# ROCKS & MINERALS

# ROCKS &
# MINERALS

**RONALD LOUIS
BONEWITZ**

**CONSULTANTS**
MARGARET CARRUTHERS,
RICHARD EFTHIM

## THE DEFINITIVE VISUAL GUIDE

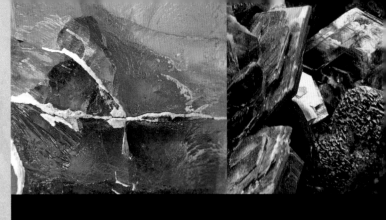

**LONDON, NEW YORK, MUNICH, MELBOURNE, AND DELHI**

**Senior Editors** Angeles Gavira, Peter Frances
**Senior Art Editor** Ina Stradins
**Designers** Paul Drislane, Peter Laws
**Production Controllers** Melanie Dowland, Luca Frassinetti, Kevin Ward
**DTP Designer** John Goldsmid

**Managing Art Editor** Philip Ormerod
**Art Director** Bryn Walls
**Publishing Manager** Liz Wheeler
**Category Publisher** Jonathan Metcalf

Produced for Dorling Kindersley by
Grant Laing Partnership
48 Brockwell Park Gardens,
London SE24 9BJ

**Managing Editor** Jane Laing
**Managing Art Editor** Christine Lacey
**Art Editors** Alison Gardner, Miranda Harvey
**Designers** Nick Avery, Vicky Short, Paul Ashby
**Project Editor** Jane Simmonds
**Editors** Frank Ritter, Helen Ridge
**Picture Researcher** Jo Walton
**Proof Reader** Becky Gee
**Indexer** Dorothy Frame
**Specially Commissioned Photography**
Linda Burgess and Gary Ombler
**Crystal Illustrations** Tim Loughhead,
Precision Illustration

**Smithsonian Project Coordinators**
Ellen Nanney, Katie Mann
**Consultants** Margaret Carruthers, Richard Efthim

First published as *Rock and Gem*
in Great Britain in 2005 by
Dorling Kindersley Limited
80 Strand, London WC2R 0RL
A Penguin Company

This edition published 2008

Colour reproduction by Colourscan, Singapore
Printed and bound in China, by Toppan

see our complete catalogue at
**www.dk.com**

# CONTENTS

# ORIGINS

EARTH HISTORY | COLLECTING SPECIMENS

# THE FORMATION OF THE UNIVERSE

WHEN WE LOOK TODAY at rocks and minerals, and at the planet Earth formed from them, it is hard for us to imagine a time when none of this existed – a time even before there were chemical elements, the building blocks from which rocks and minerals are made. Yet between 13,000 and 15,000 million years ago, the entire Universe consisted of one tiny dot of primordial energy. Then, in an instant, the Big Bang set in motion a chain of events that resulted in the creation of atoms and, over millions of years, in the formation of galaxies and stars. It was only within stars that the elements were formed that would make minerals, rocks, and planets.

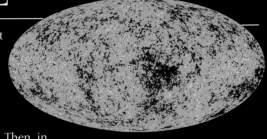

**COSMIC BACKGROUND**
*This image shows the Cosmic Microwave Radiation Background, a faint heat radiation that emanates from all points in the sky. It is thought to be left over from the fireball-like conditions prevalent in the early Universe.*

## THE BIG BANG

According to the most widely held theory of how the Universe came into being, the physical Universe and all within it emerged from an infinitely small point of pure energy in a highly compressed state and at an extremely high temperature. The rapid expansion of this point of energy through the Big Bang resulted in a relatively rapid drop in its density and temperature, and within a few seconds a number of elementary particles, such as electrons, photons, neutrons, and protons formed. This expanding mass of energy and particles was still far too hot for the formation of atoms, an event that required perhaps another 300,000 years of expansion and cooling. Only then did the first atoms appear, chiefly of hydrogen and helium.

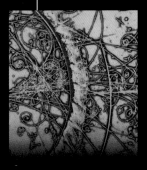

**PARTICLE TRACKS**
*Studying the tracks of subatomic particles in a particle accelerator gives scientists insight into the Big Bang's early stages.*

### EDWIN HUBBLE

American astronomer Edwin Hubble (1889–1953) is famous for proving that we live in an expanding Universe. Hubble made his most important discoveries in the 1920s. He found that other galaxies exist beyond our own galaxy, the Milky Way. By analysing changes in the wavelength of light from these other galaxies, he showed that they are moving away from the Milky Way, and that the further away a galaxy is the faster it recedes. As a result, Hubble concluded that the Universe as a whole is expanding. His crucial observations led directly to the Big Bang theory.

**AFTER THE BIG BANG**
*In the first few billionths of a second after the Big Bang, the Universe consisted of a "soup" of enormously varied particles and radiation.*

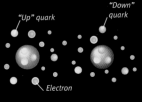

"Up" quark

"Down" quark

Electron

PROTON          NEUTRON

Neutron

Proton

Single proton

Electron

Energy released by collisions

HYDROGEN ATOM

Helium nucleus, two protons and two neutrons

Electron

Nucleus

Electron

HELIUM AT

**QUARKS BIND**
*In the next seconds quarks started binding to form protons and neutrons. One "down" and two "up" quarks made a proton, while one "up" and two "down" quarks made a neutron.*

**NUCLEI FORM**
*Between 1 and 180 seconds after the Big Bang, collisions between protons and neutrons formed the nuclei of light elements, mainly helium, though they did not yet capture electrons.*

**ATOMS APPEAR**
*After about 300,000 years, helium nuclei began to capture pairs of electrons, forming helium atoms. Protons captured one electron each, forming hydrogen atoms.*

# GALAXIES APPEAR

Hydrogen and helium gas, along with two other light elements, lithium and beryllium, were the only chemical elements in the Universe for hundreds of millions of years. Eventually, variations in the density of matter in the expanding universe caused small regions of higher gravity to occur (the denser matter becomes, the more gravitational pull it exerts). Gas was drawn into these regions by the force of gravity, creating vast, coalescing clouds of hydrogen. Within each of these huge gas clouds , or nebulae, numerous even more dense areas formed, each drawing yet more gas into itself. Eventually, in the centres of the densest areas of gas, temperatures and pressures rose to such a point that hydrogen atoms began to fuse together to form helium. This nuclear fusion generated light and yet more heat – creating the first stars to light the newly formed Universe. Each cloud of new-born stars constituted a galaxy.

**NURSERY OF STARS**
Nebulae are clouds of gas and dust that are the birthplace of stars. This nebula, known as the Keyhole Nebula, is glowing with colour because of the hot young stars within it.

**DOUBLE SPIRAL GALAXY**
Galaxies exist in five basic shapes, one of which is a spiral. A typical single galaxy will contain around 100 billion stars.

# CREATING THE FIRST MINERALS

Stars can be thought of as factories for making heavier elements from hydrogen. At the high temperatures and pressures in the cores of stars, nuclei can collide with such energy that they fuse together to become heavier elements. These heavier elements can then fuse with more hydrogen or with other freshly made nuclei to form heavier nuclei. In this way oxygen, carbon, and most of the other elements up to the mass of iron are formed. Heavier elements than iron are made towards the end of the life of large stars by a process of adding neutrons to nuclei and radioactive decay. At the end of the life of large stars, the star will explode. Such an explosion is called a supernova. Minerals are found in meteorites that formed in the outflows of Red Giant stars and others around supernovae.

**SUPERNOVA EFFECT**
The blast from an ancient supernova heated this gas cloud in the constellation Cygnus, making it glow.

**OPEN STAR CLUSTER**
The Quintuplet cluster (shown here) is the largest cluster of stars inside our Milky Way galaxy. This group of young stars formed 4 million years ago and contains the brightest star in the galaxy, called the Pistol star.

# GENERATIONS OF STARS

New stars are created from material spread through space by the explosion of other stars. Our own star, the Sun, is a third-generation star, containing elements that have been cycled through at least two other stages of star-death. Although, like all stars, second-generation stars are mainly hydrogen, they incorporate elements formed in the first generation of stars and can use these as building blocks for new isotopes and elements. We can identify these second-generation elements and isotopes in the Sun and in the planets circling it, including Earth.

**ULTRAVIOLET SUN**
Our local star, the Sun, here photographed through an ultraviolet filter, is a third-generation star.

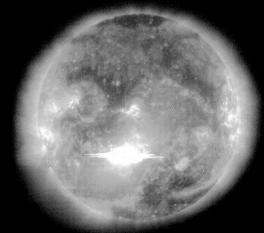

# BIRTH OF THE SUN

Our Solar System consists of the Sun, planets, moons, asteroids, comets, and assorted other debris left over from its formation. The Solar System was formed out of a rotating cloud of interstellar gas and dust (known as the Solar Nebula) that was the remnant of two earlier generations of stars. It was located towards the edge of a galaxy we know as the Milky Way. Roughly 4.6 thousand million (4.6 billion) years ago, the nebula collapsed, possibly due to shockwaves from a nearby exploding supernova. The collapse of the nebula caused its rate of rotation to increase and the collapsing material flattened into a disc, known as the protoplanetary disc. Matter continued to be drawn from this disc towards the centre. Energy from the collapse caused the temperature in the centre to rise and the density to increase. Eventually the temperature and density were high enough for hydrogen nuclei to start fusing to form helium and giving off energy in the form of heat and light. Our own star, the Sun, was born.

**MILKY WAY**
*The Milky Way is a spiral galaxy. Its nucleus consists mainly of old red and yellow stars; younger, bluer stars form the arms, where our Sun is found.*

**SUN COMPOSITION**
*The Sun, the centre of the Solar System, is roughly 75 per cent hydrogen and 25 per cent helium. Other elements are present at levels less than 1 per cent. These minor elements can only have been formed by processing through two generations of stars.*

---

# ORIGIN OF THE PLANETS

Not all of the material in the Solar Nebula ended up being dragged into the forming Sun by the force of gravity. Some gas and dust remained orbiting in the protoplanetary disc. Here dust particles stuck together into larger and larger agglomerations. Some of these dust balls were melted to form tiny rocks smaller than marbles. Known as "chondrules", these were in turn accreted, sticking together in ever larger bodies until gravitational forces could take over and keep them together. These bodies are called planetesimals. Once the planetesimals grew to around a kilometre in size, they started to attract other similar-sized objects due to gravitational attraction.

At this stage the process of accretion grew in efficiency. Many of the planetesimals continued to coalesce until they formed objects the size of Mercury or the Earth. The whole process, starting from the initial collapse of the nebula and continuing through the various stages up to the formation of planets, may well have taken about 10 million years. Not all of the planetesimals were caught up in the formation of planets. Some of them were frozen at between approximately one kilometre and tens of kilometres in size. Many of these bodies are preserved as asteroids orbiting in the asteroid belt between Mars and Jupiter.

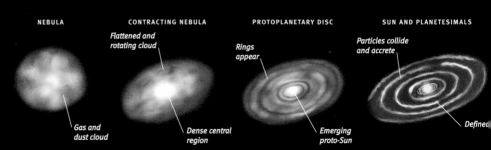

**FROM SOLAR NEBULA TO SOLAR SYSTEM**
*The Solar System formed from a huge nebula (cloud) of gas and dust, which condensed into a spinning disc with the proto-Sun at its centre. While the Sun further condensed and atomic fusion began, particles in the rest of the disc accreted, starting the process that led to the formation of planets.*

**NEBULA**
Gas and dust cloud

**CONTRACTING NEBULA**
Flattened and rotating cloud
Dense central region

**PROTOPLANETARY DISC**
Rings appear
Emerging proto-Sun

**SUN AND PLANETESIMALS**
Particles collide and accrete
Defined

# INNER AND OUTER PLANETS

The protoplanetary disc became hotter nearer the Sun than further out. In the hotter regions ices evaporated. Also, gas was driven out towards the outer parts of the Solar System, a movement helped by the solar wind, a stream of energetic particles emitted from the Sun. As a result, planets in the inner Solar System – known as the terrestrial planets – are small and rocky. But in the cold outer regions, the rocky bodies accreted ice and, because of the cold and their greater mass, could also hold on to gases such as methane, ammonia, hydrogen, and helium. The result was the formation of giant gassy planets. Pluto formed far out, along with many other similar frozen bodies. These remained small as they were sparsely distributed and so could not accrete efficiently. They also did not gain much ice and gas driven off from the inner Solar System. Some astronomers do not consider Pluto a planet, but just one of many icy bodies beyond Neptune.

**SMALL, STONY EARTH**
*The four inner planets, including Earth, are composed largely of metal and stone.*

**LARGE, GASEOUS JUPITER**
*Jupiter and the other large outer planets have rocky cores but are mostly made of frozen gases.*

Sun Mercury Venus Earth Mars Jupiter Saturn Uranus Neptune Pluto

**NEAR AND FAR**
*The distance line above gives a rough idea of each planet's distance from the Sun. The stony planets (Mercury, Venus, Earth, and Mars) cluster near the Sun, while the gas giants (Jupiter, Saturn, Uranus, and Neptune) orbit at greater distances.*

**ORBITS OF THE PLANETS**
*All of the inner planets circle the Sun in elliptical orbits, in the same direction and roughly on the same plane. A portion of Pluto's orbit crosses Neptune's, and the orbits of comets (blue line) can extend beyond that of Pluto.*

## ROCKS FROM SPACE

**ASTROBLEME**
*Meteorite craters, such as the impressive Tswaing crater near Pretoria, South Africa, are known as "astroblemes".*

Meteorites – rocks that fall to Earth from space – are of particular interest in the study of the origin of the Solar System. Nearly all meteorites are thought to be fragments of asteroids, rocky bodies that range from objects more than 900km (560 miles) in diameter to particles of microscopic dust. Mostly orbiting in a belt between Mars and Jupiter, asteroids are thought to have aggregated from material in the Solar Nebula about the same time as the Earth was formed. The larger asteroids are, in fact, planetesimals that have never accreted to form a planet. Collisions between them produce fragments called meteoroids, some of which are propelled into Earth-crossing orbits, falling to Earth as meteorites. Types of meteorite and their contribution to our understanding of the Solar System are explored on pages 74–75.

Accreted rock

**STONY-IRON METEORITE**
*Meteorites can be divided into three groups: stony, iron, and, as here, a stony-iron mix of silcates and nickel-iron alloy.*

# THE FORMATION OF THE EARTH

THE ROCKS AND MINERALS we observe on the Earth today are the result of an ongoing series of processes that began with the initial formation of the Earth from the Solar Nebula over 4.5 thousand million years ago, and which will continue for at least as long into the future.

**BULGING SPHERE**
*The Earth is nearly a perfect sphere, but because of its rotation it bulges a little at the equator and is slightly flattened at the poles.*

## THE EARTH TAKES SHAPE

As with the other planets, the Earth formed through the accretion process, sweeping up solid material from the disc of the Solar Nebula (see pp.10–11). As the mass of the Earth increased, so too did its gravity. Increasing gravity drew in ever larger planetesimals and other meteoritic debris, accelerating the growth of the planet. Initially, the accreting planet was loosely consolidated – it did not form a coherent whole. But within a relatively short time, under the impact of collisions with meteorites and planetesimals, it consolidated and developed a differentiated structure. The shock of the impacts generated heat, causing rocks to melt. The molten Earth separated into liquid iron and silicate melt, the less dense silicate material floating on top of the higher density metallic iron core. The silicate portion then also started to separate out into the higher density mantle and the lower density crust on the top. Accretion continues to this day as the Earth sweeps up tonnes of space debris each year.

*Molten core*

*Lighter eleme form mantle*

*Asteroid-size body joins others of similar size*

*Larger object forms*

*Small bodies and dust accrete to asteroid size*

**THE ACCRETION OF THE EARTH**
*Matter from the Solar Nebula was drawn together to form asteroid-size bodies. These objects then accreted to form even larger bodies. Repeating the process over time formed the Earth.*

**CORE AND MANTLE**
*As more and larger objects were drawn together, melting took place, with heavy elements migrating to the core. Lighter elements formed a mantle, which was surrounded by a rocky crust.*

**MODEST PLANET**
*Although it may seem large to us, Earth is a planet of medium size. With a diameter of 12,756km (7,928 miles), it is the largest of the four rocky inner planets, but much smaller than any of the giant gas planets.*

# THE EARTH GAINS A SATELLITE

**SURFACE OF THE MOON**
*The Moon's surface is scarred by impact craters. Some of these have been filled by flows of basaltic lava from beneath the Moon's crust, forming maria or seas.*

According to the most widely accepted current theory, the Moon was formed as the result of an awesome collision early in the Earth's history. Around 4,500 million years ago, a very large planetesimal – probably an object about the size of Mars – hit the Earth, penetrating far below the surface and splashing out a huge amount of debris. This Earth debris, along with material from the planetesimal itself, went into orbit around the Earth. There, by the process of accretion, it formed the Moon. Studies of trace elements in lunar rocks show results similar to those obtained from the Earth's mantle. This implies that at the time of the collision the structure of the Earth must have already differentiated into a lighter mantle and a denser core. Only the lighter mantle rocks were splashed out into orbit. Because lunar rocks have been dated to near the time when the Earth was formed, scientists have concluded that planet formation must be a relatively rapid process. In the space of less than a hundred million years, it is likely that the Earth formed, the Earth's mantle developed, the awesome collision with the planetesimal occurred, and the splashed-out lunar material accreted to form the Moon.

**MOON ROCK**
*The overall composition of rocks recovered from the Moon matches that of mantle rock on the Earth.*

*Basalt from the Moon's suface*

**MOON STRUCTURE**
*Like the Earth, the Moon has a mantle and a crust. It may also have an iron-rich core, but if such a core exists it is small. The Moon is less than half the size of the Earth, with a diameter of 5,562km (3,476 miles).*

**FORMATION OF THE MOON**
*The Moon was formed from material splashed out of the Earth by the impact of a large planetesimal, along with a portion of the planetesimal itself. The debris of this massive collision was held in orbit by the Earth's gravity and gradually cooled and coalesced. The process of accretion then formed the Earth's natural satellite.*

*Mars-sized impactor*

*Earth*

**BOMBARDMENT OF THE EARTH**

*Earth*

*Debris begins to coalesce*

**DEBRIS ORBITS THE EARTH**

*Earth*

*Debris accretes to form the Moon*

*Earth*

**ACCRETION FORMS MOON**

## CORE MINERALS FORM

The Earth's core seems to have formed early on in the planet's development, as heavier elements sank towards the centre. There is no way of obtaining samples of the core, at least not with current technology, but a range of ingenious methods has been used to determine the core's composition, including studies of the Earth's magnetic field, of earthquake waves as they pass through the Earth, of rock density, and of meteorites. These studies have led to the conclusion that the Earth has a solid inner core surrounded by a liquid outer core. The core is probably composed mainly of iron with a small percentage of nickel, and a lighter element, possibly sulphur. Further refinements of the core's composition are probably still taking place as a result of chemical interaction with the lower mantle. The temperatures in the inner core are estimated to be close to those on the surface of the Sun. Exactly which iron minerals might form at such temperatures and pressures is not known.

*Pyroxene*

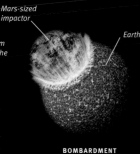

**CORE COMPOSITION**
*The composition of iron-nickel meteorites is thought to resemble that of the Earth's core.*

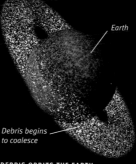

**MOLTEN IRON**
*The outer portion of the Earth's core is thought to be molten iron, as here created by human skill in a foundry.*

# MANTLE AND CRUST MINERALS

The intensity of the heat in the interior of the forming Earth encouraged chemical reactions between the iron-rich core and the surrounding material. Most of the elements that combine readily with iron formed denser minerals which accumulated in a thick shell around the core. This created what is known as the lower mantle. Lighter minerals were formed from elements with a greater affinity for oxygen than for iron. These combined as oxide compounds, predominantly silicates. Being lighter, these compounds rose upwards to form the upper mantle and the crust, in effect floating on the denser lower mantle.

### CORE, MANTLE, AND CRUST
*The Earth has three distinct layers: a core, consisting of a solid inner core and a fluid outer core, which together make up more than half the planet's diameter; the mantle, a layer of dense minerals making up most of the rest of the Earth's diameter; and a thin crust composed of rocks and minerals chemically distinct from those of the core and mantle.*

**Core–mantle boundary, where liquid outer core and solid lower mantle meet**

**Plume of hot, upwelling, mantle rock carries heat to surface**

### INNER CORE
*Solid nickel-iron; down to 6,370km (3,960 miles) below the Earth's surface*

### OUTER CORE
*Liquid nickel-iron; down to 5,150km (3,200 miles) below the surface*

**Land surface made of continental crust**

### LOWER MANTLE
*Solid, mostly silicates rich in magnesium; down to 2,990km (1,860 miles) below the surface*

### UPPER MANTLE
*Solid, primarily peridotite; down to 660km (410 miles) below the surface*

**Continental shelf drops into ocean depths**

### CONTINENTAL CRUST
*Solid, mixture of igneous, metamorphic, and sedimentary rocks; down to 70km (45 miles) below the surface*

**Sea floor made of oceanic crust**

### OCEANIC CRUST
*Solid, primarily basalt; down to 7km (4 miles) below the surface*

**Ocean surface**

**Boundary between lower and upper mantle**

### MANTLE XENOLITH
*Xenoliths are fragments of older rock incorporated in younger rock. In this xenolith from France, upper-mantle peridotite has been brought to the Earth's surface in basaltic lava.*

## DATING THE EARTH

One of the earliest attempts to calculate the age of the Earth was made in the 17th century, when Anglo-Irish Bishop Ussher, working from Biblical genealogy, estimated that the Earth and Universe were created in October, 4004BC. The first accurate estimate was obtained in 1956 by physicist Claire Patterson (shown below) who compared radioisotope measurements from meteorites and Earth minerals to give an age of 4,550 million years (more recently modified to 4,560 million years). The oldest directly measured date is for zircon crystals taken from Mount Narryer, Western Australia, dating their formation to 4,400–3,900 million years ago.

**CLAIRE PATTERSON**

### OCEAN FLOOR ROCKS
*Tectonic forces sometimes bring old ocean floor to the surface. These Newfoundland peridotites, thought to be similar to upper-mantle rocks, were once on the ocean bed.*

This bar chart illustrates the proportion of various chemical compounds that are found in the major layers of the Earth.

- OTHERS
- NICKEL OXIDE
- MAGNESIUM OXIDE
- CALCIUM OXIDE
- IRON AND IRON OXIDES
- ALUMINIUM OXIDE
- SILICON DIOXIDE

CONTINENTAL CRUST

OCEANIC CRUST

MANTLE

CORE

## THE OXYGEN FACTOR

Although the core of the Earth is denser than the mantle and crust, the distribution of the various chemical elements between the layers of the Earth's interior was not necessarily determined by the density of the actual elements themselves. The tendency of various elements to interact either with iron or with oxygen could have been the principal determining factor. For example, uranium and thorium are very dense, heavy elements, yet they are concentrated in the crust, possibly because they both have a high affinity to oxygen. The importance of the role played by oxygen in the Earth's crust is clearly illustrated by the fact that it makes up 50 per cent of the crust by weight.

## THE ATMOSPHERE AND THE OCEANS

The formation of the atmosphere and the oceans was directly connected to the formation of the crust. During the period of intense accretion, as the Earth melted and recrystallized, water vapour and other gases were released to add to the gases from the Solar Nebula accumulated during the accretion process. This constituted the Earth's first atmosphere. Then, 3.8 billion years ago, the Earth and Moon were subjected to intense meteorite bombardment which is thought to have stripped away the first atmosphere. Subsequent volcanic activity released nitrogen, carbon dioxide, and water vapour, which in turn were broken down by ultraviolet light into hydrogen, oxygen, and ozone. The water vapour collected to form oceans, where salts dissolved, increasing salinity. A second atmosphere formed, at first with little oxygen. As life developed in the oceans oxygen was released, creating today's atmosphere.

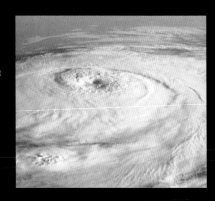

**WEATHER SYSTEM**
Earth's atmosphere extends upwards for several hundred kilometres, but most weather systems occur in the bottom 15km (10 miles).

**FIRST ATMOSPHERE**

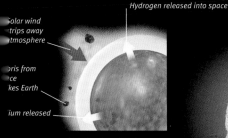

Hydrogen released into space

Solar wind strips away atmosphere

Debris from space strikes Earth

Helium released

**SECOND ATMOSPHERE**

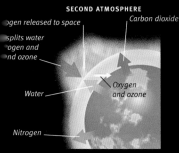

Nitrogen released to space

Carbon dioxide

UV splits water into hydrogen and oxygen and ozone

Oxygen and ozone

Water

Nitrogen

**ATMOSPHERIC CHANGE**
Earth's first atmosphere, built up from gases (nitrogen, carbon dioxide, and water vapour) released by crystallization and from the Solar Nebula, was stripped away by meteorite strikes and the solar wind. The second atmosphere formed from volcanic gases modified by ultraviolet light.

**ANCIENT LIFE**
Formed by bacteria colonies, stromatolites in Shark Bay, Australia, are "living fossils". Stromatolite mounds like these formed in shallow seas more than 3 billion years ago.

# THE EARTH'S CRUST

ONE OF THE MOST conspicuous features of the Earth's surface is its division into continents and ocean basins. Both the oceanic and the continental crust lie on top of the mantle, but the two types of crust are significantly different. The lighter, thicker continental crust floats higher in the mantle than the denser, thinner oceanic crust. About two-thirds of the Earth's surface is oceanic crust, but the continental crust has a far wider variety of rocks.

**DRIFTING ROCKS**
*The Earth's crust is fused with the upper mantle to form the lithosphere. This drifts on top of the asthenosphere.*

Continental margin of thick sediments built up on ocean crust

Oceanic crust

Lithosphere

Asthenosphere

## OCEANIC CRUST

The oceanic crust is much younger than the continental crust – even the oldest parts of the ocean floor are only 200 million years old. This is because oceanic crust is continually being recycled. It forms from mantle material rising within rifts and oceanic ridges, and spreading to either side of the ridge. This material returns to the mantle at the same rate in areas called subduction zones. The oceanic crust is composed of several layers, the topmost of which is a layer of sediments, primarily very fine muds overlying a layer of basalt. The lower layers are primarily gabbro, with increasing olivine content as the depth increases.

Abyssal plain

Continental slope

Continental crust made up of igneous, metamorphic, and sedimentary rocks

Boundary between crust and mantle, known as Mohorovicic discontinuity

Amygdule filled with zeolite

**WIDESPREAD ROCK**
*This black rock is basalt that formed on land. Basalt is also the main component of the oceanic crust.*

**BLACK SMOKER**
*Associated with volcanic activity, black smokers are underwater hot springs, expelling mineral-rich water at temperatures up to 400°C (750°F). Most occur near mid-ocean ridges, where oceanic crust is formed.*

## JOHN TUZO WILSON

In the 1960s, the Canadian geophysicist John Tuzo Wilson (1908–93) became one of the leading advocates of the theory that the sea floor is spreading from the ocean ridges. Wilson made a major contribution to the theory when he put forward the idea that volcanic mid-ocean island chains, such as Hawaii, form as a result of the sea floor moving slowly over the top of a "hotspot" – a fixed point where magma wells up inside the mantle.

**ICELANDIC RIDGE**
*Oceanic crust and mid-ocean ridges are not often seen above sea-level, but in Iceland the Mid-Atlantic Ridge rises out of the surrounding ocean.*

# CONTINENTAL CRUST

The continental crust covers about a third of the Earth's surface, forming the main landmasses and the beds of the shallow seas that surround them. It is where most of the rocks and minerals discussed in this book are found.

The continental crust varies from 25 to 70km (16 to 45 miles) in thickness, with the greatest thickness under the mountain ranges. At the heart of each continent is a stable mass of crystalline rocks from the Precambrian period (4,560 million to 543 million years ago) covering thousands of square kilometres. This is known as the continental shield. Other rocks of the continental crust have more recently been subjected to the processes of the rock cycle (see pp.28–29), such as erosion, metamorphism, and sedimentation. It is this history of transformation over billions of years that accounts for their immense variety.

Large feldspar crystal

Biotite

**CRUSTAL ROCKS**
Metamorphic rocks such as this gneiss are formed as a result of tectonic forces acting upon rocks of the continental crust.

**AUSTRALIAN SHIELD**
The central area of Western Australia is part of the ancient heart of the Australian continent – its stable continental shield.

**GRANITE OUTCROP**
These granite rock pinnacles are in the Ahaggar Massif, Algeria. Composed of silica-rich minerals, such as quartz and feldspar, granite is found in large quantities in the continental crust.

# FLOATING ROCK

The Earth's crust behaves as if it were floating on the underlying denser, flexible rock of the mantle. This concept is known as "isostasy". The crust can be visualized as resembling an iceberg floating on water, and like an iceberg it extends downwards into the medium on which it floats. Because the Earth's crust is not all of the same density or thickness, different parts of it protrude into the mantle to different depths. Thus the crust extends deeper down into the mantle under tall mountain ranges, such as the Himalayas, than it does in low-lying areas, because it requires a deeper "root" to buoy up the mountain range's additional weight. About two-thirds of the thickness of the continental crust forms the root that supports the rest.

**DEEP ROOTS**
Everest is part of the Himalayas, a relatively young mountain range created by a collision of two continental plates. Mountains such as Everest have a "root" extending many kilometres into the mantle.

17

# PLATE TECTONICS

The term "tectonics" refers to the building and movements of the Earth's crust. According to the concept of plate tectonics, the solid, brittle lithosphere, which comprises the crust and a thin section of the uppermost mantle, rests on and is carried along by the underlying layer of hot and flowing mantle rock, the asthenosphere. The lithosphere is made up of both continents and ocean basins. It is divided into relatively rigid sections known as "plates", and the movement of these plates relative to one another results in the formation and the modification of the Earth's major surface features.

### PLATE BOUNDARIES
*There are about a dozen large tectonic plates. Some of the plates consist of both continental and oceanic crust, while others are composed exclusively of oceanic crust.*

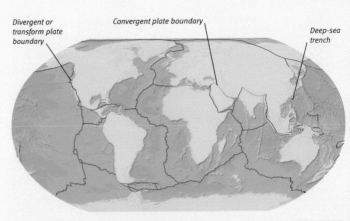

*Divergent or transform plate boundary*

*Convergent plate boundary*

*Deep-sea trench*

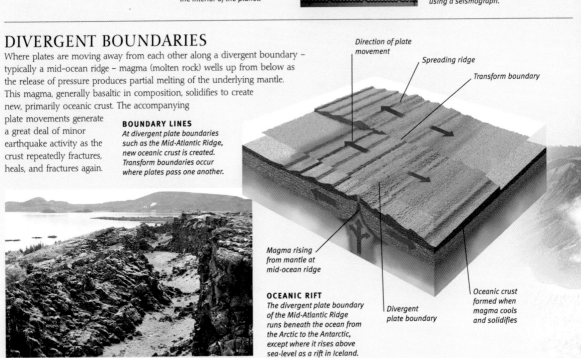

*Mantle plume rises from lower mantle to form hotspot*

*Plates move apart at oceanic rift*

*Plate dragged along by convection current*

*Upper mantle*

*Lower mantle*

*Motion of convection current*

### PLATE MOTION
*The force that drives the movement of tectonic plates is thought to be heat rising from the interior of the planet.*

*Subduction zone where tectonic plate descends into mantle*

*Core*

# MOVING FORCE

The forces that drive the lithospheric plates across the Earth's surface are as yet not fully understood, but the movement is thought to be caused by convection currents in the mantle, which are driven by heat within the planet. As the plates move relative to each other, diverging, converging, or moving past, major geological interactions take place along their boundaries. This is where many of the principal processes that shape the Earth's surface happen – for example, earthquakes, volcanism, and the deformation of the Earth's crust that builds mountain ranges.

### SEISMOGRAPH TRACE
*The energy of tectonic movements creates vibrations in the Earth's rocks that can be measured and recorded using a seismograph.*

# DIVERGENT BOUNDARIES

Where plates are moving away from each other along a divergent boundary – typically a mid-ocean ridge – magma (molten rock) wells up from below as the release of pressure produces partial melting of the underlying mantle. This magma, generally basaltic in composition, solidifies to create new, primarily oceanic crust. The accompanying plate movements generate a great deal of minor earthquake activity as the crust repeatedly fractures, heals, and fractures again.

### BOUNDARY LINES
*At divergent plate boundaries such as the Mid-Atlantic Ridge, new oceanic crust is created. Transform boundaries occur where plates pass one another.*

*Direction of plate movement*

*Spreading ridge*

*Transform boundary*

*Magma rising from mantle at mid-ocean ridge*

### OCEANIC RIFT
*The divergent plate boundary of the Mid-Atlantic Ridge runs beneath the ocean from the Arctic to the Antarctic, except where it rises above sea-level as a rift in Iceland.*

*Divergent plate boundary*

*Oceanic crust formed when magma cools and solidifies*

# CONVERGENT BOUNDARIES

A convergent boundary occurs where plates are moving towards each other. Where oceanic crust meets oceanic crust, one of the plates descends beneath the other – a process called subduction. The subducted crust is recycled into the mantle, compensating for the new crust being created continually at divergent plate boundaries. Subduction also happens where the oceanic crust of one plate meets the continental crust of another. The greater buoyancy of the continental crust prevents it from being subducted, so it is the plate carrying the oceanic crust that dives underneath. The subducted oceanic crust triggers melting at depth and molten rock rises up through the crust, erupting to form explosive volcanoes. The numerous volcanoes found along the upper Pacific coast of the USA, such as Mount Saint Helens, are formed in this manner. If both plates meet along a continental edge, then neither can be subducted. The resulting collision causes the crust to crumple, pushing mountain ranges upwards.

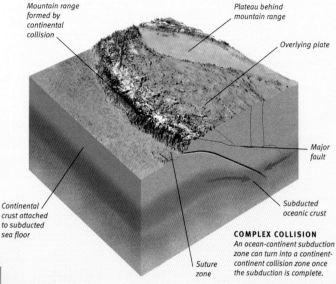

*Mountain range formed by continental collision*

*Plateau behind mountain range*

*Overlying plate*

*Major fault*

*Continental crust attached to subducted sea floor*

*Subducted oceanic crust*

*Suture zone*

**COMPLEX COLLISION**
*An ocean-continent subduction zone can turn into a continent-continent collision zone once the subduction is complete.*

**UPWARD THRUST**
*The Himalayan mountain range, the highest on the planet, is still being thrust upwards as the Eurasian and Indian-Australian tectonic plates collide.*

# TRANSFORM FAULTS

As well as divergent and convergent boundaries, there is a third type of plate boundary, where two plates move past each other without creating or destroying crust. This is called a transform boundary. Large earthquakes are caused by the continuous build-up and release of tension where the plates meet. California's San Andreas Fault and the North Anatolian fault system in Turkey are good examples of transform boundaries.

**CALIFORNIAN FAULT LINE**
*The San Andreas Fault in California is a transform boundary where the Pacific Plate and the North American Plate slide past one another.*

**RING OF FIRE**
*These volcanoes in Java are part of the "ring of fire" that surrounds the Pacific Plate. Subducted rock triggers melting, feeding volcanic activity at the Earth's surface.*

**VESUVIUS ROCKS**
*Some collections are of finds from a single locality.*

# COLLECTING ROCKS AND MINERALS

THE WORLD OF ROCKS, minerals, gems, and fossils offers virtually endless possibilities for the hobbyist. Only a very small amount of specialized knowledge – little more than the ability to identify a few common minerals – is required to open a whole new world of enjoyment of some of nature's finest creations. Probably the most popular rock hobby is mineral collecting.

## STARTING A COLLECTION

Most collectors begin by just accumulating rocks, minerals, and fossils. As their collection grows, they start being more selective, keeping only specimens with better colour, better crystallization, or more interesting crystal forms. With widening experience, most collectors begin to specialize. Some minerals come in such a wide variety of crystal forms that a collection can be made of a single mineral; other collectors may focus on ore minerals, or on minerals from a particular locality. There are practical considerations that limit the scope of most collections. Some rare minerals are very expensive, and mineral collections can take up a huge amount of space. One solution is to collect tiny specimens called micromounts, whose form and beauty can only be seen under magnification.

**MAGNIFICATION**
*Collectors of small crystals need effective microscopes to examine and enjoy their minute specimens.*

**FASCINATING FOSSILS**
*As well as minerals, fossils are a popular subject for specialized collections.*

**TREE CALCITE**

**SAND CALCITE**

**DOGTOOTH SPAR**

**CALCITE CRYSTALS**
*A mineral such as calcite, which exists in a rich variety of crystal shapes, provides suitable material for an interesting single-mineral collection.*

**FIELD EXPERIENCE**
*Rock collecting can be a hobby for a lifetime, as the collector develops knowledge and skills to enhance the activity.*

## COLLECTING SAFELY

While mineral collecting is generally a safe hobby, there are a few definite hazards that the collector needs to be aware of. The most dangerous collecting locality is around old mines and workings. Tunnels should never be entered – shoring timbers rot very quickly, and cave-ins and rockfalls are almost guaranteed to happen. In any case, there is often remarkably poor collecting inside old mines. If anything looked good, the miners usually dug it out. Mine dumps, by contrast, can be a good source of specimens, although be aware that they are often loosely piled and can be unstable. When collecting in locations such as beach cliffs, road cuttings, and rockfalls, attention must be paid not only to loose material underfoot but also to anything that may fall or roll from above. Collectors are injured every year by falling rocks. And, unless you absolutely know the locality to be safe, leave small children at home.

**TEMPTING TUNNELS**
*Old mine shafts can be tempting, but are often highly dangerous places. Better specimens are usually found in the mine dumps outside.*

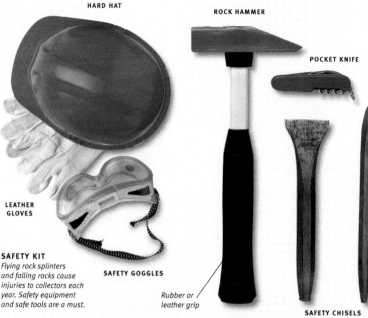

**HARD HAT**

**ROCK HAMMER**

**POCKET KNIFE**

*Trowel*

**Brushes for light cleaning**

*Sieve*

**LEATHER GLOVES**

**SAFETY GOGGLES**

**SAFETY KIT**
*Flying rock splinters and falling rocks cause injuries to collectors each year. Safety equipment and safe tools are a must.*

*Rubber or leather grip*

**SAFETY CHISELS**

**READY FOR ANYTHING**
*The experienced collector has a range of equipment for all collecting possibilities, from sieves and pans to various brushes and trowels. Most of these tools can be left in the car until they are needed.*

# ESSENTIAL EQUIPMENT

At first glance it might seem that any hammer would do for breaking rocks, but this is far from the case. Every year rock collectors are injured – including being blinded – by using the wrong hammers. Geologist's hammers are made of special steels, and the striking ends are bevelled to prevent steel splinters from flying off them. Special geological chisels are made for the same reasons. Wearing safety goggles is highly recommended when breaking or splitting stone. Access to some collecting localities requires safety equipment such as a hard hat and fluorescent vest. A mobile phone is an important piece of safety equipment – with a fully charged battery. Carry it with you even if you are only going a short distance from the car. A fall into a ravine or other low spot may take you out of sight of potential help and add hours to the time it takes to find you. In desert country, an adequate supply of water is essential, and if you are in snake country take an appropriate snake-bite kit. Clothing suitable to the weather and terrain is, of course, vital. Leave your low-cut shoes and trainers at home. Leather boots offer better protection from snake bites, cactus spines, sharp stones, jagged metal, and rolling stones, and give much better traction.

**DELICATE CHISEL**
*A collector uses a small hammer and chisel to chip away rock from an embedded fossil.*

**GEM PANNING**
*The gold pan is an essential piece of kit for a collector. In addition to gold, many gemstones can be found by panning surface sediments.*

# WHERE TO LOOK

In many countries guidebooks are available that give precise directions to localities where collectors can find rocks, minerals, and fossils. A search of the internet can also yield good information. Specialist publications and websites keep collectors abreast of new localities. In some respects, collecting your own specimens has become more difficult in recent years. Working mines and quarries are constrained by legal liability as to whom they can permit on their premises; old mines are increasingly dangerous; old mine dumps have been gone over for decades by other collectors; and public access to land is often restricted. But some traditional sites for collecting, such as road cuttings and eroded cliffs on shorelines, still provide excellent opportunities. There is also an increasing number of collecting localities open to the public on a fee-paying basis, and some clubs for enthusiasts have their own collecting sites. Those who do not wish to collect all their specimens personally can buy them from numerous dealers, many of whom offer interesting samples that become available as new localities are discovered or opened up.

**LEAF FOSSIL**

**FOSSIL HUNT**
*A rocky beach on the Isle of Wight, southern England, offers a rich hunting ground for fossil collectors.*

**LOOKING FOR GOLD**
*This mineral collector hopes to find gold nuggets by the time-honoured method of panning through gravel sediment from a river bed.*

**CORUNDUM**

**ASSOCIATED ROCKS**
*The collector needs to know in which rocks the minerals he is seeking are found. For example, chromite is found in peridotite; corundum in recrystallized limestone.*

**GOLD NUGGET**

**CHROMITE CRYSTALS**

**OLD WORKINGS**
*This collector is looking for specimens in the waste rock of an old mineshaft in Cornwall, south-west England.*

## KEEPING BASIC NOTES

When they start out, new collectors often ignore the need to write down information about their finds. But experience soon shows that investing in a notebook and devoting the minimal amount of time it takes to keep at least basic notes is essential. It is especially important to make notes about exactly where specimens were found. A considerable time may go by before you revisit the locality, and by then, in the absence of notes, you will probably be unable to find the spot again. Even if in many cases memory can be relied on, the surroundings of the location cannot. Areas overgrow; trees fall; tracks and trails move; and seasons change. With the passage of time, it is amazing how different once-familiar landmarks can look. It is far from unknown for a collector to be standing virtually on the exact spot he or she is seeking, and not recognize it. Also, should you ever decide to sell specimens at a later date, precise information on where they were found will make them more valuable. Take notes about the collecting spot itself, any hazards there may be in the area, landmarks and distances for finding the locality, map references, and the names, addresses, and telephone numbers of any people from whom you have had to get permission to collect. All of these are things easily forgotten over time. Take your camera along as well. Photos are very useful to jog memories. Also keep all your old notebooks and delete nothing. You may want to retrieve information or revisit collecting spots 20 years later.

**DRAWING LOCATIONS**
*If you have the necessary skill, make drawings in your notebook of locations and the specimens that have come from them. It may be some time before you return.*

**MAP AND COMPASS**
*A compass and a map are essential tools for finding collecting localities and relocating them at a later date. Note down map references for good locations.*

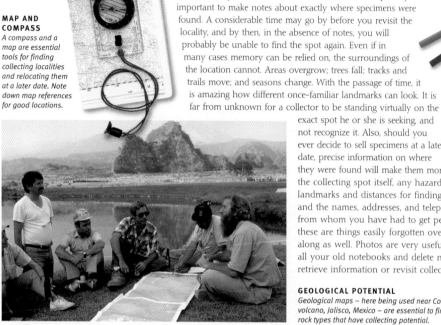

**GEOLOGICAL POTENTIAL**
*Geological maps – here being used near Colimas volcano, Jalisco, Mexico – are essential to find the rock types that have collecting potential.*

## LABELLING SPECIMENS

It is a good idea to develop the habit of labelling specimens as they are collected. Labelling should go hand in hand with the keeping of basic notes. The simplest method of labelling is to stick a piece of adhesive tape on the specimen and write a number or other identifier on it. Then write down the identifier in the notebook. Most collectors keep a separate catalogue of specimens, listing the contents of the collection itself. If you purchase specimens, carefully keep any information accompanying them. If you happen to acquire specimens from someone else's collection, save any and all labels that come with them. These may well have scientific and historical importance. In the beginning all of this may not seem so important, but as a collection grows, so too does the likelihood of mixing up specimens or forgetting the localities where they were found. Precise information enhances both the financial value of specimens and their scientific interest.

**FROZEN NOTES**
*Keeping notes of locations and of specimens found is a necessary activity even under difficult conditions, as here in freezing Arctic weather.*

**LABELLED LINE**
*A collector hangs out bagged-up, recently found specimens on a line, each one clearly identified by a label for future reference.*

# TRANSPORTING SPECIMENS

The number of specimens destroyed or damaged in the course of the journey home can be disappointingly large. Wrapping of some sort is essential for most specimens, whether they are being carried in a rucksack or a car. Some collectors use cloth or plastic specimen bags, but many specimens are resilient enough for newspaper to provide adequate protection. If you are using newspaper, though, also carry a few sheets of tissue paper in your rucksack. Delicate specimens need to be wrapped first in tissue, and only then in newspaper. After wrapping them individually, place the specimens in your rucksack with crumpled paper between them to prevent them damaging each other. If you find more specimens than you expect and your wrapping material is used up, try leaves, grass, or pine needles as a natural alternative – they are better than nothing.

If specimens are wet when they are wrapped, unwrap them and let them dry as soon as you get home. Extremely delicate samples may deteriorate if left wet for long. Also, packing material that has got wet and then dried can be difficult to remove. Some materials are totally unsuitable for wrapping. Cotton wool should be kept entirely away from specimens, as the fibres are almost impossible to remove. Cellulose wadding is not recommended either. If it remains damp for any period, the cellulose can stick to specimens and may be difficult to get off.

**IN THE BAG**
*A collector places a rock sample in a cloth specimen bag. More sensitive specimens can require elaborate wrapping to enable them to be transported safely.*

# CLEANING SPECIMENS

**MUDDY ROCKS**
*Many specimens will be muddy or dirty when collected. Most dirt is more easily removed when it is dry.*

As a general rule, clean specimens as little as possible, starting with the gentlest methods. Try out your chosen cleaning method on your least valued specimen first to see what happens. Begin by using a brush to remove loose soil and debris, possibly aided by pointed tools. Hard rock specimens, such as gneiss or granite, are unlikely to be damaged by vigorous cleaning, but with delicate minerals, such as calcite crystals, it is essential to use a fine brush. Washing a specimen may seem an obvious recourse, but remember that some minerals dissolve or disintegrate in water. Never use hot water, as the heat may cause minerals to crack or shatter. Alcohol is often used to clean borates, nitrates, and sulphates, while certain acids are suitable to clean specific minerals. Weak hydrochloric acid, for example, is good for cleaning silicates. If you do use acids, seek specific information on their use from specialized books or from another collector, and always be aware of the risks involved. Aside from the serious hazard of acid burns, even the fumes released by some of them are toxic. Soaps should be avoided, but if you must use them, choose liquid dish-washing soaps rather than hand or toilet soaps, which have additives that can penetrate specimens. The use of ultrasonic cleaners is not recommended, as they can shatter delicate specimens even at low intensities.

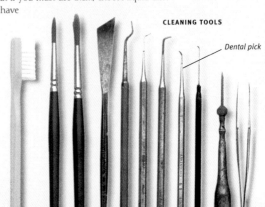

**WEAK HYDROCHLORIC ACID SOLUTION**

**CLEANING TOOLS**

Dental pick

**CLEANING UP**
*Removing rock with fine specialist tools is most often necessary when collecting fossils. The mineral collector, by contrast, is more likely to brush off dirt or to chisel out crystals.*

# PRESERVATION OF MINERALS

Once collected, some minerals are liable to experience physical and chemical effects that change or even destroy them. In some instances, they can also affect other minerals stored with them. Fortunately, these problems are well known and preventative measures can be taken.

*Deliquescence* is the absorption of water from the atmosphere into the structure of the material, causing it to dissolve. The halide minerals, especially halite, carnallite, and sylvite, are prone to deliquescence. Specimens need dry storage. They may have to be placed inside a sealed polythene bag containing a small bag of silica gel to absorb any moisture.

*Efflorescence* is the loss of the water that is incorporated into the structure of certain minerals. The water evaporates, causing the specimen to discolour or disintegrate unless it is stored in humid conditions. Chalcanthite, borax, melanterite, and laumontite are examples of minerals liable to damage by efflorescence.

*Effects of light: colour changes*. Certain minerals are noted for changing colour under strong lighting. For example, coloured topaz from Mexico and Utah, USA, eventually loses its colour entirely if left exposed to light. As many minerals may be affected by light to some degree, displayed specimens should be regularly examined with this in mind.

*Effects of light: decomposition*. Some minerals are strongly affected by light. Bright red realgar turns to golden-yellow, powdery orpiment. Other sulphides develop a surface tarnish that destroys their lustre. These minerals are often bright, colourful, and attractive, so it is tempting to light them strongly. Place them in a display case that is generally dark, and only turn on lights when they are actually being admired.

*Pyrite decay* is probably the most difficult single problem. Pyrite and marcasite are both liable to deteriorate into a heap of powder. Sulphuric acid is released as a by-product, which can damage other specimens. The heat given off can even start a fire. Examine pyrite and marcasite specimens regularly for signs of deterioration. Dry storage is one of the best solutions.

**BORAX**

*White coating results from efflorescence*

**ROUGH SORTING**
*These fossil specimens have been roughly sorted, ready for further study and preparation for display.*

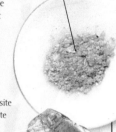

*Powdered orpiment*

**ORPIMENT**
*When exposed to the light, specimens of the red-orange sulphide realgar turn into yellow, easily powdered orpiment.*

# STORAGE AND DISPLAY

Once specimens have been collected and cleaned, they need to be stored or, in the case of the most attractive pieces, displayed. For storage, many collectors like to put specimens in card trays inside shallow drawers. Every specimen should be accompanied by a label with as much information as is feasible. Some collectors also keep a catalogue recording the complete history of the specimen, including its name, associates, exact locality, date of collection, and other relevant information. Each specimen is given a number corresponding to its listing in the catalogue.

For display, use a sturdy, preferably glass-fronted cabinet or shelves that are well out of reach of children – they find colourful minerals irresistible. Many guests will wish to handle specimens. While their appreciation is welcome, they may not be aware that handling can damage delicate examples. It is always best if the collector hands specimens to guests, as he or she will know which ones are safe to handle. It is also a good idea to have a carpet in the display area to minimize damage if specimens are dropped.

**INFORMATIVE DISPLAY**
*Some collectors not only display their prize specimens, but provide museum-style information about them.*

**STORAGE BOXES**
*These specimens are stored in individual card trays. They can usefully be numbered to link them to a catalogue listing.*

**DISPLAY CABINET**
*A part of the impressive mineral collection of the Natural History Museum, London, is displayed under glass. The same principles govern the display of a private collection – specimens should be laid out simply and clearly in a secure cabinet.*

# ROCKS

IGNEOUS | SEDIMENTARY | METAMORPHIC

# ROCK FORMATION

FROM THE GEOLOGICALLY SHORT PERSPECTIVE of a human lifetime, rocks and minerals may appear permanent and unchanging. But in reality old minerals and rocks are constantly breaking down and new rocks and minerals being formed. What we see around us is a momentary snapshot of the point rocks or minerals are now at in the process called the "rock cycle".

## THE ROCK CYCLE

Rocks are either igneous, sedimentary, or metamorphic (see pp.30–31). Igneous rock is formed as magma rises and cools. It is called extrusive if it is spewed out by volcanoes and intrusive if instead it solidifies underground. Weathering and erosion lead to the formation of sedimentary rocks, which can turn into metamorphic rock if the temperature or pressure conditions (or both) change.

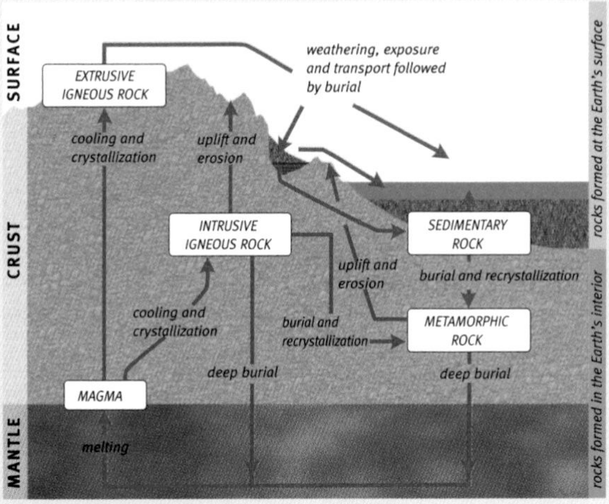

**CYCLE DIAGRAM**
*This diagram shows the various processes of the rock cycle, and their interrelationship.*

SURFACE

CRUST

MANTLE

EXTRUSIVE IGNEOUS ROCK

*weathering, exposure and transport followed by burial*

*cooling and crystallization*

*uplift and erosion*

INTRUSIVE IGNEOUS ROCK

SEDIMENTARY ROCK

*uplift and erosion*

*burial and recrystallization*

*cooling and crystallization*

*burial and recrystallization*

METAMORPHIC ROCK

MAGMA

*deep burial*

*deep burial*

*melting*

*rocks formed at the Earth's surface*

*rocks formed in the Earth's interior*

## MOLTEN ROCK

Although all the processes of the rock cycle are going on at once, and can occur in more than one order, the cycle can be thought of as starting with magma rising from the crust or upper mantle. Magma solidifies to form either intrusive or extrusive igneous rocks. If intrusive rock solidifies deep in the crust, it requires uplift and erosion (see opposite) to bring it to the surface before the next stage of the cycle can begin. If the intrusion is near the surface, the erosion of overlying rock may expose it without uplift. Extrusive volcanic rocks begin to undergo weathering and erosion immediately, whether extruded on land or in water.

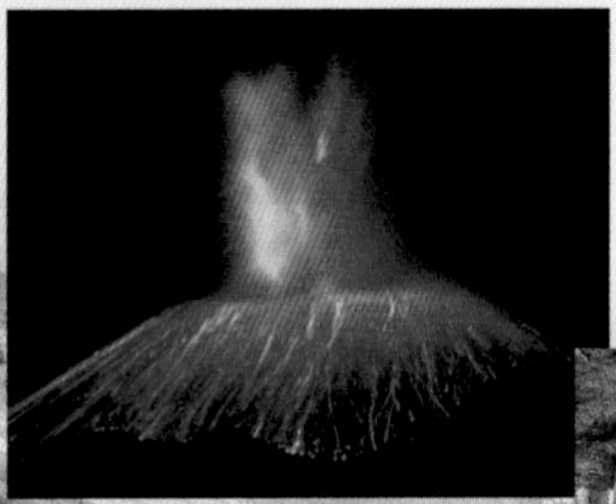

**VOLCANIC GROWTH**
*In eruptions from 1943 to 1952, Parícutin volcano in Mexico added over 300m (1,000ft) to its height.*

**PEGMATITE DYKE**
*Made of solidified magma, these sheets of pegmatite in Scotland are intrusive rocks.*

**SPANISH CANYON**
*Aniscoo Canyon in the Spanish Pyrenees shows three stages of the rock cycle: weathering of the cliffs; erosion of the slope below; and transport via the stream in the valley.*

# UPLIFT

Uplift is an upward vertical movement of the Earth's lithosphere. It can be slow and gentle, involving a broad up-warping over a long period, or sudden and dramatic, as when a fault snaps. Where uplift occurs, buried igneous, sedimentary, and metamorphic rocks are brought to the surface and exposed to weathering and erosion. Uplift can be related to the pressures of plate tectonics, or it can occur in response to the removal of a large weight, as through erosion or the retreat of glaciers.

**ANDEAN UPLIFT**
*Where tectonic plates meet, whole mountain ranges such as the Andes can be uplifted.*

# DEEP BURIAL

Sediments created by weathering and erosion are deposited in layers, which over time become compacted, forming sedimentary rocks. Because the Earth's crust moves downwards as well as upwards, great thicknesses of these sedimentary rocks can in some cases become buried deep inside of the Earth, accumulating in the

lower regions of the crust. Faulting and folding may eventually return these rocks from their deep burial back up to the surface of the Earth.

**UPLIFTED SEDIMENT**
*Originally deposited as marine sediment, nine major rock layers are visible in the Grand Canyon, Arizona, which is part of an uplifted plateau.*

# WEATHERING AND EROSION

Weathering occurs when rocks are exposed to the atmosphere, water, or living organisms at the Earth's surface. Rocks may break apart without substantial change to their chemical structure – for example, when rock is shattered by freezing water. This is called physical weathering. Chemical weathering occurs when some of the minerals within the rock break down. They may be dissolved outright, or combine with water over thousands of years to become new, easily eroded minerals. Biological weathering occurs when organisms directly attack the rock, or the growth of plants in joints and seams breaks them apart. Erosion involves the movement of rock debris away from the site where weathering has taken place. Without it, the debris would accumulate where it formed. Wind, water, glaciers, and gravity are primary agents of erosion, between them constantly reshaping the landscape. Overall, erosion's effect is to lower the Earth's elevations.

**WATER EROSION**
*This powerful, fast-flowing river in New York State, USA, transports abrasive rock fragments, which, over time, cut a gorge through the rock.*

**FROST-SHATTERED**
*Freeze-thaw action has dramatically fractured the rocks on the exposed summit of Glyder Fach in Snowdonia, Wales.*

# TRANSPORT AND DEPOSITION

Once rock has been eroded or dissolved, the products of these processes are usually transported and deposited elsewhere to become new rock (to be "lithified"). An exception is breccia, which is composed of shattered rock fragments that accumulate and lithify near their place of origin. Particles transported by the wind form dunes on land, or fall into water. Waterborne rock fragments are deposited in lakes, oceans, or old river channels. Rock material dissolved in the water is also deposited, filtering through other sediments to become cementing material for lithification. Glaciers carry rock debris in or on their ice. Also, volcanoes disgorge ash and dust that may be deposited far away.

**WIND-FORMED DUNES**
*These dunes in Colorado, USA, were formed by windborne particles that were eroded from exposed rocks elsewhere.*

**GLACIER GRINDER**
*Glaciers grind and scour rocks and carry away the frost-shattered rock debris with them as they flow downhill.*

# METAMORPHISM

Deeply buried sedimentary or igneous rocks may be subjected to tectonic forces that change their structure. Sometimes they are "baked" at high temperatures. In other cases they may be squeezed or folded by immense pressures. Or they may be subjected to both high temperature and extreme pressure. In response, rocks begin to reform, without melting, creating metamorphic rocks and new minerals that are stable under the changed conditions. If the heat and pressure increase enough, melting takes place, and the whole rock cycle starts anew – although not every rock particle is destined to undergo the full cycle.

**FOLDED ROCK**
*The folding of this sedimentary greywacke, found at Mizen Head on the coast of Ireland, shows the immense pressures to which it has been subjected in the interior of the Earth.*

# TYPES OF ROCK

A ROCK IS A NATURALLY occurring and coherent aggregate of one or more minerals – although there are a few rocks made from other substances, such as the decayed vegetation of which coal is composed. There are three major classes of rocks – igneous, sedimentary, and metamorphic – and each of these three classes is further subdivided into groups and types, principally based on differences in their mineral composition and texture.

## IGNEOUS ROCKS

The magmas (molten rocks) from which igneous rocks are formed (see p.28) either solidify underground, creating intrusive rocks such as granite – known as "plutonic" rocks – or flow onto the surface of the land or ocean bed, forming extrusive rocks such as basalt. Igneous rocks vary in texture, depending on how rapidly the magma cooled. Rapid cooling creates tiny crystals and freezes the liquid to glass, while slower cooling forms coarse–grained rocks. The oceanic crust is primarily composed of igneous rocks, including volcanic basalts and the underlying gabbros.

**Small crystals formed during rapid cooling after eruption**

**Large quartz crystal formed during slow cooling of magma before eruption**

**BASALT THIN SECTION**

**Fine-grained matrix**

**Crystals fill cavity (vug) left by gas bubble**

**BASALT WITH CRYSTAL-FILLED VUGS**

**ROCK FROM LAVA**
*Volcanic rocks, such as these on the Keanae Peninsula, Hawaii (right), are extrusive igneous rocks formed by cooled, solidified lava.*

**SIERRA GRANITE**
*The granite batholith of the Sierra Nevada range, California, is an example of intrusive igneous rock.*

**PINK GRANITE SPECIMEN**

*Pink feldspar*

*Feldspar crystallized first in this slow-cooling magma*

**GRANITE THIN SECTION**

*Mica fitted into spaces between feldspar crystals*

*Quartz formed last*

**HAWAIIAN LAVA**
*Basaltic lava flows from Kilauea volcano, Hawaii. Volcanic material varies in composition – for example, basaltic lava contains a relatively low percentage of silica, and so flows easily.*

# SEDIMENTARY ROCKS

Sedimentary rocks are generally made of deposits laid down on the Earth's surface by water, wind, or ice. These deposits are almost always laid down in layers or strata. The stratification survives compaction and cementation, and is a distinguishing characteristic of sedimentary rocks. They are also characterized by the presence of fossils, which are rarely found in any other type of rock. Some sedimentary rocks are formed by eroded fragments of other rocks in some way cemented together. They are classified by particle size, particle shape, and mineralogy. Some sedimentary rocks are of chemical origin, having been deposited in solid form from a solution, and some are of biochemical origin, composed predominantly of calcium carbonate. There are also organic deposits, such as coal, derived from the breakdown of the organic tissue of dead plants and animals.

Sand grain

Cement between grains

**SANDSTONE THIN SECTION**

Evidence of bedding

**SANDSTONE SPECIMEN**

**SEDIMENTARY LAYERS**
*This sandstone in Zion National Park, Utah, USA, displays a typical layered structure. The highest layers are the youngest, while the deepest are the oldest.*

# METAMORPHIC ROCKS

Metamorphic rocks are formed when existing rocks are subjected to extreme temperatures or pressures (or both) that alter their mineralogical composition, texture, and internal structure. Quartzite, for example, is metamorphosed sandstone, and slate is metamorphosed mudstone or shale. Such changes typically occur deep within the crust, and may result from the deformation caused by plate tectonics. But metamorphic rocks can also form virtually at the surface of the Earth under the impact of meteorites or near igneous intrusions, which create zones of high temperature around them. Metamorphic rocks include schists and gneisses, whose component minerals tend to segregate into separate bands, giving a layered appearance.

Biotite

**GNEISS SPECIMEN**

Quartz

Biotite

Feldspar

**GNEISS THIN SECTION**

**ROCKY COAST**
*These coastal rocks near Cape Breton, Nova Scotia, Canada, are mainly schists and gneisses, but marked with pink granite that has been forced into cracks in the metamorphic rocks.*

31

# IGNEOUS ROCKS

**VOLCANO VENT**
*A vent at Kilauea volcano, Hawaii, spews out lava that will cool to form igneous rock.*

IGNEOUS ROCKS ARE DEFINED as extrusive or intrusive depending on whether or not the molten magma from which they were formed emerged at the Earth's surface before crystallizing. Extrusive rocks form on the surface; intrusive rocks form below it. Intrusive rocks are categorized as plutonic if formed deep inside the crust and hypabyssal if formed at shallow depths. The silica content of the rocks is also a basis for classification. Felsic rocks have over 65 per cent silica, intermediate rocks 55–65 per cent, mafic rocks 45–55 per cent, and ultramafic rocks less than 45 per cent.

## EXTRUSIVE ROCK

Extrusive igneous rocks are also known as volcanic rocks. The principal rock types in this category include basalt, obsidian, rhyolite, trachyte, and andesite. All of these generally form from lavas – the term for a magma that has flowed onto the surface, either on land or underwater. Other extrusive rocks, such as tuff and pumice, are formed in explosive volcanic eruptions. These "pyroclastic" rocks are porous because of the frothing expansion of volcanic gases when they formed. Basalt is the commonest igneous rock, forming the floor of most oceans.

**COOLING LAVA**
*Volcanic materials vary in chemical composition and temperature. Hot and low in silica, this basaltic lava flowing on Mount Etna, Sicily, has a rough texture as it cools and hardens.*

**BASALT COLUMNS**
*Basaltic magma, the type that forms most of the oceanic crust, has cooled to create these five-sided perpendicular columns near Aldeyarfoss, Iceland.*

**GRANITE PEAKS**
*Like all batholiths, the Sierra Nevada in California is made up of many smaller intrusions which, with erosion effects, give its peaks a varied appearance.*

# PLUTONIC INTRUSION

Plutonic intrusive rocks solidify deep within the Earth's crust and are characterized by their large crystals. They occur as batholiths, plutons, and laccoliths. A batholith is a large igneous body with a surface exposure of at least 100 square km (40 square miles) and a thickness of about 10 to 15km (6 to 9 miles). Batholiths form the cores of great mountain ranges, such as the Rockies and the Sierra Nevada in North America. They are generally composites of a number of smaller intrusions called plutons, each of which may consist of several different rock types. Laccoliths are smaller scale intrusions that create a characteristic dome-shaped structure, generally with a horizontal floor. Granite, diorite, peridotite, syenite, and gabbro are all plutonic igneous rocks.

**DEVIL'S MARBLES**
*In Australia's Northern Territory, a dome of intrusive red granite has been exposed and weathered into rounded boulders.*

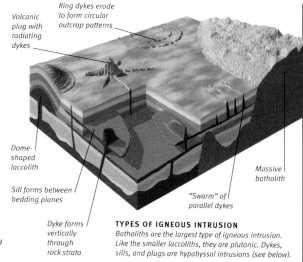

Volcanic plug with radiating dykes

Ring dykes erode to form circular outcrop patterns

Dome-shaped laccolith

Sill forms between bedding planes

Dyke forms vertically through rock strata

Massive batholith

"Swarm" of parallel dykes

**TYPES OF IGNEOUS INTRUSION**
*Batholiths are the largest type of igneous intrusion. Like the smaller laccoliths, they are plutonic. Dykes, sills, and plugs are hypabyssal intrusions (see below).*

# HYPABYSSAL INTRUSION

Formed at shallow depths, hypabyssal intrusive rocks are characterized by fine crystallization. They are found in dykes, sills, volcanic plugs, and other relatively small formations. A dyke is a sheet-like body that cuts vertically or at a steep angle to the surrounding rocks. Dykes range from less than a centimetre to many metres in thickness, and can be hundreds of kilometres in length. Sills are similar to dykes, except that they form parallel to the enclosing rocks, intruding between two strata. Plugs are formed from magma solidified inside volcanic vents.

**DOLERITE DYKE**
*A resistant dyke of igneous dolerite in Namibia projects above the surrounding rock.*

**VOLCANIC PLUG**
*The Devil's Tower, Wyoming, USA, is rock that cooled within a volcanic vent and has been exposed by erosion.*

**GRANITE TORS, HAYTOR**
*These weather-rounded granite boulders at a hilltop summit near an old granite quarry at Haytor, Devon, England, were once part of a mountain range.*

**EL CAPITAN, CALIFORNIA, USA**
*A granite buttress, El Capitan rises 1,098m (3,604ft) from the valley floor of Yosemite National Park.*

# GRANITE

THE MOST COMMON INTRUSIVE rock in the Earth's continental crust, granite is familiar as a mottled pink, white, grey, and black ornamental stone. It is coarse- to medium-grained; rhyolite is its fine-grained equivalent (see p.46). Granite's three main minerals are feldspar, quartz, and mica, which occur as dark biotite and/or as silvery muscovite. Of the three principal minerals, feldspar predominates, and quartz usually accounts for more than 10 per cent. The alkali feldspars are often pink (although they can also be white, buff, or grey), resulting in the pink granite often used as an ornamental stone for the facings of buildings and for floors.

Plagioclase feldspar also occurs in granite. The ratio between plagioclase and alkali feldspars has provided the basis for granite classifications, but if plagioclase greatly exceeds alkali feldspar, the rock is considered granodiorite. Granites from the eastern, central, and south-western United States, south-west England, the Baltic Shield area, western and central France, and Spain tend be low in plagioclase, whereas in large regions of the western United States granites in which plagioclase exceeds alkali feldspar are common. These granites in the western United States are part of a great series of batholiths –

**GRANITE PEAKS**
*A sandstone arch frames this view of the massive granite batholith that is the Sierra Nevada Mountain Range, USA.*

## GRANITE'S DURABILITY

Granite has been a favourite stone for carving and for building structures for at least four millennia. Wherever it has been available, its strength and durability have made it a first choice for uses as exotic as temples and as mundane as millstones. Modern buildings are still faced with granite, both as an ornament and an acknowledgement of its agelessness as a building material.

Granite was highly favoured by the ancient Egyptians for monumental carving, and for obelisks in particular. These needle-like spires were often tens of metres tall and weighed many tonnes. The red granite was quarried near Aswan, site of the current hydroelectric dam across the Nile. Other colossal blocks were quarried for the interior chambers of the Great Pyramid, again weighing many tonnes. The King's Chamber alone is roofed by nine full-span granite blocks, each weighing more than 50 tonnes. Another single granite block in the pyramid weighs about 70 tonnes.

**UNFINISHED OBELISK**
*A huge Egyptian obelisk lies unfinished in the granite quarry at Aswan.*

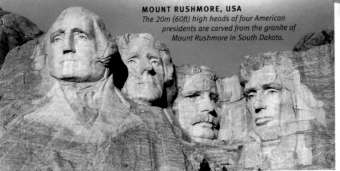

**MOUNT RUSHMORE, USA**
*The 20m (60ft) high heads of four American presidents are carved from the granite of Mount Rushmore in South Dakota.*

### PROPERTIES

**ROCK TYPE** Felsic, plutonic, igneous

**MAJOR MINERALS** Potassium-feldspar, quartz, mica, sodium

**MINOR MINERALS** Sodium-plagioclase, hornblende

**COLOUR** White, light grey, grey, pink, red

**TEXTURE** Medium to coarse

**PINK GRANITE**
*In this specimen of pink granite the three essential components of all granites can be seen: quartz, alkali feldspar, and mica. Pink granite is used as a building stone, and its extensive use in Aberdeen, Scotland, gives the city a notable pink hue. Aberdeen is the largest granite exporting area in the UK.*

huge rock masses often hundreds or thousands of square kilometres in area – stretching from Alaska and British Columbia southward through Idaho and California into Mexico. Amphiboles and pyroxenes are minor essential minerals of granite, and many others occur as accessories.

## FORMATION AND USES OF GRANITE

Granite crystallizes from silica-rich magmas tens of kilometres deep in the Earth's crust. Because it is formed at depth, the exposure of granite at the surface is evidence that the area has been uplifted and that the great thickness of rock overlying the granite has been eroded away. Many important mineral deposits are

**GRANITE QUARRY**
*Towering blocks of granite stand in the Rock of Ages Quarry, a long-producing granite quarry in Barre, Vermont, USA.*

formed in the vicinity of crystallizing granite bodies from the hydrothermal solutions that are released by them. For example, pegmatite gemstones such as topaz, rock crystal, tourmaline, and aquamarine, as well as many metalliferous ores, including those of gold, silver, lead, and titanium, are deposited by these solutions.

The quarrying of granite was once a major industry due to demand for it as a building stone. Today, granite is used rather less, mainly as a veneer for commercial buildings and, to a lesser extent, to make tombstones and kitchen worktops. In some countries, granite is also quarried for use as kerbstones.

Grey quartz crystals

**TRIUMPHAL ARCH**
*This triumphal arch in North Korea was recently constructed from granite.*

Black biotite mica

**THIN SECTION**
*An intergrowth of feldspar and quartz can be seen in this thin section of granite. Mica fits into the spaces between other crystals.*

Mica

Feldspar

Quartz

Coarse crystallization

Pink feldspar

**AQUEDUCT**
*The aqueduct at Segovia, Spain, was built of Guadaramma granite during the reign of the Roman emperor Trajan (AD98–117). It is 728m (2,388ft) long and is still in use.*

**GRANITE VARIETIES**
*Of the numerous varieties of granite, some have names relating to their texture or the presence of minerals in addition to the three basic ones. Others have informal names based on colour.*

White orthoclase feldspar

Grey quartz

Feldspar phenocryst

Spherical phenocryst

Pink orthoclase feldspar

Pale orthoclase feldspar

Dark hornblende

**WHITE GRANITE**

**PORPHYRITIC GRANITE**

**ORBICULAR GRANITE**

**GRAPHIC GRANITE**

**HORNBLENDE GRANITE**

# PEGMATITE

PEGMATITE IS THE NAME given to very coarse-grained, igneous rocks. Most have the same major constituents – quartz and feldspar – as granites or syenites, but intermediate, mafic, and ultramafic pegmatites are sometimes found. They usually form as sheets, pods, lenses, or cigar-shaped bodies in large plutonic bodies with the same principal minerals. Although crystals can be huge – metres long – the average size is 8–10cm (3–4in). The large crystals are due to the large amount of water in the magma rather than to slow cooling. Crystals can occur in hollows in the pegmatite, or the entire pegmatite may be completely filled. Some pegmatites may consist of a single or a predominant mineral, and take their names from that mineral. Quartz and the silica-rich feldspars are the major constituents of granitic and syenitic pegmatites. These types of pegmatites are particularly important because they are the chief commercial source of feldspar and sheet mica, and also a major source of gemstones. Tourmaline, aquamarine, emerald, rock crystal, smoky quartz, rose quartz, topaz, moonstone, and garnet are all pegmatite minerals. Pegmatites are also the source of other important economic minerals, which contain beryllium, lithium, tin, titanium, molybdenum, tungsten, tantalum, niobium, and other rare elements.

**TUNGSTEN-TIPPED**
*Mined from pegmatites, tungsten adds strength to many everyday objects.*

**PEGMATITE BANDS**
*Light-coloured bands of pegmatite can be clearly seen here cutting across darker bands of gneiss.*

## MINING TUNGSTEN

Pegmatites are one of the sources of important minerals. These minerals are, in turn, the ores of less common but vitally important metals. Tungsten is one of these. It goes into the manufacture of items as exotic as a rocket nozzle (see p.217), or as mundane as a set of darts or the tip of a ballpoint pen. The pegmatite minerals wolframite (in the form of hübnerite and ferberite) and scheelite are its major mineral sources. China has about 50 per cent of the world's reserves, and in Russia deposits are located in the northern Caucasus and around Lake Baikal.

**HIMALAYA MINE**
*The Himalaya Mine in California has pegmatite veins that produce stunning tourmalines.*

**THIN SECTION**
*The crystals in this pegmatite from Scourie, Scotland, are up to 1m (3ft) in size.*

*Intergrown quartz crystal*

*Large microcline crystal*

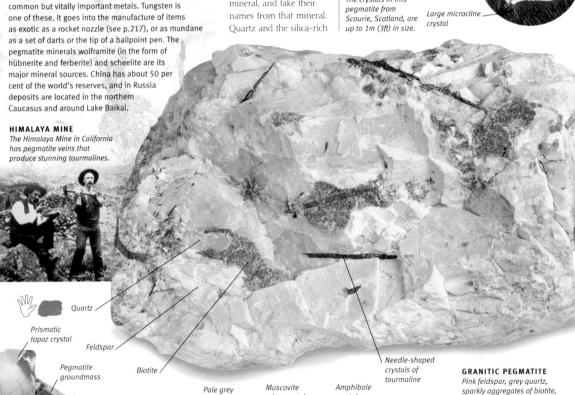

*Quartz*

*Prismatic topaz crystal*

*Feldspar*

*Pegmatite groundmass*

*Biotite*

*Feldspar*

*Quartz*

*Pale grey quartz*

*Muscovite mica crystals*

*Amphibole crystals*

*Needle-shaped crystals of tourmaline*

*White feldspar crystals*

*Mica*

**TOPAZ IN PEGMATITE**

**GROUP OF CRYSTALS IN A GRANITIC PEGMATITE**

**MICA PEGMATITE**

**FELDSPAR PEGMATITE**

**TYPES OF PEGMATITE**
*The term "pegmatite" is often qualified by either its predominant mineral composition, such as mica pegmatite, or its origin, such as granitic pegmatite.*

**GRANITIC PEGMATITE**
*Pink feldspar, grey quartz, sparkly aggregates of biotite, and black, needle-shaped crystals of tourmaline make up this specimen of pegmatite.*

**PROPERTIES**

| | |
|---|---|
| **ROCK TYPE** | Felsic, plutonic, igneous |
| **MAJOR MINERALS** | Quartz, feldspar, mica |
| **MINOR MINERALS** | Tourmaline, topaz |
| **COLOUR** | Light |
| **TEXTURE** | Very coarse |

# GRANODIORITE

GRANODIORITE IS AMONG the most abundant of intrusive igneous rocks. A medium- to coarse-grained rock similar to granite, it has more plagioclase feldspar than orthoclase feldspar. Granodiorite can be pink or white with a grain size and texture similar to granite, but the plagioclase generally makes it appear darker, and the hornblende and biotite that are

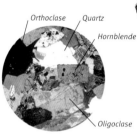

**ALPINE ROCKS**
*Greyish-pink granodiorite is seen here in the higher reaches of the Austrian Alps.*

often present give it a speckled appearance. Twinned plagioclase crystals are sometimes wholly encased by orthoclase. Its quartz can be grey to white.

**PROPERTIES**

**ROCK TYPE** Felsic, plutonic, igneous
**MAJOR MINERALS** Plagioclase, K-feldspar, quartz, mica
**MINOR MINERALS** Hornblende, augite
**COLOUR** Grey, white, or pink
**TEXTURE** Medium to coarse

Orthoclase  Quartz

Hornblende

Oligoclase

**THIN SECTION**
*Clearly visible in this section, from Strontian, Scotland, are oligoclase, quartz, orthoclase, biotite hornblende, and titanite.*

Coarse texture

Light feldspar

Dark, ferro-magnesian minerals

## INDUSTRIAL USES

Granodiorite is hard and durable, and is quarried, cut, and crushed for use as road ballast and kerbstones. In the past, it has been used for cobblestones. In modern times, it is also sawn and polished to create flooring, facing for buildings, and worktops, due to its toughness and often attractive speckling. It is one of the stones sold as "black granite".

**CUTTING STONE**
*When used as road ballast, granodiorite is often chipped into shape.*

**SPECKLED GRANODIORITE**
*"White" granodiorite has a speckled appearance because of darker minerals such as mica and hornblende.*

# DIORITE

**DIORITE CLIFFS**
*Diorite forms these cliffs in County Antrim, Northern Ireland.*

BECAUSE IT GRADES imperceptibly into granite, diorite is somewhat difficult to define. In general, though, this medium- to coarse-grained intrusive igneous rock is darker in colour than granite. It is commonly composed of about two-thirds plagioclase feldspar and one-third dark-coloured minerals, such as hornblende or biotite. In diorites the plagioclases are sodium-rich (oligoclase or andesine), distinguishing them from gabbros, in which calcium-rich plagioclases (labradorite or bytownite) predominate. Diorite can be of uniform grain size, or have large phenocrysts of plagioclase or hornblende. It can occur as large intrusions, or as smaller dikes and sills. Most diorite is intruded along the margins of continents. Rarely used as an ornamental and building material, diorite is sometimes sold as "black granite".

## ANCIENT USAGE

Diorite was one of the stones that was particularly prized in ancient Egypt. It was used in statuary, for columns, pillars, and sarcophagi, and for lining the chambers of some pyramids. In North America, a diorite bowl shaped in the form of a duck has been called "the Portland Vase (a superb glass vase from ancient Rome) of Native American culture".

**DIORITE YOKE**
*Created between 300 and 900AD, this diorite yoke is from South America.*

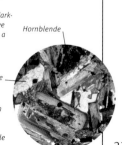

**DIORITE CHEPHREN**
*This statue of Pharaoh Chephren (2650–2134BC) is carved from diorite.*

**PROPERTIES**

**ROCK TYPE** Intermediate, plutonic, igneous
**MAJOR MINERALS** Sodium plagioclase, hornblende
**MINOR MINERALS** Biotite
**COLOUR** Mottled black/dark green and grey or white
**TEXTURE** Medium to coarse

Light plagioclase feldspar

**CONTRASTING COLOURS**
*Contrasting light-coloured plagioclase feldspar and dark-coloured hornblende give this specimen of diorite a two-tone appearance.*

Hornblende

Hornblende

Andesine

**THIN SECTION**
*This section from Glen Doll, Scotland, shows mostly andesine and hornblende with a little quartz and biotite.*

37

# ANORTHOSITE

**ROCK OUTCROP**
*This anorthosite outcrop is in South Africa.*

AN INTRUSIVE IGNEOUS ROCK, anorthosite is composed of at least 90 per cent calcium-rich plagioclase feldspar – principally labradorite and bytownite. Olivine, pyroxene, garnet, and iron oxides make up the remaining 10 per cent. Anorthosite is not a common rock on Earth, but where it does occur, it is found as immense masses, or as layers between mafic and ultramafic rocks such as gabbro and peridotite. There are large anorthosite-bearing rock bodies in New York State and Montana, USA; eastern Canada; and South Africa. Many anorthosites have an interesting "cumulate" texture, where well-formed crystals appear to have settled out of the liquid magma, in a similar way to a sediment. Anorthosite is extremely common on the surface of the Moon, especially the far side. The ancient, rough, light-coloured highlands of the Moon are made of anorthosite and similar rocks.

Pyroxene

Plagioclase crystals

**THIN SECTION**
*A small amount of pyroxene is trapped between this accumulation of plagioclase crystals.*

**PROPERTIES**

**ROCK TYPE** Ultramafic, plutonic, igneous

**MAJOR MINERALS** Calcium plagioclase

**MINOR MINERALS** Olivine, pyroxene, garnet

**COLOUR** Light grey to white

**TEXTURE** Medium to coarse

Light plagioclase feldspar crystals

Coarse grain size

**COARSE ANORTHOSITE**
*This example of anorthosite from the Bushveld Complex of South Africa has a coarse texture.*

**ANORTHOSITE FROM THE MOON**
*Fine-grained anorthosite is thought to be the first rock that crystallized from the "magma ocean" that covered the Moon's surface early in its history.*

# SYENITE

**LAYERED SYENITE**
*This layered syenite is from South Africa.*

SYENITES MAY BE VISUALLY SIMILAR to granite, and are often confused with it, but they can be distinguished from it by the absence or scarcity of quartz. A syenite is any one of a class of rocks essentially composed of an alkali feldspar or sodic plagioclase (or both), a ferromagnesian mineral – usually biotite, hornblende, or pyroxene – and little or no quartz. The alkali feldspars can include orthoclase, albite, or less commonly, microcline. Other minerals that can occur in small amounts in syenite include sphene, apatite, zircon, magnetite, and pyrite. When syenites contain quartz, they are quartz syenites. In the absence of quartz, feltspathoid minerals such as leucite, cancrinite, sodalite, or particularly nepheline, may be present.

**PINK SYENITE**
*The pink colouring of syenite is due to the presence of alkali feldspar, which predominates in syenite.*

**PROPERTIES**

**ROCK TYPE** Intermediate, plutonic, igneous

**MAJOR MINERALS** Potassium feldspar

**MINOR MINERALS** Sodium plagioclase, biotite, amphibole, pyroxene, feldspathoids

**COLOUR** Grey, pink, or red

**GRAIN SIZE** Medium to coarse

Amphibole

Coarse grain size

Grey feldspar crystals

Orthoclase

Hornblende

**THIN SECTION**
*Large, simple-twinned orthoclase crystals are present in this section, with hornblende and a little quartz.*

Quartz

Feldspar

**LARVIKITE**
*This iridescent variety of syenite contains feldspar, pyroxene, mica, and amphibole.*

**DOLERITE DYKE**
*A darker dyke of dolerite cuts through a body of lighter granite.*

# DOLERITE

DOLERITE IS EQUIVALENT to basalt and gabbro in composition, although it is intermediate between them in grain size. A fine- to medium-grained rock, dolerite is from one-third to two-thirds calcium-rich plagioclase feldspar, the remainder being principally pyroxene. It has a silica content of less than 55 per cent, and its quartz content is usually less than 10 per cent. Magnetite and olivine may be present: if olivine is present, the rock is called olivine-dolerite. Extremely hard and tough, it occurs in dykes and sills intruded into fissures in other rocks.

*Plagioclase feldspars*

*Medium texture*

**DARK GREY DOLERITE**
*Dolerite's medium texture and dark colour can be seen in this specimen. It is one of the dark-coloured rocks sold commercially as "black granite".*

## ANCIENT BUILDING STONE

The dolerite "bluestones" of the inner circle of Stonehenge were transported 385km (240 miles) from Wales to Wiltshire in England. They were brought by sea, river, and over land. Exactly what significance was attached by the builders to this particular stone is unknown, but it must have been considerable to justify the enormous effort required to get the bluestones to the site. Stonehenge was constructed in 3000–2000BC.

**STONEHENGE**
*The bluestone used for making the stone circle of Stonehenge originated at Carn Meini Quarry in Wales.*

### PROPERTIES

| | |
|---|---|
| **ROCK TYPE** | Mafic, plutonic, igneous |
| **MAJOR MINERALS** | Calcium plagioclase , pyroxene |
| **MINOR MINERALS** | Quartz, magnetite, olivine |
| **COLOUR** | Dark grey to black, often mottled white |
| **TEXTURE** | Fine to medium |

**DOLERITE OUTCROP**
*The Tees River in north-east England forms waterfalls such as this one on the hard dolerite outcrop of Whin Sill.*

# GABBRO

**GABBRO HILLS**
*The Black Cuillin Hills of the Isle of Skye are formed entirely from gabbro.*

MEDIUM- OR COARSE-GRAINED rocks, gabbros consist principally of calcium-rich plagioclase feldspar and pyroxene, usually augite. They are low in silica – usually less than 55 per cent – and quartz is rarely present. Gabbro is essentially the intrusive equivalent of basalt, but where basalt is often relatively homogeneous in mineralogy and composition, gabbros are highly variable. They are often found with layering of light and dark minerals (layered gabbro), with a significant amount of olivine (olivine gabbro), and with a high percentage of coarse crystals of plagioclase (leucogabbro). Gabbros are widespread, but not common, on the Earth's surface. Commercially, they are sold as "black granite", although they owe their major economic significance to the ores contained within them. Some gabbros are mined for their nickel, chromium, and platinum content. Those containing magnetite and ilmenite are mined for their iron or titanium.

**CHRISTIAN LEOPOLD VON BUCH (1774–1852)**
*Geologist von Buch, famed for creating the first geological map of Germany, named gabbro after a town in Italy.*

**THIN SECTION**
*This is an olivine gabbro (labradorite feldspar, augite, and olivine) from the Duluth Complex, USA.*

*Olivine*

*Augite*

### PROPERTIES

| | |
|---|---|
| **ROCK TYPE** | Mafic, plutonic, igneous |
| **MAJOR MINERALS** | Calcium plagioclase feldspar, pyroxene |
| **MINOR MINERALS** | Olivine, magnetite |
| **COLOUR** | Dark grey to black |
| **TEXTURE** | Medium to coarse |

*Light plagioclase feldspar*

*Dark pyroxene*

**LAYERED GABBRO**
*Alternating bands of light plagioclase feldspar and dark ferromagnesian minerals form this specimen.*

**COARSE GABBRO**
*This coarsely crystallized specimen of gabbro contains light-coloured crystals of plagioclase and dark crystals of pyroxene.*

**PERIDOTITE HILLS**
*The Semail ophiolite crops out of the mountain belt near Muscat, Oman. It is composed of layered peridotite and gabbro.*

## PROPERTIES

| | |
|---|---|
| **ROCK TYPE** | Ultramafic, plutonic, igneous |
| **MAJOR MINERALS** | Olivine, pyroxene |
| **MINOR MINERALS** | Garnet, chromite |
| **COLOUR** | Dark green to black |
| **TEXTURE** | Coarse |

# PERIDOTITE

ONE OF THE MOST ECONOMICALLY important rocks, peridotite is the ultimate source of all chromium ore, and its variety kimberlite (see opposite) is the source of all naturally occurring diamonds. Fresh peridotite contains olivine, pyroxene, and chromite. Peridotite that has been altered by weathering becomes serpentinite, containing chrysotile, other serpentines, and talc. Formerly, peridotite was a major source of magnesite (see p.179), an important economic mineral. It is also the major source of chrysotile asbestos (see p.258), now of declining economic importance.

An intrusive igneous rock, peridotite is coarse-grained, dark-coloured, and dense. It contains at least 40 per cent olivine. It is related to the rock pyroxenite in that they form in similar environments and both are olivine-pyroxene rocks. Pyroxenite contains at least 60 per cent pyroxene. Another related rock is dunite, a yellowish-green, intrusive igneous rock composed almost entirely of olivine. Like peridotite, dunite is an important source of the chromium ore chromite.

Peridotite is found interlayered with iron- and magnesia-rich rocks in the lower parts of layered, igneous rock bodies, where its denser crystals formed first through selective crystallization and then settled to the bottom of

**THIN SECTION**
*This is harzburgite from New Zealand, with olivine and orthopyroxene visible.*

Orthopyroxene

Olivine

## THE EARTH'S MANTLE

Peridotite is a major structural component of the Earth. It is the major component of the asthenosphere, the upper portion of the mantle upon which the plates rest. At depths of 70–200km (40–125 miles), the mantle rocks are thought to exist at temperatures slightly above their melting point, and a small percentage may be molten. This gives the mantle the plasticity for continental movement.

**LAVA FLOW**
*The deepest magmas originate in the upper layer of the mantle.*

The basaltic magmas of the oceanic crust are generated in the asthenosphere, their compositions determined to an extent by the mineral make-up of the peridotite.

**MANTLE XENOLITH**
*Fragments of mantle peridotite are brought to the surface by deep-seated magmas.*

still-fluid or semi-solid crystallizing mushes. Within peridotite bodies, layers of chromite (the principal ore of chromium), an early crystallite and the first to settle, can reach several metres thick. Peridotite is also found in mountain belts as irregular, olivine-rich masses, sometimes with related gabbro, and in dykes (vertical or steeply inclined sheet-like bodies, which often cut existing rock structure). Garnet is often present as an accessory mineral, and weathered peridotites are commonly a source of nickel. Peridotite's other important occurrence is in volcanic pipes as kimberlite.

**EXPOSED PERIDOTITE**
*Gros Morne National Park in Newfoundland, Canada contains peridotite rocks from the Earth's mantle.*

**GREEN PERIDOTITE**
*This specimen from Lanzarote in the Canary Islands takes its green colouring from the mineral olivine present within it.*

Green olivine

Dark pyroxene

Coarse texture

Greenish colouring from olivine

**DUNITE**

**PYROXENITE**

**RELATED ROCKS**
*Both pyroxene and dunite are related to peridotite. They are deep-seated in origin, and contain a large proportion of olivine.*

# KIMBERLITE

THE MAJOR SOURCE OF DIAMONDS, kimberlite is a form of peridotite. It is mica-rich, often with well-formed crystals of brown phlogopite mica. Other abundant minerals include chromium- and pyrope-rich garnet and chrome-bearing diopside; lesser amounts of ilmenite, serpentine, pyroxene, calcite, rutile, perovskite, magnetite, and diamond can also be present.

Kimberlite occurs in pipes – intrusive igneous bodies roughly circular in cross section with vertical sides, and usually less than 1km ($^3$/$_4$ mile) in diameter – and other igneous bodies that are also steep-sided and intrusive. These tend to be found in the uplifted centres of continental platforms, and appear to be roughly the same age, having formed during the Late Cretaceous Period (100 to 65 million years ago). It appears that the kimberlite was injected into zones of weakness in the crust as a relatively cool "mush" of crystals rather than as a liquid magma that crystallized within the pipes themselves, as there is usually little contact metamorphism of surrounding rocks. Xenoliths (fragments) of mantle rock are often brought to the surface in kimberlites, making them a valuable source of information about the mantle.

**KIMBERLITE PIPE**
*The mined-out diamond-bearing pipe at Kimberley, South Africa, has now filled with water. Kimberlite pipes do not have a great lateral extent, but they are very deep.*

## PROPERTIES

**ROCK TYPE** Ultramafic, volcanic, igneous
**MAJOR MINERALS** Olivine, pyroxene, mica
**MINOR MINERALS** Garnet, ilmenite, diamond
**COLOUR** Dark grey
**TEXTURE** Fine to coarse/porphyritic

## MINING KIMBERLITE

Diamonds were first discovered at De Kalk Farm on the banks of the Orange River, South Africa, in 1867, but the host rock was not discovered until two years later. At first, diamonds were extracted by panning from weathered kimberlite in open pits, but as the pits became deeper and the rock harder it was necessary to mine underground. After the rocks have been blasted, they are brought to the surface and the diamonds are separated out. Wet, crushed kimberlite is passed across grease-covered tables: water does not stick to diamond, so the dry stones stick to the grease, while the kimberlite washes away (see p.107). The diamond crystals are scraped off. Now many of the mines are worked out and South Africa is no longer one of the top diamond-producing nations.

**DRILLING**
*Drill-holes packed with explosives are blasted to break up the rocks.*

**ROUGH DIAMOND**
*Most gem-quality diamonds are found as octahedral crystals.*

Diamonds need pressures of over 50,000 atmospheres to form. This corresponds to a depth below the Earth's surface of at least 150km (90 miles), well into the upper peridotite mantle. An examination of minute mineral grains trapped inside growing diamonds reveals that most diamonds formed one to three billion years ago. Once diamonds formed in the upper mantle, they remained there for millions of years before being blasted to the surface at supersonic speeds in devastatingly volatile eruptions. Fortunately, no kimberlite volcanoes are known to have erupted for over 60 million years.

Kimberlite is also called blue ground, a reference to its colour. Weathered kimberlite is yellowish in colour, and is called yellow ground. It is in areas of yellow ground that diamonds were first discovered in South Africa, and later in the USA in Arkansas, Colorado, and Wyoming.

*Crystal of ferro-magnesian minerals*

*Dark matrix*

*Diamond*

*Altered pyroxene*

*Coarse texture*

**BRECCIATED KIMBERLITE**
*This dark, coarse-grained, heavy, and brecciated (fragmented) specimen is typical of kimberlite. The mineral is named after Kimberley in South Africa, the old centre of the South African diamond industry.*

**THIN SECTION**
*Larger, corroded crystals of phlogopite and pyroxene sit in a heavily altered matrix of olivine, calcite, and mica in this section of kimberlite from De Beer's Mine, Kimberley.*

*Phlogopite*

**WEATHERED KIMBERLITE**
*This heavily weathered specimen of kimberlite from Kimberley, South Africa, holds an octahedral diamond crystal.*

41

# BASALT

**COLUMNAR BASALT**
*This outcrop of basalt exposed by coastal weathering shows the columnar jointing that can occur as the basalt cools and shrinks at different rates.*

THIS VOLCANIC ROCK is the most common igneous rock at the Earth's surface. However, most basalt is not seen because it forms the ocean floors. Volcanoes in the ocean basins erupt basalt, and volcanic islands are made from it. On land, basalts have built enormous plateaux, such as in the Deccan of India, the Columbia River region of north-western USA, and the Paraná Basin of South America. The large, dark maria on the Moon are basalt, and it is quite likely that the volcanoes on Mars and Venus are, too.

Dark in colour and comparatively rich in iron and magnesium, basalts are mainly composed of plagioclase, pyroxene, and olivine. Most are very fine grained and compact, even glassy, but it is also common for them to be porphyritic – having distinct crystals set in a fine-grained base or groundmass. The larger crystals are called phenocrysts, and commonly consist of olivine, augite, or plagioclase feldspar. Feldspathoids (see pp.242–48), such as nepheline and leucite, occur in many mafic basalts. Nepheline-basalts are fairly common in Germany, Libya, Turkey, and in New Mexico, USA. Leucite-basalts are principally found in Italy, Germany, eastern Africa, Australia, and in Montana, Wyoming, and Arizona, USA.

Basaltic lavas are frequently charged with water, the resulting froth making their rocks spongy. Basalts are very interesting to geologists because they form by partial melting of the Earth's mantle. Unmelted fragments of the mantle (known as xenoliths) can be carried upwards in basalts, giving geologists their only direct observation of the composition of the mantle.

**BASALT PILLOWS**
*Pillow-basalt is formed when basaltic magma flows into water during underwater eruptions.*

Fine matrix

**BASALT**

Fine matrix
Pyroxene phenocryst

**PORPHYRITIC BASALT**

Empty cavities

**VESICULAR BASALT**

**VARIETIES OF BASALT**
*Several varieties of basalt are found, taking their names from their texture and crystallization.*

**STUDYING LAVA FLOW**
*Vulcanologists often take considerable personal risks in order to sample unadulterated lava.*

Fine-grained matrix
Amygdule filled with zeolite

**AMYGDALOIDAL BASALT**
*Amygdules are the small, round cavities left by gas bubbles and filled with minerals. The steam cavities left in this specimen of basalt are filled with zeolite crystals.*

Mafic minerals

## THE ROSETTA STONE

Made of black basalt, the Rosetta Stone was uncovered by Napoleon's troops in 1799 at Rosetta, near Alexandria, Egypt. Its identical inscriptions in three scripts – Egyptian hieroglyphs, demotic script (cursive hieroglyphs), and Greek – led to the decipherment of ancient Egyptian hieroglyphics by Jean-François Champollion of France. Carved in the reign of Ptolemy V Epiphanes (205–180BC), it currently resides in the British Museum, London. It was acquired as war booty when Napoleon was defeated in Egypt by the British in 1801.

Egyptian hieroglyphs
Demotic script

Pigeonite crystal
Fine-grained base
Augite crystal

**THIN SECTION**
*This section is of a lunar mare basalt collected from the edge of the Imbrium basin.*

## PROPERTIES

**ROCK TYPE** Mafic, volcanic, igneous

**MAJOR MINERALS** Sodium plagioclase, pyroxene, olivine

**MINOR MINERALS** Leucite, nepheline

**COLOUR** Dark grey to black

**TEXTURE** Fine-grained to porphyritic

# OBSIDIAN

A NATURAL VOLCANIC GLASS, obsidian forms when lava solidifies so quickly that mineral crystals do not have time to grow. The name obsidian simply refers to the glassy texture, and technically obsidian can have any chemical composition. However, most obsidian is similar in composition to rhyolite, and is commonly found on the outer edges of rhyolite domes and flows. Like rhyolite, obsidian can show flow-banding. It is also found along the rapidly cooled edges of dykes and sills. Obsidian is typically jet-black, although the presence of hematite (iron oxide) produces red and brown varieties, and the inclusion of tiny gas bubbles can create a golden sheen.

**NEWBERRY CRATER, USA**
*Obsidian is abundant in this crater in Oregon, resulting from numerous lava flows. Basalt, pumice, and rhyolite are also widespread in this huge area.*

**FABERGÉ HORSE**
*This obsidian horse has gold hooves, ruby eyes, and an enamelled bridle decorated with diamonds.*

Tiny crystals of feldspar and phenocrysts (visible crystals) of quartz can also be present. Most obsidian is relatively young because volcanic glass devitrifies – minerals crystallize from the glass – over time. Spherical clusters of radially arranged, needle-like crystals called spherulites in snowflake obsidian can form through devitrification after the rock has cooled. Pitchstone obsidian has a resinous lustre that forms as the rock absorbs water over time. Mount Hekla in Iceland, the Eolie Islands off the coast of Italy, and Obsidian Cliff in Yellowstone National Park, Wyoming, USA, are all well-known occurrences of obsidian.

## ANCIENT USES

Because obsidian is slightly harder than window glass and has a conchoidal fracture, allowing it to be chipped with razor-sharp edges, it was used by Native Americans and many other ancient peoples to create weapons, tools, and ornaments. The Aztecs and Greeks used it for mirrors because it was so reflective. It was widely traded along ancient routes, and by studying the properties of obsidian artefacts, archaeologists can determine both the source of the material and the routes along which it was traded. It is still cut as a semi-precious stone today.

**OBSIDIAN KNIFE BLADE**
*This image of a face from Mexico City, dating to 1325–1521, includes white flint and green obsidian.*

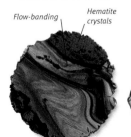

**ANCIENT ARTEFACTS**
*This obsidian Olmec mask and Aztec knife demonstrate two ancient methods of working obsidian to create a smooth and conversely, a razor-sharp finish.*

## PROPERTIES

| | |
|---|---|
| **ROCK TYPE** | Felsic, volcanic, igneous |
| **MAJOR MINERALS** | Glass |
| **MINOR MINERALS** | Hematite, feldspar |
| **COLOUR** | Black, brown, red |
| **TEXTURE** | Amorphous |

**BLACK OBSIDIAN**
*This specimen of black obsidian clearly shows the excellent conchoidal fracture that made it perfect for chipping into tools. Its bright, vitreous lustre shows why it was favoured for mirrors.*

*Conchoidal fracture*

*Glass rather than mineral crystals*

*Flow-banding*

*Hematite crystals*

*White spherulites*

**OBSIDIAN WITH RED COLOURING**

**SNOWFLAKE OBSIDIAN**

**VARIETIES OF OBSIDIAN**
*The presence of hematite (iron oxide) produces the red colouring in the thin section on the far left. The "snowflakes" in snowflake obsidian (left) are spherulites, spherical aggregates of needle-like crystals.*

**"APACHE TEAR"**
*Tumble-polished nodules of obsidian are sold as "apache tears".*

# THE GIANT'S CAUSEWAY

A PROMONTORY OF BASALT COLUMNS THAT JUT OUT OF A CLIFF FACE AS IF THEY WERE STEPS CREEPING INTO THE SEA, THE GIANT'S CAUSEWAY HAS BEEN INTENSIVELY STUDIED BY GEOLOGISTS SINCE THE 17TH CENTURY.

**JOINTED COLUMNS**
*The columnar jointing in the Causeway's basalt extends through the entire thickness of some of its component lava flows. The result of such effective jointing is columns extending up to a height of 25m (82ft).*

The Giant's Causeway extends along 6km (4 miles) of the coast of County Antrim, Northern Ireland, about 40km (25 miles) north-east of Londonderry. Formed 50 to 60 million years ago, it consists of about 40,000 stone pillars, each typically with five to seven irregular sides that vary from 40 to 50cm (15 to 20in) in diameter and measure up to 25m (82ft) in height.

## ANCIENT BASALT FLOWS

Its structure is the result of successive surges of lava that flowed towards the coast and cooled when they met the sea. The basaltic lava that welled up through fractures in the local chalk beds can still be seen as dark bands jutting out into the sea where erosion has cut away the chalk. The layers of basalt formed columns that today have the appearance of a huge bundle of tightly standing hexagonal fenceposts. Each column exhibits "columnar jointing", which develops as the cooling lava contracts at different rates. Similar columnar jointing can be seen at the Devil's Tower, Wyoming, USA (see p.48).

The tops of the columns form stepping stones that disappear under the sea. Its appearance led to the local legend that the Causeway was the creation of the giant Finn MacCoul, the ancient Ulster warrior. According to legend, he fell in love with a female giant on Staffa, an island in the Hebrides, and built this highway so that she could walk to meet him in Ulster.

In 1692 the Bishop of Derry made one of the first recorded visits to the Causeway, and a debate raged for nearly a century afterwards about whether the Causeway was the work of giants, a natural occurrence, or man-made. It was not until 1774 that French vulcanologist Nicolas Desmarest (1725–1815) proved that columnar basalts were formed by the cooling of molten lava.

**CHIMNEY STACKS**
*In some areas of the Causeway, the basalt has eroded and weakened. Some of the columns have toppled, leaving a few columns still standing like chimney stacks.*

**GIANT'S CAUSEWAY HEAD**
*The distinctive jointed columns of the Causeway are very similar in appearance to those of Fingal's Cave on the Scottish island of Staffa. Most are hexagonal, although some have four, five, seven, or eight sides.*

# RHYOLITE

**RHYOLITE MOUNTAIN**
*This mountainous outcrop in Snowdon, North Wales, is rhyolite.*

A RELATIVELY RARE volcanic rock, almost exclusively confined to the interiors and margins of continents, rhyolite is the volcanic equivalent of granite (see p.34). Most rhyolites are porphyritic, with larger crystals (phenocrysts) in a fine-grained matrix of crystals too small to be seen with the naked eye, indicating that crystallization began before the lava flowed out onto the surface. The phenocrysts are most commonly quartz and sanidine, but can also include mafic minerals such as biotite, amphibole, and pyroxene. In banded rhyolites, there are few or no phenocrysts. Rhyolite commonly occurs with obsidian and pumice, and with intermediate volcanic rocks such as andesite.

*Dark, larger crystals*
*Pale, fine-grained matrix of crystals*

**PORPHYRITIC RHYOLITE**

## PROPERTIES

**ROCK TYPE** Felsic, volcanic, igneous
**MAJOR MINERALS** Quartz, potassium feldspar
**MINOR MINERALS** Glass, biotite, amphibole, plagioclase
**COLOUR** Very light to medium grey, light pink
**TEXTURE** Fine or porphyritic

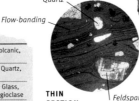

*Quartz*
*Flow-banding*
*Feldspar*

**THIN SECTION**
*Feldspar and quartz phenocrysts can be seen here.*
*Fine texture*

## RHYOLITE DOMES

Because rhyolitic lava is so silica rich, it is also very sticky, or viscous, and rarely reaches the Earth's surface. When it does, because it is so thick, the lava piles up over its vent without flowing away, and tends to form steep-sided domes rather than flows. Lava domes can be several hundred metres high, with a diameter of more than 900m (3,000ft). A lava dome is currently forming in the explosion crater at Mount St. Helens, Washington, USA.

**VOLCANIC RHYOLITE PLUG**
*Plugs such as this one at Church Rock, Arizona, USA, are the solidified vents of ancient volcanoes, the surrounding cinder-cones of which have eroded away.*

*Flow-banding*

**BANDED RHYOLITE**
*In banded rhyolite, flow-banding, with swirling layers of colours and textures, is prominent.*

# ANDESITE

**ANDES**
*Andesite is named for the Andes Mountains of South America, a subduction zone, and therefore a characteristic locality.*

THE VOLCANIC EQUIVALENT of diorite (see p.37), andesite is fine grained or porphyritic, and primarily consists of the plagioclase feldspar minerals andesine and oligoclase, plus one or more of the dark, ferromagnesian minerals, such as pyroxene or biotite. Unlike rhyolites, andesites do not contain quartz. Amygdaloidal andesite occurs when the voids left by gas bubbles in the solidifying magma are later filled in, often with zeolite minerals. Porphyritic andesite occurs when larger phenocrysts of feldspar and pyroxene form in a fine-grained matrix. Andesite erupts from explosive volcanoes and is commonly found interbedded with volcanic ash and tuff. The steep-sided volcanoes of the Pacific Rim, the Caribbean, and the Mediterranean are composed in large part of andesites.

## PROPERTIES

**ROCK TYPE** Intermediate, volcanic, igneous
**MAJOR MINERALS** Plagioclase feldspars
**MINOR MINERALS** Pyroxene, amphibole, biotite
**COLOUR** Light to dark grey, reddish-pink
**TEXTURE** Fine, porphyritic

## VOLCANOES

Andesitic volcanoes form on continental or ocean crusts above subduction zones where one oceanic plate is sinking beneath another. Ancient andesites can therefore be used to map ancient subduction zones. The Soufrière Hills in Montserrat, West Indies, Krakatau in Indonesia, Popocatépetl in Mexico, and Mounts Shasta, Hood, and Adams in the USA have all expelled large quantities of andesitic rock.

**MOUNT FUJI**
*Mount Fuji, near Tokyo, Japan, is a classic andesitic volcano.*

**PORPHYRITIC ANDESITE**
*In this specimen, light phenocrysts are in a dark matrix.*

*Lighter-coloured plagioclase crystals*
*Dark, fine-grained matrix*

**THIN SECTION**
*This andesite from Hakone volcano in Japan, shows large plagioclase and pyroxene crystals in a fine matrix.*

*Pyroxene*
*Plagioclase*

**FINE-GRAINED**
*This fine-grained andesite specimen contains small phenocrysts of light plagioclase feldspar.*

# PHONOLITE

**CRIPPLE CREEK, USA**
*Phonolite is found at Cripple Creek, In the Pike's Peak area of Colorado, USA.*

PHONOLITE IS A TYPE OF TRACHYTE (see below) that contains nepheline or leucite rather than quartz. They are commonly fine grained and compact, and split into thin, tough plates, which make a ringing sound when struck, hence the rock's name. The principal dark-coloured mineral is pyroxene, usually in the form of aegirine or augite. Phonolites are common in Europe, especially on the Eifel plateau and the Laacher See, Germany; in the Czech Republic; and in the Mediterranean area, particularly in Italy. In the USA phonolites are found in Colorado, and the Black Hills, South Dakota. The spectacular Devil's Tower, Wyoming, is probably phonolite's most famous occurrence (see p.48).

### PROPERTIES

**ROCK TYPE** Intermediate, volcanic, igneous

**MAJOR MINERALS** Sanidine, oligoclase

**MINOR MINERALS** Feldspathoids, hornblende, pyroxene, biotite

**COLOUR** Medium grey

**TEXTURE** Fine to medium, porphyritic

**COMPACT PHONOLITE**
*This phonolite specimen shows characteristic fine grains.*

Dark matrix crystals

# TRACHYTE

**TENERIFE**
*Trachytes are prominent and widespread on Tenerife, in the Canary Islands.*

THE VOLCANIC EQUIVALENT of syenite, trachyte is commonly porphyritic. Some porphyritic trachytes contain phenocrysts of the mineral sanidine that are up to 5cm (2in) across. Dark, mafic minerals such as biotite, amphibole, and pyroxene can be present in small quantities. Trachyte is similar to rhyolite in colour and occurrence, but contains very little or no quartz. Quartz-free trachytes, such as phonolite (see above), can contain feldspathoids such as leucite, nepheline, and sodalite. Unlike andesite, trachyte contains no phenocrysts of plagioclase feldspar. Trachyte has a characteristically rough texture, and its name comes from the Greek for rough.

Trachyte occurs with other alkali-rich felsic, intermediate, and mafic volcanics on both continents and ocean islands. Localities include Ascension Island; Cripple Creek, Colorado, USA; the Auvergne in France; and the Rhine district of central Europe.

## USE IN PAVING

Tough and resistant volcanic rocks such as trachyte have been used for millennia as paving stones. In addition to trachyte, granite, basalt, and gabbro can still be found forming the surfaces of streets and roads. Individual stones may have been used and reused for centuries. In some places the cobbled surfaces of Roman roads still remain, grooved by the passage of countless wheels of all varieties. Although labour-intensive to mine and shape, the resulting surface has unsurpassed wear-resistance.

**COBBLED STREET**
*The trachyte cobbles in this street have been polished smooth by years of wear.*

Larger crystals visible within matrix

Dark matrix crystals

Sanidine

Dark matrix crystals

**THIN SECTION**
*The typical texture of aligned sanidine crystals is seen here. The larger crystals show simple twinning.*

**GREY TRACHYTE**
*A fine- to medium-grained and fairly uniform structure can be seen in this trachyte specimen.*

### PROPERTIES

**ROCK TYPE** Intermediate, volcanic, igneous

**MAJOR MINERALS** Sanidine, oligoclase

**MINOR MINERALS** Feldspathoids, quartz, hornblende, pyroxene, biotite

**COLOUR** Off-white, grey, pale yellow, pink

**TEXTURE** Fine to medium, porphyritic

# DEVIL'S TOWER

THE NECK OF AN ANCIENT VOLCANO, THIS STRIKING STRUCTURE
STANDS HIGH ABOVE THE GREAT PLAINS OF WYOMING, USA. IT IS
KNOWN TO MANY AS THE LANDING STRIP USED BY THE SPACESHIP IN
STEVEN SPIELBERG'S FILM, *CLOSE ENCOUNTERS OF THE THIRD KIND.*

**VERTICAL STRIATIONS**
*As the phonolite remaining in the volcanic neck cooled and solidified, it developed vertical jointing. It is this that gives the Tower its vertically striated appearance today.*

Located in north-eastern Wyoming, USA, 14.5km (9 miles) south of the town of Hulett, the Devil's Tower is the neck, or "throat", of an ancient volcano, the pipe-like conduit through which magma flowed to the surface. After the volcano erupted, much of the magma remained in the underground plug. The magma cooled and solidified, eventually forming into uniform columns. Over millions of years, the cone of the volcano and the sedimentary rock layers around it were removed by weathering and the Belle Fourche River, which meanders around the base of the Tower, leaving the neck exposed.

The Tower is composed of phonolite, a type of silica-deficient rhyolite magma (see p.47). Phonolites are rich in nepheline and potash feldspar, and are typically fine-grained and compact. They split into thin, tough plates

**AERIAL VIEW**
*The top of the Tower stands 386m (1,267ft) above the level of the Belle Fourche River, and 1,560m (5,115ft) above sea level.*

**NATIONAL MONUMENT**
*The Devil's Tower was proclaimed a National Monument by President Theodore Roosevelt on 24 September 1906. It has become a favourite among serious rock climbers, who enjoy climbing the long crack-lines in the rock.*

that make a ringing sound when struck, giving the rock its name. The principal dark-coloured mineral in the Devil's Tower phonolite is the pyroxene aegirine.

The name "Devil's Tower" was given to the structure in 1875 by Colonel Richard Dodge. He was escorting a scientific team through the area, looking for gold, even though this was a violation of Indian treaty rights. Dodge translated one of the names given by the Indians to the area, meaning "Bad God's Tower", as "Devil's Tower". The most common Native American name used for the Tower today is Mateo Tepee or "Bear Lodge". This name is derived from a Kiowa legend in which seven girls playing in the woods were chased by bears, and found refuge on the Tower. The bears tried to climb up after them, but failed, leaving their claw marks (the vertical striations of the jointing) down the sides. It is still a sacred site of worship for many Native Americans today.

# TUFF

TUFF IS A BROAD TERM used to include any relatively soft, porous rock made of lithified pyroclastic minerals (ash and other sediments ejected from volcanic vents). Tuff should not be confused with the sedimentary rock tufa (see p.55). Tuffs originate when foaming magma wells to the surface as a mixture of hot gases and incandescent particles, and is ejected from a volcano. After the loose pyroclastic material is deposited, it eventually lithifies to become a volcanic tuff. The conditions under which the ejected pyroclastics lithify determine the final nature of the tuff. If the pyroclastic material is hot enough to fuse, a welded tuff forms at once. Other tuffs lithify slowly through compaction and cementation with calcite or silica-like sedimentary rocks. Because of variations in the conditions of their formation and of the ejected material, tuffs can vary greatly both in texture and in chemical and mineralogical composition. Most tuff formations include a range of fragment sizes and varieties. These range from fine-grained dust and ash (ash tuffs), to medium-sized fragments called lapilli (lapilli tuffs), to large volcanic blocks and bombs (bomb tuffs). Vitric tuffs are mainly composed of ash-size fragments of volcanic glass. Lithic tuffs contain a variety of crystalline rock fragments, which may be of rhyolitic, trachytic, or andesitic composition. In crystal tuffs, crystal fragments originating from partially solidified magmas are more abundant than lithic or vitric fragments.

**VOLCANO WALL**
*Tuff covers the inner wall of the volcanic caldera at Santorini on the Greek island of Thira. The island's main town, Fira, perches on the crest.*

## PROPERTIES

**ROCK TYPE** Volcanic, igneous

**FOSSILS** Generally moulds of nonmarine plants and animals, including humans

**MAJOR MINERALS** Glassy fragments

**MINOR MINERALS** Crystalline fragments

**COLOUR** Light to dark brown

**TEXTURE** Fine

**MOUNT ST. HELENS, USA**
*In 1980, Mount St. Helens erupted thousands of tonnes of pyroclastic debris that later formed tuff.*

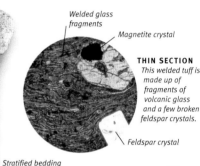

*Welded glass fragments*

*Magnetite crystal*

**THIN SECTION**
*This welded tuff is made up of fragments of volcanic glass and a few broken feldspar crystals.*

*Feldspar crystal*

*Stratified bedding*

*Fine texture*

**TUFF WITH GRADED BEDDING**
*In this specimen from Iceland, graded bedding is visible. This is as a result of formation in standing water, when the tuff can stratify exactly like a sedimentary rock. Extensive tuff deposits occur in New Zealand, Guatemala, Peru, and in the USA in Nevada and Yellowstone National Park, Wyoming.*

**ROW OF MOAI**
*The moai statues of Easter Island are carved from yellow-grey tuff.*

## NUÉE ARDENTE

A nuée ardente (French for "fiery cloud") is an incandescent mass of volcanic particles suspended in a cloud of volcanic gases. It can move very quickly down even slight inclines and may gain speeds of 160km (100 miles) per hour. A nuée ardente can reach temperatures of 700°C (1,300°F), and virtually no living thing in its path survives. It is characteristic of Pelean eruptions – violent, explosive volcanic eruptions, which most often occur in the Pacific Ring of Fire. A nuée ardente finally sealed the fate of Pompeii, two millennia ago.

**BODY FROM POMPEII**
*Archeologists fill hollows left in the ash with plaster to form casts of bodies.*

**ST. PIERRE**
*In 1902 a nuée ardente swept down from the erupting Mount Pelée in the West Indies, killing all but two of a population of 30,000 in the town of St. Pierre, Martinique.*

# PUMICE

A VERY POROUS, froth-like volcanic glass, pumice is created when gas-saturated liquid magma erupts like a fizzy drink released from a shaken bottle, and cools so rapidly that the resulting foam solidifies into a glass full of gas bubbles. The hollows in the froth (called vesicles) can be rounded, elongated, or tubular, depending on the flow of the solidifying lava. The glassy material that forms pumice can be in threads, fibres, or thin partitions between the vesicles. Pumice has a very low density due to its large number of air-filled pores. As a result, it can easily float in water, and it frequently does when erupted into the sea. After the eruption of Krakatau in Indonesia in 1883, banks of pumice covered the surface of the sea up to a depth of 1.5m (5ft). Pumice differs from tuff in that it seldom forms thick deposits, but instead tends to form thin layers on the surface.

Although pumice is mainly composed of glass, small crystals of various minerals do occur, the most common being feldspar, augite, hornblende, and zircon. In older pumices, the vesicles are more often than not filled with deposits of minerals such as zeolites brought in by water percolating through them. Under favourable conditions, most types of lava can form pumice – but silica-rich lavas are more prone to froth than silica-deficient ones because they are more viscous and tend to hold their volatiles until they are erupted. Pumices are frequently accompanied by obsidian, which forms under similar conditions but results from greater pressure. Pumices formed from silica-rich lavas are white, while those from lavas with intermediate silica content are often yellow or brown, and the rarer silica-poor pumices (like those in the Hawaiian Islands) are black. Pumice is important economically as an abrasive in cleaning, polishing, and scouring compounds.

**PUMICE LANDSCAPE**
In desert regions, beds of pumice can be sculpted into fantastic shapes by erosion from sand particles carried in the wind.

**PUMICE MINE**
This old pumice mine is in mountains on the island of Lipari, Italy. Pumice has been mined here for many centuries.

**PANTHEON, ROME**
The dome of the Roman Pantheon was made from concrete mixed with pumice. This gave the mixture lightness and strength. Pumice is still used as a lightweight building aggregate.

## PROPERTIES

| | |
|---|---|
| **ROCK TYPE** | Volcanic, igneous |
| **ORIGIN** | Extrusive |
| **MAJOR MINERALS** | Glass |
| **MINOR MINERALS** | Feldspar, augite, hornblende, zircon |
| **COLOUR** | White, yellow, brown, black |
| **TEXTURE** | Fine |

**PUMICE STONE**
Scrubbing stones made of pumice, used to remove rough skin, are an important bath accessory for many people. Pumice is soft and easily shaped.

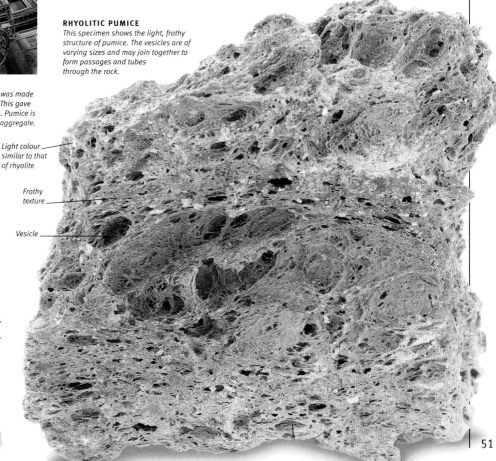

**RHYOLITIC PUMICE**
This specimen shows the light, frothy structure of pumice. The vesicles are of varying sizes and may join together to form passages and tubes through the rock.

Light colour similar to that of rhyolite

Frothy texture

Vesicle

# SEDIMENTARY ROCKS

SEDIMENTARY ROCKS MAKE UP between 80 and 90 per cent of the rock exposed at the Earth's surface. Yet they are only a minor constituent – about 8 per cent by volume – of the entire crust, which is predominantly composed of igneous and metamorphic rocks. Sedimentary rock is formed at or near the Earth's surface by either accumulation of grains or precipitation of dissolved material.

**WATER TRANSPORT**
*Swift-moving water, as at Lower Yellowstone Falls in Wyoming, USA, is both an agent of erosion and a prime means of transporting sediments to be deposited elsewhere.*

## SOURCES OF SEDIMENT

The formation of sedimentary rock begins with the weathering of other rocks. Physical weathering is the breakdown of rock into smaller rock fragments. In chemical weathering, portions of the weathered rock are dissolved. These two different weathering products create two fundamentally different kinds of sediment and sedimentary rock: clastic and chemical (see opposite). The products created by weathering are transported by running water, gravity, glaciers, or wind and eventually deposited. In many cases, during transport the clasts – rock fragments and mineral grains – are sorted by size, so when sedimentary rock is eventually formed from them it is relatively uniform in grain size. In chemical sedimentary rocks, the dissolved rock–matter is transported in solution and precipitated in a new location. Precipitation may take place through evaporation.

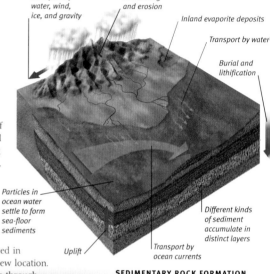

Transport by water, wind, ice, and gravity

Weathering and erosion

Inland evaporite deposits

Transport by water

Burial and lithification

Particles in ocean water settle to form sea-floor sediments

Uplift

Transport by ocean currents

Different kinds of sediment accumulate in distinct layers

**SEDIMENTARY ROCK FORMATION**
*The formation of sedimentary rocks begins with weathering and erosion on land. This creates sediments that may ultimately be transported to the sea, where deposition, burial, and lithification take place.*

**ERODED CANYON**
*At the Grand Canyon in Arizona, USA, water erosion by the Colorado River has cut through 1.6km (1 mile) of accumulated sedimentary rock.*

# LITHIFICATION

The transformation of loose grains of sediment into sedimentary rock is known as lithification. Binding the grains together often requires a cementing agent, especially in the case of sandstones and conglomerates. The cementing agent is generally precipitated from solutions filtering through the sediment, although in some cases it is created at least in part by the breakdown of some of the rock particles of the sediment itself. The most common cement is silica (usually quartz), but calcite and other carbonates also form cements, as do iron oxides, baryte, anhydrite, zeolites, and clay minerals. The cementing agent becomes an integral and important part of the sedimentary rock once it is formed. In some cases, clastic sedimentary rocks (see below) can also be formed by simple compaction, a process in which the grains are forced to bind together under extreme pressure.

Lithification can sometimes take place almost immediately after the grains have been deposited as sediment, but in other cases hundreds or even millions of years may pass before lithification occurs. In other words, at any given time there are large amounts of rock fragments produced by weathering that have not been lithified, simply existing as a form of sediment.

**BANDED ROCK**
*These cliffs in Zion Canyon, Utah, USA, show distinct layers of sedimentary rock deposited between 10 million years ago at the top and more than 250 million years ago in the lowest strata.*

Settling grains
Water
Substrate
Grains
Cement

**SEDIMENTARY AGGREGATIONS**
*Deposited by wind and water, mineral grains form loose sediments. Over time these grains are compacted or cemented until they form solid rock.*

**SANDSTONE THIN SECTION**
Sand grains
*Under magnification the individual mineral grains that make up sandstone can be seen. The grains are partially rounded as a result of abrasion during their transport.*

# CLASTIC ROCKS

Clasts are rock fragments ranging in size from boulders to microscopic particles. Clastic sedimentary rocks can be grouped according to the size of the clasts from which they are formed. Larger clasts such as pebbles, cobbles, and boulder-size gravels form conglomerate and breccia; sand becomes sandstone; and finer silt and clay particles form siltstone, claystone, mudstone, and shale. The mineral composition of clastic rock can be subject to considerable change over time. Chemical reactions may take place between clasts of different minerals, between clasts and cementing agents, or between clasts, cement, and ground water. Minerals in the rock may also be dissolved and redistributed.

Large pebble

**CONGLOMERATE**
*Conglomerates are the most coarse-grained sedimentary rocks. They consist of grains the size of pebbles, or even larger, bound together by a cementing matrix.*

Cementing agent

**STONE BEACH**
*The loose, rounded stones on this beach in Maine, USA, could provide the raw material for a sedimentary conglomerate rock if buried at depth and cemented together.*

# CHEMICAL ROCKS

Chemical sedimentary rocks are formed by precipitation of the transported, dissolved products of chemical weathering. In some cases the dissolved constituents are directly precipitated as solid rock. Examples include banded iron formations, some limestones, and bedded evaporite deposits – that is, rocks and mineral deposits of soluble salts such as halite, gypsum, and anhydrite, resulting from the evaporation of water. In the formation of other sedimentary rocks, the solid material first precipitates into particles, which are then deposited and lithified. Limestones and cherts are principally formed in this manner.

**STONE FOREST**
*These limestone deposits in Yunnan Province, China, have been weathered into strange pinnacles.*

**EVAPORITE DEPOSITS**
*In Death Valley, California, USA, evaporation is constantly forming new sedimentary deposits. Growing salt crystals create rimmed pans as they press against each other.*

53

**LIMESTONE PAVEMENT**
*The criss-cross joints of exposed limestone are further etched by acidic rainfall to create this formation.*

# LIMESTONE

COMPOSED MAINLY OF CALCITE, limestone is abundant, and occurs in thick, extensive, multiple layers. It generally forms in warm, shallow seas from the precipitation of calcium carbonate from sea water or the accumulation of the shells and skeletons of calcareous marine organisms. Its texture ranges from coarse and fossil-rich to fine and micritic. Some limestones are very compact, while others are grainy or friable. Many have sedimentary structures such as cross bedding and ripple marks.

## INDUSTRIAL USES

Limestone is very important commercially and has a number of different uses: as a building stone; when burned to form lime for cement; as a raw material in the manufacture of glass; as a flux in metallurgical processes; and in agriculture. Because calcite dissolves in acidic water, limestone facades, gravestones, and other structures weather very quickly, especially in cities. (See also Caves and caverns, pp.56–57.)

**CEMENT FACTORY**
*Limestone is used to make cement at this factory in Rugby, UK.*

### PROPERTIES

**ROCK TYPE** Marine, chemical, sedimentary
**FOSSILS** Marine and freshwater invertebrates
**MAJOR MINERALS** Calcite
**MINOR MINERALS** Aragonite, dolomite, siderite, quartz, pyrite
**COLOUR** White, grey, pink
**TEXTURE** Fine to medium, angular to rounded

Fine texture

Rounded grains of calcium carbonate

**THIN SECTION**
*Oolitic limestone contains masses of calcareous spheres, called oolites. This section is from Loch Eishort, Isle of Skye, Scotland.*

Fossil of shell

Oolitic grains

**OOLITIC LIMESTONE**

**FOSSILIFEROUS LIMESTONE**
*This limestone specimen, from Oxfordshire, England, includes several fossils.*

---

# DOLOMITE

ALSO CALLED DOLOSTONE, dolomite is the rock formed exclusively from the mineral dolomite (see p.181). Most dolomites are thought to be limestones in which calcite has been replaced by dolomite. This process is known as dolomitization, and it may occur soon after limestone is deposited by exchange with sea water, or after lithification by exchange with magnesium-bearing solutions. Dolomites tend to be less fossiliferous than limestones because fossils and other features are destroyed by the dolomitization process. It is distinguished from limestone in that it does not fizz as violently in hydrochloric acid. Dolomite occurs as massive layers and as thin layers or pods within limestone. Fresh dolomite looks very similar to limestone; weathered, it is a yellowish-grey.

**DOLOMITES**
*The Eastern Tyrol area of the Dolomites is the type locality.*

Compact carbonate rock

Fine to medium texture

**REGULAR FEATURES**
*This specimen has a fairly uniform texture, typical of dolomite. Dolomitization destroys original features, such as fossils, in the original limestone.*

### PROPERTIES

**ROCK TYPE** Marine, chemical, sedimentary
**FOSSILS** Invertebrates
**MAJOR MINERALS** Dolomite
**MINOR MINERALS** Calcite
**COLOUR** Grey to yellowish-grey
**TEXTURE** Fine to medium, crystalline

Dark calcite matrix

Dolomite rhomb

**THIN SECTION**
*Rhomb-shaped crystals of secondary dolomite form by replacing fine-grained calcite in a limestone.*

## STEEL-MAKING

In recent years, dolomite has been used in steel-making. It is used as a flux, a substance added to a furnace to combine with impurities to form slag, which can then be removed. Dolomite is more environmentally friendly than other fluxes because the slag it produces is itself valuable and therefore likely to be reused. Dolomite slag does not break down when wet, so it can be used as a lightweight building aggregate. It is used in breeze blocks, and in poured concrete whose surfaces are exposed to weathering. Raw dolomite is also used as an aggregate for cement and bitumen mixes in place of limestone.

**BLAST FURNACE**
*Dolomite helps to increase the fluidity of the metal in the steel-making process.*

# CHALK

**WHITE CLIFFS**
*These spectacular chalk cliffs are near Dover in Kent, England.*

A SOFT, FINE-GRAINED, easily pulverized, white to greyish variety of limestone, chalk is composed of the calcite shells of minute marine organisms. Small amounts of other minerals are commonly present, such as glauconite, apatite, and clay minerals. Silica can also be present in small quantities from sponge spicules, diatom and radiolarian skeletons, and nodules of chert and flint. In some localities, the beds of flint nodules are thick enough to have been mined in ancient times. Extensive chalk deposits were formed during the Cretaceous Period (142 to 65 million years ago), the name being derived from the Latin *creta*, meaning "chalk".

## INDUSTRIAL USES

Like other limestones, chalk is used for making lime and cement and as a fertilizer. It is used as a filler, extender, or pigment in a wide variety of materials, including ceramics, putty, cosmetics, crayons, plastics, rubber, paper, paints, and linoleum. Modern-day blackboard chalk is a manufactured substance rather than natural chalk. In past times chalk was mined extensively to be burned for making quicklime for mortar. There are buildings lime-mortared 500 years ago that are still standing today.

**WELLINGTON BOOTS**
*These loose-fitting, waterproof rubber boots are based upon a design worn by the first Duke of Wellington (1769–1852).*

Soft, white, powdery texture

### PROPERTIES

**ROCK TYPE** Marine, organic, sedimentary
**FOSSILS** Invertebrates, vertebrates
**MAJOR MINERALS** Calcite
**MINOR MINERALS** Quartz, glauconite, clays
**COLOUR** White, grey, buff
**TEXTURE** Very fine, angular to rounded

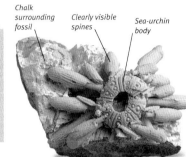

Chalk surrounding fossil

Clearly visible spines

Sea-urchin body

**MICROSCOPIC FOSSILS**
*Tiny fossils of marine organisms make up chalk, which is almost entirely calcite. A microscope is often needed to see the fossils, although larger fossils can also be present.*

**FOSSIL IN CHALK**
*This well-preserved sea urchin (Tylocidaris) fossil is a common large fossil in chalk.*

# TUFA

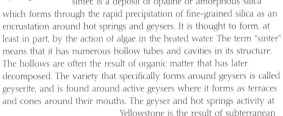

**TUFA CONE**
*A tufa cone has formed under this geyser in Yellowstone Park, Wyoming, USA.*

TUFA IS A NAME GIVEN TO two different sedimentary rocks that precipitate from water. Calcareous tufa, or calc-tufa, is a soft, porous deposit composed principally of calcium carbonate (calcite) that precipitates from hot springs, lake water, and ground water. It is often stained red by the presence of iron oxides. Siliceous tufa, also called siliceous sinter, is a deposit of opaline or amorphous silica which forms through the rapid precipitation of fine-grained silica as an encrustation around hot springs and geysers. It is thought to form, at least in part, by the action of algae in the heated water. The term "sinter" means that it has numerous hollow tubes and cavities in its structure. The hollows are often the result of organic matter that has later decomposed. The variety that specifically forms around geysers is called geyserite, and is found around active geysers where it forms as terraces and cones around their mouths. The geyser and hot springs activity at Yellowstone is the result of subterranean magma generating silica-rich hydrothermal solutions, which periodically rise to the surface. Silica is considerably more soluble in hot water under pressure, and as it cools on the surface much of the silica is precipitated as geyserite. Geyserite is found in Iceland, New Zealand, and Yellowstone National Park, Wyoming, and Steamboat Springs, Colorado, USA.

### PROPERTIES

**ROCK TYPE** Continental, chemical, sedimentary
**FOSSILS** Rare
**MAJOR MINERALS** Calcite or silica
**MINOR MINERALS** Aragonite
**COLOUR** White
**TEXTURE** Fine, crystalline

Highly porous

Irregular shapes

**CALCAREOUS TUFA**
*Formed by the evaporation of water, this tufa is full of holes and irregular shapes.*

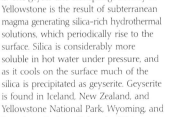

**TUFA FORMATIONS**
*These distinctive tufa towers stand above Mono Lake, California.*

**CAVE FORMATIONS**
*The evaporation of lime-bearing waters dripping into caves can produce a dramatic assortment of cave formations. These include stalactites, which grow downwards from the roof of a cave, and stalagmites, which grow upwards from the floor.*

**LAVA IN A TUBE**
*Molten lava continues to flow through this lava tube on Kilauea Volcano, Hawaii. The lava is visible through a hole in the roof of the tube at a weak point in the crust.*

**INSIDE A STALACTITE**
*Stalactites build up a layer at a time, and resemble tree rings in cross section. Differences in mineral content on deposition can cause colour banding.*

*Calcite stained with iron oxide*

*Unstained calcite layer*

# CAVES AND CAVERNS

CARVED OUT OF ROCK BY NATURAL FORCES, CAVES ARE AMONG THE MOST SPECTACULAR OF EARTH'S GEOLOGICAL FEATURES.

Naturally formed subterranean cavities, caves form as a result of many geological processes. The largest and most common caves are called solution caves. These are formed by chemical reaction between circulating ground water and underlying soluble or otherwise easily eroded rock. Other types of caves include lava tubes and marine grottos. A cave consists either of a single chamber or of an extensive system of many chambers connected by smaller passageways.

## SOLUTION CAVES

Solution caves are generally formed by dissolution of limestone (see p.54), and, in some cases, dolomite (see p.54). Limestone contains calcium carbonate (calcite), which is slightly soluble in the weak acid, known as carbonic acid ($H_2CO_3$), which is formed as carbon dioxide dissolves in ground water. This acidic water percolates down through the overlying soils. Some collects in sinkholes that are already dissolved out of the underlying limestone or dolomite, and some trickles down through cracks in the rock. As dissolution continues, conduits expand to become interconnected drains for the movement of underground water.

Solution caves are simply parts of this underground drainage system. Some are completely water-filled, while others are dry passages left behind by streams that went on to cut through to lower levels, or that disappeared entirely. The longest known solution cave is Mammoth Cave in Kentucky, USA, which extends 500km (more than 300 miles).

## VOLCANIC CAVES

Volcanic (or lava) caves are tubes or hollows formed when a stream of lava flows over the ground. The surface of the flowing lava cools as it makes contact with the air and solidifies into rock. Beneath this solid crust, molten lava continues to flow. When the supply of lava ceases, the remaining molten material drains out, leaving long, cylindrical tunnels called lava tubes.

**HARRISON'S CAVE**
*An underground stream flows through this spectacular solution cave in Barbados, filled with still-forming stalagmites and stalactites.*

# ROCK SALT

OCCURRING IN BEDS that range from a metre or so to more than 300m (1,000ft) in thickness, rock salt is the massive rock form of the mineral halite (see p.170), familiar as common table salt. It forms as a result of the evaporation of saline water in partially enclosed basins, and is commonly interlayered with beds of limestone, dolomite, shale, and other evaporites such as gypsum and anhydrite. It is found on all continents, with particularly large deposits in south-eastern Russia, France, India, Canada, and western and central USA.

Salt beds have been exploited since Neolithic times in Europe, where huge salt caverns have been produced by underground mining. Beds of rock salt are still mined or quarried by traditional excavation methods, depending on the depths and thicknesses of the deposits. In some deposits (such as in Poland and the United States), the rock salt is pure enough to be marketed without further processing. Modern extraction of salt often takes place through solution mining where high-pressure water is injected into deeply buried salt beds and the salt is then returned to the surface as brine.

**SALT MOUND**
*This mound of raw salt has been scooped from a salt pan after the evaporation of sea water.*

**ART IN OLD SALT MINE**
*During World War II, the Nazis concealed much of their looted art in abandoned salt mines.*

**EVAPORATION**
*A considerable amount of raw salt is recovered from the evaporation of sea water in shallow basins called salt pans.*

Rock salt often occurs in salt domes. These consist of a core of salt and an envelope of surrounding strata. Their formation is complex, but they occur in the subsurface, and can be 2km (1 mile) or more thick, and up to 10km (6 miles) in diameter. They occur in abundance around the Gulf of Mexico and in the Middle East, and are often associated with petroleum deposits. Oil migrates up from oil-rich shales and gets caught on the underside of a salt dome. Salt's low density makes it easy to detect by geophysical methods, making possible oil reserves easy to spot.

## PROPERTIES

**ROCK TYPE** Marine, evaporite, sedimentary
**FOSSILS** None
**MAJOR MINERALS** Halite
**MINOR MINERALS** Sylvite
**COLOUR** White, orange-brown, blue
**TEXTURE** Coarse to fine crystalline

**MASSIVE ROCK SALT**
*This massive specimen of rock salt has coarse white and blue crystals.*

Blue from colour centres (see p.92)

**CAMELS PULLING SALT**
*In this photograph from 1926, camels are being used to haul trucks of salt along tracks by the Red Sea.*

Massive habit

## SALT'S COMMERCIAL VALUE

Rock salt is a valuable and widely used economic mineral, and its mining is a large-scale undertaking. Aside from the human dietary requirements for salt, livestock also require large quantities of salt, which farmers often provide in large blocks. Salt is used in the chemical industry in the manufacture of sodium bicarbonate (baking soda), sodium hydroxide (caustic soda), hydrochloric acid, chlorine, and many other chemicals. Large quantities of salt are also used to make soaps, glazes, and porcelain enamel, and as a flux in the melting of metals.

**SALT FOR SALE**
*Cakes of salt produced by the evaporation of sea water in salt pans are stacked to await buyers.*

# ROCK GYPSUM

ALSO KNOWN AS GYPROCK, rock gypsum is the sedimentary rock formed principally from the mineral gypsum (see p.212). Rock gypsum occurs in extensive beds formed by the evaporation of ocean water, as well as in saline lakes and salt pans. It also occurs in limestones, dolomitic limestones, and some shales. It is commonly interlayered with other evaporites such as rock salt and anhydrite. Commonly granular, it can also occur in fibrous bands. Unaltered rock gypsum is used as a fluxing agent, fertilizer, filler in paper and textiles, and retardant in Portland cement. However, most of the gypsum that is extracted – about three-quarters of the

total production – is calcined (the mineral is heated to drive off some of its water) for use as plaster of Paris and as other building materials in plaster. Plaster of Paris is so called because gypsum from near Paris, France, was used from early times to make plaster and cement. It is used to cast ornamental plasterwork for ceilings and cornices, and in medicine it is used to make plaster casts to immobilize broken bones while they heal. When the calcined material is mixed with water, it recrystallizes as gypsum and hardens. Calcined gypsum is a major component of other plasters as well, usually with special retarders or hardeners added to give it particular properties.

**DEVIL'S GOLF COURSE**
*Thick deposits of evaporites such as rock gypsum can form rugged crusts. This example is at the Devil's Golf Course in Death Valley, California, USA.*

Massive habit

Iron oxide impurities

**MASSIVE ROCK GYPSUM**
*This specimen of rock gypsum is from the Marblaegis Mine, Nottinghamshire, England.*

## PROPERTIES

**ROCK TYPE** Marine, evaporite, sedimentary

**FOSSILS** None

**MAJOR MINERALS** Gypsum

**MINOR MINERALS** Anhydrite

**COLOUR** White, pinkish, yellowish, grey

**TEXTURE** Medium to fine crystalline

**PLASTER DECOR**
*Intricately carved plasterwork by James Pettifer graces the ceiling above the Great Staircase at Sudbury Hall, Derbyshire, UK.*

Granular texture

## USE IN WALLBOARD

The term "wallboard" is applied to large, rigid sheets of finishing material used to face the interior walls of dwellings and other buildings. The sheets can be made of various materials, but are commonly made from a gypsum plaster core sandwiched between layers of heavy paper. In the USA in particular, this wallboard is called – for obvious reasons – sheet rock. It is fire-resistant and, to improve this property, glass fibres are mixed with the gypsum base. Other types of gypsum boards can be used for exterior sheathing. Gypsum is a poor conductor of heat, so in addition to providing a fireproof wall covering, sheet rock is also effective for insulation purposes.

Gypsum in wallboard

**PLASTER STATUE**
*This plaster cast of Theseus Battling the Centaur was made by Antonio Canova in 1805. The "alabaster" used by sculptors is in fact rock gypsum.*

# CLASTIC ROCKS

CLASTIC ROCKS ARE those derived from material released by the weathering and breakdown of igneous, metamorphic, and other sedimentary rocks. The resulting clasts (pieces of rock debris) are transported by water, wind, ice, or gravity, before being deposited and then cemented to form a coherent rock. Clastic sedimentary rocks are classified on

**LAYERED SANDSTONE**
*In Arizona, USA, layers of sandstone have been contorted and then worn smooth by glacier erosion.*

the basis of physical and chemical properties. Those classified principally by grain size, from largest to smallest, are: conglomerate, breccia, sandstone, shale, siltstone, and mudstone. Those classified primarily on the basis of their mineral makeup or cementation (or both) are shown on pages 68–71. Clastic sedimentary rocks provide important information about the rocks from which they originated, and their environment of deposition.

## CONGLOMERATE

**RIVER DEPOSITS**
*This conglomerate in South Africa was deposited by a river.*

CONGLOMERATES ARE formed by the lithification of rounded rock fragments over 2mm (0.08in) in diameter. They can be further classified by the average size of their constituent materials: pebble (fine), cobble (medium), and boulder (coarse). The name can also include a description of the rock or mineral fragments of which a conglomerate is composed; for example, a quartz pebble conglomerate. Conglomerates fall into two categories, indicative of their depositional environments. First are those whose pebbles are well sorted (having a small size variation), generally of only one rock or mineral type, and with few small particles between the pebbles. Second are those with poorly sorted pebbles (of varying sizes), of mixed rock and mineral types, and with a number of small particles between the pebbles. Well-sorted conglomerates result from normal water flow over a long period; poorly sorted conglomerates from rapid deposition.

### PROPERTIES

**ROCK TYPE** Marine, freshwater, and glacial detrital sedimentary

**FOSSILS** Very rare

**MAJOR MINERALS** Any hard mineral can be present

**MINOR MINERALS** Any mineral can be present

**COLOUR** Varies

**TEXTURE** Very coarse, rounded clasts

**PUDDINGSTONE**
*The pebbles cemented together to form this rock from Hertfordshire, England, have been previously rounded by water.*

*Rounded clasts*

*Fine-grained sediment*

*Large, partially rounded fragment*

**POLYGENETIC CONGLOMERATE**
*This type of conglomerate can contain fragments of igneous, metamorphic, and sedimentary rocks, and particles of various minerals.*

## BRECCIA

**BRECCIA IN WALES**
*This example of breccia is found in Pembrokeshire.*

**POORLY SORTED BRECCIA**
*This specimen shows large and small angular fragments with no clear pattern of orientation.*

*Angular fragments*

*Grey siliceous fragments*

BRECCIAS ARE LITHIFIED SEDIMENTS with the same size clasts as conglomerates, but in breccias the clasts are angular or only slightly rounded. The lack of rounding indicates little or no transportation has taken place. Breccias can form in several ways: firstly, a rock body can break and then become cemented in its original position; secondly, rock fragments accumulating at the base of a cliff can become cemented where they fell; thirdly, in areas of active faulting, newly broken material along the line of the fault can be cemented in place; and fourthly, where faulting takes place underwater, newly shattered material can move in underwater landslides and become cemented.

### PROPERTIES

**ROCK TYPE** Marine, freshwater, and glacial detrital sedimentary

**FOSSILS** Very rare

**MAJOR MINERALS** Any hard mineral can be present

**MINOR MINERALS** Any mineral can be present

**COLOUR** Varies

**TEXTURE** Very coarse, angular clasts

# SHALE

ONE OF THE MOST ABUNDANT sedimentary rocks, shale consists of silt- and clay–sized particles deposited by gentle transporting currents, and laid down on deep-ocean floors, basins of shallow seas, and river floodplains. Shales differ from mudstones in that they easily split into thin layers. Most shales occur in widespread sheets up to several metres thick, although they are also found thinly interbedded with layers of sandstone or limestone. Shales consist of a high percentage of clay minerals, substantial amounts of quartz, and smaller quantities of feldspars, iron oxides, carbonates, fossils, and organic matter. Reddish and purple shales result from the presence of hematite and goethite; blue, green, and black from ferrous iron; grey or yellowish from calcite. A valuable raw material, shale is used to make tiles, bricks, and pottery.

**GRAPHITOLITE SHALES**
*These shales are in Pembrokeshire, Wales.*

## PROPERTIES

**ROCK TYPE** Marine, freshwater, and glacial detrital sedimentary

**FOSSILS** Invertebrates, vertebrates, plants

**MAJOR MINERALS** Clays, quartz, calcite

**MINOR MINERALS** Pyrite, iron oxides, feldspar

**COLOUR** Grey

**TEXTURE** Fine

## OIL SHALES

Oil shales are organic-rich shales containing kerogen, a complex mixture of solid hydrocarbons derived from plant and animal matter. In oil shales these are present in high enough quantities to yield oil when subjected to intense heat. Scientists have attempted to extract the oil in an economically viable way, but have had limited success. The detonation of small nuclear devices within the shale layers has even been suggested, but has been rejected as too dangerous.

**RIFLE, COLORADO, USA**
*This is one of several centres where attempts are being made to extract oil from shale.*

**SPLITTING SHALE**
*Fissile shale splits easily into layers that are usually parallel to the bedding planes.*

Fossilized ammonite

Fossilized bivalve

Layers of shale visible

**POLISHED OIL SHALE**
*This polished section of oil shale shows darker, kerogen-rich layers. Darker colours come from the organic matter.*

**FOSSILIFEROUS SHALE**
*This shale specimen from Runswick Bay, Yorkshire, England, contains fossils of bivalves and ammonites.*

# SILTSTONE

**CROSS BEDDING**
*Cross bedding is visible in this siltstone.*

SILTSTONE IS LITHIFIED SILT-SIZED particles, 0.0039–0.063mm (0.00015–0.0025in) in diameter, smaller than fine sand but larger than clay. Siltstones tend to be hard and durable, and are not easily split into thin layers. They can show cross bedding, ripple marks, and internal layering. Siltstones are much less common than shales or sandstones, and rarely form thick deposits.

**GREY SILTSTONE**
*The tiny grains that make up siltstone are too small to be seen without a microscope.*

Uneven fracture

## PROPERTIES

**ROCK TYPE** Marine, freshwater, and glacial detrital sedimentary

**FOSSILS** Invertebrates, vertebrates, plants

**MAJOR MINERALS** Quartz, feldspar

**MINOR MINERALS** Mica, chlorite, micaceous clay minerals

**COLOUR** Grey to beige

**TEXTURE** Fine, rounded to angular

Fine-grained sediment

# MUDSTONE

**MUDSTONE MOUND**
*This mound in Namibia is under a sandstone column.*

MUDSTONE IS MADE UP primarily of a mix of clay- and silt-sized particles, like shale, but deposited and lithified so that it is not laminated or easily split into thin layers. Mudstones generally occur in the same colour range as shales with similar associations between colour and mineral content. Those containing a substantial amount of calcite are known as calcareous mudstones. Mudstones can form deposits several metres thick.

## PROPERTIES

**ROCK TYPE** Marine, freshwater, and glacial detrital sedimentary

**FOSSILS** Invertebrates, vertebrates, plants

**MAJOR MINERALS** Clays, quartz

**MINOR MINERALS** Calcite

**COLOUR** Grey, brown, black

**TEXTURE** Fine, microscopic

**FINE-GRAINED MUDSTONE**
*Plant debris is visible in this specimen of mudstone, which is fine-grained.*

Fine-grained texture

Plant debris

# THE TERRACOTTA ARMY

THE SHAPING OF CLAY EVOLVED FROM UTILITARIAN NEED INTO AN ART FORM THOUSANDS OF YEARS AGO. NOWHERE HAS THIS BEEN MORE SPECTACULARLY EXPRESSED THAN IN THE TERRACOTTA ARMY OF THE FIRST CHINESE EMPEROR, QIN SHIHUANGDI.

In 1974 some farmers digging a well near Xi'an, China, discovered what would prove to be three pits containing over 7,000 life-size clay warriors and horses adjacent to the burial tomb of the first Chinese Emperor. The first pit held over 6,000 soldiers and a few horses and chariots; the second pit housed 1,400 soldiers and cavalry; and the third pit contained 68 officers. They were interred as part of the mausoleum of Qin Shihuangdi, who died in 210BC. An empty fourth pit was also discovered, apparently unfinished due to Qin's untimely death.

Qin Shihuangdi was a hated tyrant who sacrificed thousands of lives on his personal projects. His tomb, constructed by 700,000 forced labourers and not yet discovered, was said to have a ceiling inlaid with pearls, a stone floor forming a map of the Chinese kingdom with 100 rivers of mercury flowing across it, and vast quantities of treasure protected by deadly booby-traps. The discovery of high levels of mercury in the soil surrounding the site suggests that there may be some truth to the tale.

**SOLDIERS AND HORSES**
*Each warrior stands poised to attention, as if awaiting the command to attack. Many weapons were also found in the tomb: bows and arrows, spears, and swords, as well as bronze and leather bridles, silk, linen, jade, and bone artefacts, and iron farm implements.*

## CREATING THE ARMY

Each of the otherwise hollow statues carved out of clay stood unaided upon solid legs. This had never been achieved before. The remaining hollow body parts were made from a variety of moulds, and then assembled. Pieces, such as ears, beards and armour, were modelled separately and attached. The entire figure was coated with thin layers of clay, and the final details, giving each soldier a unique appearance and personality, were sculpted. The whole figure was fired at a high temperature and then painted and coated with lacquer. For two millennia the clay figures lay buried in water-soaked soil. The moment the figures were unearthed, their water-soaked glaze started to dry out and the lacquer, with its coloured pigments, began to flake and fall off. However, a chemical treatment developed recently at the University of Munich promises to preserve the crumbling warriors for future generations.

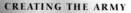

**CONSERVATION**
*To preserve the figures, each one is being bathed in a solution containing hydroxyethyl methacrylate (HEMA), which is able to penetrate tiny pores in the glaze. It is then bombarded with electrons in a particle accelerator, converting the liquid into a strong polymer that bonds the lacquer together from the inside without altering the figure's appearance.*

63

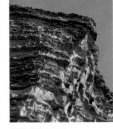

**DORSET CLIFF**
*Sandstones interlayered with shales are visible in this sea cliff in Dorset, England.*

# SANDSTONE

THE SECOND MOST COMMON sedimentary rock after shale, sandstone makes up about 10 to 20 per cent of the sedimentary rocks in the Earth's crust. It is the lithified accumulation of sand-sized grains, 0.063–2mm (0.0025–0.08in) in diameter, and is of great interest to geologists because it is abundant, well exposed, and with a wide range of textures and mineralogy, is an important indicator of erosional and depositional processes.

Sandstones consist of two parts: the grains themselves; and the space between the grains which, in the lithification process, fills with a chemical cement of silica or calcium carbonate. Most sandstones have a grain component that is principally quartz, feldspar, and lithic fragments. Other minerals can be present in varying amounts. The relative proportions of all of these are indicators of both the source area and the rate of deposition.

**PAINTED DESERT, ARIZONA**
*Cross-bedded sandstone is exposed to dramatic effect in the Painted Desert of Arizona, USA.*

**FOSSIL RECORD**
*Fossils, such as the ammonoid above right, are found in sandstone, although preservation is often not as good as in finer grained rocks. The dinosaur footprint (above) is preserved in fine-grained sandstone in Namibia.*

## CLASSIFYING SANDSTONE

Sandstones are classified according to texture and mineralogical properties. Several important types are orthoquartzite (nearly pure quartz); arkose (feldspar-rich, with only a minor fine-grained component); and greywacke (feldspar- and quartz-rich, but with a significant fine-grained component). Sedimentary structures preserved in sandstone can be important indicators of the depositional environment. These include bedding (cross bedding occurs due to the force of wind or water at a high velocity); ripple marks, left by shallow, relatively

**STRUCTURES IN SANDSTONE**
*This sandstone specimen from Yorkshire, England, shows particularly well-preserved sedimentary structures of bedding and slumping.*

### PROPERTIES

**ROCK TYPE** Continental, detrital, sedimentary
**FOSSILS** Vertebrates, invertebrates, plants
**MAJOR MINERALS** Quartz, feldspar
**MINOR MINERALS** Silica, calcium carbonate
**COLOUR** Cream to red
**TEXTURE** Fine to medium grained, angular to rounded

**CONSTRUCTING THE RED FORT AT OLD DELHI**
*Named after the colour of its sandstone, the Red Fort (below) was constructed in the reign of the Mogul emperor Shah Jahan in the mid-17th century. This 17th-century painting (left) shows the Red Fort under construction, supervised by Shah Jahan's master masons.*

*Red colouring due to iron oxides*

*Evidence of slumping*

*Fine-grained texture*

*Evidence of bedding*

slow-flowing water, showing the direction of flow; fossil tracks and trails; mud cracks, which are an indicator of dry or desert conditions; and slumping, indicating that the sandstone slipped down a slope, especially a cliff, usually with a rotational movement.

Sandstone formations are some of nature's most spectacular, and sandstone has long been a source of natural wonder. Most sandstones are resistant to erosion and, where they are exposed, especially with softer rock beneath, form prominent ridges, bluffs, and mesas. Canyonlands National Park in Utah, USA, is a spectacular wilderness of wind- and water-eroded sandstone spires, canyons, and mesas.

## BUILDING IN SANDSTONE

Sandstone has long been an important building material. Its durability as both a building and as a sculptural material is unsurpassed among sedimentary rocks. A superb example of building in sandstone is the Red Fort in Old Delhi, India, so called because of its red sandstone walls. It encloses palaces, gardens, barracks, and other buildings of red sandstone.

*Semirounded grains*

### THIN SECTION
*This section of Permian desert sandstone is formed from wind-blown sand with clearly outlined, but not particularly well-rounded, sand grains.*

*Sand grains*

*Cement between grains*

**PETRA**
*The Treasury, one of the finest and most widely recognized of Petra's buildings.*

# CARVING OUT OF SANDSTONE

Two locations in the world illustrate the ageless possibilities of sandstone. Both are carved from the rock as it stands in its natural setting, and the youngest is two millennia old.

## PETRA

The ancient city of Petra, in modern Jordan, was the centre of the Arab kingdom of the Nabataeans in Greek and Roman times. From the 4th century BC, Petra prospered as a centre of the spice trade, which stretched as far as China, Egypt, and India. The city was conquered by the Romans in AD106, and was mostly abandoned by the 7th century. What remained were the buildings cut into the sandstone of the Wadi as-Sik. The well-preserved buildings are a major tourist attraction today.

## ABU SIMBEL

Abu Simbel, Egypt, is the site of two temples built by the Egyptian king Ramses II (who reigned from 1279 to 1213BC), and consists, in part, of four 20m (65ft) colossal statues of Ramses carved out of a sandstone cliff on the west bank of the Nile. In the 1960s the temples were threatened by the rising waters of the Nile caused by the erection of the Aswan High Dam. The entire temple complex was cut into some 16,000 blocks, and reassembled on high ground above its previous site.

**ABU SIMBEL, EGYPT**
*In one of the engineering feats of the late 20th century, the colossal statues of Ramses II were relocated 60m (200ft) above the Aswan reservoir.*

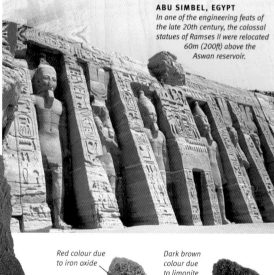

*Red colour due to iron oxide*

*Dark brown colour due to limonite*

*Rounded grains*

**RED SANDSTONE**

**LIMONITIC SANDSTONE**

# THE PAINTED DESERT

ONE OF THE MOST DRAMATIC AND COLOURFUL
EXPOSURES OF SEDIMENTARY ROCK ON THE EARTH
IS FOUND IN THE HIGH DESERT AREA OF NORTH
AMERICA. IT IS KNOWN AS THE PAINTED DESERT.

**NAVAJO SAND PAINTING**
*This contemporary sand painting of a yei (holy person), was created by a Navajo medicine man. Sand painting is an art form designed to heal the sick. The patient lies on top of the sand painting and the medicine man rubs sand over him. At the end of the healing ceremony, the sand painting is destroyed, destroying the illness at the same time.*

Government explorer Lieutenant Joseph C. Ives named the high desert area in the State of Arizona the Painted Desert, in 1858. He thought it aptly described the rainbow of colourful layers of sedimentary rock exposed in the stark desert landscape.

The Painted Desert is a narrow, crescent-shaped arc, about 240km (150 miles) long and varying in width from 16km (10 miles) at its northern end near the Grand Canyon to 80km (50 miles) at its southern end near Holbrook. Elevations within the Painted Desert range from 1,370 to 1,980m (4,500 to 6,500ft). The land is arid, with only 125 to 230mm (5 to 9in) of annual rain, and temperature extremes of –31 to +41°C (–25 to +105°F). It is sparsely vegetated, and very heavily eroded in parts. Within the Painted Desert at the very southern tip is the Petrified Forest National Park (see pp.334–35).

## THE CHINLE FORMATION

The rock exposed in the Painted Desert is the Chinle Formation, of Late Triassic Period (200–250 million years ago), composed of sandstones, mudstones, and shales. During the Triassic Period, north-eastern Arizona was located near the equator. About 60 million years ago, after the Chinle Formation had been deeply buried by younger strata, the region was uplifted as part of the massive Colorado Plateau, through which the Colorado River eroded the Grand Canyon. Other erosion also took place, exposing the rocks that are now seen in the Painted Desert.

The varied colours of the formation are a result of the varying mineral content in the sediments when they were laid down, and the rate of their deposition. With slow deposition, oxides of iron and (hematite) aluminium are concentrated in the sediment, creating the red, orange, and pink colours seen at the north end of the Desert. In the southern portion of the Desert, deposition was rapid, leading to removal of oxygen from the sediment, and resulting in blue, grey, white, and lavender layers.

The most heavily eroded areas of the Painted Desert contain many deep, tortuous gullies with intervening saw-toothed divides, combined with flat-topped mesas and buttes. These areas are known as badlands, and are the result of periodic torrential rains during thunderstorms and flash flooding. The soft mudstone erodes rapidly, creating gullies, while hills capped with sandstone or shale are more resistant, causing the development of hills and buttes with a step-like profile.

**MULTI-COLOURED LAYER CAKE**
*Blue, grey, lavender, and white sedimentary layers run through the eroded rock in the southern portion of the Painted Desert. These coloured rocks have been compared to a multi-coloured layer cake.*

**RED AND WHITE LAYERS**
White horizontal lines of mudstone strata run
through red and orange sandstone formations in
the Wilderness area alongside the Blue Canyon
in the Painted Desert of Arizona.

# GREENSAND

**GREENSAND CLIFFS**
*These cliffs are on the Isle of Wight, off England's south coast.*

A QUARTZ SANDSTONE with a high percentage of the green mica mineral glauconite, greensand is also called glauconitic sandstone. It is thought to form in marine environments by the replacement of calcite or by primary deposition. Some glauconite pellets are biogenic, originating as faecal pellets. The potassium in glauconite is useful in radiometric age-dating. Greensand is used as fertilizer for its high potassium and phosphorus levels and as a water softener.

*Well-sorted sediment*

*Glauconite*

*Quartz grains*

*Green colouring from glauconite*

**THIN SECTION**
*This thin section of greensand shows rounded grains of quartz and green glauconite.*

**MARINE ORIGIN**
*Greensand and other marine sedimentary rocks are common on the North American Atlantic coastal plain.*

### PROPERTIES

| | |
|---|---|
| **ROCK TYPE** | Marine, detrital, sedimentary |
| **FOSSILS** | Vertebrates, invertebrates, plants |
| **MAJOR MINERALS** | Quartz, glauconite |
| **MINOR MINERALS** | Feldspar, mica |
| **COLOUR** | Green |
| **TEXTURE** | Medium, angular |

# GREYWACKE

**CORNWALL**
*Greywacke in Cornwall, south-west England, is laid down in deep marine waters.*

GREYWACKE IS ALSO CALLED dirty sandstone, and is easily confused with basalt. It is composed of coarse- to fine-grained quartz, feldspar, and dark-coloured mafic minerals such as amphibole and pyroxene, in a matrix of clay, quartz, or calcite. Greywackes seem to result from rapid deposition in a turbulent marine environment, and can be several thousand metres thick.

### PROPERTIES

| | |
|---|---|
| **ROCK TYPE** | Marine, detrital, sedimentary |
| **FOSSILS** | Rare |
| **MAJOR MINERALS** | Quartz, feldspar, mafic minerals |
| **MINOR MINERALS** | Chlorite, biotite, clay, calcite |
| **COLOUR** | Grey, greenish-grey |
| **TEXTURE** | Fine to medium, angular |

**ANCIENT EGYPT**
*In ancient Egypt, greywacke was used to make sarcophagi, vessels, and statues, such as this Pharaoh's head.*

*Coarse-grained quartz*

*Fine-grained matrix*

**THIN SECTION**
*This section shows a variety of grain sizes and rock fragments, and also single mineral grains.*

*Feldspar*

*Rock fragment*

*Angular quartz grains*

**DIRTY SANDSTONE**
*This specimen from New World Island, Canada, shows a typical dirty appearance and poor sorting.*

# MARL

*MULTI-COLOURED*
*Arizona's painted desert contains marl. The various colours are a result of varied mineral content in the sediments.*

MARL, OR CALCAREOUS MUDSTONE, is a term applied to a variety of rocks that have a considerable range of compositions, but are generally earthy mixtures of fine-grained minerals. Consisting of clay minerals and calcium carbonate, they form under both marine and freshwater conditions. The calcium carbonate content is frequently made up of shell fragments of marine or freshwater organisms, or calcium carbonate precipitated by algae. Greensand marls contain the green mineral glauconite (see p.261); red marls contain iron oxide. Both marls are commonly whitish-grey or brownish in colour.

**BRICK FACTORY**
*Marl is sometimes ground, moulded, and fired for the manufacture of bricks.*

### PROPERTIES

| | |
|---|---|
| **ROCK TYPE** | Marine or freshwater, detrital, sedimentary |
| **FOSSILS** | Vertebrates, invertebrates, plants |
| **MAJOR MINERALS** | Clays, calcite |
| **MINOR MINERALS** | Glauconite, hematite |
| **COLOUR** | Various |
| **TEXTURE** | Fine, angular |

*Fine-grained sediment*

*Curved fracture*

**GREEN MARL**
*Marls differ in colour, depending on their mineral content. Green can be caused by glauconite or chlorite.*

# ARKOSE

**SEDIMENTS**
*These Torridonian sediments containing arkose were deposited in north-west Scotland.*

A RELATIVELY COARSE SANDSTONE, arkose consists primarily of quartz and feldspar grains with small amounts of mica. It has a high feldspar content (more than 25 per cent of the sand grains), setting it apart from other sandstones. Its grains tend to be moderately well sorted, angular or slightly rounded, and usually cemented with calcite, although they can sometimes be cemented with iron oxides or silica. Arkoses are presumed to derive from granites and, like them, are pink or grey. Sandstones with a feldspar content of 5–25 per cent are called subarkoses.

**GREY ARKOSE**
*The development of arkoses is thought to indicate either a climatic extreme or rapid uplift and high relief.*

Epidote

Feldspar

Quartz

Pinkish feldspar

Quartz grains

**THIN SECTION**
*This medium-grained feldspathic red sandstone is from Pre-cambrian sediments at Stoer, Scotland.*

## PROPERTIES

**ROCK TYPE** Terrestrial, marine, or freshwater detrital, sedimentary

**FOSSILS** Rare

**MAJOR MINERALS** Quartz, feldspar

**MINOR MINERALS** Mica

**COLOUR** Pinkish to pale grey

**TEXTURE** Medium, angular

# IRONSTONE

**HAMMERSLEY RANGE**
*These Australian mountains are coloured red by iron in the rocks.*

IRONSTONE IS A GENERAL TERM applied to sedimentary and metamorphic rocks that contain over 15 per cent iron. Very old, Precambrian sedimentary iron formations typically consist of iron minerals such as hematite and magnetite interlayered with chert. There are a number of younger large ironstone formations, from the Paleozoic Era, that consist of oolites (small spheres) of hematite and are fossiliferous. There are also small ironstone formations in bogs. With the exception of bog iron, ironstone does not appear to be forming any more. Partly because of this, the formation of ironstone is something of a mystery. Some seem to have formed early in Earth's history when oxygen was not as abundant in the atmosphere as it is now.

## PROPERTIES

**ROCK TYPE** Marine or continental, chemical, sedimentary

**FOSSILS** None or invertebrates

**MAJOR MINERALS** Hematite, goethite, chamosite, magnetite, siderite, limonite, jasper

**MINOR MINERALS** Pyrite, pyrrhotite

**COLOUR** Red, black, grey, striped

**TEXTURE** Fine to medium, crystalline to angular, oolitic

**BANDED IRONSTONE**
*These banded strata of ironstone rock are in Ontario, Canada.*

Magnetite

Jasper

Red colour from iron

Rounded ooliths

**OOLITIC IRONSTONE**
*The small, rounded ooliths that make up this rock are often cemented by calcite.*

# MICACEOUS SANDSTONE

**GRAND CANYON, USA**
*Micaceous sandstone is found in the Grand Canyon area.*

MICACEOUS SANDSTONE not only contains quartz, but also a large amount of mica and feldspar. The mica is commonly muscovite, and less commonly biotite. It is particularly visible where the rock is broken along bedding planes, revealing many sparkling flakes. The presence of mica is a good indicator of the depositional environment of the sandstone: small flakes of mica are very light, and are blown away easily in sediments deposited on the land surface, indicating that micaceous sandstones are likely to have been deposited in water.

**ANGULAR GRAINS**
*Micaceous sandstone is a well-sorted, medium-grained rock. The majority of the grains are angular with the mica occurring typically as flakes.*

**THIN SECTION**
*Slightly crumpled detrital flakes of muscovite are apparent in this thin section from Raasay, Inner Hebrides.*

Mica flakes

Mica flakes

Iron oxide patches

## PROPERTIES

**ROCK TYPE** Marine or freshwater, detrital, sedimentary

**FOSSILS** Invertebrates, plants, vertebrates

**MAJOR MINERALS** Quartz, feldspar, mica

**MINOR MINERALS** None

**COLOUR** Buff, green, grey, pink

**TEXTURE** Medium, angular to flattened

**NODULES IN MUDSTONE**
*These nodules with calcite cement are in Tertiary calcareous mudstone in Sicily, Italy. Their shape, composition, and texture differentiate them from the surrounding rock.*

*Crystals radiating from centre*

**PYRITE "SUN"**
*The pyrite in this nodule has formed a flat, coin-like disc, but still maintains its radial crystal structure. These nodules are often marketed as pyrite "suns".*

# NODULES AND CONCRETIONS

NODULES AND CONCRETIONS are features that develop during or after the formation of a sedimentary rock. Most form within the sedimentary layer, but sea-floor nodules are found above accumulating sediments.

## NODULES

A nodule is a rounded mineral accretion that differs in composition from its surrounding rock. Most are formed by the accumulation of silica in the sediments, which then solidifies. Nodules are commonly elongated, with a knobby irregular surface, and are usually oriented parallel to the bedding of their enclosing sediment. One mineral commonly found as nodules is pyrite. It occurs as spheres or as rounded cylinders of radiating crystals, and in a few localities, as flat, radiating discs, or "suns". Clay ironstone, a mixture of clay and siderite, sometimes occurs as layers of dark grey to brown nodules overlying coal seams. Nodules containing manganese, phosphorous, titanium, chromium, and other valuable metals develop on the sea floor, but their depth and relative lack of concentration make them uneconomical to mine. Other nodules form around plant and animal remains, and become part of the fossilization process.

## CHERT AND FLINT NODULES

Chert and flint often occur as dense and structureless nodules of nearly pure cryptocrystalline quartz within beds of limestone or chalk. They seem to have formed through the replacement of the carbonate rock by silica, and often replace or include fossils and trace fossils such

*Weathered outer surface*

**SPHERICAL NODULE**
*This nodule of pyrite is made up of characteristic radiating pyrite crystals.*

*Internal radiating structure*

**PROPERTIES**

| | |
|---|---|
| **ROCK TYPE** | Continental, marine, detrital, sedimentary |
| **FOSSILS** | Vertebrates, invertebrates, plants |
| **MAJOR MINERALS** | Quartz, feldspar |
| **MINOR MINERALS** | Silica, calcium carbonate |
| **COLOUR** | Cream to red |
| **TEXTURE** | Fine to medium grained, angular to rounded |

**FLINT NODULE**
*This flint nodule shows variation in its internal colour due to the differing extraneous material that was enclosed as the silica making up the nodule solidified. Also seen clearly is its conchoidal fracture, which made it a useful material for tool-making.*

*Conchoidal fracture*

*Sharp edges*

## FLINT TOOLS AND WEAPONS

Chert and flint were the main sources of tools and weapons for Stone Age peoples. Because of the uniformly fine grain, brittleness, and conchoidal fracture of flint and chert, it was relatively straightforward to flake off chips to shape them, leaving razor-sharp edges. Modern experiments have shown that the edge of a single flint flake is as sharp as that of a razor blade, and can be used to butcher meat as effectively as a modern knife.

Flint quarrying was probably the earliest human mining venture. Archaeologists can trace prehistoric trade routes by studying the distribution of flint mined from a certain area. During the Neolithic period, or New Stone Age (about 8000–2000BC), the soft chalk deposits

in France and Britain were mined for their flint nodules in shafts sunk to a depth of up to 100m (330ft). Until the first quarter of the 19th century, chipping flints for flintlock firearms remained an important cottage industry.

**FLINTLOCK PISTOL**
*A flint striking steel produced a spark that ignited the gunpowder in a flintlock pistol.*

**FLINT ARROWHEADS**
*Flint can be chipped into delicate, intricate, and very sharp forms.*

*Crust of weathered flint*

as burrows. Because they are hard and chemically resistant, they become concentrated in residual soils as the surrounding carbonate rock weathers away – in some places, the soils are composed of more flint than soil.

Flints and cherts usually have a certain amount of other minerals minutely intermixed, often making them opaque. In general, flints are found in nodules but cherts are also found as bedded deposits. Flints are more whitish to grey, cherts pale brown to black.

## CONCRETIONS

Concretions differ from nodules in being made of the same material as their host sediment, cemented by other minerals. Concretions are often much harder and more resistant to erosion than their surrounding rock, and can be concentrated by weathering. In some localities, concretions have formed around plant and animal remains, perfectly fossilizing them.

**IRONSTONE CONCRETION**
*From the famous fossil locality of Mazon Creek, Illinois, USA, this ironstone concretion has formed around a fossil fern.*

# BUILDING WITH FLINT

In areas where flint is common, it is a favoured and durable building material. Consisting of individual flints set in mortar, flint walls are virtually indestructible by the elements. In the southern counties of England, flints cleared from agricultural fields provide an easily accessible and plentiful supply of cheap material. City walls of flint set in concrete still stand from Roman times in numerous parts of Britain. The walls of many medieval cathedrals and castles were built with an inner and outer layer of shaped stone, but with a core of flint set in mortar. Flints can be set as gathered, with the white rind of weathered material intact, or with a face broken off to reveal the colourful, fresh material inside. Gravel-sized flint pebbles are also flung onto still-wet exterior plaster, creating a "pebble-dash" finish.

**FLINT WALL**
*Flint walls can be made from flints randomly arranged, or can be built with regularly shaped flints set in attractive repeating patterns.*

**FLINT COTTAGE**
*Flint cottages are found in large areas of the southern counties of England that are underlain by chalk that contains flint nodules.*

Compact silica

**SEPTARIAN CONCRETION**
*An unusual type of concretion is the septarian concretion (or septarian nodule). This is formed when cracks appear in the nodule due to shrinkage, and are filled with crystals deposited from percolating brines.*

Dark calcareous material

Pale calcite in cracks

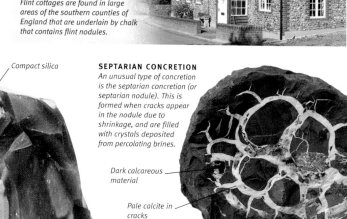

**EXPOSED NODULES**
*The Whaleback in Pennsylvania, USA, is an area where hundreds of large concretions have been exposed by the weathering of the surrounding rock. The weathering followed upwarping of the area.*

Subconchoidal fracture

**CHERT**

Fine-grained texture

# STONEHENGE

ONE OF THE MOST FAMOUS STRUCTURES BUILT BY HUMAN HANDS,
STONEHENGE WAS CREATED OVER SEVERAL CENTURIES, DURING A PERIOD
SOMETIMES CALLED THE NEOLITHIC REVOLUTION. THIS WAS A TIME
OF TRANSITION FROM STONE CONSTRUCTION USING ROUGH, UNHEWN
STONES TO CONSTRUCTION USING SHAPED AND JOINTED STONES.

**MORTISE AND TENON**
*The lintels of the sandstone circle and trilithons are held on their uprights by mortise-and-tenon joints. The tenon can just be seen at the top of the far upright.*

Located about 30km (18 miles) south of Avebury, Wiltshire, England, Stonehenge represents a significant step in the use of stone as a building material. The structure was among the first monumental structures in Europe to be built using shaped stones, apparently in imitation of already-developed woodworking techniques. The stone circles at Stonehenge may have been used either as a centre for religious ritual or as an observatory for predicting astronomical events.

Although opinions about the purpose of the megalithic monument differ, archeologists are agreed about its construction. Stonehenge was built in several stages, the first beginning in about 3100BC. In the first phase, a circular ditch and bank about 98m (320ft) in diameter was constructed and a circle of 56 shallow holes dug just inside the bank. Beyond the circular ditch to the north-east, two parallel stones of sandstone were erected, marking the entrance of the earthwork. One of these remains and is known as the Heel Stone. Cremated remains have been found in the 56 holes (known as the Aubrey Holes), indicating that funerary rites were performed here at this time.

The site was used until about 2600BC, when it was abandoned. Then, in about 2100BC, the site was reoccupied and radically remodelled. About 80 dolerite

bluestone pillars (see p.39) were brought to the site from the Preseli Mountains in south-west Wales – a distance of some 385km (240 miles). They weighed up to 4 tonnes each, and were transported directly across sea, river, and land. These dolerite pillars were to be erected in the centre of the site to form two concentric circles. However, the stone circles were never completed.

## MAIN PERIOD OF CONSTRUCTION

Around 2000BC, a new phase of construction began. The dolerite bluestones were dismantled, and a lintelled circle and a horseshoe of large sarsen sandstones was erected. Up to 9m (30ft) in height and 50 tonnes in weight, these gigantic sandstones dominate the site today. They are composed of Tertiary sandstone and were brought from the Marlborough Downs, 30km (20 miles) to the north of the site. The circle consisted of 30 uprights capped by a continuous ring of stone lintels. Within this circle, a horseshoe formation of five trilithons was erected, each of which comprised a pair of large stone uprights supporting a stone lintel.

The uprights of the sarsen ring were dressed and tapered by pounding with stone hammers. The lintels are held on their uprights by mortise-and-tenon (dovetail) joints. They are dressed into a curve to produce a circle, and are joined to each other with tongue-and-groove joints. The lintels of the trilithon sandstones are similarly held on their uprights.

Between 2000 and 1550BC some of the dismantled dolerite bluestones were re-erected in a horseshoe within the sandstone horseshoe. Then, between 1550 and 1100BC a circle of dolerite pillars was erected between the sandstone circle and horseshoe. Finally, a large block of green sandstone, now known as the Altar Stone, and also brought from south Wales, was erected within the inner horseshoe.

## POPULAR THEORY

In 1963 American astronomer Gerald Hawkins concluded in an article for the journal *Nature* that the stones provided sightlines for the rising and setting of the Sun and Moon at the summer and winter solstices, and that the Aubrey Holes were used to predict eclipses of the Moon. His theory that Stonehenge was used to establish a basic calendar and to chart the movement of the Sun, Moon, and stars remains the most popular today.

**AERIAL VIEW**
*This overhead view encompasses the still-standing remains of the inner and outer circles and the trilithons, but does not show the outer ditch.*

**REMAINING OUTER CIRCLE SEGMENT**
*The largest still-standing segment of the outer stone circle consists of four sarsen sandstones, still capped by their lintels. There were originally 30 of these lintel-capped uprights, forming a continuous outer circle.*

# METEORITES

METEORITES ARE ROCKS that formed on other bodies in our Solar System, were chipped from their parent bodies, and orbited the Sun until colliding with Earth. Much of the material that makes this journey is travelling so fast when it hits the Earth's atmosphere that friction causes the rock to evaporate completely in the upper atmosphere, forming meteors (shooting stars). Pieces that are relatively large (weighing more than 10g) or very small (weighing less than a milligram) often survive to fall to the surface. Meteorites can be classified into three types based on properties visible to the naked eye: iron, stony-iron, and stony meteorites. Although some meteorites have been referred to as "cosmic sediments", meteorites as a whole are not regarded as being igneous, metamorphic, or sedimentary rocks.

## IRON AND STONY-IRON METEORITES

As their name suggests iron meteorites are formed mostly of metallic iron, mixed with varying amounts of nickel to produce two minerals, kamacite and taenite, which are rare on Earth outside meteorite collections. Other minerals present in very minor amounts include troilite (FeS), some silicates (most commonly olivine and pyroxene), phosphides, and graphite. Iron meteorites are thought to originate in the cores of large asteroids. Because iron meteorites are so different from most terrestrial rocks, they are recognized more often than other kinds of meteorite.

Stony-iron meteorites are composed of a mixture of iron and silicate minerals. The most common type of stony irons are called pallasites; these look like iron meteorites with centimetre-sized translucent olivine (less commonly, pyroxene) crystals scattered throughout. Pallasites are thought to originate along the boundary of the iron core and silicate mantle of large asteroids.

Widmanstätten pattern    Pyroxene

**IRON METEORITE**
*The intergrowth of large kamacite and taenite crystals forms this Widmanstätten pattern, which is revealed by chemical etching, and is characteristic of some iron meteorites. These large crystals can form only through very slow cooling. This specimen was found in Namibia.*

**FALLING TO EARTH**
*The accretion of the Earth is still going on to a small degree as debris from the solar nebula (see p.11) still falls to Earth as meteorites. Several tonnes reach Earth each year.*

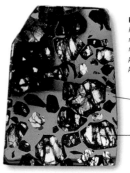

Ida

Dactyl

Irregular shape

Cratered surface

**IDA AND DACTYL**
*Meteorites are derived from asteroids. The asteroid Ida and its satellite, Dactyl, are in the asteroid belt between Mars and Jupiter. Irregularly shaped and heavily cratered, Ida is 56km (35 miles) long.*

**PALLASITE**
*Pallasites are stony-iron meteorites and are relatively rare. They are typically one part olivine crystals to two parts metal.*

Iron-nickel matrix

Olivine crystal

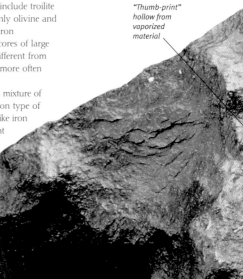

"Thumb-print" hollow from vaporized material

**BARRINGER CRATER**
*This impact crater, named for its owner, Daniel Barringer, is in the Arizona Desert, USA. It has a diameter of 1.2km (¾ mile). The meteorite that created it fell about 50,000 years ago, and was about 50m (150ft) across. The landing of such large meteorites is so explosive that they almost entirely vaporize, leaving large craters but little meteorite debris.*

## STONY METEORITES

Stony meteorites fall into two categories, chondrites and achondrites. The majority of the meteorites falling to Earth are chondrites. These are named for the small igneous, silicate spheres called chondrules, which are the dominant component in most chondritic meteorites. The largest chondrules are no more than 1cm (1/2in) across, and are made largely of olivine and pyroxene with lesser amounts of glass and in some cases feldspar. Chondrules are miniature igneous rocks that were melted and then crystallized while floating in space during the birth of the Solar System. Many chondrites have been altered on their parent bodies by aqueous fluids or metamorphism. However, the geological processing is relatively light compared with that of terrestrial rocks or other meteorites. Achondritic meteorites have been geologically processed. Most represent the crust or mantle of differentiated asteroids, and are very similar to igneous rocks and impact breccias on Earth. The majority of achondrites appear to come from asteroids, but about 20 have been identified as Martian and another 20 are Lunar.

**Moldavite**

### MOLDAVITE PENDANT
This moldavite tektite has been mounted as a pendant, together with a topaz crystal.

### PROPERTIES OF STONY METEORITES

| | |
|---|---|
| **ROCK TYPE** | Igneous, metamorphic, sedimentary |
| **FOSSILS** | No macroscopic |
| **MAJOR MINERALS** | Olivine, pyroxene, plagioclase |
| **MINOR MINERALS** | Various |
| **COLOUR** | Grey, greenish, tan, or black |
| **TEXTURE** | Fine to medium |

**Medium-grained texture**

**Surface worn away by heat-friction**

### STONY METEORITE
This stony meteorite was found in Haryana, India.

## TEKTITES

When a large meteorite hits the Earth, the terrestrial rock is sometimes melted, thrown in the air, then quickly cools to form glass. These pieces of glass, called tektites, are typically a few millimetres to centimetres across and dumbell- or disc-shaped. Tektites have been found on every continent except Antarctica and South America. They are usually found in limited but large areas (of hundreds to thousands of square km) and can, in some cases, be related to specific impact craters. For example, the beautiful green glass tektites found in and around the Czech Republic have been shown by their age and chemical composition to come from the Reis Crater hundreds of kilometres away in Germany.

### CONE-IN-CONE

### GREEN GLASS (MOLDAVITE) TEKTITE

### OTHER IMPACTITES

Impacts often pulverize rather than melt rock. This can lead to the formation of an impact breccia called suevite. The pressure wave produced by the impact can also lead to the reorientation of mineral grains within the rock, producing shatter cones. These may be millimetre- to metre-sized structures, depending upon the shock intensity, rock type, and grain size, but the nose of the cone always points towards the crater.

### DISC-SHAPED TEKTITES

### DUMBELL TEKTITE

### STONY-IRON METEORITE
This stony-iron meteorite, found in the Atacama Desert, Chile, shows the concave "thumb print" surface of many meteorites, caused by the vaporization of material as it passed through the atmosphere. Its surface is also weathered from its time on Earth.

**Weathered surface**

## MARTIAN METEORITES

A group of meteorites known as SNCs (Shergottites, Nakhlites, and Chassignites) comes from Mars. The composition of small pockets of gas inside the rocks matches that of the Martian atmosphere almost perfectly. These are our only physical samples of Mars. One Martian meteorite, ALH84001, is thought by some to contain evidence of Martian life. Microbe-like, rod-shaped tubules were found within it that some took to be fossil bacteria. Others believe that this structure was the result of high-temperature mineral formation.

### MARTIAN METEORITE
The ALH84001 meteorite is believed by some to contain evidence of life on Mars.

# METAMORPHIC ROCKS

METAMORPHISM OCCURS WHEN an existing rock (a protolith) is subjected to pressures or temperatures very different from those under which it originally formed. This causes its atoms and molecules to rearrange themselves into new minerals while still in the solid state, without melting taking place. The result of this transmutation is called a metamorphic rock. There are three different ways in which metamorphic rocks are formed: dynamic metamorphism, contact metamorphism (or thermal metamorphism), and regional metamorphism.

## DYNAMIC PRESSURE

Metamorphism may be the result of large-scale movements in the Earth's crust, especially along fault planes and at continental margins where tectonic plates are colliding. Rock masses are crushed as great pressures come to bear. The resulting mechanical deformation, with little temperature change, is known as dynamic metamorphism. The rocks that are produced by this form of metamorphism range from angular fragments to fine-grained, granulated or powdered rocks, such as mylonite. They are characterized by foliation – the alignment of mineral grains in parallel plates. They are often distorted as a result of the stresses to which they have been subjected.

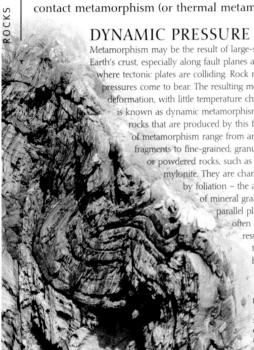

Section of rock deep in the Earth's crust

Tectonic compression

Folded strata

Vertical, slaty cleavage forms at right angles to forces of compression

*Tectonic compression*

**DYNAMIC METAMORPHISM**
Tectonic forces applying pressure to sedimentary rocks transform them by dynamic metamorphism. Rock strata fold and slaty cleavages form.

**FOLDED ROCK**
The pressures that create folding of sedimentary rocks such as this greywacke at Mizen Head, Ireland, can result in metamorphism even at low temperatures.

## THERMAL CONTACT

Contact metamorphism occurs mainly as a result of increases in temperature with little or no contribution from pressure. It is common in rocks near an igneous intrusion. Heat from the intrusion alters rocks in the surrounding area to produce an "aureole" of metamorphic rock. Since the rocks nearest to the intrusion are subjected to higher temperatures than those farther away, they exhibit different characteristics. Thus the temperature gradient, from high to low, creates concentric zones of distinctive metamorphic rocks. The new minerals that form due to contact metamorphism depend on the composition of the host rocks. For example, aluminium-bearing minerals, such as the feldspar in arkosic sandstones, are changed to micas and garnets, but when carbonate minerals, such as calcite, are present, hornblende, epidote, and diopside are formed.

Existing rock changed by hot intrusion

Eroded landscape

Zones of decreasing heat and metamorphism

Granite intrusion

**CONTACT METAMORPHISM**
A large intrusion of an igneous rock such as granite releases heat into the surrounding rocks, altering their mineralogy. This is known as contact (thermal) metamorphism.

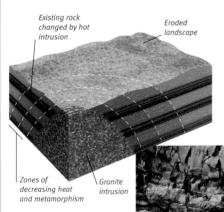

**BAKED SEDIMENT**
At Salisbury Crag, Edinburgh, Scotland, a dark intrusion of igneous dolerite has "baked" neighbouring paler sandstones.

**SHALE**

**SLATE**

**PHYLLITE**

**SCHIST**

**GNEISS**

**SHALE TO GNEISS**
This sequence shows how shale, a sedimentary rock, can be metamorphosed by the application of heat and pressure. Relatively low pressures will suffice to turn shale into slate. Increasing temperature and pressure will create phyllite and schist. Finally, with extreme heat and pressure, gneiss forms. Gneiss is primarily associated with regional metamorphism.

## CHANGING MINERALS

Metamorphism is said to be low grade if it occurs at relatively low temperature and pressure and high grade at the intense end of the temperature and pressure range. The assemblages of minerals in rocks are affected differently depending on the grade of metamorphism and the relative importance of pressure and temperature in the reaction. In some low-grade reactions, the components of existing mineral assemblages are simply redistributed. For example, the iron-rich garnet almandine and magnesian-biotite may be metamorphosed into the magnesium-rich garnet pyrope and iron-biotite. The iron from the almandine garnet migrates to the biotite, and the magnesium from the biotite migrates to the garnet. In other reactions at higher temperatures and pressures, the biotite and garnet may disappear entirely, their chemical components combining with other components present in the rock to form an entirely new set of minerals or a liquid. Some metamorphic reactions are more dependent on pressure than temperature. For example, undergoing the high pressures found in subduction zones, albite breaks down to form jadeite and quartz. Jadeite is much denser than albite, reflecting the closer packing of atoms under pressure. The number of possible reactions found in metamorphism is vast because of the chemical and mineralogical complexity of the Earth's crust.

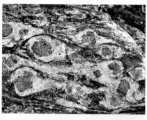

**GARNET AUGEN GNEISS**
*Almandine, an iron-rich garnet, is a mineral often found in metamorphic rocks such as gneiss. The garnet can grow as large, eye-shaped grains called "augens".*

**QUARTZOSE SCHIST**
*It is often far from obvious that a rock is metamorphic. This quartzose mica schist at Kintyre, Scotland, could easily be confused with sandstone.*

## REGIONAL CHANGE

The formation of regional metamorphic rocks is associated with the process of mountain-building through the collision of tectonic plates. This process generates increases in temperature and pressure that may extend over an area of thousands of square kilometres, producing widespread metamorphism. The most important regional metamorphic rocks include slates, schists, and gneisses. Which one of these is produced depends on the temperature and pressure to which the existing rocks have been subjected, as well as the amount of time they have spent under those conditions. Slates are produced in areas of relatively low temperature and pressure. Where temperatures and pressures are at their highest, gneisses are produced. Schists are formed in zones of intermediate temperature and pressure.

**ANTARCTIC MARBLE**
*These folded rocks in Antarctica's Royal Society Range began as beds of limestone and have been subsequently metamorphosed – deformed and recrystallized to form marble.*

**SCOTTISH MARBLE**
*From Ledmore North Quarry, Assynt, Sutherland, this marble source is one of only a few in a country better known for its quarrying of granite.*

# MARBLE

MARBLE IS A GRANULAR METAMORPHIC rock derived from limestone or dolomite (see p.54). Marbles are formed under the influence of heat and pressure, and consist of a mass of interlocking calcite or dolomite grains. The latter are called dolomitic marbles. Marbles rich in serpentine minerals are known as ophicalcites.

Marbles form in two ways: as a result of the deep burial of limestone in the older layers of the Earth's crust with consequent heat and pressure from the thick layers of overlying sediments; and as a result of contact metamorphism near igneous intrusions. Pure marble is nothing but calcite; impurities in the original limestone often recrystallize during metamorphism to give mineral impuritites in the marble. The most common impurities are quartz, mica, graphite, iron oxides, and small pyrite crystals. Silicates of lime or magnesia are found in other marbles, and can include diopside, tremolite, labradorite, albite, anorthite, garnet, vesuvianite, spinel, zoisite, epidote, tourmaline, and sphene. Many of these will be present in sufficient amounts to give a characteristic texture or colour (or both) to the marbles in which they occur.

**MARBLE QUARRY**
*Huge blocks of sparkling white marble are sliced from the rock-face at Carrara, Italy, and removed by heavy machinery.*

**SWITCHBACKS**
*The haulage road at the Carrara marble quarry in Italy climbs back and forth across the steep rock-face.*

## VARIETIES OF MARBLE

Pure marble is white. Some other marbles take their common names from their colour or mineral impurities. Because the impurities were originally layers of other minerals thinly interbedded in the original limestone, they sometimes occur as bands in the marble. Impurities of serpentine, tremolite, brucite, and olivine may all impart a greenish tinge. Grey marble usually forms from relatively pure limestone and contains few mineral impurities. Small amounts of graphite colour

**HAZRATBAL MOSQUE**
*This mosque, with its distinctive dome and minaret, is on the shores of Lake Dal in Kashmir. It is made of white marble.*

*Marble fragment*

**ARCH OF CONSTANTINE**
*This marble arch, in Rome, Italy, was built in the 4th century AD. It is the model for many modern arches, including London's Marble Arch.*

"The **marble** not yet carved can hold the form of every **thought** the greatest **artist** has."

**MICHELANGELO**

some grey marbles; some black marbles have higher graphite concentrations. Brown marble is coloured by garnet and vesuvianite. Veined and patterned marbles are often created when an existing marble is cracked or shattered, and the spaces between the marble fragments filled in with calcite or other minerals. In marbles that have not been fully metamorphosed, fossils from the original limestone can sometimes still be seen. Marbles from Carrara, Italy, and at Bergen, Norway, are examples of these. Where complete recrystallization has taken place, all fossil remains are completely obliterated.

## PROPERTIES

**ROCK TYPE** Regional or contact metamorphic
**TEMPERATURE** High
**PRESSURE** Low to high
**STRUCTURE** Crystalline
**MAJOR MINERALS** Calcite
**MINOR MINERALS** Diopside, tremolite, actinolite
**COLOUR** White, pink
**TEXTURE** Fine to coarse
**PROTOLITH** Limestone, dolomite

## CARVING IN MARBLE

Michelangelo notwithstanding, the golden age of marble sculpture was arguably in the Classical Period of ancient Greece, from about 480 to 330BC. Two sculptors epitomize the age: Praxiteles and Phidias. The statues of Praxiteles transformed the detached, majestic style of his immediate predecessors into his own, expressing gentle grace and sensuous charm. His most famous work was the *Aphrodite of Cnidus*, celebrated by the Roman author Pliny the Elder as the finest statue in the world. In this statue, the goddess is depicted naked, a bold innovation at the time, which prompted his model to exclaim, "Alas, where did Praxiteles see me naked?" Phidias undertook marble sculpture on a grand scale: he was the artistic director of the construction of the Parthenon in Athens. Phidias and his assistants also carved the marble figures that ornamented the Parthenon frieze. Most of the remains of these – often called the Elgin Marbles – are in the British Museum.

**DAVID**
Michelangelo's David (created 1501–04) was carved from marble from Carrara, Italy.

Iron oxide cement

Hematite vein

Limonite vein

**SCULPTOR AT WORK**
The fine grain and flawless colouring of some marble allows intricate carvings, such as this example by a Vietnamese sculptor.

**MARBLE BRECCIA**
From the Isle of Skyros, Greece, this specimen of marble has been cracked and shattered, the veins infilled, and fragments recemented with iron-rich calcareous cement. Colour variations in the veining are a result of the decomposition products of mineral impurities in the original marble, such as reddish hematite, brown limonite, and pale green talc.

**THIN SECTION**
This is forsterite marble (forsterite and calcite), a typical, high-grade metamorphic marble from South Africa.

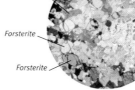

Calcite

Forsterite

Forsterite

**TYPES OF MARBLE**
The many variations in colouring of marble is determined by the other mineral impurities that are enclosed within it.

Colour from graphite traces

Colour from olivine

Visible olivine grains

Colour from serpentine or tremolite

Colour from diopside

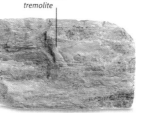

**GREY MARBLE**

**OLIVINE MARBLE**

**GREEN MARBLE**

**POLISHED SLAB OF BLUE MARBLE**

# THE TAJ MAHAL

A MARBLE MAUSOLEUM IN AGRA, INDIA, THE TAJ MAHAL WAS
BUILT FOR THE MOGUL EMPEROR SHAH JAHAN TO MEMORIALIZE
HIS FAVOURITE WIFE, MUMTAZ MAHAL,
WHO DIED IN CHILDBIRTH IN 1631.

**PIETRA DURA**
*Precious and semiprecious
stones were meticulously cut
and shaped into tendrils and
floral arabesques, and then set
in rows in the white marble.*

**LOVING COUPLE**
*Shah Jahan was heartbroken
when his favourite wife, Mumtaz
Mahal, died giving birth to
their 14th child. It is said that
he locked himself in his rooms
and refused food for eight
days. When he finally emerged,
observers noted that his black
beard had turned white.*

**THE CENOTAPHS**
*The central chamber of the
mausoleum houses the public
tomb of Mumtaz Mahal (below
right), which rests beneath the
centre of the dome and in line
with the entrance to it. The
tomb of Shah Jahan was later
squeezed in beside it.*

Renowned as the finest example of Mogul
architecture, the Taj Mahal combines symmetry
and balance between all parts of the building
with superb delicacy of detail in ornamentation. The
chief architect was probably an Indian of Persian
descent, Ustad Ahmad Lahawri.

Building began around 1632, with more than 20,000
workers employed from India, Persia, the Ottoman
Empire, and Europe. Construction of the mausoleum
complex spanned 22 years and cost between four and
five million rupees. The central mausoleum itself was
completed by about 1639, the adjunct buildings by 1643,
and the decoration work in 1647.

## MAKRANA MARBLE

The mausoleum itself is built of unusual white marble
called Makrana marble, which changes hue according to
the intensity of sunlight or moonlight. It rests on a
plinth 7m (25ft) high, with four nearly identical facades,
each with a wide central arch that rises to 33m (110ft).
Each of the four corners is slanted, and incorporates
smaller arches. The central dome reaches a height of
73m (240ft). It is surrounded by four lesser domes. The
superb acoustics inside the main dome cause the single
note of a flute to resound five times.

The interior is organized around an octagonal
marble chamber, which is ornamented with low-relief
carvings and hard-stone inlay called *pietra dura*. Within
this are the cenotaphs of Mumtaz Muhal and Shah
Jahan, both of which are inlaid with semiprecious stones.
They are enclosed by a finely carved filigree marble
screen. The true sarcophagi lie beneath, at garden level.

**TAJ MAHAL, INDIA**
*Framed by four tall minarets, the central mausoleum is
set at the north end of a formal garden quartered by
long pools of water in which the building is reflected.*

# SLATE

A FINE-GRAINED metamorphic rock, slate has a characteristic "slaty cleavage", allowing it to be split into relatively thin, flat sheets. True slates split along the foliation planes formed during metamorphism rather than along the original sedimentary layers. Its cleavage is a result of microscopic mica crystals that have all grown oriented in the same plane. Slate forms when mudstone, shale, or felsic volcanic rocks are buried and subjected to low temperatures and pressures. Common around the world in regionally metamorphosed terrains, such as the Appalachians, USA, the Rhenish mountains, western Germany, and the Taihang mountains, Hebei Province, China, slate occurs in a number of colours, depending on the mineralogy and oxidation conditions of the original sedimentary environment. Black slate, for example, forms in a very oxygen-poor environment; red in an oxygen-rich environment. Many slates are mottled or spotted. Fossils of plants and animals can be preserved in slate. Slate is quarried in large pieces for use in electrical panels, worktops, blackboards, and flooring, and in smaller pieces for roofing.

## PROPERTIES

| | |
|---|---|
| **ROCK TYPE** | Regional metamorphic |
| **TEMPERATURE** | Low |
| **PRESSURE** | Low |
| **STRUCTURE** | Foliated |
| **MAJOR MINERALS** | Quartz, mica, feldspar |
| **MINOR MINERALS** | Pyrite, graphite |
| **COLOUR** | Various |
| **TEXTURE** | Fine |
| **PROTOLITH** | Mudstone, siltstone, shale, or felsic volcanics |

**THIN SECTION**
*This slate contains a folded band of coarser silt. The slaty cleavage crosses the bedding at a high angle.*

Band of silt

Slaty cleavage

Fine grain size

## SLATE ROOFING

The fissile nature of slate – its ability to split into thin sheets with parallel faces – makes it ideal as a durable roofing material. Slate quarrying and preparation was once a huge industry, now largely supplanted by modern materials. Fissile limestones are also used as roofing slates and known as "slate", though they are not true slates.

**PIEDMONT SLATES**
*Local slates, such as these in Italy, are often part of a region's typical architecture.*

Fissile splitting

**WELSH SLATE**
*This specimen is from Wales, Britain's principal source of slate. In the USA, slate is quarried in Pennsylvania and Vermont.*

Foliated structure

---

**PERU**
*Many phyllites can be found in the Peruvian Andes, a vast mountain chain.*

# PHYLLITE

LIKE SLATE, PHYLLITE is a fine-grained metamorphic rock that forms when fine-grained sedimentary rocks, such as mudstones or shales, are buried and subjected to relatively low temperatures and pressures for a long period of time. It has a similar tendency to split into sheets or slabs because of the parallel alignment of mica minerals, but its grains are larger than those of slate. Phyllite also has a shinier sheen than slate because of its larger mica crystals.

Many phyllites have a scattering of large crystals called porphyroblasts, which grow during metamorphism. Tourmaline, cordierite, andalusite, staurolite, biotite, and pyrite are all minerals that are commonly found as porphyroblasts in phyllites. Phyllite occurs in regionally metamorphosed terrains, in both young and old eroded mountain belts. It is sometimes used as a paving stone.

**THIN SECTION**
*This section shows the dense mat of aligned muscovite that gives rise to the sheen on the specimen's surface.*

Muscovite

Sheen on surface

## PROPERTIES

| | |
|---|---|
| **ROCK TYPE** | Regional metamorphic |
| **TEMPERATURE** | Low to moderate |
| **PRESSURE** | Low |
| **STRUCTURE** | Foliated |
| **MAJOR MINERALS** | Quartz, feldspar, muscovite mica, graphite, chlorite |
| **MINOR MINERALS** | Tourmaline, analusite, cordierite, pyrite, magnetite |
| **COLOUR** | Silvery to greenish-grey |
| **TEXTURE** | Fine |
| **PROTOLITH** | Mudstone, shale, siltstone, or felsic volcanics |

**GEOLOGICAL INDICATOR**
*Phyllite has little economic use, except as a paving stone, but it is a valuable indicator of geological conditions.*

Wavy foliation

# SCHIST

**FOLDED SCHISTS**
*This schist outcrop is in the Austrian Alps.*

SCHISTS ARE foliated metamorphic rocks with visible mineral crystals. They commonly show distinct layering of light- and dark-coloured minerals. Most schists are composed largely of platy minerals such as muscovite, chlorite, talc, biotite, and graphite.

Schist's characteristic schistose fabric (it splits easily along planes) is a result of the parallel orientation of these minerals. There are a number of different schists. For example, greenschist is a schist rich in the green minerals chlorite, actinolite, and epidote; blueschist is rich in blue glaucophane. The specific mineral composition of schist depends both on its original rock (protolith) and on its metamorphic environment. The mineral assemblage can thus be used to help determine both the environment in which the original rock formed and its metamorphic history.

**THIN SECTION**
*This typical garnet-mica schist comes from the Austrian Alps.*

- Garnet
- Kyanite
- Muscovite

**MUSCOVITE SCHIST**
*The silvery appearance of muscovite schist is due to the abundance of muscovite mica.*

- Silvery mica

## PROPERTIES

**ROCK TYPE** Regional metamorphic
**TEMPERATURE** Low to moderate
**PRESSURE** Low to moderate
**STRUCTURE** Foliated
**MAJOR MINERALS** Quartz, feldspar, mica
**MINOR MINERALS** Garnet, actinolite, hornblende, graphite, kyanite
**COLOUR** Silvery, green, blue
**TEXTURE** Medium
**PROTOLITH** Mudstone, siltstone, shale, or felsic volcanics

Wavy foliation

Garnet porphyroblast

**GARNET SCHIST**

**KYANITE SCHIST**

- Biotite
- Bladed blue kyanite

**SCHIST VARIETIES**
*Many schists contain porphyroblasts: large crystals of minerals such as garnet that grow during metamorphism.*

---

# GNEISS

**DOMES OF GNEISS**
*These granitic gneiss domes are in South Africa.*

DISTINCT BANDS of minerals of different colours and grain sizes characterize this metamorphic rock. In most gneisses the bands are folded, although the folds may be too large to see in a hand specimen. Gneiss is medium- to coarse-grained, and differs from schist in that its foliation is well developed, but it has little or no schistosity (tendency to split along planes).

Most gneisses contain abundant quartz and feldspar, with porphyroblasts of metamorphic minerals such as garnet. They form at very high temperatures and pressures from sedimentary or granitic rocks, and make up the cores of mountain ranges. Pencil gneiss has rod-shaped individual minerals or aggregates of minerals. Being crystalline, gneisses do not split easily, and are a popular building and facing stone. In the building industry, they are classified as granites.

**GNEISS BOWL**
*This Egyptian diorite-gneiss bowl dates from the period 2890–2686BC.*

**ABRAHAM WERNER**
*Mineralogist Werner (1749–1817) adopted and popularized the use of the ancient terms gneiss and schist.*

## PROPERTIES

**TYPE** Regional metamorphic
**TEMPERATURE** High
**PRESSURE** High
**STRUCTURE** Foliated, crystalline
**MAJOR MINERALS** Quartz, feldspar
**MINOR MINERALS** Biotite, hornblende, garnet, staurolite
**COLOUR** Grey, pink, multi-coloured
**TEXTURE** Coarse
**PROTOLITH** Granite, shale, granodiorite, mudstone, siltstone, or felsic volcanics

**THIN SECTION**
*Alternating bands of light and dark minerals give this granodioritic gneiss from the Hohe Tauern, in the Austrian Alps, its gneissic texture.*

- Quartz
- Biotite
- Feldspar

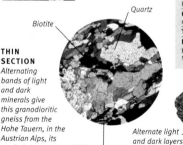

Alternate light and dark layers

**FOLDED GNEISS**

Eye- ("augen"-) shaped patches

**BIOTITE GNEISS**
*Biotite is one of the minerals that form under metamorphic conditions, and occurs in quantity in biotite gneiss.*

**AUGEN GNEISS**

**GNEISS VARIETIES**
*Gneisses can have a huge range of forms and mineral characteristics, and are named for them.*

- Biotite
- Pale feldspar

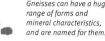

83

# AMPHIBOLITE

**AMPHIBOLITE IN SCHIST**
*Layers of amphibolite run through schist in the Austrian Alps.*

SCHISTOSITY AND FOLIATION are well developed in most amphibolites. Made up predominantly of amphibole minerals, such as actinolite (see Tremolite, p.279), many samples contain microscopic grains of feldspar, pyroxene, and calcite. Some also contain macroscopic crystals of minerals such as garnet, diagnostic of regional metamorphic rocks formed under low to moderate temperatures and pressures. Amphibolites form from the metamorphism of mafic igneous rocks such as gabbros, and can also form from sedimentary rocks such as greywacke.

**MINERAL COMPOSITION**
*Amphibolite comprises mainly amphibole minerals, but may also contain feldspar, garnet, pyroxene, and epidote.*

Amphibole crystals

Coarse texture

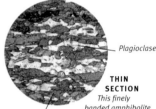

## PROPERTIES

**ROCK TYPE** Regional metamorphic
**TEMPERATURE** Low to moderate
**PRESSURE** Low to moderate
**STRUCTURE** Foliated, crystalline
**MAJOR MINERALS** Hornblende, actinolite
**MINOR MINERALS** Feldspar, calcite, pyroxene
**COLOUR** Grey, black, greenish
**TEXTURE** Coarse
**PROTOLITH** Basalt, greywacke, dolomite

## ROAD BUILDING

Amphibolite is used in road building and in other aggregates where a high degree of strength and durability is required. It is also sawn into sheets for use as ornamental stone in the facings of buildings, as flagstones, or as flooring. It is sometimes sold for use as worktops and for ornamental purposes under the name of "black granite".

**HIGH-GRADE AGGREGATE**
*When crushed, amphibolite makes a high-grade aggregate for road building.*

Plagioclase

**THIN SECTION**
*This finely banded amphibolite is from the Lewisian Complex at Achmelvich, Scotland.*

Hornblende

# HORNFELS

**PRECAMBRIAN HORNFELS**
*These hornfels are in Charnwood Forest, UK.*

HORNFELS FORMS BY contact metamorphism close to igneous intrusions at temperatures as high as 700°–800°C (1,300°–1,450°F). Hornfels can form from almost any parent rock, and its composition depends on both that rock and the exact temperatures and fluids to which the rock is exposed. It is notoriously difficult to identify. The rock is usually dense, hard, very fine grained, relatively homogeneous, and has a conchoidal, flint-like fracture. It may be banded. Hornfels is often categorized by the mix of minerals present in the sample. Garnet hornfels, for example, is characterized by porphyroblasts (large crystals set into a rock matrix) of garnet. Dark hornfels is easily confused with basalt.

Dark pyroxene crystals

Rhombic chiastolite

Bladed chiastolite

Fine-grained texture

**PYROXENE HORNFELS**     **CHIASTOLITE HORNFELS**     **CORDIERITE HORNFELS**

**HORNFELS VARIETIES**
*Varieties of hornfels take their names from the presence of porphyroblasts of various minerals in the rock, such as pyroxene hornfels, chiastolite hornfels, and cordierite hornfels.*

Garnet porphyroblasts

Hornblende and plagioclase

**GARNET HORNFELS**
*The red colour of this specimen of garnet hornfels is a result of the number of garnet crystals present in the rock.*

Cordierite

Spinel

Biotite

## PROPERTIES

**ROCK TYPE** Contact metamorphic
**TEMPERATURE** Moderate to high
**PRESSURE** Low to high
**STRUCTURE** Crystalline
**MAJOR MINERALS** Hornblende, plagioclase, andalusite, cordierite, and others
**MINOR MINERALS** Magnetite, apatite, titanite
**COLOUR** Dark grey, brown, greenish, reddish
**TEXTURE** Microcrystalline to fine
**PROTOLITH** Almost any rock

**THIN SECTION**
*This rock was a mudstone heated to very high temperature adjacent to an intrusion.*

# QUARTZITE

**QUARTZITE CAP**
*The peak of Stob Ban in Scotland is capped with white quartzite.*

THE QUARTZ-RICH ROCK is formed by the burial, heating, and squeezing of sandstone. The name quartzite can also refer to rock formed by precipitation of cement in pore spaces. Such rock contains rounded quartz grains. Quartzite is very hard and brittle, and shows conchoidal fracture.

### PROPERTIES

| | |
|---|---|
| **ROCK TYPE** | Regional metamorphic |
| **TEMPERATURE** | High |
| **PRESSURE** | Low to high |
| **STRUCTURE** | Crystalline |
| **MAJOR MINERALS** | Quartz |
| **MINOR MINERALS** | Mica, kyanite, sillimanite |
| **COLOUR** | Almost any |
| **TEXTURE** | Medium |
| **PROTOLITH** | Sandstone |

*Crystalline quartz*

**METAQUARTZITE**
*The metamorphic variety of quartzite is a crystalline rock with over 90 per cent quartz.*

# FULGURITE

**LIGHTNING**
*As lightning strikes sand in the desert , some of the sand fuses into a fulgurite.*

NAMED AFTER the Latin for "thunderbolt", fulgurites form when lightning strikes. They form either as crusts or as tubes, some of which branch out like trees. The tubes are lined with fused sand, or glass, with partially melted sand on the outside. In theory, fulgurites can form in any type of easily melted rock, but most are found in desert regions where sand has been melted, and the surrounding loose material has since eroded away. The largest fulgurite ever recorded was about 5m (16ft) long. Fulgurites have no practical use, although they are decorative.

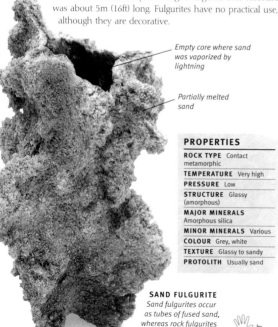

*Empty core where sand was vaporized by lightning*

*Partially melted sand*

### PROPERTIES

| | |
|---|---|
| **ROCK TYPE** | Contact metamorphic |
| **TEMPERATURE** | Very high |
| **PRESSURE** | Low |
| **STRUCTURE** | Glassy (amorphous) |
| **MAJOR MINERALS** | Amorphous silica |
| **MINOR MINERALS** | Various |
| **COLOUR** | Grey, white |
| **TEXTURE** | Glassy to sandy |
| **PROTOLITH** | Usually sand |

**SAND FULGURITE**
*Sand fulgurites occur as tubes of fused sand, whereas rock fulgurites are glassy crusts.*

# GRANULITE

**FAULT BLOCK**
*The Tetons in Wyoming, USA, contain granulite.*

GRANULITE IS THOUGHT to form at very high temperatures and pressures near the base of the crust. It usually has a high concentration of pyroxene, with diopside or hypersthene, garnet, calcium plagioclase, and quartz or olivine. Granulites are tough, massive, and coarse-grained. The "granulite facies" is one of the major divisions in the classification of metamorphic rocks.

### PROPERTIES

| | |
|---|---|
| **ROCK TYPE** | Regional metamorphic |
| **TEMPERATURE** | High |
| **PRESSURE** | High |
| **STRUCTURE** | Crystalline |
| **MAJOR MINERALS** | Feldspar, quartz, garnet |
| **MINOR MINERALS** | Spinel, corundum |
| **COLOUR** | Grey, pinkish, brownish, mottled |
| **TEXTURE** | Medium to coarse |
| **PROTOLITH** | Felsic igneous and sedimentary rocks |

*Pale crystals* — *Fine matrix*

**UNFOLIATED ROCK**
*Unlike many other regionally metamorphosed rocks, granulite is characterized by a lack of foliation.*

# MIGMATITE

**MIXED ROCK**
*The granitic rock is light against the dark bands of schist or gneiss.*

MIGMATITE, MEANING "mixed rock", consists of schist or gneiss interlayered, streaked, or veined with granite rock. The bands may be tightly folded. Migmatites often occur in areas of high-grade metamorphism. The granite streaks are the result of partial melting of the parent rock, at temperatures below the melting point of the schist or gneiss. Migmatites also form near large intrusions of granite when some of the magma has intruded into surrounding metamorphic rocks.

### PROPERTIES

| | |
|---|---|
| **ROCK TYPE** | Regional metamorphic |
| **TEMPERATURE** | High |
| **PRESSURE** | High |
| **STRUCTURE** | Foliated, crystalline |
| **MAJOR MINERALS** | Quartz, feldspar, mica |
| **MINOR MINERALS** | Various |
| **COLOUR** | Banded light and dark grey, pink, white |
| **TEXTURE** | Coarse |
| **PROTOLITH** | Various, including granite and gneiss |

*Light quartz and feldspar*

*Dark gneissose component*

**MIXED ROCK**
*Folds of granite are interlayered with gneiss in this sample of migmatite.*

**BANDED MIGMATITE**
*The interlayering of schist and gneiss with lighter-coloured granite is visible in this specimen.*

# MINERALS

## MORE THAN 350 OF THE EARTH'S MINERALS

# WHAT IS A MINERAL?

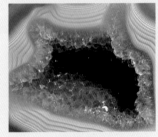

**QUARTZ GEODE**

MINERALS ARE THE STUFF of which the Earth's rocks are made. A mineral is defined as a naturally occurring solid with a specific chemical composition and a distinctive internal crystal structure. Minerals are usually formed by inorganic processes, although there are organically produced substances, such as hydroxylapatite in teeth and bones, that are also considered minerals. Certain substances, including opal and glass, resemble minerals in chemistry and occurrence but do not have a regularly ordered internal arrangement. These are known as mineraloids.

## WHAT ARE MINERALS MADE OF?

Most minerals are chemical compounds, composed of two or more chemical elements, although copper, sulphur, gold, silver, and a few others occur as single "native" elements. A mineral is defined by its chemical formula and by the atomic arrangement of its crystallization. For example, iron sulphide has the chemical formula $FeS_2$ (where Fe is the element iron, and S is sulphur). It can crystallize in two different ways. When it crystallizes in the cubic crystal system (see p.100), it is called pyrite; but when instead it crystallizes in the orthorhombic crystal system (see p.101), it becomes a different mineral, marcasite.

**COPPER DUCK**

**NATIVE ELEMENT**
Native copper was probably the first metal used by humankind, either hammered into tools or cast into artefacts. This copper duck's head was made in North Africa about 1,900 years ago.

**MARCASITE**

**PYRITE**

**IRON SULPHIDES**
Pyrite and marcasite have exactly the same chemical composition – they are both iron sulphide – but their different crystal structures make them different minerals.

## ELECTRICAL CHARGE

A mineral compound is based on an electrical balance between a positively charged part, often a metal, and a negatively charged part. These charged particles are known as ions. In many minerals the negative electrical charge is carried by a "radical" which is a combination of atoms, rather than by single atom. For example, the sulphur in pyrite is present as the disulphide ($S_2$) group, with two sulphur atoms, which has a negative charge the same as the positive charge of an iron (Fe) ion. The sulphur radical balances the iron ion and the formula for pyrite is $FeS_2$. To take another example, one atom of carbon (C) and three atoms of oxygen (O) combine to give the $CO_5$ radical, which acts as a single, negatively charged unit. The element or elements that carry the negative electrical charge determine which chemical group a mineral is assigned to (see p.90). For example, minerals formed with two sulphur atoms are known as sulphides, those formed with oxygen alone are oxides, those made with the carbon and oxygen ion are called carbonates, and those formed with silicon and oxygen are silicates.

*Carbon atom*

*Oxygen atom*

**CARBONATE**
Carbonates are formed with one carbon atom and three oxygen atoms.

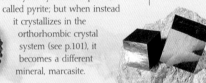

**NATIVE SULPHUR**
Sulphur is mined at Kawah Ijen, Java. Volcanic gases escaping from small openings in the ground (fumaroles) carry dissolved sulphur to the surface, where it deposits as a yellow crust.

*Each silicon atom is bonded to four oxygen atoms that form a tetrahedral shape*

*ca tetrahedra a at the ners to form elix*

## COMMON MINERALS

There are more than 4,000 known minerals, but only about 100 of these are common. Silicon and oxygen are by far the most abundant elements in the Earth's crust. They make up about three-quarters of the crust by weight, and silicates, such as quartz, feldspar, and olivine, are by far the most common minerals in rocks, making up about 90 per cent of the rocks at the Earth's surface. Carbonates are important in forming sedimentary rocks such as limestone. Many sulphides, including pyrite and galena, and oxides, such as hematite, are also relatively common. So are some of the native elements, such as copper.

*Crystal face*

**QUARTZ CRYSTAL**

**QUARTZ STRUCTURE**
*Silica tetrahedra link to form quartz and other silicate minerals.*

**HIGH SIERRA**
*The peaks of the Sierra Nevada, California, are made of granite, a rock typically composed of about 70 per cent silica.*

## MINERAL OR NOT?

The term mineral is commonly applied to certain organic substances, such as coal, oil, and natural gas (as when referring to a nation's mineral wealth), although these particular materials are more accurately referred to as hydrocarbons. Gases and liquids are not, in the strict sense, minerals. Although ice, the solid state of water, is considered a mineral, liquid water is not. Nor is liquid mercury, which is sometimes found in mercury ore deposits. Synthetic equivalents of certain minerals, such as emeralds and diamonds, are produced in the laboratory, but these are not minerals, in that they do not occur naturally. The "minerals" referred to when talking of foods are also not minerals in the geological sense, as they refer to single elements, such as calcium, zinc, or iron.

**SYNTHETIC GEMS**
*Emeralds and rubies that are grown synthetically are not classified as minerals.*

**SYNTHETIC RUBY BOULE**

**SYNTHETIC GILSON EMERALD**

**CRUDE OIL**
*Although called a "mineral" in economic terms, oil is classed as a hydrocarbon.*

**ICEBERG**
*The crystallized water that forms an iceberg is a mineral, but the liquid water it floats in is not.*

# CLASSIFYING MINERALS

Mineralogists classify minerals according to their chemical composition (see pp.88–89). Shown below are the major chemical groups, with an example of each. Within chemical groups, minerals are further classified into sub-groups, taking their name from their most characteristic mineral. For example, gold, silver, platinum, and copper are all members of the native elements chemical group, but they are also classified as part of the gold group of minerals, because they have an identical arrangement of their atoms.

Bonds extend between atoms in all directions

Gold atom

## GOLD GROUP

Metals of the gold group all have a cubic arrangement of close-packed atoms with bonds in all directions. The atoms stack up in a sequence of layers, with the atoms of one layer fitting into the hollows in the layers above and below, so that each atom has 12 nearest neighbours.

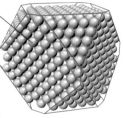

Topmost layer of gold atoms

**OCTAHEDRAL GOLD CRYSTAL**

**COPPER**

**CHALCOCITE**

## OXIDES

When oxygen alone combines with a metal or semi-metal, an oxide mineral is formed. Aluminium oxide is corundum, the red variety of which is called ruby.

**RUBY**

## NATIVE ELEMENTS

Native elements are minerals formed of a single chemical element. They include metals such as gold and copper and nonmetals such as sulphur and carbon.

## SULPHIDES

The sulphides are formed when a metal or semi-metal combines with sulphur. In chalcocite the metal element is copper.

## HYDROXIDES

Hydroxide minerals contain a hydroxyl (hydrogen and oxygen) radical combined with a metallic element – manganese in the case of brucite.

**BRUCITE**

## HALIDES

A halogen element (chlorine, bromine, fluorine, or iodine) combined with a metal or semi-metal makes a halide. Sylvite is a compound of chlorine and potassium.

**SYLVITE**

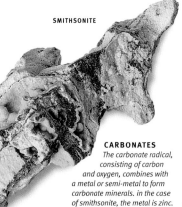

**SMITHSONITE**

**COLEMANITE**

## ARSENATES, PHOSPHATES, AND VANADATES

In these minerals a radical of oxygen and either arsenic, phosphorus, or vanadium combines with a metal or semi-metal. Apatite is a phosphate.

**APATITE**

## BORATES AND NITRATES

These minerals contain respectively radicals of boron and oxygen, and of nitrogen and oxygen. Colemanite is a borate in which boron and oxygen combine with calcium and water.

## CARBONATES

The carbonate radical, consisting of carbon and oxygen, combines with a metal or semi-metal to form carbonate minerals. in the case of smithsonite, the metal is zinc.

**AMETHYST**

**AMBER**

## SULPHATES, CHROMATES, TUNGSTATES, AND MOLYBDATES

Sulphur, molybdenum, chromium, or tungsten, form a radical with oxygen that combines with a metal or semi-metal. Celestine is a sulphate.

**CELESTINE**

## SILICATES

In this group, silicon and oxygen form a tetrahedral silica radical that combines with various metals or semi-metals. Silica also occurs on its own as quartz. Amethyst is a variety of quartz.

## ORGANIC MINERALS

Organic compounds with well-defined crystal structures are classified as minerals. Amber, however, which originates as a resin and is amorphous, is not a mineral.

**OLIVINE SANDS**
*The green sand in this Hawaiian cove consists mainly of olivine, a mineral that can have a range of compositions which form a solid-solution series.*

## CHEMICAL FORMULAE

The chemical formula identifies which atoms are present in a mineral and in what proportions. In some crystalline structures the components and their proportions are fixed. The formula for pyrite, for example, is always $FeS_2$, denoting iron (Fe) and sulphur (S) in a 1:2 ratio. In other cases, such as solid solutions, components may be variable. Thus in olivine that has complete substitution possible between iron and magnesium (Mg), the formula is written as $(Fe,Mg)_2SiO_4$, indicating that the iron and magnesium are found in varying amounts.

### CHEMICAL ELEMENTS

| Symbol | Name | Symbol | Name |
|---|---|---|---|
| Ac | Actinium | Mn | Manganese |
| Ag | Silver | Mo | Molybdenum |
| Al | Aluminium | N | Nitrogen |
| Am | Americium | Na | Sodium |
| Ar | Argon | Nb | Niobium |
| As | Arsenic | Nd | Neodymium |
| At | Astatine | Ne | Neon |
| Au | Gold | Ni | Nickel |
| B | Boron | No | Nobelium |
| Ba | Barium | Np | Neptunium |
| Be | Beryllium | O | Oxygen |
| Bi | Bismuth | Os | Osmium |
| Bk | Berkelium | P | Phosphorus |
| Br | Bromine | Pa | Protactinium |
| C | Carbon | Pb | Lead |
| Ca | Calcium | Pd | Palladium |
| Cd | Cadmium | Pm | Promethium |
| Ce | Cerium | Po | Polonium |
| Cf | Californium | Pr | Praseodymium |
| Cl | Chlorine | Pt | Platinum |
| Cm | Curium | Pu | Plutonium |
| Co | Cobalt | Ra | Radium |
| Cr | Chromium | Rb | Rubidium |
| Cs | Cesium | Re | Rhenium |
| Cu | Copper | Rh | Rhodium |
| Dy | Dysprosium | Rn | Radon |
| Er | Erbium | S | Sulphur |
| Es | Einsteinium | Sb | Antimony |
| F | Fluorine | Sc | Scandium |
| Fe | Iron | Se | Selenium |
| Fm | Fermium | Si | Silicon |
| Fr | Francium | Sm | Samarium |
| Ga | Gallium | Sn | Tin |
| Gd | Gadolinium | Sr | Strontium |
| Ge | Germanium | Ta | Tantalum |
| H | Hydrogen | Tb | Terbium |
| He | Helium | Tc | Technetium |
| Hf | Hafnium | Te | Tellurium |
| Hg | Mercury | Th | Thorium |
| Ho | Holmium | Ti | Titanium |
| I | Iodine | Tl | Thallium |
| In | Indium | Tu | Thulium |
| Ir | Iridium | U | Uranium |
| K | Potassium | V | Vanadium |
| Kr | Krypton | W | Tungsten |
| La | Lanthanum | Xe | Xenon |
| Li | Lithium | Y | Yttrium |
| Lu | Lutetium | Yb | Ytterbium |
| Lw | Lawrencium | Zn | Zinc |
| Md | Mendelevium | Zr | Zirconium |
| Mg | Magnesium | | |

## SOLID SOLUTIONS

Some minerals do not have specific chemical compositions. Instead, they are homogenous mixtures of two mineral species. These homogenous mixtures are known as solid solutions. For example, the olivine group of silicates includes the minerals forsterite and fayalite. Forsterite is a magnesium silicate (chemical formula $Mg_2SiO_4$), while fayalite is an iron silicate ($Fe_2SiO_4$).

But in fact, pure forsterite and pure fayalite are both rare. Most specimens are homogenous mixtures of the two, with the relative content of magnesium and iron varying from specimen to specimen. These minerals are described as part of a "solid-solution series" in which forsterite and fayalite are the "end members". Another example of an important solid-solution series is the plagioclase feldspars (see p.234).

**FAYALITE**
*The olivine mineral fayalite forms in igneous rocks with a relatively high silica content. It is an iron silicate.*

**FORSTERITE**
*The olivine mineral forsterite forms in igneous rocks with low silica content. It is a magnesium silicate.*

## PRIMARY AND SECONDARY MINERALS

A primary mineral is one that has crystallized directly through some igneous, sedimentary, or metamorphic process. A secondary mineral is one that has been produced through the alteration of a primary mineral after its formation. For example, when copper-bearing primary minerals come into contact with carbonated water they can be turned into bright blue azurite or bright green malachite. The azurite and malachite are secondary minerals.

**MALACHITE**
*Malachite, the green mineral coating the rocks in this mine in Zambia, is a secondary mineral formed from copper ore.*

**BRAZILIANITE**
*The phosphate mineral brazilianite is a primary mineral, mainly formed in pegmatites.*

# IDENTIFYING MINERALS

EVEN FOR EXPERTS, MINERAL IDENTIFICATION is often far from easy. But there are certain physical properties – determined by the crystalline structure and chemical composition of the mineral – that can help with identification without requiring the use of expensive equipment. Some of these properties, such as hardness and density, can be measured objectively, while others, such as colour and lustre, demand a more subjective assessment. Where minerals are found as distinct crystals, there is an additional set of properties that can be used for identification (see pp.98–103).

## COLOUR

Some minerals have characteristic colours and others do not. The bright blue of azurite, the yellow of sulphur, and the green of malachite allow an almost instant identification. At the other end of the scale, fluorite occurs in virtually all colours, so it can only be identified by observing other of its properties.

Colour in minerals is caused by the absorption or refraction of light of particular wavelengths. This can happen for a number of reasons. One is the presence of foreign atoms – atoms not part of the chemical makeup of the mineral – in the crystal structure. These are called trace elements. As few as three or four atoms per million can absorb enough of certain parts of the visible light spectrum to give colour to a mineral. The colour produced by a particular trace element varies according to the mineral it inhabits. For example, chromium is the colouring element in both red ruby and green emerald. Colour can also result from the absence of an atom or ionic radical from a point that it would normally occupy in a crystal. These types of defects are called vacancies, and their result is called a colour centre. The violet colour of some fluorite is produced by a vacancy.

The structure of the mineral itself, without any defects or foreign elements, may also cause colour. For example, opal is composed of minute spheres of silica that diffract light, while the colour and sheen of moonstone is determined by the thin interlayering of two different feldspars. In some crystals, light vibrates in different planes within the crystal, with the result that, whatever the initial cause of its colour may be, it appears as different colours when observed along different axes. This is called pleochroism.

**GREEN FLUORITE**

**YELLOW FLUORITE**

**PURPLE FLUORITE**

**COLOUR RANGE**
*These specimens show only a few of the many colours that can occur in fluorite. Different coloration depends on a range of factors, such as, the presence of traces of hydrocarbons.*

**COLOUR VARIATION**
*The multiplicity of colours in opal (above) is produced by the arrangement of microscopic spheres of silica in its structure. An image made using an electron microscope (left) shows the opal's fractured surface.*

# LUSTRE

A mineral's lustre is the general appearance of its surface in reflected light. There are two broad types of lustre: metallic and nonmetallic. Metallic lustre is that of an untarnished metal surface, such as gold, steel, or copper. Minerals with metallic lustre are opaque to light, even on thin edges. By contrast, minerals with nonmetallic lustre are generally lighter in colour, and show some degree of transparency or translucency, even though this may only be on a thin edge. There are a number of terms which describe nonmetallic lustres: vitreous, having the lustre of a piece of broken glass; adamantine, having the brilliant lustre of diamond; resinous, having the lustre of a piece of resin; pearly, having the lustre of pearl or mother-of-pearl; greasy, appearing to be covered with a thin layer of oil; silky, appearing as the surface of silk or satin; dull, producing little or no reflection; and earthy, having the nonlustrous appearance of raw earth. With practice, the main types of lustre can be distinguished easily by eye, but the difference between them cannot be precisely quantified.

**METALLIC**
*The sulphide galena has a metallic lustre and a distinctive pattern of cleavage.*

**VITREOUS**
*Many silicate minerals, such as this quartz crystal, have a vitreous lustre and appear similar to glass surfaces.*

**GREASY**
*Orpiment can have a greasy or a resinous lustre. Making the distinction requires a subjective judgement.*

**RESINOUS**
*Crystals of native sulphur are transparent or translucent with a resinous lustre.*

**SILKY**
*The borate ulexite exhibits a silky lustre, having the surface sheen of a bolt of satin.*

**DULL**
*A dull lustre, such as that of this sample of hematite, is nonreflective, but not as granular as earthy lustre.*

**EARTHY**
*Minerals with an earthy lustre, such as this fine-grained calcite, have the look of freshly broken, dry soil.*

# DIAGNOSTIC STREAK

A streak is the colour of the powder produced when a specimen is drawn across a surface such as a piece of unglazed porcelain – the reverse side of a kitchen tile, for example. This technique is an extremely useful diagnostic because a mineral's streak is far more consistent than its colour, which tends to vary from specimen to specimen. It can allow you to distinguish between minerals that are otherwise easily confused. For example, the iron oxide hematite gives a red streak, whereas magnetite, another oxide of iron, gives a black streak. If a mineral is too hard to mark a streak plate, the colour of its powder can be determined by filing or crushing a small sample.

ORPIMENT

CINNABAR

CROCOITE

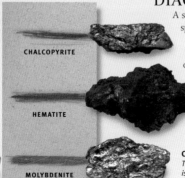

CHALCOPYRITE

HEMATITE

MOLYBDENITE

**CONSISTENCY**
*The streak of a mineral is consistent from specimen to specimen, as long as you test an unweathered surface .*

93

# CLEAVAGE

Cleavage is the ability of a mineral to break along flat, planar surfaces. It occurs at places within the mineral's crystal structure where the forces bonding atoms are weakest. Cleavage surfaces are generally smooth and reflect light evenly. Cleavage is described by its direction relative to the position of crystal faces (using the terms cubic, prismatic, and basal) and by the ease with which it is produced. If cleavage produces smooth, lustrous surfaces with great ease, it is called perfect. Terms for lesser degrees of ease include distinct, imperfect, and difficult, while some minerals have no cleavage at all. A mineral can have cleavages in different directions, each of which may be of a different quality. For example, one may be perfect, another may be imperfect. Cleavage is consistent property and thus useful for identification.

**BERYL CLEAVAGE**
*Under a microscope, a beryl crystal shows structural breaks, which coincide with its potential cleavages. The direction of cleavages and the angles between them offer clues to aid identification.*

**CLEAR BREAKS**
*The parallel cleavage planes of this baryte crystal are clearly visible. Baryte has perfect cleavage.*

*Cleavage plane*

**PERFECT SHAPE**
*When Iceland spar calcite (left) is broken along its cleavage planes, a perfect rhombic shape results. Iceland spar is also identified by its transparency and double refraction (see opposite).*

# FRACTURE

Some minerals will break in directions other than along cleavage planes. These breaks are known as fractures and may also help in mineral identification. For example, hackly fractures, with jagged edges, are often found in metals, while shell-like conchoidal fractures are typical of quartz. Other terms for fractures include even (rough but more or less flat), uneven (rough and completely irregular), and splintery (with partially separated fibres).

*Conchoidal fracture*

**CONCHOIDAL**
*This Bronze Age chert axe shows conchoidal fracture, with cavities shaped like a bivalve seashell.*

*Hackly fracture surface*

**HACKLY**
*This gold nugget shows hackly fracture, involving sharp edges and jagged points.*

*Irregular surface*

**UNEVEN**
*This specimen of chalcopyrite shows uneven fracture. Its broken surface is rough and irregular, with no pattern evident.*

# TENACITY

Tenacity is a term used for a set of physical properties – such as malleability, ductility, or brittleness – that depend on the cohesive force between atoms in mineral structures. Gold, silver, and copper are good examples of malleable minerals, capable of being flattened without breaking or crumbling. Acanthite is sectile, able to be smoothly cut with a knife. Talc is flexible, bending easily and staying bent after the pressure is removed. Other terms used include ductile (capable of being drawn into a wire), brittle (showing little resistance to breakage), and elastic (capable of being bent or pulled out of shape but returning to the original form when relieved).

**WORKING SILVER**
*High malleability is one of the distinguishing characteristics of silver. It is exploited in the making of silverware.*

# HARDNESS

Testing hardness is an extremely useful aid to mineral identification. The hardness of a mineral is the relative ease or difficulty with which it can be scratched. A harder mineral will scratch a softer one, but not vice versa. Any mineral can be allotted a number on the Mohs scale, which measures hardness relative to ten minerals of increasing hardness, from 1 (as soft as talc) to 10 (as hard as diamond). Hardness should not be confused with toughness or strength. Very hard minerals (including diamond) can be quite brittle. There is a general link between hardness and chemical composition. Most hydrous minerals – that is, minerals containing water molecules – are relatively soft, as are halides, carbonates, sulphates, phosphates, and most sulphides. Most anhydrous oxides – those not containing water molecules – and silicates are relatively hard (above 5 on the Mohs scale).

## THE MOHS SCALE OF HARDNESS

| Hardness | Mineral | Other materials for hardness testing |
|---|---|---|
| 1 | Talc | Very easily scratched by a fingernail |
| 2 | Gypsum | Can be scratched by a fingernail |
| 3 | Calcite | Just scratched with a copper coin |
| 4 | Fluorite | Very easily scratched with a knife but not as easily as calcite |
| 5 | Apatite | Scratched with a knife with difficulty |
| 6 | Orthoclase | Cannot be scratched with a knife, but scratches glass with difficulty |
| 7 | Quartz | Scratches glass easily |
| 8 | Topaz | Scratches glass very easily |
| 9 | Corundum | Cuts glass |
| 10 | Diamond | Cuts glass |

**TESTING METHODS**
*Specialist tools are available to test hardness, but you can use a fingernail (scratches a mineral less than 2¹/₂ on the Mohs scale) or a knifeblade (scratches under 5¹/₂).*

**FINGERNAIL TEST**

Hardness point

**SPECIALIST TOOLS**

# REFRACTIVE INDEX

When light passes through a transparent or translucent mineral, it changes velocity and direction. The extent of this refraction is measured by the refractive index, the ratio of the velocity of light in air to its velocity in the crystal. A high refractive index is linked to the dispersion of the light into its component colours, which gives minerals such as diamond their fire. For the amateur, refractive indices can be found using specialized liquids or with relatively inexpensive equipment.

Rhombohedral calcite

Twin image due to double refraction

**DOUBLE REFRACTION**
*A calcite rhomb refracts light to two different degrees, thus creating a double image.*

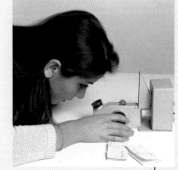

**LIGHT INDEX**
*A refractometer is used to determine the refractive index of an unknown mineral.*

# SPECIFIC GRAVITY

Specific gravity (SG) is a measure of the density of a substance. It is the ratio of the mass of the substance to the mass of an equal volume of water. A mineral with an SG of 2 is twice as heavy as water of the same volume. Specific gravity can be determined by the use of specialized balances or by employing liquids formulated to allow minerals of below a given SG to float. Bromoform, for example, can be used to distinguish quartz from topaz – quartz floats in it while topaz sinks. With experience, you will find that the "feel" of a specimen begins to relate instinctively to its specific gravity. For example, quartz, with an SG of 2.6, feels "normal", galena, SG 7.6, feels "very heavy", and so on.

**GALENA**

**QUARTZ**

**LIGHT BUT HARD**
*Although quartz is much harder than the sulphide galena, it has a much lower specific gravity.*

# FLUORESCENCE

Some minerals are characterized by fluorescence. This means that they emit visible light of various colours when subjected to ultraviolet radiation. To test for fluorescence, all that is needed is an ultraviolet light, which can be readily obtained from dealers selling lapidary and collectors' equipment. On its own, fluorescence is generally one of the less reliable indicators of a mineral's identity because it lacks consistency. Some specimens of a mineral will exhibit fluorescence while others do not, even where the specimens are from the same location and appear identical. Fluorescence is discussed further on pages 160–61.

**FLUORESCENT GYPSUM**
*This white gypsum specimen is fluorescing rose-pink when lit by ultraviolet light. It contains layers of other nonfluorescent minerals.*

# MINERAL ASSOCIATIONS

MINERALS NORMALLY OCCUR TOGETHER in groups or "associations". The study of how different minerals came to be associated with each other provides important information on how rocks were formed. Just as fossils give a record of life on Earth, mineral associations hold a record of the Earth's geological history. Patterns of association or assemblage can also help with the identification of minerals.

Apophyllite

Stilbite

**ZEOLITE ASSOCIATES**
*Minerals belonging to the zeolite group of silicates, such as apophyllite and stilbite, are often found in association with one another.*

## LOCAL ASSOCIATIONS

Some minerals are consistently found together over large areas because they are found in the same rock type (see mineral assemblages, opposite). Other associations occur less extensively in veins, cavities, encrustations, or thin layers. Most such localized associations have built up through geological time – the associated minerals were not formed simultaneously. For example, within a hollow left by a gas bubble in solidified lava (a geode), you may find agate, amethyst, and calcite. First, dissolved silica was deposited as an agate lining on the inside of the bubble. Later, other manganese-bearing silica solutions deposited the layer of amethyst, followed at yet another time by carbonate-rich waters, which deposited the calcite. By studying such associations, it is possible to build up a picture of a region's geological history.

## IDENTIFICATION

The fact that certain minerals are likely to be found together can help both with the discovery and identification of minerals. For those in search of metals, it is useful to know that lead and zinc ore minerals, such as galena and sphalerite, are often associated with calcite and baryte, while gold is frequently found in association with quartz. Also if you have difficulty identifying a mineral but recognize another that it is associated with, this can offer a valuable clue to its identity. For example, bertrandite is an obscure silicate that may be hard to identify, but it is found in association with the more familiar beryl. Again, if you have identified one mineral as a zeolite, it is likely that an associated mineral will be a zeolite too.

Baryte

Smoky quartz

**CRYSTAL ORDER**
*Associations can be used to determine in which order minerals crystallized. In this specimen the baryte must have formed after the smoky quartz.*

## COPPER DEPOSIT

Important mineral ores, such as this copper deposit at Bingham, Utah, USA, are often discovered through their associated minerals. In the case of copper, the original (primary) ore minerals – for example, chalcopyrite – are oxidized near or at the Earth's surface into secondary minerals, such as malachite or azurite. When prospectors find these secondary minerals on the surface, they know that they will lead to a rich concentration of copper ore below. Similarly, vivianite forms near or at the surface from the alteration of iron or manganese ores and indicates their presence at greater depth.

Calcite

Amethyst

**AMETHYST GEODE**
*The amethyst in this geode, a cavity in volcanic rock, was deposited by manganese-bearing silica solutions. Calcite was deposited later by carbonate-rich waters.*

**WALLS OF SHALE**
*Water erosion at this canyon on the San Juan River, Utah, USA, has exposed layers of shale. Differences in the assemblage of minerals in various shale layers can reveal much of the geological history of the region.*

# MINERAL ASSEMBLAGES

An association of minerals that forms more or less simultaneously and is usually present in a specific rock type is called an assemblage. For example, the association of orthoclase, albite, quartz, and biotite is a mineral assemblage for granite, while plagioclase, augite, olivine, and magnetite is an assemblage for gabbro. The minerals present in any specific rock vary depending on the forces present when it was formed. A shale formed at relatively low temperature, for example, will have the assemblage muscovite-kaolin-dolomite-quartz-feldspar; another shale formed at a higher temperature will have the assemblage garnet-sillimanite-biotite-feldspar. This effect is especially striking in metamorphic rocks. A basalt that is metamorphosed at high pressures and low temperatures recrystallizes into a rock containing glaucophane and albite. These minerals have a bluish coloration and the resulting rocks are known as blueschist. The same rock type metamorphosed at more moderate pressures and temperatures would contain abundant chlorite and actinolite, both of which are green, giving this rock the name greenschist.

**SEDIMENTARY SHALE**
*Shale is an example of a sedimentary assemblage. It is a mixture of clay minerals, quartz, feldspar, and mica.*

Feldspar

Quartz

Biotite mica

**GRANITE ASSEMBLAGE**
*Granite is characterized by the familiar assemblage of quartz, feldspar, and mica.*

Mica gives silvery sheen

Garnet

**METAMORPHIC MIX**
*The assemblage of garnet, quartz, and mica in this specimen tells geologists that this metamorphic rock formed under conditions of relatively moderate pressure and relatively low temperature.*

**SCHIST EXPOSURE**
*Greenschist, as seen in this exposure on the Tibetan Plateau, is a metamorphic rock formed at moderate pressures and temperatures.*

97

# WHAT IS A CRYSTAL?

ALL MINERALS ARE CRYSTALLINE. A crystalline material is a solid in which the component atoms are arranged in a particular, repeating, three–dimensional pattern. This is expressed externally as flat faces arranged in geometric forms. Some crystals are large enough to see, even reaching dimensions measured in metres, and others are so small they can be seen only with the most powerful microscopes. But from the tiniest to the largest, crystals of the same mineral are built to the same atomic pattern. Crystallography – the study of the geometric properties and internal structure of crystals – is a fundamental part of mineralogy.

**STRUCTURE OF GOLD**
The highly magnified image (far right) shows the individual atoms that make up the specimen of gold (right).

**MAGNIFIED GOLD**

**GOLD SPECIMEN**

Gold plates

## ATOMIC STRUCTURE

A crystal is built up of individual, identical, structural units of atoms or molecules called unit cells. A crystal can consist of only a few unit cells, or billions of them. The unit cell is reproduced over and over in three dimensions, constructing the larger scale internal structure of the crystal, which is called the lattice. The shape of the unit cell and the symmetry of the lattice determine the position and shape of the crystal's faces. The crystals of many different minerals have unit cells that are the same shape but are made of different chemical elements. The final development of the faces that appear on any given crystal is determined to a large extent by the geological conditions at the time that the crystal is forming. Certain faces may be emphasized, while others disappear altogether. The final form a crystal takes is known as its habit (see pp.102–103).

Unit cells combine to form lattice

Atom

**CELL LATTICE**
A lattice is the three-dimensional representation of the crystal's internal structure.

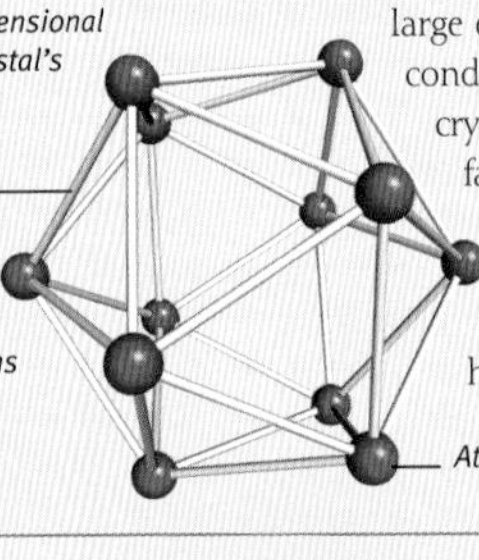

Bonding energy

**CRYSTAL STRUCTURE**
"Stick-and-ball" diagrams show how each atom in the structure of a crystal is bonded.

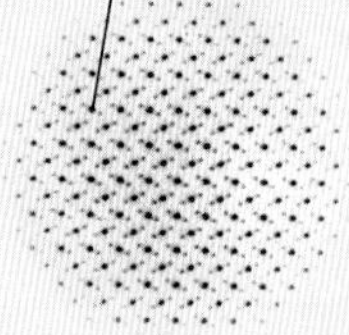

Atom

Iron atom flanked by two smaller sulphur atoms

**UNIT CELLS**

**STRUCTURE OF MARCASITE**
Marcasite crystals are created from repeating unit cells of one atom of iron flanked by two atoms of sulphur.

**MARCASITE CRYSTALS**

Chalk groundmass

## CRYSTAL SYMMETRY

Because a crystal is built up of repeating geometric patterns, all crystals exhibit symmetry. Patterns of crystal symmetry fall into six main groups, called crystal systems (see pp.100–101). The first of these is the cubic system in which all crystals exhibit cubic symmetry. The characteristics of cubic symmetry may be explained as follows. If opposite points of a cube-shaped cubic crystal (such as halite) are held between the thumb and forefinger and the crystal is rotated through 360 degrees, it will appear identical three times as the different faces and edges come into view. All cubic crystals have four axes of threefold symmetry. They all have other axes of symmetry, but these differ among classes within the cube system. For example, halite has three axes of fourfold symmetry, in addition to its four axes of threefold symmetry.

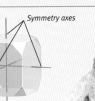

Symmetry axes

**CUBIC SYMMETRY**
All cubic crystals, such as those of halite (right), have four, threefold axes of symmetry.

Rock groundmass

**HALITE**

Cubic halite crystals

Sodium atom

Chlorine atom

**HALITE LATTICE**
This diagram shows the cubic arrangement of sodium and chlorine atoms in the halite lattice.

# TWIN CRYSTALS

When two or more crystals of the same species, such as gypsum or fluorite, form a symmetrical intergrowth, they are referred to as twinned crystals. Twins can form by contact or by interpenetration. Penetration twinning may occur with the individual crystals at an angle to one another, as in the staurolite cross, or with them parallel to one another, as in the Carlsbad twin of orthoclase. If a twin involves three or more individual crystals, it is referred to as a multiple, or repeated, twin. Albite often forms multiple twins. Many minerals form twins, but they are particularly characteristic of some species (a group of minerals that are chemically similar), such as the "fishtail" contact twins of gypsum, or the penetration twins of fluorite.

Centre twinning

**FISHTAIL TWIN**
*The selenite variety of the mineral gypsum sometimes forms "fishtail" contact twins.*

Twin plane

**ST. ANDREW'S CROSS**
*Staurolite commonly forms two types of penetration twin, both of which resemble crosses: a St. Andrew's Cross (left), and a 90-degree cross.*

**CYCLIC TWIN**
*A group of crystals that forms radially from a common centre, as in this cerussite specimen, is called a cyclic twin.*

Centre of twinning

Twin plane

Area of intergrowth

**CARLSBAD PENETRATION TWIN (ORTHOCLASE)**

**ALBITE CONTACT TWIN**

**FLUORITE TWINS**
*Many of the cubic crystals of fluorite in the cluster shown here are penetration twinned.*

# CRYSTAL SYSTEMS

CRYSTALS ARE CLASSIFIED INTO SIX DIFFERENT SYSTEMS according to the maximum symmetry of their faces. The systems are cubic, tetragonal, hexagonal and trigonal (regarded as a single system), monoclinic, orthorhombic, and triclinic. Each system is defined by the relative lengths and orientation of its three crystallographic axes (indicated by the letters a, b, and c if they are all of different lengths). These axes are imaginary lines that pass through the centre of an ideal crystal. Crystal systems, and subgroups called classes, are also defined by the crystals' axes of symmetry (see pp.98–99). Crystals in any single crystal system can assume a variety of shapes. One shape is shown for each system below.

**HESSONITE**
*The individual crystals of hessonite garnet in this cluster are in the cubic system.*

## CUBIC

Cubic crystals have three crystallographic axes at right angles and of equal length ($a_1$, $a_2$ and $a_5$), and four threefold axes of symmetry. The main forms within this system are: cube, octahedron, and rhombic dodecahedron. Minerals that crystallize in the cubic system include halite, copper, gold, silver, platinum, iron, fluorite, leucite, diamond, garnet, spinel, pyrite, galena, and magnetite. The cubic system is also sometimes known as the isometric system.

*$a_3$ axis*
*$a_1$ axis*
*$a_2$ axis*

*Cubic habit*

**PYRITE CUBES**
*Pyrite crystals form commonly as cubes, but can also be found as pentagonal dodecahedra and octahedra, or combinations of all three forms.*

## TETRAGONAL

Tetragonal crystals have three crystallographic axes at right angles – two are equal in length ($a_1$ and $a_2$), and the third (c) is longer or shorter. They have one principal, fourfold axis of symmetry. Tetragonal crystals have the look of square prisms. Some of the minerals that crystallize in the tetragonal system are rutile, calomel, cassiterite, zircon, chalcopyrite, and wulfenite.

*c axis*
*$a_2$ axis*
*$a_1$ axis*

*Pyramid face*
*Prism face*

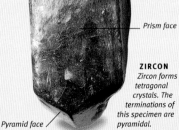

**ZIRCON**
*Zircon forms tetragonal crystals. The terminations of this specimen are pyramidal.*

*Pyramid face*

## HEXAGONAL AND TRIGONAL

Some crystallographers consider that there are seven crystal systems rather than six, separating the hexagonal from the trigonal crystals. Both hexagonal and trigonal crystals have three crystallographic axes of equal length ($a_1$, $a_2$, and $a_5$) set at 120 degrees to one another, and a fourth (c), perpendicular to the plane of the other three axes. They differ from one another in that trigonal crystals have only threefold symmetry, whereas hexagonal crystals have sixfold symmetry. Minerals that crystallize in the hexagonal system include beryl (emerald and aquamarine) and apatite. Some of the minerals that crystallize in the trigonal system are calcite, quartz, and tourmaline.

*c axis*
*$a_1$ axis*
*$a_2$ axis*
*$a_3$ axis*

*Hexagonal prism*

**APATITE**
*Apatite forms hexagonal-shaped prisms, and its overall symmetry is hexagonal.*

**TOURMALINE**
*Minerals in the tourmaline group form trigonal prisms.*

## MONOCLINIC

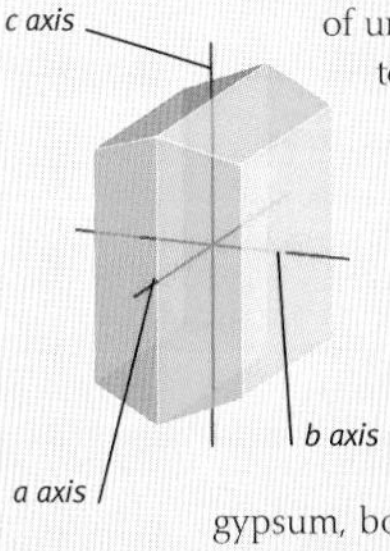

Monoclinic crystals have three crystallographic axes of unequal length. One (c) is at right angles to the other two (a and b); these two axes are not perpendicular to each other, although they are in the same plane. The crystals have one twofold axis of symmetry. More minerals crystallize in the monoclinic system than in any other. The term "monoclinic" means "one incline". Minerals that crystallize in the monoclinic system include gypsum, borax, orthoclase, muscovite, clinopyroxene, jadeite, azurite, malachite, orpiment, and realgar.

c axis

b axis

a axis

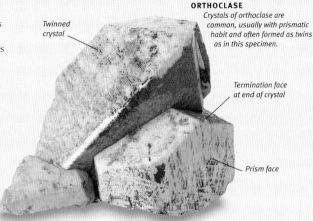

**ORTHOCLASE**
*Crystals of orthoclase are common, usually with prismatic habit and often formed as twins as in this specimen.*

Twinned crystal

Termination face at end of crystal

Prism face

**GOLDEN BARYTE**
*These orthorhombic crystals of golden baryte are from the Barrick Goldstrike Mine, in Elko, Nevada, USA.*

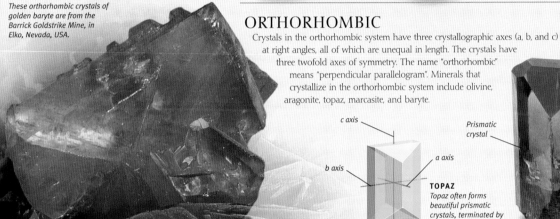

## ORTHORHOMBIC

Crystals in the orthorhombic system have three crystallographic axes (a, b, and c) at right angles, all of which are unequal in length. The crystals have three twofold axes of symmetry. The name "orthorhombic" means "perpendicular parallelogram". Minerals that crystallize in the orthorhombic system include olivine, aragonite, topaz, marcasite, and baryte.

c axis

b axis

a axis

Prismatic crystal

**TOPAZ**
*Topaz often forms beautiful prismatic crystals, terminated by bipyramids or other prisms.*

## TRICLINIC

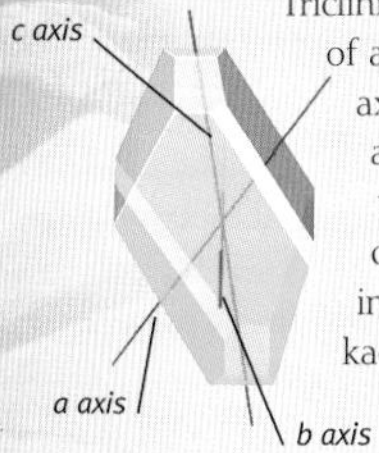

Triclinic crystals have the least symmetrical shape of all crystals. They have three crystallographic axes of unequal length (a, b, and c), which are inclined at angles of less than 90 degrees to one another. The orientation of a triclinic crystal is arbitrary. Minerals that crystallize in the triclinic system include albite, anorthite, kaolin, kyanite, and microcline.

c axis

a axis

b axis

Bladed crystals

**KYANITE**
*Kyanite is an important triclinic mineral that often forms masses of blue, bladed crystals.*

101

# CRYSTAL HABITS

A COMPLETE DESCRIPTION of the external shape of a crystal is known as its habit. This description includes all the crystal's visible characteristics. It incorporates the names of the crystal's faces, for example prismatic or pyramidal, and the name of its form, for example cubic or octahedral. It also includes more general descriptive terms, such as bladed or dendritic. The description of crystal habit can relate to a single crystal, or to an assemblage of intergrown crystals, known as an aggregate. Various terms are used to describe aggregates, for example massive or radiating.

**GLOBULAR**
*Aggregates of calcite crystals with a globular habit rest on top of tiny quartz crystals in this specimen.*

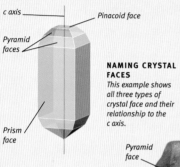

**NAMING CRYSTAL FACES**
*This example shows all three types of crystal face and their relationship to the c axis.*

c axis

Pinacoid face

Pyramid faces

Prism face

## CRYSTAL FACES

There are three types of crystal face: prism, pyramid, and pinacoid, which are determined by their relationship to the crystallographic axes. Prism faces are those parallel to the c axis; pyramid faces cut through the c axis at an angle; and pinacoid faces are at right angles to the c axis. A crystal may have numerous sets of pyramid faces, each at a different angle to the c axis, as well as major and minor prism faces, with edges parallel to each other. However, in most crystals some faces are more developed than others. As a crystal grows, the faces that grow most quickly eventually eliminate themselves, while those that grow more slowly become prominent. Where prism faces predominate in a crystal, the crystal habit may be described as prismatic; where pyramid faces predominate, it may be described as pyramidal; and where pinacoid faces predominate, the habit is described as platy. A face at the end of a crystal is called a termination face.

Prism face

Long prismatic habit

Pyramid face

Dipyramidal habit

Platy habit

**PLATY**
*These yellow wulfenite crystals could be described as having a platy habit.*

**PRISMATIC**
*In this long specimen of beryl, prism faces predominate. Its habit can therefore be described as long prismatic.*

**PYRAMIDAL**
*If pyramid faces dominate in both directions, as in this specimen of sapphire, the habit is dipyramidal.*

Wulfenite crystals

## CRYSTAL FORMS

Some crystal habits derive their names purely from their crystal forms: for example, cubic, crystallizing in the form of cubes; dodecahedral, crystallizing in the form of dodecahedrons; and rhombohedral, crystallizing in the form of rhombohedrons. If crystals of one system crystallize in forms that appear to be the crystals of another system, the habit name is preceded by the word "pseudo". For example, if cyclic twins of orthorhombic aragonite appear to form hexagonal prisms, they are described as pseudohexagonal. If the terminations (end faces) of the crystal are different from each other, the habit is known as hemimorphic.

Cube face

Crystal forms as cube, with 6 plane faces

**CUBIC**
*Crystals that form in the cubic system can take the form of perfect cubes, as in this pyrite specimen.*

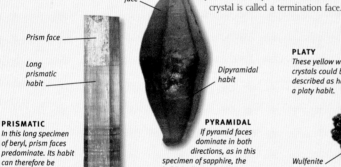

Crystal forms as octahedron, with 8 plane faces

**OCTAHEDRAL**
*This magnetite specimen has crystallized as an octahedron, and can be described as having an octahedral habit.*

Octahedron face

Garnet crystal

Crystal forms as dodecahedron, with 12 plane faces

**DODECAHEDRAL**
*These garnets on schist have crystallized as dodecahedrons, and can be described as having a dodecahedral habit.*

## FIBROUS

*A fibrous aggregate, such as this tremolite, consists of slender, parallel or radiating fibres.*

*Fibrous strands*

# AGGREGATES

Aggregates are groups of intimately associated crystals. They differ from clusters in that in crystal clusters there are a number of individuals growing together, but there is not an intimate intergrowth as in aggregates. The type of aggregation is often typical of the particular mineral species. Terms used to describe aggregates include: granular, fibrous, radiating, botryoidal, stalactitic, concentric, geode, oolitic, and massive.

## STALACTITIC

*Stalactitic minerals, such as this rare green aragonite from a cave in France, form in cylinders or cones resembling icicles.*

## BOTRYOIDAL

*This hematite has formed in globular aggregates, resembling a bunch of grapes, and is described as botryoidal.*

*Globular*

*Radiating crystals*

*Aragonite crystal*

## GEODE

*A geode is a partially filled rock cavity lined by mineral material, such as agate or amethyst.*

## CONCENTRIC

*This rhodochrosite is a concentric aggregate, forming roughly spherical layers around a common centre.*

*Concentric layers*

## RADIATING

*The crystals of this malachite specimen radiate from a common centre. Malachite can sometimes form star-like groups.*

*Massive habit*

## MASSIVE

*A mass of crystals that cannot be seen individually makes up this massive specimen of dumortierite.*

# CRYSTAL APPEARANCE

Some crystal habits are descriptions of the general appearance of a crystal. For example, tabular is used to describe a crystal with predominantly large, flat, parallel faces; bladed describes elongated crystals flattened like a knife blade; and stalactitic describes crystal aggregates that have grown in the shape of a stalactite.

*Lenticular crystals*

*Fern-like shape*

## LENTICULAR

*The crystals of this selenite rose have formed in lens shapes, and are described as being lenticular.*

## LAMELLAR

*These mica crystals are described as lamellar. They are flat, plate-like individual crystals arranged in layers.*

*Layered crystals*

## DENDRITIC

*Dendritic aggregates, such as this copper, form in slender, divergent, somewhat plant-like branches.*

*Long, thin crystals*

## BLADED

*These kyanite crystals are elongated and flattened like knife blades. They have a bladed habit.*

## ACICULAR

*This mass of slender, radiating mesolite crystals can be described as acicular, meaning needle-like.*

*Flat crystals*

103

# WHAT IS A GEMSTONE?

THE DEFINITION OF A GEMSTONE is broad: the term can be used to describe any mineral that is highly prized for its beauty, durability, and rarity. A gem is a mineral enhanced in some manner by altering its shape, usually by cutting and polishing. Most gems begin as crystals of minerals (such as diamonds or sapphires), or as aggregates of crystals (such as malachite or jadeite). A limited number of noncrystalline materials of organic origin (such as pearl and amber) are also classified as gemstones, and are usually referred to as organic gems.

**ANTIQUE BROOCH**
*This finely made gold antique brooch is set with faceted diamonds and rubies, both precious gems.*

**STRING OF PEARLS**
*These beautiful and lustrous pink pearls were cultured in fresh waters over several years before polishing and threading into this lovely necklace.*

## THE BEAUTY OF GEMS

The prime requisite for a gem is that it must be beautiful.
A gemstone can owe its beauty to various properties: its depth of colour or transparency; its colour pattern, as seen in opal or agate; the intensity of its brilliance; or the pattern light makes within it, as seen in star sapphires or cat's-eye chrysoberyl. It must also remain beautiful, withstanding wear and preserving its polish or other finish.

A number of otherwise beautiful gemstones are too soft or too brittle to wear, and are cut only for collectors. More than 4,000 minerals have been identified, but fewer than 100 are used as gemstones. Of these, only a minority are of major importance: diamond, corundum (sapphire and ruby), beryl (emerald and aquamarine), chrysoberyl, feldspars (sunstone, moonstone, and labradorite), garnets, jadeite and nephrite (jade), lazurite (lapis lazuli), olivines (peridot), opal, aragonite (pearl), quartz (in all its varieties), spinel, topaz, tourmalines, turquoise, and zircon.
Gems are usually divided into two categories: precious and semiprecious. Diamond and the two colour varieties of corundum, sapphire and ruby, are considered precious, as is the deep green variety of beryl, emerald.

**HINDU WEDDING**
*Conspicuous display in the form of gem-encrusted jewellery is an essential part of the Hindu wedding ritual.*

**MOONSTONE EARRINGS**
*These contemporary earrings are set with moonstone cabochons, which exhibit a blue schiller or sheen.*

**SAPPHIRES**
*Although it is normally thought of as blue, the variety of corundum known as sapphire actually forms in a wide range of colours.*

# EARLY USES OF GEMS

The use of gemstones goes far back in human history: people were adorning themselves with shells, pieces of bone, teeth, and pebbles by at least the Upper Paleolithic period (25,000–12,000BC). Bright colours or beautiful patterns drew people initially, and when the shaping of stones for adornment began, the stones chosen were opaque and soft. As the techniques of shaping improved – in essence, the first gem cutting – other, harder stones were employed. Carnelian and rock crystal beads, both varieties of quartz, which is relatively hard, were fashioned at Jarmo in Mesopotamia (now Iraq) in the 7th millennium BC. The next technical leap in cutting also took place in Mesopotamia, with the engraving of cylinder seals: finger-sized, engraved stone cylinders used as a means of identifying goods. When the seal was rolled on damp clay, a unique imprint resulted. The cylinder seals were also valued as adornment and possibly a symbol of status. Records of the time are the first indicating the belief that stones themselves have a "mystic" value – beliefs that persist today in the wearing of birthstones, and in New Age uses. The Egyptians, Babylonians, and Assyrians all believed that coloured stones had healing properties. The choice of stones and other healing materials was dictated by the colours that the disease under treatment caused in the body: yellow jaundice, blue lips, or fever-red skin, for example. Other coloured materials, such as plants, were considered equally effective, but stones retained their colours over time.

**QUARTZ BEADS**
*These roughly shaped quartz beads from Ghana were threaded on a necklace and used by traders as a form of currency.*

**FRESCO AT SIGIRIYA**
*The beautiful 6th-century frescoes in the temples of the fortress city of Sigiriya in central Sri Lanka illustrate some of the sumptuous jewels created during that period.*

**SUMERIAN EARRINGS**
*These gold and lapis lazuli earrings are from one of the graves in the Royal Cemetery of Ur, Iraq, and date from about 2500BC.*

**IRAQI CARNELIAN NECKLACE**
*This necklace comes from what is now Iraq and was made in about 2500BC from lapis lazuli, carnelian, and etched carnelian.*

**SEAL AND IMPRINT**
*In ancient Mesopotamia, intricate cylinder seals (right) were carved from semiprecious stones, such as serpentine and amethyst. When rolled on damp clay (far right), the seal made an imprint that identified the owner.*

# NEW AGE BELIEFS ABOUT CRYSTALS

Gemstones have been associated with particular months of the year since ancient times. In astrology, each of the 12 signs of the zodiac is associated with one or two gemstones, which are thought to resonate with the essential character of the person born under that sign, and consequently to bring the person luck. A hugely contentious subject between scientists and New Age practitioners is that of "crystal healing". Crystal healers claim that certain crystals – particularly quartz – give off "healing energy" when placed on the body, a claim contested by mineralogists, who recognize the impossibility of this given the minutely bound and intricately balanced energies that create a crystal in the first place. Yet anecdotal evidence from many sources suggests that when patients are treated by crystal healers, healing does occur in many instances. A new field of medical research suggests that the state of the immune system (and hence the state of health) is profoundly influenced by both conscious and unconscious mental and emotional factors. Perhaps it is the belief that the crystal will heal the body that causes it to heal, rather than the crystal itself.

| JANUARY GARNET | FEBRUARY AMETHYST | MARCH AQUAMARINE | APRIL DIAMOND |
| --- | --- | --- | --- |

| MAY EMERALD | JUNE PEARL | JULY RUBY | AUGUST PERIDOT |
| --- | --- | --- | --- |

| SEPTEMBER SAPPHIRE | OCTOBER OPAL | NOVEMBER TOPAZ | DECEMBER TURQUOISE |
| --- | --- | --- | --- |

**BIRTHSTONES**
*The particular birthstones associated with each month vary from country to country and from one decade to the next. A modern European set of associations is shown here.*

**HEALING CRYSTALS**
*New Age practitioners believe that crystals have healing powers when placed against the body. This belief can be traced back to the ancient Egyptians, Babylonians, and Assyrians.*

# GEM MINING

MINING TECHNIQUES AND OPERATIONS have varied little
over the millennia until very recent times. In common with other
types of mining, gems are recovered in one of two ways: from the
rocks in which they formed; and in weathered rock debris where
they have been released from their original rock. The first type is
often called "hard–rock" mining; the second is called placer mining.
Gemstone mining of both kinds can be done by an individual or
by huge corporations. Many gem mines have consisted of no
more than a few individuals working cooperatively.

**MOUNT MICA**
*The tourmaline deposit at Mount
Mica, Maine, USA, has produced
thousands of carats of gem
tourmaline (see p.288).*

## HARD–ROCK MINING

Gemstone deposits form in several different geological environments.
Perhaps the best known are the "pipes" of kimberlite from which most
diamonds are recovered (see p.41). Solid kimberlite is mined by the usual
hard-rock methods of drilling and blasting. The rock is then crushed, and
the diamonds extracted, sorted, and graded.

Diamonds are also found in lamproite pipes. The first commercial
extraction of diamond from lamproite rock was at the Argyle Mine
in Western Australia. Diamonds found in nearby placer deposits in
1979 led to the discovery of the lamproite pipe, and the mine opened
in 1985. Argyle is an open-pit operation producing more diamonds
than any other operation in the world. Placer deposits there are
almost exhausted, and most diamonds are extracted from the solid
rock. About 5 per cent are gem quality, and they include 90–95
per cent of the world's natural pink diamonds. These are
sold by tender for as much as $100,000 per carat.

Many other gems are also sufficiently
concentrated in their original rock to be
mined economically. Among these are
various quartz varieties, tourmaline,
topaz, opal, emerald, aquamarine,
some sapphire and ruby,

turquoise, lapis lazuli, and chrysoberyl. Few of these require – or are
valuable enough to justify – the extensive and expensive methods of
mining diamond pipes. Even the largest pegmatite deposits, which are
a rich source of gemstones and other valuable minerals, rarely justify
highly mechanized extraction operations, such as the one used to
extract tourmaline at Mount Mica (see above). Most are worked by
hand using basic tools. Drilling and blasting are avoided for most gems,
because they are shock sensitive. Additionally, many gem veins are in
deteriorated rock and the veins are often deteriorated themselves. In
these deposits, mining is done with little more than a pick and shovel.

**DRILLING KIMBERLITE PIPES**
*Kimberlite that has not deteriorated
through weathering is a hard rock
that has to be drilled and
blasted. Vast quantities
of rock are processed
in the search for
diamonds.*

**SIBERIAN DIAMOND MINE**
*The kimberlite in this Siberian diamond
pipe is deeply weathered, permitting
open-pit mining operations.*

# PLACER MINING

Because many gemstones are hard, dense, and impervious to chemical weathering, once released by weathering they can be carried considerable distances by water to concentrate in river beds, beaches, and on the ocean floor. These concentrations are called placer deposits.

Placer mining uses techniques that mimic the creation of the placer in the first place: the separation of the denser minerals – including the gemstones – in running water. The simplest method is panning (see right). In sieving, the separated denser minerals end up in the centre of the sieve, and the gemstones are hand-sorted from this concentrate. Another method is to run gravel through a trough of flowing water with baffles on the bottom. The lighter material is washed away, and the denser gemstones are retained by the baffles. Jade and some quartz gemstones can be found by people simply walking up

**PANNING**
In Sri Lanka and Laos rubies and sapphires are mined by scooping up gravel in pans or baskets and hand-washing it to separate out the gems.

**SAPPHIRE GRAVELS**
These sapphires were recovered from stream gravels in Montana by placer mining.

and down river beds or across deserts and picking up material from the surface. Some of the most productive diamond placers occur where diamond-bearing gravels have been dumped into the sea at the mouths of rivers, and then redistributed by wave action along beaches. Within recent decades, new underwater mining techniques have permitted the large-scale exploitation of sea-floor diamond deposits (see p.123).

**SAND SHIFTING, NAMIBIA**
In large-scale operations, such as the beaches of Namibia, earth-movers shift millions of tonnes of sand, recovering tens of thousands of carats of diamonds from the bedrock.

# SORTING AND GRADING

After drilling and crushing rock containing diamonds, the diamonds are recovered using grease tables. Crushed kimberlite is wetted and made to flow in water across a sloping table covered with thick grease. The wet rock particles do not stick, but the dry diamonds do (water does not stick to diamonds). Periodically, the diamond-covered grease is scraped from the table, and the grease is removed from the diamonds before they are sent for sorting and grading.

Grease tables are sometimes used in the recovery of placer diamonds as well. In the simplest placer operations, the gemstones are first concentrated by placer mining methods, and then picked out by eye.

**AFGHANISTAN**
The mountains of Afghanistan have been producing gemstones for at least 7,000 years.

**GREASE TABLE**
The table is coated with grease, to which only the dry diamonds stick.

**ROUGH DIAMONDS**
Uncut diamonds such as these are graded according to their properties, for example, colour and clarity.

## MINING AND CUTTING TECHNIQUES

Ancient miners were far from inefficient – some of their methods are still used today. Panning and basket-sieving of stream gravels have changed little. Digging into decomposed gem veins differs only in the tools used: antler picks were used in ancient times rather than steel. When copper and bronze tools came into use in the 2nd century BC, true hard-rock mining was possible. Emeralds were mined at Cleopatra's mines from 1500BC (see p.293). Gem cutting differed little in principle from cutting today: the gem was abraded on a series of stones of greater and greater hardness, becoming ever finer in grain size.

**ANCIENT MASTERPIECE**
This scarab pectoral from the tomb of Tutankhamun (c.1361–52BC) is inlaid with gold, lapis lazuli, carnelian, and other semiprecious gems.

# ANCIENT MINING

The mining of precious stones has a long history: lapis lazuli was mined at Badakhshan, Afghanistan, up to 7,000 years ago, and turquoise mining on the Sinai Peninsula began around 5,000 years ago. At about the same time, the Egyptians were mining emeralds near Aswan. Gemstones were widely valued commodities and were traded over long distances: for example, lapis lazuli from Badakhshan reached Egypt before 3000BC, and China, India, and Greece by 2500BC. Baltic amber was traded all around the Mediterranean, and Egyptian emeralds are found in Roman jewellery. With the conquests of Alexander the Great (356–323BC), trade from the East increased, as did the number of gemstones available. In North and South America, turquoise was extensively mined in pre-Columbian times.

**EGYPTIAN GOLDSMITHS**
Goldsmiths weigh and smelt gold, and present jewellery in this wall painting in the tomb of sculptors Nebamun and Ipuky (c.1390–36BC).

# GEM CUTTING

THERE ARE A NUMBER OF WAYS of shaping gemstones. Opaque or translucent semiprecious stones, such as agate or jasper, are tumble–polished, carved, engraved, or cut *en cabochon*. Transparent stones, like amethyst, diamond, and sapphire, are generally faceted to maximize the brilliance and "fire" of the stone, or in some instances to enhance its colour. The process of shaping stones is called "cutting", although shaping is in fact most often done by grinding.

**CARVING JADE**
*This Chinese woman is carving nephrite (jade), a gemstone that has been worked in China for more than 2,000 years. It is the toughness of nephrite that makes it so suitable for carving.*

**INSPECTING THE CUT**
*Here, a lapidary is using a magnifying glass to assess his progress as he facets a diamond.*

## POLISHING AND CARVING

Many semiprecious stones are rounded and polished by placing them in a rotating cylinder with progressively finer abrasive grits and water, which is rotated about its long axis. Called tumble-polishing, this is essentially the same process that rounds beach pebbles. Layered gemstones, such as onyx and sardonyx, have been carved to create cameos and intaglios for 2,000 years. In a cameo the stone is cut around the design, so that it stands in relief against a differently coloured background. In an intaglio, the subject is cut away to create a recessed image that could be used as a seal.

**ROUGH AND POLISHED UNAKITE**
*Unakite is one of many inexpensive stones shaped by tumble-polishing.*

**CAMEO OF VENUS**
*This 19th-century cameo plaque depicting the birth of Venus was carved from banded agate by Jean Louis- Francois.*

## CUTTING CABOCHONS

The best way to display colours and other optical effects in opaque or translucent stones is to cut them *en cabochon* - with a rounded upper surface and a flat underside. Cabochons are cut and polished on abrasive wheels. The stone is first ground to its outline shape, and then the top is ground and polished into a dome, using progressively finer abrasives. Some gemstones have oriented inclusions or structures within them: chatoyancy, or cat's-eye effect, is caused by microscopic fibres or hollow tubes within the gem; asterism, or star effect, occurs when there are sets of fibres at angles of nearly 90 degrees or 60 degrees to each other. These are revealed by cutting the stones *en cabochon*. The soft sheen produced in moonstone when cut *en cabochon* is called the "schiller" effect.

**BEAD**

**CABOCHON**

Soft play of colours

**MOONSTONE SCHILLER**

**OPTICAL EFFECTS**
*These three cabochon gems demonstrate the different optical effects produced by inclusions within the gemstones.*

White "eye" effect

**CAT'S-EYE CABOCHON**

Star effect

**STAR CABOCHON**

# FACETING

The best way to maximize the beauty of a transparent gem is to cut the surface into a series of flat, reflective faces called facets. The gem is glued to a holder called a dop, and held against a horizontally rotating wheel charged with an abrasive powder, which grinds each face. Facets are placed in specific geometric positions and at specific angles. During the cutting process, much of the gemstone material is ground away – often as much as half the stone or more – but the value of the finished gem is greatly increased. Faceting is a relatively complex operation, since care must be taken in the orientation of the stone to produce the best colour (or minimize the worst colour in pleochroic stones), preserve the maximum amount of material, and produce the best brilliance. The stages in faceting a round, brilliant-cut diamond are shown in the sequence below.

**1** *A gem-quality piece of rough diamond – an octahedral crystal – is selected for cutting.*

Crown

**2** *The point at the top of the octahedron is sawn off the crystal to form the large top facet known as the crown.*

Girdle (outer edge)

Table facet

**3** *The stone is rounded on a lathe by another diamond and the top (table) facet is cut.*

Bezel facet

Bezel facet

**4** *Pairs of facets are cut top and bottom to centre the stone, beginning with the bezel facets.*

Upper girdle facet

**5** *The 16 "main" facets are then cut above the girdle and below it.*

Star facet

**6** *Eight small "star" facets are cut around the table facet.*

Pavilion

**7** *Thirty-two "girdle" facets are cut around the lower girdle and the pavilion (the underside).*

Total of 58 facets

**FINISHED BRILLIANT CUT**
*The brilliant cut emphasizes the internal colours of the stone, making it sparkle. It is particularly popular for colourless stones.*

## THE VALUE OF GEMS

Gemstones are valued according to the "four Cs": colour, clarity, cut, and carats. In the case of colourless diamonds, the total lack of colour is the highest grade; in coloured stones, such as ruby or sapphire, the purity and intensity of the colour determine value. Clarity refers to the lack of visible or invisible foreign matter within the stone. The cut is graded on the basis of its technical perfection, and the brilliance it produces. The carat is a unit of measurement of weight, equivalent to one-fifth of a gram; the finished weight also affects a stone's value.

When all these criteria are met, one other factor determines the final price: rarity. For example, a superb garnet will never command the same price as an equivalent ruby, because the ruby is hugely rarer. A synthetic ruby will never command the same price as an identical natural one, for exactly the same reason. In general, larger stones are much rarer than smaller ones. With some stones, an increase in weight is associated with a disproportionately large increase in price, so when a gemstone doubles in weight, its price may go up by four or five times.

RUBY

GARNET

**THE ART OF THE JEWELLER**
*This work by Giovanni Antonio Guardi (1698–1760) shows the cutting and trading of gemstones.*

GILSON STEP CUT

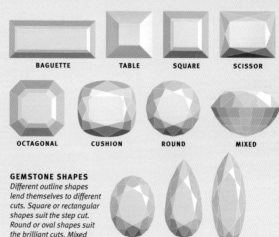

---

# GEM CUTS

Faceting as we know it today began with the cutting of diamond, probably in France and the Netherlands, in the 15th century. At first only the natural faces of diamond's octahedral crystals were polished, but in the 17th century the rose cut was developed, and by about 1700 the brilliant cut (today's favourite for diamonds and other colourless gems) was created. Not long after, as emeralds poured in from the Americas, the emerald cut was developed, primarily to save valuable emerald material, as the rectangular cut conforms to the shape of emerald crystals. Other gem cuts were soon developed, and today there are hundreds of possible shapes. Some are cut totally free form, and others in the more familiar cuts shown here.

STEP CUT

CUSHION MIXED CUT

BRILLIANT CUT

**THREE CUTS**
*These gems show the three basic types of facet cuts: step (rectangular facets), brilliant (triangular facets), and mixed (a mixture of the two).*

**BAGUETTE** **TABLE** **SQUARE** **SCISSOR**

**OCTAGONAL** **CUSHION** **ROUND** **MIXED**

**GEMSTONE SHAPES**
*Different outline shapes lend themselves to different cuts. Square or rectangular shapes suit the step cut. Round or oval shapes suit the brilliant cuts. Mixed cuts can be used on both.*

**OVAL** **PENDELOQUE** **MARQUISE**

# MEGAGEMS

IN THE USA, THE SMITHSONIAN INSTITUTION'S
NATIONAL GEM COLLECTION CONTAINS SOME OF THE
LARGEST, FINEST, AND MOST BEAUTIFUL GEMSTONES
EVER FOUND. THEY ARE HOUSED AT THE NATIONAL
MUSEUM OF NATURAL HISTORY IN WASHINGTON, DC.

The term "megagem", although not an official one, nonetheless represents a group of gemstones that stand out from others of their type in size and beauty. Many gems in the Smithsonian's National Gem Collection take their names from people who donated them to the collection, or who once figured in their histories. Some have no name, but are no less remarkable.

But what makes a gem a megagem? The first consideration for many is the gem's size, but in the case of very common materials, such as fluorite or calcite, large gems may also be commonplace. In such instances, the megagems are large but also particularly high-quality examples that best represent their type of mineral. The second consideration in defining a megagem is the gem's value, which may not be governed by size. The Blue Heart Diamond (see right) is small in comparison with the fluorite or the calcite gems also shown, but it is thousands of times more valuable. Rarity is a key factor. Intensely blue diamonds are seldom found, and those attaining the size, and also the shape, of the Blue Heart are much rarer still. Thus a relatively small stone cut from a very rare material is a megagem, whereas a larger stone cut from a more common material still may not be. All the stones illustrated here are highly prized, and therefore valuable, megagems in that they represent superb examples, both in size and quality, of their respective gemstone species.

Diamond consists of almost pure carbon

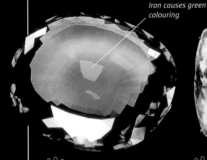
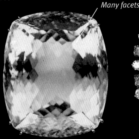

**AMERICAN GOLDEN TOPAZ**
At 22,892.5 carats, the 4.6kg (10.1lb) topaz (centre) was cut from a Brazilian cobble in the late 1980s. Alongside it are two of the largest high-quality topaz crystals ever found, also from Brazil.

Pear-shaped cut

Cut draws out gem's natural heart shape

**GROSSULAR GARNET**
Just one of 15 minerals known as garnet, grossular may be pink, orange, red-orange, yellow, brown, or green. This 30-carat specimen was found in Sri Lanka in 1991.

Iron causes green colouring

Many facets

Bevelled square cut

Fine facets

**PERIDOT**
This 311.8-carat peridot from Zabargad, Egypt, is one of the largest peridot gems known. Peridot is the gem variety of the mineral forsterite. The green colouring is caused by iron; the more iron, the deeper the colour.

**HELIODOR**
The common name for heliodor is golden beryl. There are larger known examples than this 216-carat stone, but few can rival the beauty of its colour or the brilliance of its cut. It came from Minas Gerais, Brazil.

**HOOKER EMERALD**
Weighing 75.47 carats, the Hooker Emerald is unusually flawless for an emerald of its size. Mined in Colombia and taken to Europe in the 17th century by the conquistadores, it was donated by Mrs J.A. Hooker in 1977.

**CALCITE**
The normally soft and fragile structure of calcite restricts its use as a gemstone. Originating at Balmat in New York State, USA, this specimen weighs 1,800 carats and is probably the world's finest calcite gem.

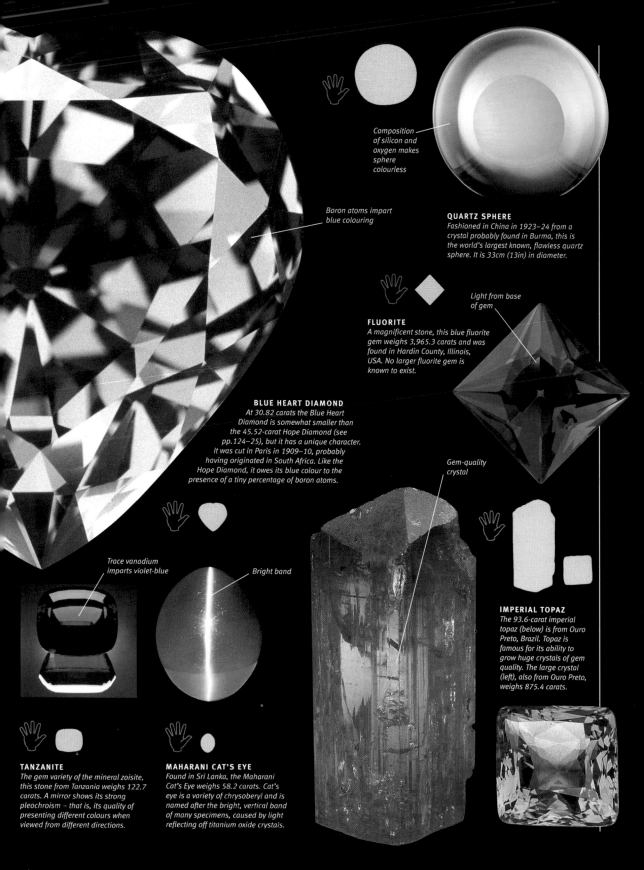

Composition of silicon and oxygen makes sphere colourless

Boron atoms impart blue colouring

**QUARTZ SPHERE**
Fashioned in China in 1923–24 from a crystal probably found in Burma, this is the world's largest known, flawless quartz sphere. It is 33cm (13in) in diameter.

**FLUORITE**
A magnificent stone, this blue fluorite gem weighs 3,965.3 carats and was found in Hardin County, Illinois, USA. No larger fluorite gem is known to exist.

Light from base of gem

**BLUE HEART DIAMOND**
At 30.82 carats the Blue Heart Diamond is somewhat smaller than the 45.52-carat Hope Diamond (see pp.124–25), but it has a unique character. It was cut in Paris in 1909–10, probably having originated in South Africa. Like the Hope Diamond, it owes its blue colour to the presence of a tiny percentage of boron atoms.

Gem-quality crystal

Trace vanadium imparts violet-blue

Bright band

**IMPERIAL TOPAZ**
The 93.6-carat imperial topaz (below) is from Ouro Preto, Brazil. Topaz is famous for its ability to grow huge crystals of gem quality. The large crystal (left), also from Ouro Preto, weighs 875.4 carats.

**TANZANITE**
The gem variety of the mineral zoisite, this stone from Tanzania weighs 122.7 carats. A mirror shows its strong pleochroism – that is, its quality of presenting different colours when viewed from different directions.

**MAHARANI CAT'S EYE**
Found in Sri Lanka, the Maharani Cat's Eye weighs 58.2 carats. Cat's eye is a variety of chrysoberyl and is named after the bright, vertical band of many specimens, caused by light reflecting off titanium oxide crystals.

# NATIVE ELEMENTS

THE NATIVE ELEMENTS are those chemical elements that occur in nature uncombined with other elements. They are commonly divided into three groups: metals (copper, platinum, iron, gold, and silver); semimetals (arsenic); and nonmetals (sulphur and carbon). Other native elements that occur more rarely and are not represented here are cadmium, mercury, nickel, chromium, tin, iridium, palladium, osmium, antimony, bismuth, tellurium, and selenium. These mostly occur in minute amounts, often alloyed with other native elements. The native elements form under greatly contrasting physical and chemical conditions and in a variety of rock types.

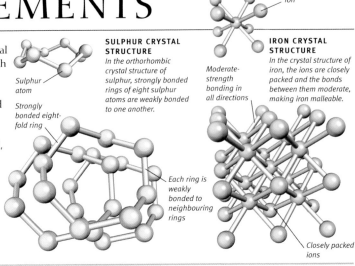

**Sulphur atom**

**Strongly bonded eight-fold ring**

**SULPHUR CRYSTAL STRUCTURE**
*In the orthorhombic crystal structure of sulphur, strongly bonded rings of eight sulphur atoms are weakly bonded to one another.*

**Each ring is weakly bonded to neighbouring rings**

**Ion**

**Moderate-strength bonding in all directions**

**IRON CRYSTAL STRUCTURE**
*In the crystal structure of iron, the ions are closely packed and the bonds between them moderate, making iron malleable.*

**Closely packed ions**

# COPPER

COPPER IN ITS NATURALLY occurring free metallic state was probably the first metal to be used by humans. Neolithic people used copper as a substitute for stone by 8000BC. It was first cast in moulds by the Egyptians in about 4000BC, and was alloyed with tin to produce bronze in around 3500BC. It takes its name from the Latin *aes Cyprium*, meaning "metal of Cyprus", shortened to *cyprium* and later corrupted to *cuprum*. Copper in industrial quantities is usually found in massive aggregates, some weighing up to several tonnes. When well crystallized, it is found in cubic or dodecahedral crystals, often arranged in tree-like shapes. Native copper seems to be a secondary mineral, a result of the reduction of copper-bearing solutions by iron-bearing minerals. The Keweenaw Peninsula, Michigan, USA, has probably the world's largest concentration of native copper. Other localities include Corocoro, Bolivia; the Ural Mountains, Russia; Cornwall, England; Broken Hill, NSW, and Mount Isa, Queensland, Australia; Rheinland-Pfalz, Germany; and Bisbee, Arizona, USA.

**KEWEENAW PENINSULA**
*Here, in Michigan, USA, single masses of native copper weighed 420 tonnes.*

**VIKING AMULET**
*This copper locket, crafted in Grotland, Sweden, contains a decomposed snake, believed to possess magical powers.*

**Crystalline copper**

**COPPER FIGURE**
*Created in AD1000–1400, this copper male figure is from South America.*

**Massive copper**

**Accessory quartz**

**NATIVE COPPER**
*This specimen of native copper is part crystalline and partly massive. It is accompanied by accessory quartz.*

## PROPERTIES

| | |
|---|---|
| **GROUP** | Native elements |
| **CRYSTAL SYSTEM** | Cubic |
| **COMPOSITION** | Cu |
| **COLOUR** | Copper-red to brown |
| **FORM/HABIT** | Massive |
| **HARDNESS** | $2\frac{1}{2}$–3 |
| **CLEAVAGE** | None |
| **FRACTURE** | Hackly, ductile |
| **LUSTRE** | Metallic |
| **STREAK** | Rose |
| **SPECIFIC GRAVITY** | 8.9 |
| **TRANSPARENCY** | Opaque |

# PLATINUM

**BUSHVELD COMPLEX**
*This complex in South Africa is a major source of platinum.*

ALTHOUGH PLATINUM has been used for thousands of years, it was not recognized as a distinct metal until 1735. The first documented discovery was by the Spaniards in the 1500s in the alluvial gold mines of the Río Pinto, Colombia. They called it *platina del Pinto*, *platina* meaning "little silver", in the belief that it was an impure ore of silver. Platinum is found as flakes or grains, and only rarely in nuggets. Crystals are rare as well. Native platinum almost always contains some iron and other metals such as iridium, rhodium, and palladium. Platinum occurs in mafic and ultramafic igneous rocks and in quartz veins associated with hematite, chlorite, and pyrolusite. It concentrates in placers, but most commercial recovery is from primary deposits. Important sources are the Bushveld Complex, Transvaal, South Africa; the Stillwater Complex, Montana, and Goodnews Bay, Alaska, USA; Ontario, Canada; and Norlisk, Russia.

## USES OF PLATINUM

The use of platinum in jewellery began about 1900, but its high melting point (it is higher than that of gold) meant that it was not until the 1920s that a technology was developed sufficiently to work it easily. Today platinum is more important in industry than in ornamentation. It is used in aircraft spark plugs, molecular converters in the refining of petroleum, and in catalytic converters for cars. It is also used to make razor blades.
Platinum compounds have been incorporated recently in chemotherapy drugs.

**JEWELLERY**
*The making of platinum jewellery requires specialist skills because of the element's high melting point.*

**PLATINUM NUGGET**
*Although most of the platinum mined from placer gravels is in small grains, sizeable nuggets are sometimes found.*

**PLATINUM CRYSTALS**
*Platinum crystals are rare, but are found as cubes when they do occur.*

*Rounded surface*

**PLATINUM GRAINS**
*Found in placers, platinum is usually recovered as small grains.*

### PROPERTIES

| | |
|---|---|
| **GROUP** | Native elements |
| **CRYSTAL SYSTEM** | Cubic |
| **COMPOSITION** | Pt |
| **COLOUR** | Whitish steel-grey |
| **FORM/HABIT** | Cubic |
| **HARDNESS** | 4–4½ |
| **CLEAVAGE** | None |
| **FRACTURE** | Hackly |
| **LUSTRE** | Metallic |
| **STREAK** | Whitish steel-grey |
| **SPECIFIC GRAVITY** | 14.0–19.0 |
| **TRANSPARENCY** | Opaque |

# IRON

**DISKO ISLAND**
*Large masses of kamacite occur here in Greenland.*

IRON MAKES UP FIVE PER CENT of the Earth's crust and is fourth in abundance behind oxygen, silicon, and aluminium. It is also relatively plentiful in the Sun and other stars. Native or free iron is rare in the crust, and it is invariably alloyed with nickel to some degree. Low-nickel iron (up to 7.5 per cent nickel) is called kamacite, and high-nickel iron (up to 50 per cent nickel) is called taenite. Kamacite takes its name from the Greek for "bar", and taenite from the Greek for "band". Both crystallize in the cubic system. A third form of iron-nickel, mainly found in meteorites and crystallizing in the tetragonal system, is called tetrataenite. All three are generally found as disseminated grains or rounded masses. Important localities for kamacite are Taimyr, Russia; and Bühl, Germany. It is the major component of most iron meteorites, is found in most chondritic meteorites, and occurs as microscopic grains in some lunar rock. Taenite and tetrataenite are mainly found in meteorites, often associated or intergrown with kamacite.

**IRONWORK BRACELET**
*Made in Berlin in the late 19th century, this ironwork bracelet is one of a pair.*

### PROPERTIES

| | |
|---|---|
| **GROUP** | Native elements |
| **CRYSTAL SYSTEM** | Cubic |
| **COMPOSITION** | Fe |
| **COLOUR** | Steel-grey to iron-black |
| **FORM/HABIT** | Crystals rare |
| **HARDNESS** | 4 |
| **CLEAVAGE** | Basal |
| **FRACTURE** | Hackly |
| **LUSTRE** | Metallic |
| **STREAK** | Steel-grey |
| **SPECIFIC GRAVITY** | 7.3–7.9 |
| **TRANSPARENCY** | Opaque |

**NATIVE IRON**
*The majority of native iron is in the Earth's core, but iron from meteorites was used by early man from about 3000BC.*

*Nickel-iron composition*

*Previously melted surface*

# GOLD

THROUGHOUT HUMAN HISTORY, gold has been the most prized of metals. Its colour and brightness are highly attractive, it is extremely malleable, and is usually found in nature in a relatively pure form – all qualities that have made it exceptionally valuable. It is also remarkably inert, and neither bonds nor reacts with most chemicals, so resists tarnishing. People have been using gold for at least 6,000 years, since the civilizations of ancient Egypt and Mesopotamia. In ancient times gold was almost exclusively recovered from river and stream gravels – placer deposits – where particles of gold weathered from its original rock were concentrated.

## SOURCES OF GOLD

Gold is rarely found in well-formed octahedral and dodecahedral crystals, but more commonly it occurs as dendritic growths, and as grains and scaly masses. Crystals about 2.5cm (1in) across have been found in California, and masses of over 90kg (200lb) have been recovered in Australia. Virtually all igneous rocks contain gold in low concentrations, where it occurs mostly as invisible, disseminated grains. Within the Earth's crust, its abundance is estimated at about 0.005 parts per million. Because gold does not easily bond with other elements, there are only a few relatively rare gold-bearing minerals, in which it combines with tellurium and selenium.

One unusually rich concentration of these gold minerals was discovered at the famous gold-rush town of Cripple Creek, Colorado, USA. Large masses of rock rich enough to mine only for the gold are rare, but fall into

**WITWATERSRAND, SOUTH AFRICA**
*Witwatersrand means "ridge of white waters". Its fabulously rich gold deposits occur in conglomerate beds.*

**MIXTEC PENDANT**
*This gold pendant of a Mexican deity was crafted by the Mixtecs of ancient Mexico.*

**ALFRED JEWEL**
*Made for Alfred the Great in the 9th century AD, the Alfred Jewel is a reading pointer of gold, enamel, and rock crystal.*

**SNAKE ARMLET**
*This pure gold armlet was found in the ruined Roman city of Pompeii, Italy. It was crafted between the 1st century BC and the 1st century AD.*

**EGYPTIAN AMULET**
*This ancient Egyptian amulet in the form of the eye of Horus is crafted from gold, lapis lazuli, and enamel.*

**EGYPTIAN GOLD**
*Gold was a highly valued metal in ancient Egypt, worked with great skill. This relief (right) from the Tomb of Rekhmire depicts workers engraving and polishing gold and silver vases (c.1567–1320BC).*

*Scaly gold*

**SCALES OF GOLD**
*This specimen of native gold is from Baita, Transylvania, Romania. The gold has formed as thin plates set in a groundmass of quartz. Gold is also often found in association with sulphides.*

## GOLD RUSHES

The enormous gold discoveries in the western USA and Australia in the 1840s and 1850s – aside from creating more destitution than wealth among the miners in general – poured so much gold into the world marketplace that all but a few countries decided to use their gold reserves to back their currencies. Concerning the situation of the average miner and prospector, two sayings emerged: "more money goes into the ground than ever comes out", and "the way to make a small fortune in mining is to start with a large one"! Two positive benefits (except perhaps for the native inhabitants of gold rush areas) were that the population shifts that accompanied the gold rushes opened new areas for agriculture, and that other valuable minerals were discovered aside from gold.

**PANNING FOR GOLD**
*The gold pan is a prospector's tool. Panning is a slow but thorough method.*

**GOLD GRAINS**
*Most gold recovered from river and stream placers is in the form of scales and grains.*

**NEVADA GOLD RUSH**
*In 1902 a major gold rush occurred in Nevada after discoveries were made at a place later called Goldfield.*

*Massive quartz*

two types: hydrothermal veins, where it is associated with quartz and pyrite (fool's gold); and placer deposits. Some placer deposits are unconsolidated, and the gold there can be recovered by panning. Smaller quantities of gold often occur in copper and lead deposits, where it is recovered as a by-product in the refining of those metals.

The principal ancient sources of gold were southern Egypt and Nubia (Sudan). Major sources in Europe during the Middle Ages were the mines of Saxony and Austria. The largest single gold-ore body known to the modern world is in South Africa's Witwatersrand. Half the planet's known gold reserves are in South Africa, with other large deposits in Russia, Brazil, Canada, Australia, and the USA.

## THE COLOUR OF GOLD

In its native state, gold is always golden-yellow, but it is too soft in its pure state to wear well. To increase its hardness for use in jewellery, gold is alloyed with other metals. Pale or white gold results when silver, platinum, nickel, or zinc are added. Copper yields red or pink gold, and iron gives a blue tinge. The purity of alloyed gold is expressed as its carat (Ct) value, defined as the proportion of pure gold metal present. Twenty-four carats equates to 100 per cent gold. So a piece of gold jewellery labelled, 9-carat is 9 parts out of 24 (or 37 per cent) pure gold.

## GOLD IN INDUSTRY

By far the most popular use of gold is in jewellery, but it has important industrial applications as well. It has high electrical conductivity (71 per cent that of copper) and is chemically inert, so it finds an important application in the electronics industry for plating contacts, terminals, printed circuits, and semiconductor systems. The salvaging of obsolete computer electronics for their gold content has become a major recycling enterprise. Thin films of gold no more than a single atom thick reflect up to 98 per cent of incident infrared radiation. These have been employed on satellites to control temperature and on space-suit visors to afford protection. Large office buildings often have their windows similarly coated, reducing the need for air-conditioning.

Gold compounds also have important uses. Sodium aurichloride is used in the treatment of rheumatoid arthritis, and gold mercaptide compound is used for decorating china and glass articles.

**GOLD WINDOWS**
*A thin gold coating on window glass in hot climates reduces the need for air-conditioning.*

**SATELLITE**
*Gold coatings on satellite components are vital for temperature control.*

"What do you not drive **human hearts** into, cursed **craving** for **gold**!"

**VIRGIL, AENEID**

**HABITS OF GOLD**
*Gold occurs in several habits: as crystalline masses, as flat plates, and as nuggets and grains recovered from stream and river placers.*

**GOLD CRYSTALS**

**FLAT GOLD PLATES**

### PROPERTIES

| | |
|---|---|
| **GROUP** | Native elements |
| **CRYSTAL SYSTEM** | Cubic |
| **COMPOSITION** | Au |
| **COLOUR** | Golden-yellow |
| **FORM/HABIT** | Octahedral, dodecahedral, dendritic |
| **HARDNESS** | $2^{1}/_{2}$–3 |
| **CLEAVAGE** | None |
| **FRACTURE** | Hackly |
| **LUSTRE** | Metallic |
| **STREAK** | Golden-yellow |
| **SPECIFIC GRAVITY** | 19.3 |
| **TRANSPARENCY** | Opaque |

*Other mineralization in quartz*

*Actual size of nugget*

**LARGE GOLD NUGGET IN SMITHSONIAN COLLECTION**

115

**TUTANKHAMUN'S SECOND GOLD COFFIN**
*Each of the three gold coffins is inlaid with lapis lazuli, turquoise, and carnelian, and shows Tutankhamun in the form of Osiris, god of the dead.*

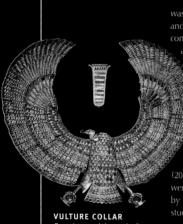

**GOLDEN THRONE**
*Tutankhamun's wooden throne is overlaid with gold and inlaid with lapis lazuli, carnelian, and turquoise. Depicted on the back of the throne is Queen Ankhesenamun as she adjusts the king's clothing.*

# EGYPTIAN GOLD

## A SYMBOL OF THE SUN IN A CULTURE THAT WORSHIPPED THE SUN GOD RA, GOLD WAS HIGHLY PRIZED BY THE ANCIENT EGYPTIANS, WHO USED IT FOR ADORNMENT IN LIFE AND DEATH.

I n 1924 English archeologist Howard Carter (1874–1939) lifted the lid of the last shrine in the burial chamber in the tomb of Tutankhamun. He removed the layers of linen shrouds and "a golden effigy of the young boy-king of most magnificent workmanship, filled the whole interior of the sarcophagus!" This coffin proved to be the outermost of three. The third and final coffin was made of 110kg (243lb) of solid gold. Within this final coffin lay the mummy of Tutankhamun, his face covered with a superb mask of gold inlaid with lapis lazuli and coloured glass. The tomb as a whole contained the greatest treasure ever discovered by an archeologist.

The ancient Egyptians had prized gold and deeply coloured precious stones long before the reign of Tutankhamun (18th dynasty – 1539–1292BC). They probably first came across nuggets of gold in the waters of the Nile, where they had been washed downstream from the mountains of Nubia (modern Sudan). Gold, silver, and copper artefacts were discovered in the predynastic graves of Naqada near Nubt, which was a trading centre for gold from about 3100BC. *Nub* is the Egyptian word for "gold". From the beginning of the Old Kingdom (2575–2134BC) pharoahs sent expeditions to Nubia to mine the quartz lodes there for gold. They sunk deep shafts into the rock and lit fires around the quartz to shatter it. Then, they crushed it, extracting the gold within. The mines were exhausted by the late 18th dynasty, and the Egyptians were then forced to mine the much harder rocks of the eastern desert for gold.

Virtually indestructible and highly malleable, gold was either hammered into shape while cold or melted and then cast. Goldsmiths' workshops were strictly controlled by the vizier on behalf of the pharaoh, and any gold objects made there belonged to him. By the fourth dynasty (c.2575–2465BC), gold was worked into vessels, furniture, and funerary equipment, and sophisticated jewellery, such as settings, chains, chokers, and armlets, was created. Precious stones such as lapis lazuli, turquoise, and carnelian were favoured for inlay work, beads and cabachons, which were mounted in gold settings. Broad collars and earrings were made from the Middle Kingdom (2040–1640BC) onwards. Elaborate and heavy pectorals were created in the New Kingdom (1550–1070BC). Worn by men as well as women from this time, earrings were stud-mounted for use in pierced ear lobes.

### TUTANKHAMUN'S GOLD FUNERARY MASK
*The magnificent gold funerary mask of Tutankhamun was inlaid with strips of lapis lazuli and coloured glass. The gold symbolized Ra, the sun god, and the lapis lazuli represented the god Horus, who, it was believed, guided the dead through the underworld.*

**VULTURE COLLAR**
*Found around the neck of Tutankhamun's mummy, this gold collar contains 250 inlaid segments of glass and obsidian. Each claw grasps a "shen", a symbol of totality.*

**ASPEN, COLORADO**
*Now known as an exclusive ski resort and recreation mecca, the town of Aspen in the Rocky Mountains of Colorado, USA, was originally founded in 1878 by silver miners. Rich veins of silver were found here.*

# SILVER

SILVER IS WIDELY DISTRIBUTED in nature, but in comparison with other metals it is found relatively rarely. Silver crystals are uncommon; they occur occasionally as indistinct cubes or wiry aggregates. Scaly, dendritic, and massive habits are more common. Although much of the world's silver production is a by-product of refining lead, copper, and zinc, deposits of native silver are also commercially important. As a metal for jewellery or coinage, it is harder than gold but softer than copper. Next to gold, it is the most malleable and ductile metal. The chemical symbol for silver, Ag, comes from the Latin word for silver, *argentum*, which in turn derives from a Sanskrit word meaning "white" and "shining".

## USES OF SILVER

The earliest silver ornaments and decorations have been found in tombs dating back as far as 4000BC, and silver coinage appeared not far behind that of gold, around 550BC. But it is in modern times that silver's physical properties have brought it into its own as a metal. By 1960, the demand for silver for industrial purposes exceeded the total world production. With its superior electrical and thermal conductivity, it finds use in electrical circuits, and it is alloyed with nickel or palladium for use in electrical contacts. As a catalyst, silver has a unique ability to convert ethylene to ethylene oxide, a precursor of many organic compounds. But the largest single use of silver is in the photographic industry, which uses 60 per cent of all silver production. Its use in silverware, ornaments, and jewellery continues to be important, although few countries retain silver coinage. Pure silver is too soft to wear well, so it is alloyed with other

**MEXICAN PENDANT**
*The ancient Mexicans valued silver as highly as gold for their sacred artifacts.*

**ANCIENT EGYPTIAN RING**
*This silver ring belonged to Rameses IV (1153–1147BC).*

**ROMAN BRACELET**
*This silver bracelet was made at the end of the 1st century AD. It was discovered at Great Chesters, England.*

> "Genius without **education** is like **silver** in the mine."
>
> **BENJAMIN FRANKLIN**

*Accessory quartz*

*Tarnished surface*

**COINS**
*Silver has been a companion metal to gold for coinage since ancient times. Shown here are a silver coin of English King William the Conquerer (top), a Jewish silver shekel from the Roman period (centre), and a 4th-century Persian coin with a profile of Shapur II (bottom).*

**DAGUERROTYPE**
*Invented in 1839, the daguerrotype used silver in the first photographic process.*

## SILVER IN PHOTOGRAPHY

In the predigital age, photography would have been impossible without the sensitivity of two silver halides to light – silver chloride and silver bromide. Even today, between 25 and 40 per cent of industrial silver is consumed in the production of these photosensitive chemicals. To make film, one or other of these silver halides is suspended in a layer of gelatine. When exposed to light in the camera, changes take place in the silver halide, allowing it to be reduced to silver atoms. Developing agents attack only the silver halide crystals that have had sufficient exposure to light, resulting in a negative image formed by grains of dark metallic silver. A fixing solution removes the unaffected silver halide crystals.

**WIRE SILVER**
*This example comes from the mines at Kongsberg, Buskerud, Norway, which have been worked for several centuries. Exceptional specimens of wire silver are found there. Tarnishing can be seen on the exposed surfaces.*

metals to increase its durability. Sterling silver is 92.5 per cent silver with another metal, usually copper, making up the other 7.5 per cent.

The earliest known silver mines of any size were those of the pre-Hittites of Cappadocia in Anatolia (Turkey), where it may have been mined as early as 4000BC. Today the major silver mining areas are Peru, the USA, Canada, Australia, Russia, and Kazakhstan; but the greatest single producer of silver is probably Mexico, where silver has been mined from about AD1500.

## SILVER LORE

Silver is said to relate to the Moon, which is itself related to the female principle. It is also associated with the dignity of kingship: in legend, when the Irish King Nuada lost his arm in battle and through the loss of a limb became disqualified from kingship, the god of healing, Dian Cécht, made him an artificial arm of silver, allowing him to return to the throne. In many cultures, silver is the symbol of purity, and in Christian symbolism silver stands for divine wisdom. Mystics assert that the soul is connected to the body by a silvery thread, which is sundered at death.

**MODERN MINE, MEXICO**
*Mexico is still the world's leading producer of silver. Here workers remove rock and debris from a silver mine in Taxco, Mexico.*

*Native copper*

**POLISHED SILVER AND COPPER SLICE**

**HABITS OF SILVER**
*In these two specimens silver can be seen in two habits: mixed with native copper (above), and in a finely crystallized, dendritic habit (below).*

*Metallic lustre*

**TIFFANY TEAPOT**
*This striking 1920s silver Art Deco teapot with red handle was designed by Louis Comfort Tiffany as part of a tea set.*

*Crystal growth stages visible as ridges*

**DENDRITIC SILVER**

**PROPERTIES**

| | |
|---|---|
| **GROUP** | Native elements |
| **CRYSTAL SYSTEM** | Cubic |
| **COMPOSITION** | Ag |
| **COLOUR** | Silver-white |
| **FORM/HABIT** | Cubic, octahedral, dodecahedral, wiry, arborescent |
| **HARDNESS** | $2\frac{1}{2}$–3 |
| **CLEAVAGE** | None |
| **FRACTURE** | Hackly |
| **LUSTRE** | Metallic |
| **STREAK** | Silver-white |
| **SPECIFIC GRAVITY** | 10.1–11.1 |
| **TRANSPARENCY** | Opaque |

*Dendritic crystals*

**ELECTRICAL CIRCUIT BOARD**
*Silver's conductivity of heat and electricity is second to none. It is used in printed electrical circuits as a coating for electronic conductors.*

*Quartz crystal*

**SULPHUR CRUST**
*Crusts of sulphur build up around fumaroles, where gas is venting around volcanoes.*

# SULPHUR

THE NINTH MOST ABUNDANT element in the entire universe, according to estimates, sulphur constitutes about 0.03 per cent of the Earth's crust. After oxygen and silicon, it is the most abundant constituent of minerals, in the form of sulphides, sulphates, and elemental sulphur. Elemental or native sulphur has been known and used since ancient times, when it was called brimstone. The name sulphur is derived from the Latin word for brimstone, meaning "burning stone". Crystals may exhibit as many as 56 different habits. It can also be massive, stalactitic, and reniform. Well-crystallized sulphur is usually formed as a sublimate from volcanic gases and is found encrusting volcanic vents and fumaroles. Massive sulphur is often found in thick beds in sedimentary rocks, particularly those associated with salt domes. These are common along the coastal region of Texas and Louisiana, USA. The sulphur from these deposits is recovered by the Frasch process – holes are drilled into the sulphur beds, and superheated water is injected, melting the sulphur, which is then pumped to the surface to be collected in reservoirs and transferred to vats or bins to solidify. The United States, Canada, Poland, France, Russia, Mexico, and Japan are all major producers of elemental sulphur.

**SULPHUR MINE**
*These piles of raw sulphur recovered from underground sources stand at Vancouver Harbour, Canada, ready for shipment and further processing.*

*Acicular crystals*

*Resinous lustre*

**ACICULAR CRYSTALS**
*Sulphur can form masses of fine, elongated crystals with a needle-like appearance.*

Sulphur is considered to be one of the four most important basic chemical commodities. Its principal use is in the production of sulphuric acid, which is itself a major chemical in the production of dozens of other chemical products.

Sulphur is easy to identify by its distinctive yellow colour. It is also a poor heat conductor, which means that it feels warm. It should be handled and stored with care, however. Some people have contact allergies to sulphur, and fine-grained sulphur reacts quickly with atmospheric water to form sulphuric acid, an irritant and, in high concentrations, a poison.

## INDUSTRIAL USES

Sulphur compounds are very important industrial chemicals, sulphuric acid being the most important. It is used in the manufacture of fertilizers, detergents, dyes, pigments, drugs, and explosives. Synthetic organic sulphur compounds are used in drugs and skin treatments; others are used as insecticides, solvents, and in the production of rubber and rayon.

**DYEING FABRIC**
*Sulphur compounds are widely used in the preparation of dyes.*

**SULPHUR CRYSTALS**
*Superbly formed orthorhombic crystals of sulphur up to 4cm (2in) in length adorn this specimen from Conil, Andalucía, Spain.*

*"...brimstone shall be scattered upon his habitation"*

**THE BIBLE, JOB 18**

*Resinous lustre*

*Orthorhombic crystals*

*Rock groundmass*

### PROPERTIES

| | |
|---|---|
| **GROUP** | Native elements |
| **CRYSTAL SYSTEM** | Orthorhombic |
| **COMPOSITION** | S |
| **COLOUR** | Yellow |
| **FORM/HABIT** | Bipyramidal, thick tabular |
| **HARDNESS** | $1^1/_2$–$2^1/_2$ |
| **CLEAVAGE** | Indistinct |
| **FRACTURE** | Conchoidal to uneven, brittle |
| **LUSTRE** | Resinous to greasy |
| **STREAK** | White |
| **SPECIFIC GRAVITY** | 2.1 |
| **TRANSPARENCY** | Transparent to translucent |

# ARSENIC

**SAXONY**
*St. Andreasberg and Annaberg-Buchholz in Saxony, Germany, are prime sources of arsenic.*

THE CHEMICAL ELEMENT arsenic has been known since antiquity, and derives its modern name from the Greek *arsenicon*, which was used to describe orpiment, an arsenic sulphide (see p.136). It is found widely in nature, although much less commonly in its uncombined form as native arsenic. It is classified as a semimetal – an element that possesses some of the properties of metals and some of nonmetals. On fresh surfaces, native arsenic is tin-white, but it quickly tarnishes to dark grey. Crystals are rare; it is usually massive, reniform, or stalactitic, often with concentric layers. Native arsenic is found in hydrothermal veins, generally associated with antimony, silver, cobalt, and nickel-bearing minerals. Notable localities include Fukui Prefecture, Japan; and Washington Camp, Arizona, USA.

When heated, arsenic quickly sublimates (turns into a gas without melting first) to form a vapour that smells distinctly of garlic. It is known for being highly poisonous to humans, although it is used in some medicines to treat

## PROPERTIES

| | |
|---|---|
| **GROUP** | Native elements |
| **CRYSTAL SYSTEM** | Hexagonal/trigonal |
| **COMPOSITION** | As |
| **COLOUR** | Tin-white |
| **FORM/HABIT** | Massive |
| **HARDNESS** | 3½ |
| **CLEAVAGE** | Perfect, fair |
| **FRACTURE** | Uneven, brittle |
| **LUSTRE** | Metallic or dull earthy |
| **STREAK** | Grey |
| **SPECIFIC GRAVITY** | 5.7 |
| **TRANSPARENCY** | Opaque |

## USE IN PESTICIDES

Arsenic compounds are widely used in pesticides. Arsenic acid, arsenic pentoxide, lead arsenate, and calcium arsenate are all variously used in crop sprays and as soil sterilizers. They are a considerable danger to those who regularly handle them, and leave residues on crops that are potentially hazardous to the consumer. The use of arsenic is one reason for always washing fruit and vegetables before eating them.

**CROP SPRAYING**
*Arsenic compounds are a hazard to the pilots of crop-dusters.*

infections. Elemental arsenic is used in certain alloys to increase high-temperature strength, in bronzing, in pyrotechnics, and as a herbicide and pesticide.

**TARNISHED ARSENIC**
*This massive specimen of native arsenic from St. Andreasberg, Lower Saxony, Germany, shows a darkly tarnished surface.*

*Dull, earthy lustre*

# GRAPHITE

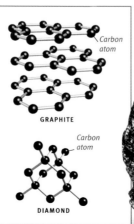

**MADAGASCAR**
*Madagascar is a significant source of the mineral graphite.*

GRAPHITE TAKES ITS NAME from the Greek verb *graphein*, meaning "to write", a reference to the black mark it leaves when rubbed against paper. It occurs as hexagonal crystals, flexible sheets, scales, or large masses, and it can be granular, compact, or earthy. Graphite is formed by the metamorphism of carbonaceous sediments, such as limestones rich in organic material, and by the reaction of carbon compounds with hydrothermal solutions. It is mined commercially in Sabaragamuwa, Sri Lanka; Sonora, Mexico; Ontario, Canada; North Korea; Madagascar; and New York, USA. Graphite is used as a lubricant and in nuclear reactors. Because it conducts electricity without melting, it is also used for arc lamps, batteries and brushes for electric motors.

## PROPERTIES

| | |
|---|---|
| **GROUP** | Native elements |
| **CRYSTAL SYSTEM** | Hexagonal |
| **COMPOSITION** | C |
| **COLOUR** | Black |
| **FORM/HABIT** | Hexagonal plates, foliated masses |
| **HARDNESS** | 1 |
| **CLEAVAGE** | Perfect basal |
| **FRACTURE** | Uneven |
| **LUSTRE** | Metallic or dull earthy |
| **STREAK** | Black to steel-grey, shiny |
| **SPECIFIC GRAVITY** | 2.2 |
| **TRANSPARENCY** | Opaque |

**GRAPHITE PENCIL**
*The familiar pencil "lead" in fact contains graphite. Graphite's first use in pencils was described in 1575.*

*Perfect cleavage*

*Metallic lustre*

**MASSIVE GRAPHITE**
*This massive specimen of graphite has a soapy or greasy feel when touched.*

## THE STRUCTURE OF GRAPHITE

Like diamond, graphite is a form of native carbon. Not only does it look dramatically different from diamond, it is at the opposite end of the hardness scale. Graphite's softness is a result of the way that the carbon atoms are bonded to each other. Graphite consists of rings of six carbon atoms arranged in widely spaced horizontal sheets. The atoms are strongly bonded within the rings, but very weakly bonded between the sheets. In diamond, the carbon atoms are strongly bonded in three dimensions, giving its hardness.

*Carbon atom*

**GRAPHITE**

*Carbon atom*

**DIAMOND**

# DIAMOND

MARIE LOUISE
DIADEM
*This diadem contains
1,006 diamonds and 79
turquoise stones, which
replaced emeralds in
the original 1810 design.*

**NAMIBIA**
*Some beach deposits in Namibia
are rich in alluvial diamonds.
Huge earth-moving operations
take place to remove the
overlying sand to reach diamond
concentrations on bedrock.*

THE HARDEST MINERAL ON EARTH, diamond
is pure carbon. Diamond crystals are usually well
formed because of the highly uniform arrangement
of their component carbon atoms, occurring as
octahedrons and cubes with rounded edges and slightly
convex faces. Its name is from the Greek *adamas*,
meaning "I take" or "I subdue", a reference to its superior
hardness. In addition to its crystalline form, diamond
occurs in two other forms: bort, or boart, is irregular or
granular black diamond; carbonado occurs as
microcrystalline masses.

Crystals may be transparent, translucent, or opaque,
and range from colourless to black, with brown and
yellow being the most common colours. Colourless
or pale blue gemstones are the most often used in
jewellery. Red and green have long been considered
the rarest colours, but pure orange and violet are
much rarer and so are more valuable. Industrial
diamonds tend to be grey or brown and are
translucent or opaque. The colour of diamonds can
be changed by artificial exposure to intense radiation or
by heat treatment; many of the "fancy" coloured stones
on the market today are the result of such treatments.

In the atomic arrangement of diamond, each
carbon atom is linked to four equidistant
neighbours, creating a close-knit, dense,
strongly bonded structure – the source
of its unsurpassed hardness and many
other properties.

### DIAMOND LOCALITIES
Most diamonds come from two rare
kinds of volcanic rocks, lamproite
and kimberlite (see p.41), but they
are much older than the rocks in
which they are found. The right

conditions for diamond to crystallize occur in the mantle
of the Earth generally more than 150km (95 miles) deep,
below ancient continental masses. Kimberlite magmas
originate particularly deep and, when they erupt,
diamonds and other fragments scoured from mantle rocks
are forced up to the Earth's surface. As the magma cools,
it forms a steeply conical pipe-shaped body. Most natural
diamonds are mined from these kimberlite pipes.

**MARIE ANTOINETTE EARRINGS**
*A gift from Louis XVI of France to
Marie Antoinette, these diamond
earrings disappeared in the
French Revolution, resurfacing
in Russia. They are now in the
Smithsonian Institution in
Washington, D.C., USA.*

*Rounded habit*

*Adamantine lustre*

**YELLOW BRILLIANT**
*The colour of most yellow
natural diamonds is
produced by traces of
nitrogen in the structure.*

**BROWN BRILLIANT**
*Natural brown diamonds
are found, but brown is
also one of the colours
produced by artificial
irradiation.*

**ORANGE BRILLIANT**
*Natural orange diamonds
are very rare, but can be
produced artificially. If the
colour is artificial it must
be declared to the buyer.*

**COLOURLESS BRILLIANT**
*The high colour
dispersion of diamond
makes it one of the
most exquisite and fiery
of gems.*

## INDUSTRIAL DIAMONDS

Although it has been replaced in large
part by synthetic diamonds, natural
diamond still finds wide use as an
industrial abrasive. Powdered
diamond is the only material that will
cut gem diamond, and it is also used
for cutting other hard gems like
sapphire. Diamond-
tipped drills are used
for boring holes in hard
materials; the diamonds are either set
in the surface of the bit or the bit is
impregnated with diamond grit
or abrasive. Diamond-edged
saws are used for slicing
through rock.

**DRILL BIT**
*Diamond abrasive
covers this oil-
drilling bit.*

**TOOL MAKER**
*A diamond-edged
industrial cutting
tool is being
produced here.*

### PROPERTIES
| | |
|---|---|
| **GROUP** | Native elements |
| **CRYSTAL SYSTEM** | Cubic |
| **COMPOSITION** | C |
| **COLOUR** | White to black, colourless, yellow, pink, red, blue, brown |
| **FORM/HABIT** | Octahedral, cubic |
| **HARDNESS** | 10 |
| **CLEAVAGE** | Perfect octahedral |
| **FRACTURE** | Conchoidal |
| **LUSTRE** | Adamantine |
| **STREAK** | Will scratch streak plate |
| **SPECIFIC GRAVITY** | 3.4–3.5 |
| **TRANSPARENCY** | Transparent to opaque |
| **R.I.** | 2.42 |

*Gem-quality
octahedron*

**DIAMOND CRYSTALS**
*This group of diamond crystals shows the
range of diamond forms, from transparent
and perfectly formed octahedrons to
the dark, irregular, grainy clumps of
industrial-grade diamond known as bort.*

*Octahedral
crystal*

Diamonds are also found in alluvial gravels and glacial tills as placer deposits. These have been eroded from a kimberlite matrix and redeposited. Because they are hard and impervious to chemical weathering, once released, they can be carried considerable distances by water to settle in river beds and oceans. Since they are denser than typical silicate rock, they concentrate in stream gravels and beach sands, where they can be mined.

India was the earliest source of diamonds, the Golconda field having been worked for centuries – possibly as early as 800BC. World production shifted to Brazil after the Spanish Conquest, with the Brazilian fields still productive to a minor extent. Today the major gem diamond reserves are in Botswana, Australia, Russia, Congo, and Angola. Many of the South African diamond mines are worked out, and South Africa is no longer one of the top diamond-producing nations.

Australia, Congo, and Russia are the main producers of industrial diamonds. Canada has several diamond mines in kimberlite pipes still to come to full production; once this happens, Canada will be producing at least 15–20 per cent of the total world diamond output. Small quantities have also been found in the USA. Synthetic diamonds have been produced commercially since 1960.

### NAPOLEON I NECKLACE
*A gift to Empress Marie-Louise from Emperor Napoleon I in 1811 to celebrate the birth of their son, this superb necklace is constructed of old mine-cut diamonds weighing a total of about 263 carats.*

### KOH-I-NOOR DIAMOND
*Originating in the Indian mines in 1304, the Koh-I-Noor came to the British when they annexed the Punjab in 1849. The poorly cut 191-carat stone was recut in 1852 to an oval brilliant, weighing 109 carats. The Koh-I-Noor is now set in the British Queen Mother's crown and is on display in the Tower of London.*

*The Koh-I-Noor diamond*

**QUEEN MOTHER'S CROWN**

### DIAMOND RINGS
*Small, brilliant-cut diamonds are set all the way round this platinum Victorian eternity ring (left). A single brilliant-cut diamond set in gold is the focus of the elegant 1920s' ring (right).*

Modified octahedral habit

Modified octahedron

Twinned crystal

Industrial-grade diamond bort

# DIAMOND MINING

Diamonds are found in kimberlite and lamproite pipes and as placer deposits. Kimberlite rock is named after Kimberley in South Africa, the centre of the South African diamond-mining industry. At first, diamonds were extracted by panning from weathered kimberlite in open pits, but as the pits became deeper and the rock harder, underground mining became necessary (see p.41). Today drill-holes are packed with explosives and blasted to break up the rock. It is brought to the surface and the diamonds separated out using grease tables (see p.107).

### BULTFONTEIN, SOUTH AFRICA
*The mine at Bultfontein is one of several around Kimberley, South Africa. Its high-quality diamonds are extracted from kimberlite pipes.*

## MINING PLACERS

Placer mining uses versions of a mining technique that mimics the creation of the placer in the first place: the separation of heavier minerals – in this case diamond – in running water. On an industrial scale, diamond-bearing gravel is run through a trough of flowing water with baffles on the bottom to retain the heavier material, which sinks. The lighter material is washed away, and the baffles are periodically emptied of their contents: diamonds and other heavy minerals. The concentrate is then hand-sorted to remove the diamonds. Some of the most productive diamond placers occur where diamond-bearing gravels have been dumped into the sea at the mouths of rivers, and then redistributed by wave action along the beaches (see p.106). Sea-floor placers are formed in a similar manner to beach placers, only by ocean currents instead of waves.

### TRAWLING FOR DIAMONDS
*Floating dredges suck up sea-floor sediments using giant vacuum hoses, and the diamonds are then recovered as with other placer mining. Tens of thousands of carats of high-quality diamond are recovered annually by this method.*

# HOPE DIAMOND

## WEIGHING 45.52 CARATS, THE HOPE DIAMOND IS AN EXCEEDINGLY RARE GEMSTONE. IT ALSO HAS A FASCINATING AND MYSTERIOUS HISTORY.

The Hope diamond was probably discovered at the Kollur Mine in the Golconda area of India. Precisely when it was found is not known, but a French gem merchant, Jean-Baptiste Tavernier, sold a 112-carat blue diamond from India to King Louis XIV of France in 1668. It was recut from its original Indian-style cut to a 67-carat, more brilliant heart shape in 1673. In 1749, Louis XV had the stone – now known as the French Blue – set into a piece of ceremonial jewellery for the Order of the Golden Fleece, a decoration worn only by the King.

During the French Revolution in 1792, the Hope diamond disappeared. Then, in 1812, a London jeweller, John Francillon, noted a 44-carat (45.5 modern metric carats) blue diamond in the possession of London diamond merchant Daniel Eliason. This diamond was probably cut from the French Blue.

**MARIE ANTOINETTE**
In this portrait of Marie Antoinette (1750-93), the wife of Louis XVI may be wearing the French Blue on her bust.

### FROM FRENCH BLUE TO HOPE

The diamond again mysteriously disappeared until 1820 when it was bought by the British King George IV. In the King's possession at the time of his death in 1830, it was purchased by a London banker and gem collector, Henry Philip Hope, whose name it bears today. After Hope's death in 1839, the diamond passed through several hands, until bought in 1909 by Pierre Cartier. He in turn sold it to Evalyn Walsh McLean in 1910 for $180,000. He told her that the diamond brought bad luck to anyone who wore it, but this was probably because she had told him that she felt objects that were bad luck for others were good luck for her. Mrs McLean wore the Hope diamond everywhere she went. Two years after her death in 1947, her entire collection of jewellery was purchased by New York jeweller Harry Winston. On 10 November, 1958, he presented the diamond to the Smithsonian Institution as the foundation of a National Gem Collection. It arrived in a plain brown package by registered mail, insured for $1 million. The diamond has remained on continuous exhibition there ever since.

White cushion-cut diamond

Antique brilliant-cut blue diamond

**FRONT AND BACK**
The Hope diamond measures 25.6mm (1in) long by 21.8mm (⁴/₅in) wide. An asymmetrical cushion antique brilliant cut, it is a fancy diamond with 58 facets plus two extra facets on the pavilion and additional facets on the girdle.

**OLD DIAMOND IN A NEW SETTING**
The Hope diamond's present-day setting was designed by Pierre Cartier in 1920. The gem is in a platinum setting surrounded by 16 pear-shaped and cushion-cut diamonds, and suspended from a chain of 45 diamonds.

# SULPHIDES

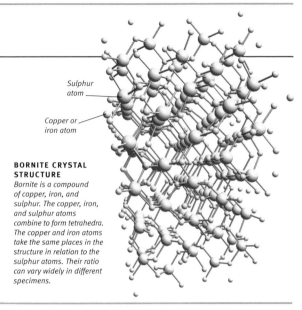

SULPHIDE MINERALS ARE THOSE in which sulphur is combined with one or more metals. They tend to have simple structures, highly symmetrical crystal forms, and many of the properties of metals, including metallic lustre and electrical conductivity. Many of the sulphides are brilliantly coloured, and most have low hardness and high specific gravity. They are the ore minerals of most metals used by industry, such as antimony, bismuth, and copper, and are also an important source of gold, silver, and platinum. Sulphides occur in all rock types, and most are quite common. Related mineral families are the selenides, tellurides, antimonides, and arsenides. In these families, selenium, tellurium, antimony, and arsenic respectively combine with metals or semimetals. These mineral families tend to occur in association with sulphide minerals, and most are relatively rare.

Sulphur atom

Copper or iron atom

**BORNITE CRYSTAL STRUCTURE**
Bornite is a compound of copper, iron, and sulphur. The copper, iron, and sulphur atoms combine to form tetrahedra. The copper and iron atoms take the same places in the structure in relation to the sulphur atoms. Their ratio can vary widely in different specimens.

# ACANTHITE

**ZACATECAS**
Acanthite is found in this Mexican state.

ACANTHITE TAKES ITS NAME from the Greek *akantha*, meaning "thorn", a reference to the spiky appearance of some of its crystals. It is a form of silver sulphide. Above 177°C (350°F), silver sulphide crystallizes in the cubic system, and it used to be assumed that cubic silver sulphide – known as argentite – was a separate mineral from acanthite. It is now known that they are the same mineral with acanthite crystallizing in the monoclinic system at lower temperatures. Acanthite is the most important ore of silver. It forms in hydrothermal veins with native silver, pyrargyrite, proustite, and other sulphides such as galena. It also forms as a secondary alteration product of primary silver sulphides. It occurs in most silver deposits, including the Harz Mountains, Germany; and Kongsberg, Norway.

## COMSTOCK LODE

The Comstock Lode was an immensely rich silver deposit that was discovered in Nevada, USA, in June, 1859. It was so rich that a branch of the US mint was established at nearby Carson City to coin its output. However, by 1882, the deposit was exhausted, and its pit flooded. The lode was formed by a shallow type of hydrothermal deposit, known as an epithermal deposit. As well as acanthite, other important silver-bearing minerals, stephanite and pyrargyrite, were found there.

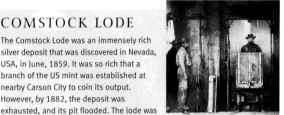

**COMSTOCK MINE**
The mine was named for Henry Comstock, part-owner of the property on which it was discovered.

**SILVER FEATHER**
This modern Native American piece features a silver feather with turquoise beads.

**SILVER BUCKLE**
This silver belt buckle has a central turquoise surrounded by a design stamped in the silver.

**PSEUDOMORPH**
This specimen is an argentite pseudomorph with the outward form of cubic symmetry.

Metallic lustre

Uneven fracture

Darkened, weathered surface

Pseudomorphic cubic form

## PROPERTIES

| | |
|---|---|
| **GROUP** | Sulphides |
| **CRYSTAL SYSTEM** | Monoclinic |
| **COMPOSITION** | $Ag_2S$ |
| **COLOUR** | Black |
| **FORM/HABIT** | Pseudocubic |
| **HARDNESS** | 2–2½ |
| **CLEAVAGE** | Indistinct |
| **FRACTURE** | Subconchoidal, sectile |
| **LUSTRE** | Metallic |
| **STREAK** | Black |
| **SPECIFIC GRAVITY** | 7.2–7.4 |
| **TRANSPARENCY** | Opaque |

# CHALCOCITE

ONE OF THE MOST important ores of copper, chalcocite is usually massive but, on rare occasions, it occurs in short prismatic or tabular crystals, or as pseudohexagonal prisms formed by twinning. The name chalcocite is derived from the Greek word for copper. Old names for chalcocite are chalcosine, copper glace, and redruthite (an allusion to its occurrence at Redruth, Cornwall, England), but all are now obsolete.

Chalcocite belongs to a group of sulphide minerals formed at relatively low temperatures, often as alteration products of other copper minerals such as bornite (see p.128). Concentrated in secondary alteration zones, it can yield more copper than the element's primary deposits. These alteration zones are often hydrothermal veins with minerals such as bornite, quartz, calcite, covellite, chalcopyrite, galena, and sphalerite in addition to chalcocite.

### INDUSTRIAL USES OF COPPER

Most copper is used for electrical applications. The windings of massive electrical generators are made of copper, as are the millions of miles of transmission lines. Most telephone calls worldwide still go through copper wires. Massive copper-alloy propellers drive ships, and aircraft are made from duralumin, an aluminium-copper alloy. Copper is also used in smaller but important applications such as roofing and cladding, brewer's vats, water and gas valves, copper-alloy fittings for water and gas pipes, and often for the pipes themselves. It would not be an exaggeration to say that the modern industrial age would be virtually impossible without copper. Copper also has many domestic uses (see p.128).

**COPPER WIRE**
*Most copper goes for electrical applications.*

**COPPER ROOFING**
*Copper's malleability allows it to be used for curves.*

**SMELTING COPPER**
*Sparks fly in a foundry in Krasnoyarsk, Russia, during the smelting of copper.*

**CORNWALL, ENGLAND**
*Cornwall has been a storehouse of minerals which have been mined for three millennia. Excellent chalcocite crystals were extracted and used here.*

## PROPERTIES

| | |
|---|---|
| **GROUP** | Sulphides |
| **CRYSTAL SYSTEM** | Monoclinic |
| **COMPOSITION** | $Cu_2S$ |
| **COLOUR** | Blackish lead-grey |
| **FORM/HABIT** | Short prismatic, thick tabular |
| **HARDNESS** | $2\frac{1}{2}$–3 |
| **CLEAVAGE** | Indistinct |
| **FRACTURE** | Conchoidal |
| **LUSTRE** | Metallic |
| **STREAK** | Blackish lead-grey |
| **SPECIFIC GRAVITY** | 5.5–5.8 |
| **TRANSPARENCY** | Opaque |

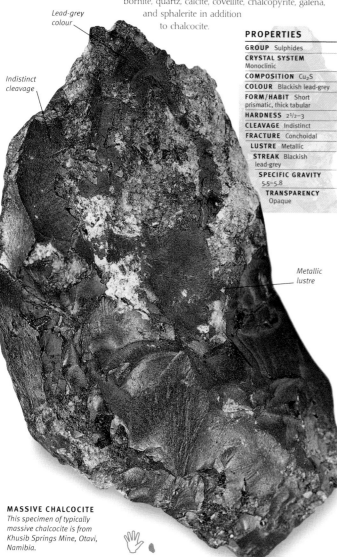

*Lead-grey colour*

*Indistinct cleavage*

*Metallic lustre*

**MASSIVE CHALCOCITE**
*This specimen of typically massive chalcocite is from Khusib Springs Mine, Otavi, Namibia.*

Chalcocite itself alters to native copper and other copper ores. Superbly crystallized specimens have come from the Ural Mountains of Russia, and Cornwall, England. The Cornish deposits have been worked since the Bronze Age. Valuable ore deposits occur in the USA at Ely, Nevada, Morenci, Arizona, and Butte, Montana; and excellent crystals are found in Bristol, Connecticut. Chalcocite is also found in Australia, Chile, the Czech Republic, Norway, Peru, Russia, and Spain.

*Prismatic crystal*

*Metallic lustre*

*Pseudohexagonal crystals*

*Dolomite groundmass*

**LONG PRISMATIC CRYSTALS**

**SHORT PRISMATIC CRYSTALS**

**CHALCOCITE CRYSTALS**
*These specimens show two forms of relatively rare chalcocite crystals.*

**TASMANIA**
*The large island of Tasmania, which lies off the south-eastern coast of Australia, is a source of numerous minerals, including bornite.*

## PROPERTIES

| | |
|---|---|
| **GROUP** | Sulphides |
| **CRYSTAL SYSTEM** | Tetragonal |
| **COMPOSITION** | $Cu_5FeS_4$ |
| **COLOUR** | Coppery red, brown |
| **FORM/HABIT** | Usually massive |
| **HARDNESS** | 3 |
| **CLEAVAGE** | Poor |
| **FRACTURE** | Uneven to conchoidal, brittle |
| **LUSTRE** | Metallic |
| **STREAK** | Pale greyish-black |
| **SPECIFIC GRAVITY** | 5.1 |
| **TRANSPARENCY** | Opaque |

# BORNITE

ONE OF NATURE'S MOST COLOURFUL minerals, bornite is a copper iron sulphide, and a major ore of copper. It can show iridescent purple, blue, and red splashes of colour on broken, tarnished faces, resulting in its common name, "peacock ore". Its formal mineral name comes from the Austrian mineralogist Ignaz von Born (1742–1791). Other informal names are "purple copper ore" and "variegated copper ore".

Bornite crystals are relatively uncommon, but when found they are pseudocubic, dodecahedral, or octahedral, frequently with curved or rough faces, although bornite is more often compact, granular, or massive. Its natural colour can be coppery red, coppery brown, or bronze. It alters readily upon weathering to chalcocite (see p.127) and other copper minerals. Bornite forms principally in hydrothermal veins, with minerals such as quartz, chalcopyrite, marcasite, and pyrite. It also forms in some silica-poor intrusives, pegmatite veins, and contact metamorphic zones. Major deposits of bornite occur in Tasmania, Chile, Peru, Kazakhstan, Canada, and in Arizona and at Butte, Montana, USA. Good-quality crystals are found in deposits in England.

**BORNITE CRYSTALS**
*This group of well-developed bornite crystals shows both curved and rough crystal faces.*

*Iridescent surface*

*Massive habit*

*Purple oxidation*

## DOMESTIC USES OF COPPER

Copper and its alloys surround us in our everyday lives. It is highly durable, can be easily formed by casting, drawing, pressing, and forging, and is an unsurpassed conductor of heat and electricity. At home, our electricity flows along copper wires, and water flows through copper pipes and out through copper-alloy taps.

**RADIATOR COIL**
*Copper is an unseen component of many domestic appliances.*

Water heaters have coils made from copper and most are copper lined. Most domestic appliances have copper somewhere in their construction. Pots and pans made entirely of copper or with a copper base provide even heating. Many of the coins in our pockets are made of copper or copper alloy. Gold jewellery usually has copper alloyed with the gold to give it durability. Even getting into your house relies on copper – most keys and locks are made of copper alloys. Copper is also widely used in industry (see p.127).

**BRACELET**
*Copper bracelets are worn by many to relieve the pain of arthritis.*

> "Gold is for the mistress – silver for the maid – **Copper** for the craftsman, cunning at his **trade**"
>
> **RUDYARD KIPLING,**
> "COLD IRON"

**MASSIVE BORNITE**
*This specimen of massive, tarnished bornite with yellow chalcopyrite, found in Russia, shows the rich oxidation colours that give it the names "purple copper ore" and "peacock ore".*

*Uneven fracture*

# SPHALERITE

SPHALERITE TAKES ITS NAME from the Greek *sphaleros*, meaning "treacherous", referring to its tendency to occur in a number of forms, which can be mistaken for other minerals, especially galena (see p.130). Red or reddish-brown transparent crystals are sometimes called ruby zinc or ruby blende, and opaque black crystals have sometimes been referred to as black jack. Although these are not formal mineral names, they can still be seen on labels in old collections. Sphalerite is also called blende from the German word for "blind", a name alluding to the fact that sphalerite does not yield lead.

Well-crystallized specimens are common in most zinc deposits. Iron may also be present in sphalerite, and can make up a quarter of its metallic content. Sphalerite is the most common zinc mineral and the principal ore of zinc; it is found associated with galena in most important lead-zinc deposits. It occurs in hydrothermal vein deposits, contact metamorphic zones, and high-temperature replacement deposits, where it is also associated with chalcopyrite and pyrite. Sphalerite is also found in minor amounts in meteorites and lunar rock. There are important deposits in the Mississippi River Valley, USA, and in Canada, Mexico, Spain, and Russia.

Sphalerite is cut only as an unusual collectors' stone since it is notoriously difficult to cut into faceted gems because of its softness and tendency to cleave.

**MISSISSIPPI RIVER VALLEY**
*One of the most abundant deposits of sphalerite is found where the US States of Missouri, Arkansas, and Kansas meet.*

## PROPERTIES

| | |
|---|---|
| **GROUP** | Sulphides |
| **CRYSTAL SYSTEM** | Cubic |
| **COMPOSITION** | ZnS |
| **COLOUR** | Brown, black, yellow |
| **FORM/HABIT** | Tetrahedral, dodecahedral |
| **HARDNESS** | 3–4 |
| **CLEAVAGE** | Perfect in six directions |
| **FRACTURE** | Conchoidal |
| **LUSTRE** | Resinous to adamantine, metallic |
| **STREAK** | Brownish to light yellow |
| **SPECIFIC GRAVITY** | 3.9–4.1 |
| **TRANSPARENCY** | Opaque to transparent |
| **R.I.** | 2.36–2.37 |

*Quartz*
*Sphalerite crystal*

*Resinous lustre*

**MASSIVE SPHALERITE**

*Sphalerite crystals*

*Sphalerite crystals*

*Pale dolomite and rock groundmass*

*Resinous lustre*

**CRYSTALLINE SPHALERITE**

**CRYSTALS IN MATRIX**

**SPHALERITE CRYSTALS**
*These superbly formed sphalerite crystals, with pyrite and quartz, are from Casapalca, Lima, Peru.*

*Pyrite*
*Quartz*

**SPHALERITE VARIETIES**
*Sphalerite is found in a variety of different habits and colours, including a red "ruby blende" (right).*

*Dark red crystals*

**RUBY BLENDE SPHALERITE**

## USES OF ZINC

Sphalerite is the principal ore of zinc. Zinc is used mainly for galvanizing iron and steel, in alloy with copper to make brass, and in alloys to make die casts (metal moulds). Zinc oxide is used in the production of phosphors for TV tubes and fluorescent lamps, in cosmetics, plastics, paints, and printing inks, and as a catalyst in the manufacture of synthetic rubber.

**DIE PRODUCTION**
*Zinc is a major metal used in die-cast components of numerous items.*

**BRILLIANT**
*This yellow-brown sphalerite has been faceted into a round brilliant cut. Its high fire shows rainbow colours.*

**OVAL**
*This oval cut shows off the golden-brown colour of sphalerite. Such stones are cut for collectors.*

**EMERALD**
*Faceting this yellow-green sphalerite in an emerald cut is a supreme test of lapidary skill.*

# GALENA

THERE ARE MORE THAN 60 known minerals that contain lead, but by far its most important ore is galena, or lead sulphide. It is possible that galena was the first ore to be smelted to release its metal: lead beads found in Turkey have been dated to around 6500BC. Smelting required nothing more complex than heating the galena in the embers of a campfire, and retrieving the metal from beneath it when it cooled.

In addition to lead, galena often contains small amounts of silver, zinc, copper, cadmium, arsenic, antimony, and bismuth. Galena can contain so much silver that it is the principal ore of that metal in a particular locality, and it can also be a source of the other metals it may contain. The Romans efficiently separated the silver from smelted lead – some Roman ingots bear the inscription *ex arg*, indicating that they are made of lead from which the silver has been removed. Galena's name comes from the Latin, and means "lead ore", or "dross from melted lead".

Galena forms cubic crystals, predominantly cubes and cubo-octahedrons. Crystals exceeding 2.5cm (1in) are common. It weathers easily to form secondary lead minerals, such as cerussite, anglesite, and pyromorphite. Banded nodules of anglesite and

**GALENA MINE, IDAHO**
*The Galena Mine is one of several in the Coeur d'Alene area in Shoshone County, Idaho, USA. Lead, silver, zinc, and gold are all mined here.*

## PROPERTIES

| | |
|---|---|
| **GROUP** | Sulphides |
| **CRYSTAL SYSTEM** | Cubic |
| **COMPOSITION** | PbS |
| **COLOUR** | Lead-grey |
| **FORM/HABIT** | Cubes, cubo-octahedrons |
| **HARDNESS** | 2½ |
| **CLEAVAGE** | Perfect |
| **FRACTURE** | Subconchoidal |
| **LUSTRE** | Metallic |
| **STREAK** | Lead-grey |
| **SPECIFIC GRAVITY** | 7.6 |
| **TRANSPARENCY** | Opaque |

## USES OF LEAD

Mystic associations are connected to lead rather than galena: the metal was associated with Saturn in alchemical practice, and efforts to transmute lead into gold were as much about symbolic personal transformation from the base into the pure as creating one metal from the other – although no one would have been disappointed if that had actually occurred! Lead was widely used in ancient times for everything from water pipes to cooking pots. In some localities such as the hot springs in Bath, England, Roman lead pipes are still in place and carry water exactly as they did 1,600 years ago.

**ANCIENT USES**
*This Roman ingot (top) and Papal bulla (document seal) are made of lead.*

**MODERN USES**
*Now supplanted by electronic technology, lead was until recently used to make metal type (above). It is still used in car batteries.*

cerussite with a galena core are frequently found. Galena occurs in many different types of deposits: in hydrothermal veins, which contain the highest silver values; as replacements of limestone or dolomite; and in deposits of contact-metamorphic origin. In the extensive Mississippi River valley deposits of the USA, where 90 per cent of US production is mined, galena occurs in brecciated (fractured) zones in limestone and chert. It occasionally occurs as a replacement of organic matter, and sometimes appears in coal beds. Commonly it is found in association with sphalerite, pyrite, and marcasite. Its occurrences are widespread, but major deposits occur in Canada, Mexico, Germany, England, Serbia, Italy, Russia, Australia, and Peru.

*Accessory dolomite*

*Cubic crystal cut by octahedral faces*

*Metallic lustre*

**GALENA CRYSTALS**
*Galena is usually found in cube-shaped crystals, but the crystal shape sometimes also incorporates the faces of octahedrons, as in this specimen.*

*Metallic lustre*

**CUBIC CRYSTALS**

# PENTLANDITE

**KOLA PENINSULA**
*This part of Russia is a major commercial source of pentlandite.*

NAMED IN 1856 for the Irish scientist Joseph Pentland, its discoverer, pentlandite is a nickel and iron sulphide. It principally has a massive or granular habit, and its crystals cannot be seen by the naked eye. Pentlandite occurs in silica-poor, intrusive igneous rocks, and has also been found in meteorites. It is almost always accompanied by pyrrhotite, and commonly with other sulphides such as chalcopyrite and pyrite. Pentlandite is the chief source of nickel. It is relatively widespread, but commercial deposits are scarce. The major ones are at Sudbury, Ontario, Canada; Lillehammer and Bodø, Norway; the Kola Peninsula of Russia; the Bushveld Complex of South Africa; and Stillwater, Montana, USA.

## PROPERTIES

| | |
|---|---|
| GROUP | Sulphides |
| CRYSTAL SYSTEM | Cubic |
| COMPOSITION | $(Ni, Fe)_9S_8$ |
| COLOUR | Bronze-yellow |
| FORM/HABIT | Massive |
| HARDNESS | $3^{1}/_2$– 4 |
| CLEAVAGE | None |
| FRACTURE | Conchoidal |
| LUSTRE | Metallic |
| STREAK | Bronze-brown |
| SPECIFIC GRAVITY | 4.6–5.0 |
| TRANSPARENCY | Opaque |

**CANADIAN NICKEL**
*This monument celebrates the Canadian five-cent piece, made from nickel.*

**NICKEL SMELTING**
*Nickel ores require extensive refining and smelting to release the metal. This smelting plant is in Russia.*

**MASSIVE PENTLANDITE**
*This typically massive specimen of pentlandite also contains pyrrhotite.*

*Massive habit*

*Uneven fracture*

**HOT NICKEL**
*Molten nickel pours into ingot moulds at this nickel smelter.*

# COVELLITE

**BUTTE, USA**
*This mineral-rich area of Montana is a key source of covellite.*

COVELLITE IS COPPER SULPHIDE, named in 1832 for the Italian Nicolas Covelli, who first described the mineral. It is generally massive and foliated in habit. When crystals form they occur as thin, tabular, hexagonal plates, which, when thin enough, are flexible. Its colour is indigo-blue, often tinged with purple iridescence. Covellite fuses very easily when heated, producing a blue-coloured flame. It typically occurs as an alteration product of other copper sulphide minerals such as chalcopyrite, chalcocite, and bornite, and is a primary mineral in some places. It rarely occurs as a volcanic sublimate, as on Mount Vesuvius, where Nicolas Covelli collected it. Occurrences include Dillenburg, Germany; Bor, Serbia; Moonta, South Australia; Alghero, Sardinia; Butte, Montana, USA; and at numerous copper mines in Arizona, USA. Covellite is a minor ore of copper in some localities.

## PROPERTIES

| | |
|---|---|
| GROUP | Sulphides |
| CRYSTAL SYSTEM | Hexagonal |
| COMPOSITION | CuS |
| COLOUR | Indigo-blue to black |
| FORM/HABIT | Foliated |
| HARDNESS | $1^{1}/_2$–2 |
| CLEAVAGE | Perfect basal |
| FRACTURE | Uneven |
| LUSTRE | Submetallic to resinous |
| STREAK | Lead-grey to black, shiny |
| SPECIFIC GRAVITY | 4.6–4.7 |
| TRANSPARENCY | Opaque |

*Iridescence*

*Tabular crystals*

*Oxidized material*

**IRIDESCENT COVELLITE**
*This spectacular, massive covellite specimen showing purple iridescence is from the Leonard Mine at Butte, Montana, USA.*

**TABULAR COVELLITE CRYSTALS**

*Foliated habit*

**MOUNT VESUVIUS**
*Covellite was first collected and identified at Mount Vesuvius, near Naples, Italy.*

# GREENOCKITE

NAMED IN 1840 for Lord Greenock, the British army officer who discovered it in Scotland, greenockite is a rare cadmium sulphide. It forms single-ended pyramidal crystals, and can also be prismatic and tabular, or occur as earthy coatings. Greenockite is an alteration product of cadmium-bearing minerals, and forms coatings on sphalerite and other zinc minerals, which often contain significant amounts of cadmium. It is one of two minerals that contain a substantial amount of cadmium; the other is hawleyite, and the two can be hard to distinguish from one another. Greenockite is commonly found associated with calcite, pyrite, quartz, prehnite, chalcopyrite, and wavellite. Localities include Renfrew, Scotland; Příbram, the Czech Republic; and New Jersey and Missouri, USA.

## USES OF CADMIUM

Cadmium is primarily used for plating steel and other easily corroded metals. It also finds use in special solders because of its low melting point. It is alloyed with nickel to make nickel-cadmium (Ni-Cad) rechargeable batteries. Cadmium efficiently absorbs thermal neutrons, making it ideal for use in the control rods for certain nuclear reactors.

**SOLDERING**
*Special solders are used in the production of electrical circuits.*

*Greenockite coating*

**EARTHY GREENOCKITE**
*On this specimen, a coating of yellow-orange greenockite can be seen on a rock groundmass.*

*Conchoidal fracture*

*Rock groundmass*

### PROPERTIES

| | |
|---|---|
| **GROUP** | Sulphides |
| **CRYSTAL SYSTEM** | Hexagonal |
| **COMPOSITION** | CdS |
| **COLOUR** | Yellow to orange |
| **FORM/HABIT** | Pyramidal |
| **HARDNESS** | $3–3\frac{1}{2}$ |
| **CLEAVAGE** | Distinct, imperfect |
| **FRACTURE** | Conchoidal |
| **LUSTRE** | Adamantine to resinous |
| **STREAK** | Yellow, orange, brick-red |
| **SPECIFIC GRAVITY** | 4.8–4.9 |
| **TRANSPARENCY** | Nearly opaque to translucent |

# PYRRHOTITE

**BUTTE, USA**
*This part of Montana is a well-known pyrrhotite locality.*

THE MOST COMMON MAGNETIC mineral after magnetite, pyrrhotite is an iron sulphide in which the ratio of iron to sulphur atoms is variable but is usually slightly less than one. Its magnetism varies inversely with its iron content and serves to distinguish it from other brassy-looking sulphide minerals.
Pyrrhotite generally has a massive or granular habit, but forms fine pseudohexagonal crystals at a number of localities. Widespread, pyrrhotite is found in silica-poor igneous rocks, pegmatites, high-temperature hydrothermal veins, and in some metamorphic rocks. It is often found as a product of magmatic separation – the settling of heavy minerals to the bottom of a crystallizing magma. It is commonly found with pentlandite, pyrite, and quartz. Localities include Dalnegorsk, Russia; Herja Mine, Romania; St. Andreasberg, Germany; Kambalda, Western Australia; Trentino, Italy; Sudbury, Ontario, Canada; many mines in Japan; and New York, Franklin, New Jersey, and Pennsylvania, USA.

### PROPERTIES

| | |
|---|---|
| **GROUP** | Sulphides |
| **CRYSTAL SYSTEM** | Monoclinic |
| **COMPOSITION** | $Fe_{1-x}S$ |
| **COLOUR** | Bronze-yellow |
| **FORM/HABIT** | Massive |
| **HARDNESS** | $3\frac{1}{2}–4\frac{1}{2}$ |
| **CLEAVAGE** | None |
| **FRACTURE** | Subconchoidal to uneven |
| **LUSTRE** | Metallic |
| **STREAK** | Dark grey-black |
| **SPECIFIC GRAVITY** | 4.6–4.7 |
| **TRANSPARENCY** | Opaque |

*Tabular crystal*

**TABULAR**
*This specimen of pyrrhotite is composed of tabular yellow crystals stacked one on top of the other.*

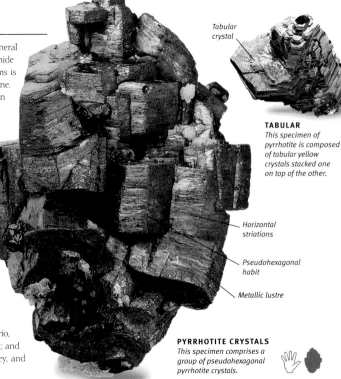

*Horizontal striations*

*Pseudohexagonal habit*

*Metallic lustre*

**PYRRHOTITE CRYSTALS**
*This specimen comprises a group of pseudohexagonal pyrrhotite crystals.*

**HUNAN PROVINCE, CHINA**
*The finest specimens of cinnabar are found in Hunan Province in south-central China.*

# CINNABAR

THE MAJOR SOURCE OF MERCURY, cinnabar is mercury sulphide. Crystals are rare; it usually forms massive or granular aggregates. Widely distributed, cinnabar commonly occurs associated with pyrite, marcasite, and stibnite in veins near recent volcanic rocks, and is also found in deposits around hot springs. The mineral's name is derived from the Persian *zinjirfrah*, and the Arabic *zinjafr*, which both mean "dragon's blood". The mining and use of cinnabar is implied, but not authenticated, in the early 2nd millennium BC in Egypt.

Cinnabar has been mined for at least 2,000 years at Almadén in Spain, which still produces excellent crystals, making it one of the oldest, if not the oldest, continuously mined deposits in the world. It is still the world's most important deposit. Smaller amounts come from Peru, Italy, Slovenia, Uzbekistan, and California, USA. It used to be widely used as the source of the pigment vermilion, and was highly valued by the Olmecs of ancient Mexico. Ancient Peruvian silversmiths used cinnabar in place of enamel for inlay. In recent times, mercury was used in the mining industry. It dissolves numerous metals to form compounds called amalgams. The amalgam is recovered in the refining process, and boiled away to recover the metal it has dissolved. This process was used in gold mining in the 19th and early 20th century, and is in large part responsible for mercury pollution in old mining areas.

**LACQUER ORNAMENT**
*This Japanese lacquer Manju netsuke (carved ornament) is from the Meiji period (1867– 1912). Cinnabar was used as a pigment to colour the foliate scrolling decoration.*

**CINNABAR VASE**
*This Chinese cinnabar porcelain baluster vase is decorated with figures in a landscape.*

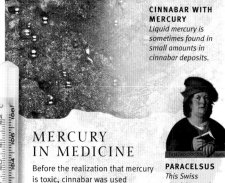

**CINNABAR WITH MERCURY**
*Liquid mercury is sometimes found in small amounts in cinnabar deposits.*

## MERCURY IN MEDICINE

Before the realization that mercury is toxic, cinnabar was used medicinally – paradoxically as a drug of immortality. It was both rubbed on the skin and taken internally. These practices occurred throughout China, India, and Europe, where it was recommended by Paracelsus. The separation of mercury from cinnabar was an alchemical operation that was regarded as symbolizing rebirth. The term "mercurial", meaning a tendency to rapid, extreme mood changes, may derive from observation of those with mercury poisoning.

**PARACELSUS**
*This Swiss physician and alchemist lived from 1493 to 1541.*

**THERMOMETER**
*Mercury expands and contracts with small changes in temperature, making it ideal for thermometers.*

## PROPERTIES

| | |
|---|---|
| **GROUP** | Sulphides |
| **CRYSTAL SYSTEM** | Hexagonal/trigonal |
| **COMPOSITION** | HgS |
| **COLOUR** | Cochineal-red |
| **FORM/HABIT** | Trigonal |
| **HARDNESS** | 2–2½ |
| **CLEAVAGE** | Perfect |
| **FRACTURE** | Subconchoidal to uneven |
| **LUSTRE** | Adamantine to dull |
| **STREAK** | Scarlet |
| **SPECIFIC GRAVITY** | 8.0 |
| **TRANSPARENCY** | Transparent to opaque |

**MASSIVE CINNABAR**
*This massive specimen of cinnabar from Monte Amiata, Tuscany, Italy, is unusual in that it also contains some crystals.*

*Crystalline cinnabar*

*Calcite*

*Rock groundmass*

**VERMILION**
*Powdered cinnabar has been the source of the pigment vermilion since ancient times, but tends to darken as it ages. It has now been replaced by a synthetic (and less toxic) substitute.*

133

# REALGAR

**FREIBERG**
*These outcrops in Freiberg, Germany, contain fine specimens of realgar.*

REALGAR TAKES ITS NAME from the Arabic *rahj al ghar*, "powder of the mine". Crystals are not commonly found, but when they occur they are short, prismatic, and striated. Realgar is found more frequently in coarse to fine granular masses, and as encrustations. Its bright colour is characteristic of the mineral. Specimens of realgar disintegrate on prolonged exposure to light, forming an opaque yellow powder, which is principally orpiment or pararealgar (see p.136). Specimens should be kept in darkened containers and exposed to light for only short periods.

An important ore of arsenic, realgar is typically found in low-temperature hydrothermal deposits, frequently associated with orpiment and other arsenic minerals. It also forms as a sublimate around volcanoes, around hot spring and geyser deposits, and as a weathering product of other arsenic-bearing minerals. It is often found with stibnite and calcite. Significant occurrences are at Freiberg, Saxony, Germany; Zacatecas, Mexico; Corsica; Cavnic and Sacaramb, Romania; Utah, Nevada, and Washington, USA.

**PIGMENT**
*Powdered realgar was once used as a red pigment, and to colour fireworks.*

## USES OF REALGAR

Realgar has been known since ancient times – the Roman historian Pliny the Elder described it in his encyclopedia *Historia Naturalis*. The Chinese have used it for carvings, but these deteriorate under light. It was once used as a red pigment. More recently it has been used to create the red colour in fireworks, but has been replaced by strontium to avoid the hazards associated with arsenic.

**PLINY THE ELDER**

**REALGAR BOTTLE**
*This Chinese bottle (dating from 1760–1840) is carved from realgar.*

*Rare prismatic realgar crystals*

*Rock groundmass*

*Light grey quartz*

**SCARLET CRYSTALS**
*Bright red realgar crystals cluster alongside quartz crystals in this sample.*

### PROPERTIES

| | |
|---|---|
| **GROUP** | Sulphides |
| **CRYSTAL SYSTEM** | Monoclinic |
| **COMPOSITION** | AsS |
| **COLOUR** | Scarlet to orange-yellow |
| **FORM/HABIT** | Granular |
| **HARDNESS** | 1½–2 |
| **CLEAVAGE** | Good |
| **FRACTURE** | Conchoidal |
| **LUSTRE** | Resinous to greasy |
| **STREAK** | Scarlet to orange-yellow |
| **SPECIFIC GRAVITY** | 3.6 |
| **TRANSPARENCY** | Subtransparent to opaque |

---

# CHALCOPYRITE

**CORNWALL**
*Chalcopyrite is found in Cornwall, south-west England, as well as many other localities.*

WITH ITS NAME DERIVED from the Greek *khalkos*, "copper", and "pyrite", chalcopyrite is a tetragonal mineral, usually forming crystals shaped like tetrahedra. On freshly broken surfaces, it is brassy yellow. Broken surfaces often form an iridescent tarnish. Although not rich in copper, its widespread occurrences make it the most important copper ore. It forms under a variety of conditions. It is commonly found in hydrothermal ore veins deposited at medium and high temperatures, and as replacements, often associated with large concentrations of pyrite. Specimens come from Colorado, Arizona, and the Tri-State mining district, USA; England; Tasmania; Germany; Canada; Spain; Japan; and China.

## RIO TINTO

The Rio Tinto area of south-western Spain has produced metals since before 800BC. Its rich mineral deposits include chalcopyrite, as well as silver, copper, and zinc. Huge quantities of silver were recovered from the time of the Roman Empire. Mining of all three ores continues today throughout the area.

**RED RIVER**
*The basin of the Rio Tinto, which means "river of ink", is an area of huge mineral wealth.*

*Metallic lustre*

**MASSIVE CHALCOPYRITE**
*An iridescent tarnish is clearly visible on the broken surface of this specimen from Mexico.*

*Metallic yellow chalcopyrite*

### PROPERTIES

| | |
|---|---|
| **GROUP** | Sulphides |
| **CRYSTAL SYSTEM** | Tetragonal |
| **COMPOSITION** | CuFeS$_2$ |
| **COLOUR** | Brassy yellow |
| **FORM/HABIT** | Tetrahedral |
| **HARDNESS** | 3½–4 |
| **CLEAVAGE** | Distinct |
| **FRACTURE** | Uneven, brittle |
| **LUSTRE** | Metallic |
| **STREAK** | Green-black |
| **SPECIFIC GRAVITY** | 4.2 |
| **TRANSPARENCY** | Opaque |

# STANNITE

STANNITE'S NAME comes from the Latin *stannum*, "tin". Crystals are rare but, when found, they are usually pseudo-octahedral in habit. Stannite is an ore of tin, and is often associated with cassiterite, pyrite, tetrahedrite, and chalcopyrite, and in tin veins of hydrothermal origin. Important localities are Cornwall, England; New Brunswick, Canada; Tasmania, Australia; Potosí, Bolivia; Keystone, South Dakota, and Seward, Alaska, USA; Kutná Hora, Czech Republic; and Norilsk, Russia.

**TASMANIA**
*Stannite is found at Zeehan, Tasmania.*

### PROPERTIES

| | |
|---|---|
| **GROUP** | Sulphides |
| **CRYSTAL SYSTEM** | Tetragonal |
| **COMPOSITION** | $Cu_2FeSnS_4$ |
| **COLOUR** | Steel-grey to iron-black |
| **FORM/HABIT** | Massive |
| **HARDNESS** | 4 |
| **CLEAVAGE** | Indistinct |
| **FRACTURE** | Uneven |
| **LUSTRE** | Metallic |
| **STREAK** | Black |
| **SPECIFIC GRAVITY** | 4.4 |
| **TRANSPARENCY** | Opaque |

*Metallic lustre*

*Quartz*

**TIN ORE**
*This massive, tarnished specimen with pale green-grey muscovite is from Cligga Mine, Cornwall, England.*

# NICKELINE

NICKELINE TAKES ITS NAME from its nickel component and also from the German *Kupfernickel*, "devil's copper", a name given to it in the Middle Ages when it was believed to contain copper that proved impossible to extract. It is found in ore deposits with other nickel minerals, and in vein deposits containing copper and silver. The Harz Mountains, Germany, and Ontario and the Northwest Territories of Canada are among key sources. Nickeline is not considered an important ore of nickel.

**HARZ MOUNTAINS**
*These mineral-rich German mountains contain nickeline.*

### PROPERTIES

| | |
|---|---|
| **GROUP** | Arsenides |
| **CRYSTAL SYSTEM** | Hexagonal |
| **COMPOSITION** | NiAs |
| **COLOUR** | Copper-red |
| **FORM/HABIT** | Massive |
| **HARDNESS** | 5–5½ |
| **CLEAVAGE** | None |
| **FRACTURE** | Conchoidal to uneven, brittle |
| **LUSTRE** | Metallic |
| **STREAK** | Black |
| **SPECIFIC GRAVITY** | 7.8 |
| **TRANSPARENCY** | Opaque |

*Metallic lustre*

*Copper-red on fresh surfaces*

**GRANULAR NICKELINE**
*Crystals of nickeline are rare; it is usually massive, as shown here, or granular.*

# MILLERITE

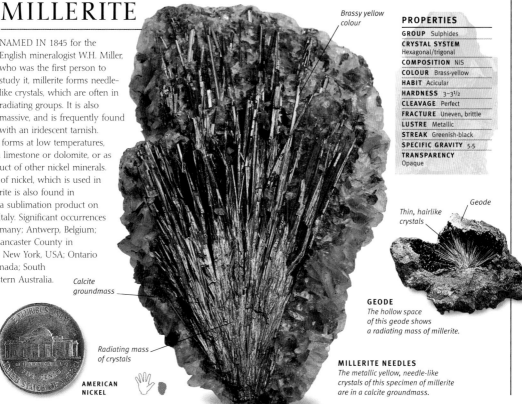

NAMED IN 1845 for the English mineralogist W.H. Miller, who was the first person to study it, millerite forms needle-like crystals, which are often in radiating groups. It is also massive, and is frequently found with an iridescent tarnish.

Millerite normally forms at low temperatures, often in cavities in limestone or dolomite, or as an alteration product of other nickel minerals. Millerite is an ore of nickel, which is used in metal alloys. Millerite is also found in meteorites and as a sublimation product on Mount Vesuvius, Italy. Significant occurrences are at Wissen, Germany; Antwerp, Belgium; Keokuk in Iowa, Lancaster County in Pennsylvania, and New York, USA; Ontario and Manitoba, Canada; South Wales; and in Western Australia.

**ONTARIO**
*One of the Canadian localities for millerite is Ontario.*

*Brassy yellow colour*

### PROPERTIES

| | |
|---|---|
| **GROUP** | Sulphides |
| **CRYSTAL SYSTEM** | Hexagonal/trigonal |
| **COMPOSITION** | NiS |
| **COLOUR** | Brass-yellow |
| **HABIT** | Acicular |
| **HARDNESS** | 3–3½ |
| **CLEAVAGE** | Perfect |
| **FRACTURE** | Uneven, brittle |
| **LUSTRE** | Metallic |
| **STREAK** | Greenish-black |
| **SPECIFIC GRAVITY** | 5.5 |
| **TRANSPARENCY** | Opaque |

*Geode*

*Thin, hairlike crystals*

**GEODE**
*The hollow space of this geode shows a radiating mass of millerite.*

*Calcite groundmass*

**NICKEL COIN**
*A nickel-bearing alloy called cupronickel is used to make coins. The American nickel takes its name from the element, althought it is only 25 per cent nickel.*

*Radiating mass of crystals*

**AMERICAN NICKEL**

**MILLERITE NEEDLES**
*The metallic yellow, needle-like crystals of this specimen of millerite are in a calcite groundmass.*

# STIBNITE

THE PRINCIPAL ORE of antimony, stibnite is antimony sulphide. Its name comes from its Latin name, *stibium*. It forms prismatic crystals, often striated parallel to the prism faces. Lead-grey to steel-grey in colour, it often has a black, iridescent tarnish. Stibnite occurs in hydrothermal veins, hot-spring deposits, and low-temperature replacement deposits. It is often associated with realgar, galena, pyrite, quartz, cinnabar, and orpiment. In gneiss and granite it is found in massive aggregates. It is a widespread mineral, with fine crystals coming from the Hunan Province of China; the island of Shikoku, Japan; the Harz Mountains, Germany; Nerchinsk, Russia; Potosí, Bolivia; Baia Sprie and Herja, Romania; Príbram, the Czech Republic; New Brunswick, Canada; and Oxaca, Mexico.

**FIREWORKS**
*Antimony provides
the intense blue colour
in fireworks.*

## USES OF ANTIMONY

Antimony is and has been widely used for hardening lead for various purposes. It adds to the strength and rigidity of the grids used in lead-acid storage batteries. It is also a major component of type metal, and is used in lead-free solders. In the electronics industry, it is used in semiconductors. Antimony compounds are widely used as flame retardants in plastics, paints, textiles, and rubber.

**SOLDERING**
*Antimony-based solders are
much safer to use than lead-
based solders.*

*Prismatic
crystals*

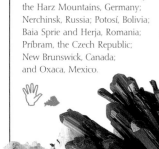

*Radiating
habit*

*Acicular crystals*

**ACICULAR CRYSTALS**
*This specimen shows a
mass of radiating, acicular
stibnite crystals.*

*Quartz and
baryte*

**STIBNITE
CRYSTALS**
*This group of long prismatic,
striated stibnite crystals is on a
quartz and baryte groundmass.*

*Striations*

### PROPERTIES

| | |
|---|---|
| **GROUP** | Sulphides |
| **CRYSTAL SYSTEM** | Orthorhombic |
| **COMPOSITION** | $Sb_2S_3$ |
| **COLOUR** | Lead-grey to steel-grey, black |
| **FORM/HABIT** | Prismatic |
| **HARDNESS** | 2 |
| **CLEAVAGE** | Perfect |
| **FRACTURE** | Subconchoidal |
| **LUSTRE** | Metallic |
| **STREAK** | Lead-grey to steel-grey |
| **SPECIFIC GRAVITY** | 4.6 |
| **TRANSPARENCY** | Opaque |

# ORPIMENT

USUALLY FOUND as foliated or columnar masses, orpiment is an arsenic sulphide. Distinct crystals are uncommon, but when found they are short prismatic, with an orthorhombic appearance. Orpiment is lemon-yellow to brownish-yellow. It takes its name from the Latin *auri*, meaning "golden", and *pigmentum*, "paint". It is a low-temperature hydrothermal mineral, and also forms as an alteration product of other arsenic-bearing minerals. It is widely distributed, with specimens coming from Sakha (Yakutia), Russia; Nye County, Nevada, USA; Copalnic, Romania; Hakkâri, Turkey; Hokkaido, Japan; St. Andreasberg, Germany; Quiruvilca Mine, Peru; and Guizhou Province, China.

*Uneven
fracture*

**USE IN PAINTING**
*Yellow pigment derived
from orpiment was used
in this painting by Sir
David Wilkie, entitled* The
Artist's Parents *(1813).*

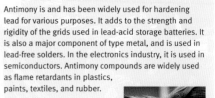

*Foliated
appearance*

*Yellow pigment*

**FOLIATED ORPIMENT**
*This specimen is a foliated mass of
orpiment. The mineral has a
resinous lustre on fresh surfaces,
but cleavage surfaces are pearly.*

*Resinous
lustre*

### PROPERTIES

| | |
|---|---|
| **GROUP** | Sulphides |
| **CRYSTAL SYSTEM** | Monoclinic |
| **COMPOSITION** | $As_2S_3$ |
| **COLOUR** | Yellow |
| **FORM/HABIT** | Massive, foliated |
| **HARDNESS** | $1^{1}/_{2}$–2 |
| **CLEAVAGE** | Perfect |
| **FRACTURE** | Uneven, sectile |
| **LUSTRE** | Resinous |
| **STREAK** | Pale yellow |
| **SPECIFIC GRAVITY** | 3.5 |
| **TRANSPARENCY** | Transparent to translucent |

# PYRITE

KNOWN SINCE ANTIQUITY, pyrite is perhaps better known to the general public by its informal name: "fool's gold". Although it is much lighter than gold, its brassy colour and relatively high density have led many novice prospectors into disappointment. Its name is derived from the Greek word *pyr*, meaning "fire", because pyrite emits sparks when struck by iron. Nodules of pyrite (see p.70) have been discovered in prehistoric burial mounds, although the suggestion that they were used as a means of producing fire is questionable, since iron, other than scarce meteoritic iron, was unknown in antiquity. The sun-like colour of pyrite would probably have been enough to assure its value. Native Americans in the American south-west set polished pyrite slices into a wooden base to construct mirrors.

Pyrite commonly forms in cubes, but octahedra and pyritohedra (pentagonal dodecahedra) are also common. Crystal faces are often deeply striated. Pyrite can also be massive, granular, nodular, or botryoidal. It forms under varied conditions, occurring in hydrothermal veins, by segregation from magmas, in contact metamorphic rocks, and in sedimentary rocks, such as shale, coal, and limestone.

Pyrite is often found in large deposits, and would be a source of iron were there not other minerals better suited to its extraction. It has been used as a source of sulphur for the manufacture of sulphuric acid, although today the acid is mostly made from the hydrogen-sulphide gas recovered from natural gas.

Pyrite is widespread worldwide. The Rio Tinto region of Spain has produced vast quantities, but splendid specimens also come from Bolivia, Brazil, Peru, Japan, Canada, Italy, Norway, Portugal, Greece, and Slovakia. In the USA, Tennessee, Virginia, Colorado, and California are all important producers.

**RIO TINTO REGION**
*The Rio Tinto region of Spain has been a source of minerals, including an enormous quantity of pyrite, which have been mined since Roman times.*

**NODULAR PYRITE**

**PYRITOHEDRAL CRYSTAL**

**PYRITE "SUN"**

**PYRITOHEDRAL CRYSTALS**

### PYRITE HABITS
*Pyrite occurs in many different shapes and forms, from massive aggregates to distinct crystals. It is also found in nodules and as attractive, coin-shaped "suns" of radiating crystals.*

**PYRITE BEADS**
*Pyrite can be ground into unusual and regularly shaped beads, such as those in this necklace.*

**WHEEL-LOCK GUN**
*The gunpowder in a wheel-lock gun is ignited by the rotation of an iron wheel against a piece of pyrite.*

**PYRITE CRYSTALS**
*Three perfectly formed cubic pyrite crystals from Navajún, La Rioja, Spain, are in a marl groundmass. The largest crystal has faces measuring 3.5cm (1¹/₂in) across.*

## PROPERTIES

| | |
|---|---|
| **GROUP** | Sulphides |
| **CRYSTAL SYSTEM** | Cubic |
| **COMPOSITION** | $FeS_2$ |
| **COLOUR** | Pale brass-yellow |
| **FORM/HABIT** | Cubic, octahedral, pyritohedral |
| **HARDNESS** | 6–6¹/₂ |
| **CLEAVAGE** | None |
| **FRACTURE** | Conchoidal |
| **LUSTRE** | Metallic |
| **STREAK** | Greenish-black to brownish-black |
| **SPECIFIC GRAVITY** | 5.0 |
| **TRANSPARENCY** | Opaque |

Metallic lustre

Cubic habit

Brass-yellow colour

Rock groundmass

# BISMUTHINITE

**MADAGASCAR**
*This island in the Indian Ocean is one source of the rare mineral bismuthinite.*

A COMPARATIVELY RARE mineral, bismuthinite is structurally related to the sulphides of antimony (stibnite) and arsenic (orpiment). Like stibnite, it forms prismatic to acicular crystals, often elongated and striated lengthways. It is usually massive with a foliated or fibrous texture. Bismuthinite forms in high-temperature hydrothermal veins and granitic pegmatites, and is sometimes found with native bismuth, as well as other sulphide minerals. It is found in Australia, Norway, Romania, Japan, Bolivia, Mexico, Canada, England, and the USA.

## USES OF BISMUTH

Since bismuth expands slightly upon solidifying, its alloys are particularly suitable for the manufacture of sharply detailed metal castings. They also have low melting points and are used in special solders and in fire-safety devices such as automatic sprinkler heads and safety plugs for compressed-gas cylinders. Another bismuth alloy is used as a catalyst in the manufacture of acrylic fibres, plastics, and paints. Bismuth oxychloride gives a pearlescent quality to eye shadow, nail varnish, and lipstick. Medicinally, bismuth compounds are used in soothing agents for the treatment of digestive disorders and to treat injuries and infections of the skin.

**FIRE SPRINKLER**
*Bismuth is used in the triggers of fire sprinklers.*

**DIE-CAST MODEL**
*Bismuth alloys are widely used in precision casting.*

### PROPERTIES

| | |
|---|---|
| **GROUP** | Sulphides |
| **CRYSTAL SYSTEM** | Orthorhombic |
| **COMPOSITION** | $Bi_2S_3$ |
| **COLOUR** | Lead-grey to tin-white |
| **FORM/HABIT** | Short prismatic to acicular |
| **HARDNESS** | 2 |
| **CLEAVAGE** | Perfect |
| **FRACTURE** | Uneven |
| **LUSTRE** | Metallic |
| **STREAK** | Lead-grey |
| **SPECIFIC GRAVITY** | 6.8 |
| **TRANSPARENCY** | Opaque |

*Metallic lustre*

*Acicular crystals*

**BISMUTHINITE CRYSTALS**
*A mass of acicular bismuthinite crystals is visible in this specimen. Crystals of this mineral are relatively rare.*

---

# HAUERITE

**URAL MOUNTAINS**
*The Russian Ural Mountains are a source of hauerite.*

HAUERITE WAS NAMED IN 1846 FOR Austrian geologists Joseph and Franz von Hauer. A manganese sulphide, it is found in both octahedral and cubo-octahedral crystals, and can also be globular or massive. It is reddish-brown to dark brown in colour. Hauerite forms in the caps of salt domes by the alteration of evaporites. Localities include the salt domes of Texas, USA; the Ural Mountains of Russia; and Radussa Mine in Sicily, Italy.

*Octahedral crystals*

*Rock groundmass*

### PROPERTIES

| | |
|---|---|
| **GROUP** | Sulphides |
| **CRYSTAL SYSTEM** | Cubic |
| **COMPOSITION** | $MnS_2$ |
| **COLOUR** | Red-brown to brown-black |
| **FORM/HABIT** | Octahedral |
| **HARDNESS** | 4 |
| **CLEAVAGE** | Perfect cubic |
| **FRACTURE** | Subconchoidal to uneven |
| **LUSTRE** | Adamantine to submetallic |
| **STREAK** | Red-brown |
| **SPECIFIC GRAVITY** | 3.5 |
| **TRANSPARENCY** | Opaque |

**HAUERITE CRYSTALS**
*In this specimen, two intergrown octahedral crystals of hauerite occur in a rock groundmass.*

---

# GLAUCODOT

**SOUTHERN SWEDEN**
*Crystals of glaucodot are found at Askersund and Lindesberg in southern Sweden.*

TAKING ITS NAME FROM the Greek for "blue", an allusion to its use in colouring blue glass, glaucodot is a cobalt iron arsenic sulphide. It forms prismatic crystals, which are frequently found in cruciform twins. It can also be massive. Glaucodot forms in hydrothermal veins, often accompanied by pyrite. Localities include Tunaberg, Sweden; Cobalt, Ontario, Canada; and Franconia, New Hampshire, and Sumpter, Oregon, in the USA.

*Massive glaucodot*

*Prismatic crystal*

### PROPERTIES

| | |
|---|---|
| **GROUP** | Sulphides |
| **CRYSTAL SYSTEM** | Orthorhombic |
| **COMPOSITION** | $(Co,Fe)AsS$ |
| **COLOUR** | Grey to white |
| **FORM/HABIT** | Prismatic |
| **HARDNESS** | 5 |
| **CLEAVAGE** | Perfect, distinct |
| **FRACTURE** | Uneven |
| **LUSTRE** | Metallic |
| **STREAK** | Black |
| **SPECIFIC GRAVITY** | 6.0 |
| **TRANSPARENCY** | Opaque |

**MASSIVE AND CRYSTALLINE**
*Here well-developed prismatic crystals of glaucodot are found on massive glaucodot.*

# MARCASITE

AN IRON SULPHIDE, marcasite is chemically identical to (or a polymorph of) pyrite (see p.137), but unlike pyrite it crystallizes in the orthorhombic system. Its light or silvery-yellow crystals are paler than those of pyrite, although they darken with exposure. It comes in a number of crystal forms, predominantly pyramidal and tabular; it is also found in characteristic twinned, curved, sheaf-like shapes that resemble a cockscomb, sometimes referred to as cockscomb pyrites. Nodules with radially arranged fibres are common; marcasite can also be massive, stalactitic, or reniform. Marcasite is found near the surface where it forms from acidic solutions percolating downwards through beds of shale, clay, limestone, and chalk. Sometimes, it is found replacing the petrifying material of fossils, and as nodules in coal. It also occurs with lead and zinc minerals in metalliferous veins, such as those at Galena, Illinois, USA, and at Clausthal–Zellerfeld and Linnich, Germany. Generally widespread in sedimentary rocks, it has notable occurrences in England, France, the Czech Republic, Uzbekistan, Mexico, Canada, Japan, and Bolivia.

Jewellery that was said to be made of marcasite, particularly in the 19th century, was, in fact, often made of other minerals, such as hematite or pyrite. Its name comes from an Arabic or Moorish word applied generally to minerals with the appearance of pyrite.

**CZECH REPUBLIC**
*Deposits of good-quality marcasite are found in the Czech Republic.*

## PROPERTIES

| | |
|---|---|
| **GROUP** | Sulphides |
| **CRYSTAL SYSTEM** | Orthorhombic |
| **COMPOSITION** | $FeS_2$ |
| **COLOUR** | Pale bronze-yellow |
| **FORM/HABIT** | Tabular, prismatic |
| **HARDNESS** | 6–6½ |
| **CLEAVAGE** | Distinct |
| **FRACTURE** | Conchoidal |
| **LUSTRE** | Metallic |
| **STREAK** | Grey to black |
| **SPECIFIC GRAVITY** | 4.9 |
| **TRANSPARENCY** | Opaque |

**MARCASITE JEWELLERY**
*Marcasites were very popular in Art Deco jewellery and, along with jet, were widely used for mourning jewellery in late-Victorian times. Marcasite jewellery often incorporated numerous small stones, as these pieces illustrate.*

> "I assured them…that the same was **marcasite**, and of no riches or **value**."

**SIR WALTER RALEIGH, ON THE RETURN OF MEN SEARCHING FOR GOLD**

Metallic lustre

Rosette-shaped aggregate

Silvery-yellow colour

**MARCASITE CRYSTALS**
*This striking group of marcasite crystals is on a groundmass of chalk. It comes from Cap Blanc-Nez, Pas-de Calais, France.*

Spear-shaped crystals

**MARCASITE CRYSTALS IN A MATRIX**
*Here, several groups of spear-shaped twinned crystals have formed within a matrix of limestone.*

Limestone matrix

Chalk groundmass

# COBALTITE

**BROKEN HILL**
*Cobaltite occurs in this part of Australia.*

COBALTITE, ALSO KNOWN AS cobalt glace, was so named in 1832 because it contains the element cobalt. It is a cobalt arsenic sulphide in which iron commonly replaces part of the cobalt. Many crystals have striated faces. It can also be massive and granular. It ranges from steel-grey to silver-white with a reddish tinge, and its crystals are pink. Found in high-temperature hydrothermal deposits, cobaltite also occurs as veins in contact-metamorphic zones. Notable localities are Ontario, Canada; Tunaberg, Sweden; New South Wales, Australia; and Bou Azzer, Morocco. Cobaltite is the principal ore of cobalt. Cobalt alloys are used wherever a combination of high strength and heat resistance is needed. It is soluble in nitric acid.

### PROPERTIES

| | |
|---|---|
| **GROUP** | Sulphides |
| **CRYSTAL SYSTEM** | Orthorhombic |
| **COMPOSITION** | CoAsS |
| **COLOUR** | Silver-white, pink |
| **FORM/HABIT** | Pseudocubic or pyritohedral |
| **HARDNESS** | 5½ |
| **CLEAVAGE** | Perfect basal |
| **FRACTURE** | Conchoidal or uneven, brittle |
| **LUSTRE** | Metallic |
| **STREAK** | Greyish-black |
| **SPECIFIC GRAVITY** | 6.3 |
| **TRANSPARENCY** | Opaque |

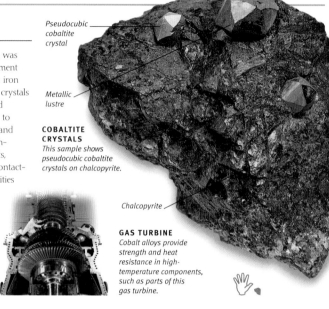

*Pseudocubic cobaltite crystal*

*Metallic lustre*

**COBALTITE CRYSTALS**
*This sample shows pseudocubic cobaltite crystals on chalcopyrite.*

*Chalcopyrite*

**GAS TURBINE**
*Cobalt alloys provide strength and heat resistance in high-temperature components, such as parts of this gas turbine.*

# MOLYBDENITE

**TASMANIA**
*Moina, on the Australian island of Tasmania, is a source of molybdenite.*

THE MOST IMPORTANT source of molybdenum, molybdenite was originally thought to be lead. This explains its name, which is derived from the Greek *molybdos*, meaning "lead". It was recognized by Swedish chemist Carl Scheele as a distinct mineral that contained a new element in 1778, but the metal was not extracted until his colleague Jacob Hjelm did so in 1782. Molybdenite forms platy, flexible, greasy-feeling hexagonal crystals that can be confused with graphite, although it has a much higher specific gravity, a more metallic lustre, and a slightly bluer tinge. It can also be massive or scaly. Molybdenite is found in granites and pegmatites, in high-temperature hydrothermal veins, and in contact metamorphic deposits. Molybdenite is widespread, and is found in commercial quantities in Japan, England, Tasmania, Canada, Norway, and Colorado, USA.

### PROPERTIES

| | |
|---|---|
| **GROUP** | Sulphides |
| **CRYSTAL SYSTEM** | Hexagonal/trigonal |
| **COMPOSITION** | MoS₂ |
| **COLOUR** | Lead-grey |
| **FORM/HABIT** | Tabular, prismatic |
| **HARDNESS** | 1–1½ |
| **CLEAVAGE** | Perfect basal |
| **FRACTURE** | Uneven |
| **LUSTRE** | Metallic |
| **STREAK** | Greenish- or bluish-grey |
| **SPECIFIC GRAVITY** | 4.7 |
| **TRANSPARENCY** | Opaque |

## USES OF MOLYBDENUM

Although a relatively rare metal, molybdenum is an important one. When added to alloys, it imparts high strength at temperatures above which many other metals and alloys are molten. It increases the hardness of both iron and steel, and provides good corrosion resistance. Unalloyed molybdenum is used for heating elements in high-temperature electric furnaces, and in some electronics.

*Granite groundmass*

**FILAMENT**
*Molybdenum is widely used for filament supports.*

**CARL SCHEELE**
*Scheele identified molybdenite in 1778 and was the first to identify the metal it contains.*

*Hexagonal foliated mass*

**LAYERED MASSES**
*The molybdenite masses in this specimen show a typical layered structure.*

*Metallic lustre*

# ARSENOPYRITE

THE MOST COMMON arsenic mineral, arsenopyrite takes its name from the contraction of its older name, arsenical pyrites. It also used to be known by its German name, *mispickel*. Its monoclinic crystals are frequently twinned, giving an orthorhombic appearance, and cruciform and multiple twins occur. Granular or compact arsenopyrite is also common. Its colour is silver-white to steel-grey on freshly broken surfaces, but it can tarnish, giving it a brownish or pink colour. Like other arsenic-bearing minerals, it yields a garlic odour when heated; these fumes are toxic. Some cobalt may substitute for the iron. Arsenopyrite is most commonly found in ore veins that were formed at moderate to high temperatures, associated with minerals such as gold and cassiterite. It is also found in contact-metamorphic deposits. Arsenopyrite weathers to scorodite. It is a widespread mineral, with noted localities in Freiberg, Saxony, Germany; Chihuahua, Mexico; Broken Hill, NSW, Australia; Panasqueira Mine, Portugal; New Jersey and New Hampshire, USA; and Ontario, Canada.

**CHIHUAHUA**
*Arsenopyrite is found in Copper Canyon in Chihuahua State, Mexico.*

## MEDICINAL USES

Although arsenic has a well-deserved reputation for being poisonous to humans, several complex organic arsenic compounds are used in the treatment of disease. Arsenic compounds are useful in the treatment of diseases caused by microorganisms, such as amoebic dysentery. Syphilis was universally treated by the arsenic compound arsphenamine until the discovery of penicillin in 1928. Arsenic is today a component of Salvarsan, an anti-syphilis drug.

**ARSENIC**

### PROPERTIES
| | |
|---|---|
| GROUP | Sulphides |
| CRYSTAL SYSTEM | Monoclinic |
| COMPOSITION | FeAsS |
| COLOUR | Silver-white to steel-grey |
| FORM/HABIT | Prismatic |
| HARDNESS | 5½–6 |
| CLEAVAGE | Distinct, indistinct |
| FRACTURE | Uneven |
| LUSTRE | Metallic |
| STREAK | Dark greyish-black |
| SPECIFIC GRAVITY | 6.1 |
| TRANSPARENCY | Opaque |

**PORTUGUESE ARSENOPYRITE**
*This specimen of arsenopyrite from Panasqueira Mine, Portugal, occurs with siderite and pyrite.*

Striated crystals · Metallic lustre · Siderite · Prismatic crystals

# SYLVANITE

FIRST IDENTIFIED IN 1835 in Transylvania, Romania, after which it is named, sylvanite is a gold and silver telluride. It forms complex prismatic, tabular, or bladed crystals, which are frequently twinned. Some twin forms are tree-like and resemble written characters. This form is known as graphic tellurium. Sylvanite occurs in minor amounts in numerous gold and silver deposits, and in a few instances in sufficient quantities to constitute an ore of gold. It is often found with the other gold tellurides calaverite and krennerite.

Important localities for sylvanite are Ontario, Canada; Kalgoorlie, Western Australia; Värmland, Sweden; Sacaramb, Transylvania, Romania; Thames District, New Zealand; Teller County, Colorado, and Calaveras County, California, USA.

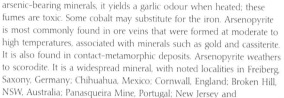

**CRIPPLE CREEK, USA**
*The mineral riches of this area of Colorado include sylvanite.*

**KALGOORLIE GOLD MINE**
*This open pit in Western Australia is famous for its gold deposits.*

### PROPERTIES
| | |
|---|---|
| GROUP | Tellurides |
| CRYSTAL SYSTEM | Monoclinic |
| COMPOSITION | (Au,Ag)₂Te₄ |
| COLOUR | Silver-white to pale yellow |
| FORM/HABIT | Prismatic, tabular, bladed |
| HARDNESS | 1–2 |
| CLEAVAGE | Perfect |
| FRACTURE | Uneven, brittle |
| LUSTRE | Metallic |
| STREAK | Grey |
| SPECIFIC GRAVITY | 8.2 |
| TRANSPARENCY | Opaque |

Sylvanite · Quartz groundmass · Metallic lustre

**GRAPHIC TELLURIUM**
*This specimen containing graphic tellurium and quartz is from Sacaramb, Transylvania.*

141

# SULPHOSALTS

SULPHOSALTS ARE A LARGE GROUP of mostly rare minerals, that contain two or more metals. They occur in small amounts in hydrothermal veins, generally in association with more common sulphides. They have a metallic lustre, high densities, and most of them are brittle. Characterized by complicated atomic and crystal structures, sulphosalts have a general chemical formula $AmBnXp$, in which m, n, and p are whole numbers; $A$ can be lead (Pb), silver (Ag), thallium (Tl), or copper (Cu); $B$ can be antimony (Sb), arsenic (As), bismuth (Bi), tin (Sn), or germanium (Ge); and $X$ can be sulphur (S) or selenium (Se). The atomic structures of many sulphosalts appear to be based on structural fragments of simpler compounds such as galena (PbS) blocks and stibnite (Sb$_2$S$_3$) sheets.

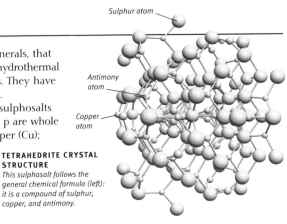

Sulphur atom

Antimony atom

Copper atom

**TETRAHEDRITE CRYSTAL STRUCTURE**
This sulphosalt follows the general chemical formula (left): it is a compound of sulphur, copper, and antimony.

## POLYBASITE

**CZECH REPUBLIC**
The area around Príbam is a key polybasite source.

Striations on crystal surfaces

Tabular crystals

Uneven fracture surface

**IRON-BLACK POLYBASITE**
This specimen exhibits the typical colouring and striated, tabular crystals of polybasite. The crystals show a bright, metallic lustre.

Tabular crystals

POLYBASITE DERIVES its name from the Greek for "many" and "base", a reference to the number of base metals in the mineral. It forms iron-black, tabular, pseudohexagonal crystals, often showing triangular striations, and can also be massive. It melts at a low temperature. Polybasite occurs in hydrothermal veins with native silver, other silver minerals, lead minerals, and other sulphides and sulphosalts such as galena and argentite. It sometimes occurs in sufficient quantities to be a significant ore mineral of silver. Important localities include Saxony, Germany; Príbram, the Czech Republic; Sabana Grande, Honduras; and many of the silver mines in Mexico and the western United States.

**PROPERTIES**

| | |
|---|---|
| **GROUP** | Sulphosalts |
| **CRYSTAL SYSTEM** | Monoclinic |
| **COMPOSITION** | (Ag,Cu)$_{16}$Sb$_2$S$_{11}$ |
| **COLOUR** | Iron-black |
| **FORM/HABIT** | Pseudohexagonal |
| **HARDNESS** | 2–3 |
| **CLEAVAGE** | Imperfect |
| **FRACTURE** | Uneven |
| **LUSTRE** | Metallic |
| **STREAK** | Black |
| **SPECIFIC GRAVITY** | 6.1 |
| **TRANSPARENCY** | Opaque |

## ENARGITE

**CHUQUICAMATA, CHILE**
Fine enargite crystals occur in this area.

REFERRING TO ITS PERFECT cleavage, enargite takes its name from the Greek *enarge*, "distinct". It is copper arsenic sulphide. Crystals are usually small, tabular or prismatic, sometimes pseudohexagonal, and sometimes hemimorphic – with the terminations different at each end. Its crystals occasionally form star-shaped multiple twins. It can also be massive or granular. On exposure to light, its lustre tends to dull from metallic silver-grey to black. Enargite forms in veins and replacement deposits associated with chalcopyrite, covellite, bornite, galena, pyrite, and sphalerite. In some places it is an important ore of copper. Localities include Chi-lung, Taiwan; Morococha and Cerro de Pasco, Peru; Chuquicamata, Chile; Calabona, Sardinia, Italy; Bor, Serbia; Tsumeb, Namibia; and Butte, Montana, Ouray, Colorado, Superior, Arizona, and Tintic, Utah, USA.

**PROPERTIES**

| | |
|---|---|
| **GROUP** | Sulphosalts |
| **CRYSTAL SYSTEM** | Orthorhombic |
| **COMPOSITION** | Cu$_3$AsS$_4$ |
| **COLOUR** | Greyish-black to iron-black |
| **FORM/HABIT** | Tabular, columnar |
| **HARDNESS** | 3 |
| **CLEAVAGE** | Perfect |
| **FRACTURE** | Uneven, brittle |
| **LUSTRE** | Metallic |
| **STREAK** | Black |
| **SPECIFIC GRAVITY** | 4.4–4.5 |
| **TRANSPARENCY** | Opaque |

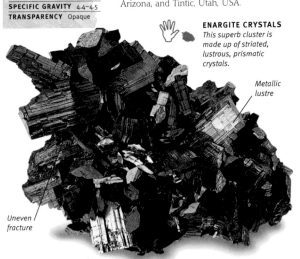

**ENARGITE CRYSTALS**
This superb cluster is made up of striated, lustrous, prismatic crystals.

Metallic lustre

Uneven fracture

# STEPHANITE

SAXONY, GERMANY
*This region of Germany is a key stephanite locality.*

SOMETIMES CALLED BRITTLE SILVER ORE, stephanite is silver antimony sulphide. Its crystals are short prismatic to tabular, but it can also be massive and granular. Stephanite generally occurs in small amounts in hydrothermal silver veins, but was found in sufficient quantity to be an ore of silver in the Comstock Lode, Nevada, USA. It is found at a number of localities in Mexico. Other occurrences include Cobalt, Ontario, and the Yukon, Canada; Cornwall, England; Espedalen, Norway; Transbaikalia, Russia; and Colquechaca, Bolivia.

## PROPERTIES

| | |
|---|---|
| **GROUP** | Sulphosalts |
| **CRYSTAL SYSTEM** | Orthorhombic |
| **COMPOSITION** | $Ag_5SbS_4$ |
| **COLOUR** | Iron-black |
| **FORM/HABIT** | Short prismatic to tabular |
| **HARDNESS** | $2–2^{1}/_{2}$ |
| **CLEAVAGE** | Imperfect |
| **FRACTURE** | Subconchoidal to uneven, brittle |
| **LUSTRE** | Metallic |
| **STREAK** | Iron-black |
| **SPECIFIC GRAVITY** | 6.2–6.5 |
| **TRANSPARENCY** | Opaque |

*Short, tabular crystal*

*Metallic lustre*

*Hexagonal outline*

**STEPHANITE CRYSTALS**
*Many of the twinned, short prismatic crystals in this cluster exhibit a pseudohexagonal outline.*

# TENNANTITE

ZACATECAS
*This mineral-rich Mexican state is a major locality for tennantite.*

NAMED IN 1819 for the English chemist Smithson Tennant, tennantite is a copper iron arsenic sulphide. It is an end member of the solid-solution series with the similar mineral tetrahedrite (see below). It forms crystals that are sometimes similar to tetrahedrite's. It can also be massive, granular, and compact. Tennantite is found in hydrothermal and contact metamorphic deposits, often associated with fluorite, baryte, galena, quartz, chalcopyrite, and sphalerite. Localities include Frieberg, Saxony, Germany; Lengenbach, Switzerland; and Butte, Montana, and Aspen and Central City, Colorado, in the USA.

## PROPERTIES

| | |
|---|---|
| **GROUP** | Sulphosalts |
| **CRYSTAL SYSTEM** | Cubic |
| **COMPOSITION** | $(Cu,Fe)_{12}As_4S_{13}$ |
| **COLOUR** | Steel-grey to black |
| **FORM/HABIT** | Cubic |
| **HARDNESS** | 4 |
| **CLEAVAGE** | None |
| **FRACTURE** | Subconchoidal to uneven, brittle |
| **LUSTRE** | Metallic |
| **STREAK** | Black |
| **SPECIFIC GRAVITY** | 4.6–4.7 |
| **TRANSPARENCY** | Opaque |

*Steel-grey colouring*

*Massive habit*

**MASSIVE TENNANTITE**
*This massive example of tennantite is from deposits in Cornwall, England.*

# TETRAHEDRITE

AUSTRIA
*Tetrahedrite is found in the Tyrol region of Austria.*

COMMONLY INCLUDING minor amounts of zinc, silver, lead, and mercury, tetrahedrite is a copper iron antimony sulphide. It is usually massive, compact, or granular. Tetrahedrite forms a solid-solution series with the similar mineral tennantite (see above), in which arsenic replaces antimony in the crystal structure. Tetrahedrite is a common mineral – probably the most common of the sulphosalts. It is an important ore of copper and sometimes of silver. It forms in low- to medium-temperature metal-bearing hydrothermal veins, often associated with galena, pyrite, chalcopyrite, baryte, bornite, and quartz. It is also found in contact metamorphic deposits. Significant occurrences are in Germany, England, Mexico, Peru, Bolivia, Chile, Australia, Romania, France, and various localities in the USA.

## PROPERTIES

| | |
|---|---|
| **GROUP** | Sulphosalts |
| **CRYSTAL SYSTEM** | Cubic |
| **COMPOSITION** | $(Cu,Fe)_{12}Sb_4S_{13}$ |
| **COLOUR** | Flint-grey to iron-black |
| **FORM/HABIT** | Massive |
| **HARDNESS** | 3–4 |
| **CLEAVAGE** | None |
| **FRACTURE** | Subconchoidal to uneven |
| **LUSTRE** | Metallic |
| **STREAK** | Brown to black to cherry-red |
| **SPECIFIC GRAVITY** | 4.6–5.1 |
| **TRANSPARENCY** | Opaque |

**BRASS ARMOUR**
*Tetrahedrite has been mined for its copper content for centuries all over the world. This 16th-century German suit of armour was made from the copper alloy, brass.*

**BRASS ASTROLABE**
*Copper alloys such as brass and bronze were used to make instruments for surveying and navigation, such as this 9th-century Arabic astrolabe.*

*Twinned tetrahedral crystals*

*Quartz crystals*

*Triangular crystal face*

**RARE CRYSTALS**
*Tetrahedrite takes its name from its relatively rare tetrahedral crystals.*

# PYRARGYRITE

ALSO CALLED DARK RUBY SILVER,
pyrargyrite takes its name from the Greek
words *pyros*, meaning "fire", and *argent*,
meaning "silver", an allusion to its colour
and its silver content. Its dark red colour
becomes darker when exposed to light.

Pyrargyrite forms in relatively low-
temperature hydrothermal veins
with other silver minerals, such as
proustite (also called, confusingly,
ruby silver) as well as galena,
sphalerite, tetrahedrite, and
calcite. Some of the best
crystallized specimens are
from St. Andreasberg in the
Harz Mountains and from
Freiberg, Saxony, both in Germany;
from Colquechaca, Bolivia;
Guanajuato, Mexico; near Silver
City, Idaho, and the Comstock
Lode, Nevada, in the USA.

## PROPERTIES

| | |
|---|---|
| **GROUP** | Sulphosalts |
| **CRYSTAL SYSTEM** | Hexagonal/trigonal |
| **COMPOSITION** | $Ag_3SbS_3$ |
| **COLOUR** | Deep red |
| **FORM/HABIT** | Prismatic, scalenohedral |
| **HARDNESS** | $2\frac{1}{2}$ |
| **CLEAVAGE** | Distinct |
| **FRACTURE** | Conchoidal to uneven |
| **LUSTRE** | Adamantine |
| **STREAK** | Purplish-red |
| **SPECIFIC GRAVITY** | 5.8 |
| **TRANSPARENCY** | Translucent |

*Twinned crystals*

**SILVER ORE**
*This specimen exhibits good
prismatic crystals and twinning.
Named in 1831, pyrargyrite is an
important source of silver.*

*Adamantine lustre*

# BOURNONITE

**CORNWALL**
*Bournonite is
one of many
minerals found in
the rocks of
Cornwall, south-
west England.*

NAMED IN 1805 for the French mineralogist Count
J.L. de Bournon, bournonite is a lead copper antimony
sulphide. It occurs as heavy, dark crystal aggregates
and masses, as well as the interpenetrating cruciform
twins that give it its informal name of cogwheel ore.
Bournonite is widespread, found in medium-
temperature hydrothermal veins and associated
with galena, sphalerite, chalcopyrite, and pyrite.
Particularly prized specimens come from the
Harz Mountains of Germany, where a few crystals exceed
2.2cm (1in) in diameter; Herodsfoot Mine,
Cornwall, England; and localities in
Italy, France, Bolivia, Peru, Canada,
Australia, Romania, Greece,
Japan, and Utah, USA.

## HARZ MOUNTAINS

The Harz Mountains which are located in
Saxony, central Germany, have been a
source of minerals, including bournonite,
for centuries. The town of Goslar was
founded in AD922 to protect the rich silver
mines nearby, and silver, lead, and iron
deposits are found in a number of other
localities. Smaller mineral deposits dot
the mountains, and many new minerals
were discovered there. A
number of its mines can
be visited by the
public today.

**NATIONAL PARK**
*Mines in the Harz Mountains
now have protected
status.*

*Quartz
groundmass*

*Metallic
lustre*

*Prismatic
crystals*

## PROPERTIES

| | |
|---|---|
| **GROUP** | Sulphosalts |
| **CRYSTAL SYSTEM** | Orthorhombic |
| **COMPOSITION** | $PbCuSbS_3$ |
| **COLOUR** | Steel-grey |
| **FORM/HABIT** | Short prismatic to tabular |
| **HARDNESS** | $2\frac{1}{2}-3$ |
| **CLEAVAGE** | Indistinct |
| **FRACTURE** | Subconchoidal to uneven |
| **LUSTRE** | Metallic |
| **STREAK** | Steel-grey |
| **SPECIFIC GRAVITY** | 5.8 |
| **TRANSPARENCY** | Opaque |

**COGWHEEL ORE**
*This specimen shows the shape
of the prismatic, orthorhombic
crystals that give bournonite its
nickname "cogwheel ore".*

# BOULANGERITE

**HARZ MOUNTAINS**
*These German mountains contain boulangerite.*

BOULANGERITE WAS NAMED in 1837 for French mining engineer Charles Boulanger. It forms flexible, prismatic to needle-like crystals, and is also fibrous or massive. Boulangerite occurs in hydrothermal veins and other environments. It is widespread in small amounts, and found in Germany, Canada, Mexico, Sweden, France, the Czech Republic, and the USA.

## PROPERTIES

**GROUP** Sulphosalts
**CRYSTAL SYSTEM** Monoclinic
**COMPOSITION** $Pb_5Sb_4S_{11}$
**COLOUR** Bluish-lead-grey
**FORM/HABIT** Long prismatic to acicular
**HARDNESS** $2^1/2$–3
**CLEAVAGE** Good
**FRACTURE** Uneven, brittle
**LUSTRE** Metallic
**STREAK** Brownish-grey to grey
**SPECIFIC GRAVITY** 6.2
**TRANSPARENCY** Opaque

*Boulangerite crystals*

**SLENDER CRYSTALS**
*This matted mass of fine acicular crystals in quartz comes from Peru.*

# PROUSTITE

**CZECH REPUBLIC**
*The Czech Republic is a source of proustite.*

AN IMPORTANT SOURCE of silver, proustite is light sensitive, turning transparent scarlet to opaque grey in strong light. Crystals are prismatic, rhombohedral, or scalenohedral. It is also massive and compact. It forms in fairly low-temperature hydrothermal veins, with other silver minerals, galena, and calcite. Notable localities are Chañarcillo, Chile; and Saxony, Germany.

## PROPERTIES

**GROUP** Sulphosalts
**CRYSTAL SYSTEM** Hexagonal/trigonal
**COMPOSITION** $Ag_3AsS_3$
**COLOUR** Scarlet, grey
**FORM/HABIT** Prismatic to short prismatic
**HARDNESS** $2–2^1/2$
**CLEAVAGE** Distinct rhombohedral
**FRACTURE** Conchoidal to uneven, brittle
**LUSTRE** Adamantine to submetallic
**STREAK** Vermilion
**SPECIFIC GRAVITY** 5.8
**TRANSPARENCY** Translucent

*Prismatic crystals*

*Striated crystal face*

**CRYSTALLINE PROUSTITE**
*Due to its colour, proustite is also called ruby silver.*

# JAMESONITE

**MEXICO**
*Jamesonite is found in Mexico's Sierra Tarahumara.*

FOUND AS ACICULAR TO FIBROUS crystals, jamesonite can also be massive and plumose. A lead iron antimony sulphide, it occurs in hydrothermal veins with other sulphosalt minerals; and in quartz with carbonate minerals, such as rhodochrosite, dolomite, and calcite. Jamesonite is widespread in small amounts, with the best specimens coming from Mexico, Serbia, Romania, England, and Bolivia.

**ROBERT JAMESON**
*Jamesonite was named for a Scottish mineralogist, Robert Jameson, in 1825.*

## PROPERTIES

**GROUP** Sulphosalts
**CRYSTAL SYSTEM** Monoclinic
**COMPOSITION** $Pb_4FeSb_6S_{14}$
**COLOUR** Steel-grey to dark lead-grey
**FORM/HABIT** Acicular, fibrous
**HARDNESS** 2–3
**CLEAVAGE** Good
**FRACTURE** Uneven to conchoidal
**LUSTRE** Metallic
**STREAK** Greyish-black
**SPECIFIC GRAVITY** 5.5–6.0
**TRANSPARENCY** Opaque

*Metallic lustre*

**FIBROUS JAMESONITE**
*This specimen shows fibrous crystallization which is common in this mineral.*

*Rock groundmass*

*Fibrous crystals*

# ZINKENITE

**HARZ MOUNTAINS**
*Zinkenite occurs in these German mountains.*

ZINKENITE, A LEAD TIN SULPHIDE, was named in 1826 for J.K.L. Zinken, a German mineralogist. It forms acicular crystals in columnar and radiating aggregates. It is often massive or forms thick mats of hair-like fibres. Zinkenite usually occurs in quartz associated with other sulphosalts. The best specimens are found in St. Pons, France; Oruro, Bolivia; and Harz, Germany. It also occurs in Canada, England, Australia, Romania, and Colorado and Nevada, USA.

## PROPERTIES

**GROUP** Sulphosalts
**CRYSTAL SYSTEM** Hexagonal
**COMPOSITION** $Pb_9Sb_{22}S_{42}$
**COLOUR** Steel-grey
**FORM/HABIT** Fibrous, massive
**HARDNESS** $3–3^1/2$
**CLEAVAGE** Indistinct
**FRACTURE** Uneven, brittle
**LUSTRE** Metallic
**STREAK** Steel-grey
**SPECIFIC GRAVITY** 5.3
**TRANSPARENCY** Opaque

**ZINKENITE CRYSTALS**
*This specimen of striated, acicular zinkenite crystals in quartz is from Wolfsberg, Lower Saxony, Germany.*

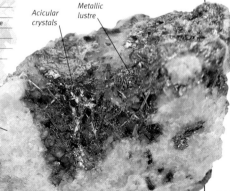

*Acicular crystals*

*Metallic lustre*

*Quartz groundmass*

145

# OXIDES

OXIDES ARE A FAMILY of minerals having structures consisting of close–packed oxygen atoms with metal or semimetal atoms occupying the spaces in between. There are two groups: simple oxides and multiple oxides. Simple oxides, such as cuprite ($Cu_2O$) are a combination of only one metal or semimetal and oxygen. Multiple oxides have two differing metal sites, both of which may be occupied by several different metals or semimetals. The spinel group of minerals are examples of multiple oxides: one site is potentially occupied by magnesium, iron, zinc, or manganese, while the second site is potentially occupied by aluminium, iron, manganese, or chromium. Oxide minerals occur in many igneous rocks, in pegmatites, and as decomposition products of sulphide minerals.

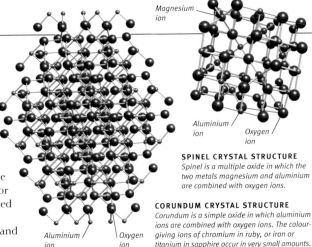

Magnesium ion

Aluminium ion

Oxygen ion

**SPINEL CRYSTAL STRUCTURE**
Spinel is a multiple oxide in which the two metals magnesium and aluminium are combined with oxygen ions.

**CORUNDUM CRYSTAL STRUCTURE**
Corundum is a simple oxide in which aluminium ions are combined with oxygen ions. The colour-giving ions of chromium in ruby, or iron or titanium in sapphire occur in very small amounts.

Aluminium ion

Oxygen ion

# ILMENITE

**BLACK SAND**
Ilmenite is a major component of many black beach sands.

NAMED AFTER THE LOCALITY of its discovery in the Il'menski Mountains, near Miass, Russia, ilmenite is a major source of titanium. Its crystals are usually thick tabular, although they sometimes occur as thin lamellae or rhombohedra. Ilmenite can also be compact or massive or occur as disseminated grains. It is iron-black, and can be mistaken for magnetite or hematite. Unlike magnetite, however, it is non-magnetic or very weakly magnetic. It can be distinguished from hematite by its streak.

Ilmenite is widely distributed as an accessory mineral in igneous rocks, such as gabbro, diorite, and anorthosite, and is also found in veins, pegmatites, black beach sands, and placer deposits. Crystals come from Bancroft, Ontario, Canada; Maderanertal and Binntal, Switzerland; Kragerø, Norway; Betafo, Madagascar; Woodstock, Western Australia; and Chester, Maine, and Tahawus and Warwick, New York, in the USA.

## PROPERTIES

| | |
|---|---|
| **GROUP** | Oxides |
| **CRYSTAL SYSTEM** | Hexagonal/trigonal |
| **COMPOSITION** | $FeTiO_3$ |
| **COLOUR** | Iron-black |
| **FORM/HABIT** | Thick tabular |
| **HARDNESS** | 5–6 |
| **CLEAVAGE** | None |
| **FRACTURE** | Conchoidal |
| **LUSTRE** | Metallic to submetallic |
| **STREAK** | Black |
| **SPECIFIC GRAVITY** | 4.7 |
| **TRANSPARENCY** | Opaque |

Ilmenite crystal

**ILMENITE CRYSTALS**
The ilmenite crystals here are with actinolite and quartz. This specimen is from Mount Painter, South Australia.

Quartz

Actinolite

# PEROVSKITE

**GREENLAND**
Crystals up to 8cm (3in) are found in Greenland's Gardiner Complex.

PEROVSKITE WAS NAMED IN 1839 for Russian mineralogist L.A. Perovski. Its crystals are pseudocubic, but can be pseudo–octahedral in varieties where niobium or cerium has replaced a large amount of the titanium. When black, it has a metallic lustre; when brown or yellow, it is adamantine.

Perovskite is thought to be a major constituent of the upper mantle of the Earth, and it is also found in carbonaceous chondrite meteorites. At the Earth's surface, it occurs in mafic igneous rocks, in contact metamorphic rocks associated with mafic intrusives, and in some schists. Greenland is a major locality for perovskite. Other localities include Trentino, Italy; Baden, Germany; Bagagem, Brazil; Leanchoil, British Columbia, Canada; and Magnet Cove, Arkansas, and Powderhorn, Colorado, USA.

Pseudocubic crystal

## PROPERTIES

| | |
|---|---|
| **GROUP** | Oxides |
| **CRYSTAL SYSTEM** | Orthorhombic |
| **COMPOSITION** | $CaTiO_3$ |
| **COLOUR** | Black, brown, yellow |
| **FORM/HABIT** | Pseudocubic |
| **HARDNESS** | 5½ |
| **CLEAVAGE** | Imperfect |
| **FRACTURE** | Subconchoidal to uneven |
| **LUSTRE** | Adamantine/metallic |
| **STREAK** | Grey to colourless |
| **SPECIFIC GRAVITY** | 4.0 |
| **TRANSPARENCY** | Transparent to opaque |

Plagioclase groundmass

**PEROVSKITE CRYSTALS**
In this specimen, two striated, pseudocubic perovskite crystals are in a groundmass of plagioclase feldspar.

Striations on crystal

# HEMATITE

A DENSE AND RELATIVELY HARD iron oxide, hematite is the most important iron ore because of its high iron content (70 per cent) and its abundance. It has a number of different habits, many of which have their own names: the steel-grey crystals and coarse-grained varieties with a brilliant metallic lustre are known as specular hematite; thin, scaly forms are micaceous hematite; and crystals in flower-like arrangements are called iron roses. Hematite can also occur as short, black, rhombohedral crystals, and may have an iridescent tarnish. Compact varieties often occur with a kidney-shaped surface (kidney ore) or a fibrous structure (pencil ore). Much hematite is a soft, fine-grained, earthy form called red ochre, which is used as a pigment. A purified form of ground hematite called rouge is used to polish plate glass and jewellery, and was in the past used to polish gems.

The most important hematite deposits are sedimentary in origin, either in sedimentary beds or metamorphosed sediments. The Lake Superior district in North America is the world's largest producer, but other major deposits are found in the Ukraine, China, India, Australia, Liberia, Brazil, and Venezuela. Hematite is also found as an accessory mineral in many igneous rocks. Cuttable material is found in England, Germany, and on Elba, Italy.

The name hematite is derived from the Greek *haimatitis*, meaning "blood-red", an allusion to the red colour of its powder. Its apparent association with blood goes back much further: the bones of Neolithic burials have been found smeared with powdered hematite, and in the 10th millennium BC, hematite was scattered around skeletal remains buried at Chou-k'ou-tien, China. In later times, it was worn to protect the wearer from bleeding.

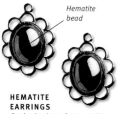

**LAKE SUPERIOR**
The iron deposits around Lake Superior, Michigan, USA, are rich in hematite.

Hematite bead

**HEMATITE EARRINGS**
Oval cabochons form a striking centre to this pair of earrings. Most cut or carved hematite is reconstituted hematite, known as "hematine".

**SPECULAR HEMATITE**
This superb group of well-formed hematite crystals is from Elba, Italy.

**SEALS**
These two carved hematite stamp seals date from c.1400BC.

**EXAMPLES OF HEMATITE**
Hematite comes in a number of different habits and lustres.

MASSIVE HEMATITE

SPECULAR HEMATITE

KIDNEY ORE

IRIDESCENT HEMATITE

Metallic lustre

Uneven fracture

Modified rhombohedral crystal

**HEMATITE FROG**
Larger pieces of hematite are a popular and often spectacular carving medium for lapidaries.

**BEAD**
Hematite is a popular material for making polished black, gleaming beads.

**OVAL CABOCHON**
The top of this cabochon of black hematite has been faceted. Hematite cabochons have been sold as "marcasites".

**MARQUISE CABOCHON**
The top of this marquise cabochon has been faceted. The brittleness of hematite makes pointed shapes vulnerable.

## PROPERTIES

**GROUP** Oxides
**CRYSTAL SYSTEM** Hexagonal
**COMPOSITION** $Fe_2O_3$
**COLOUR** Steel-grey
**FORM/HABIT** Tabular, sometimes platy, botryoidal
**HARDNESS** 5–6
**CLEAVAGE** None
**FRACTURE** Subconchoidal to uneven
**LUSTRE** Metallic to dull
**STREAK** Cherry-red or red-brown
**SPECIFIC GRAVITY** 5.3
**TRANSPARENCY** Opaque
**R.I.** 2.94–3.22

# CORUNDUM: SAPPHIRE

IT WAS NOT UNTIL the 18th century that it was clearly established that sapphires and rubies are, in fact, the same mineral - corundum, natural aluminium oxide. The name corundum is probably derived from the Sanskrit *kuruvinda*, meaning "ruby", the name given to red corundum (see pp.150–51). When it is found in other colours, it is called "sapphire". Next to diamond, corundum is the hardest mineral on Earth. It crystallizes in the hexagonal system, forming dipyramidal or rounded barrel-shapes.

## THE COLOURS OF SAPPHIRE

Although popularly thought of as being blue, sapphire can also be colourless, green, pink, and a wide range of other hues. Rare pink-orange stones are called padparadscha, and sapphire that appears blue in daylight and reddish or violet in artificial light is called alexandrine or alexandrite sapphire. Sapphires are commonly pleochroic, that is, they appear to be differently coloured when viewed from different directions.

All sapphires apart from blue ones are identified by the term "sapphire" preceded by the colour: "pink sapphire", "yellow sapphire", and rare, colourless "white sapphire". All coloured sapphires other than blue or deep red ones are collectively known as fancy sapphires. Technically, these names are reserved for gem-quality stones, but generally they are simply used for stones of the right colour. When blue, sapphire is coloured by trace titanium and iron, when pale green, yellow, or brown it is coloured by iron, and when pink by very small traces of

**SAPPHIRE CRYSTAL**
This water-worn crystal demonstrates the colour zoning that is commonly seen in sapphires.

Colour zoning

**MISSOURI RIVER**
An unusually large placer deposit of gem-quality sapphire is found by the Missouri near Helena, the capital of Montana, USA.

**LOGAN SAPPHIRE**
The 423-carat Logan Sapphire was mined in Sri Lanka and may be the world's largest blue sapphire. It is about the size of a hen's egg.

**SAPPHIRE RING**
A natural, uncut sapphire crystal is mounted in a contemporary-style silver ring.

## THE LORE OF SAPPHIRE

Because the term "sapphire" was applied only to blue corundum before the 19th century, all historic references relate to this colour. In ancient Greece, and later in the Middle Ages, there was a belief that sapphires cured eye diseases and set prisoners free. Medieval alchemists related it to the element air, and sapphires were also believed to be an antidote against poisons and endowed with a power to influence the spirits at that time. In the 11th century the French bishop Marbodius described blue sapphire: **"Sapphires possess a beauty like that of the heavenly throne; they denote the hearts of the simple, of those moved by a sure hope and of those whose lives shine with their good deeds and virtuousness."** In the East, the sapphire is regarded as a powerful charm against the evil eye.

**SAPPHIRE BUDDHA**
This Buddha representation is carved from a superb single sapphire crystal.

**SAPPHIRE GRAVELS**
*These 3–5mm (1/8–1/4in) uncut sapphires from Phillipsburg, Montana, USA, show the wide range of colours to be found in corundum.*

chromium. With increasing amounts of chromium, pink sapphire forms a continuous colour range with ruby. Vanadium, nickel, and cobalt can also alter colours. Most corundum contains nearly 1 per cent iron oxide. Much sapphire is unevenly coloured, with concentric zones of colour; for example, green sapphire is often found to be alternating bands of yellow and blue, but skilful cutting of unevenly coloured stones yields gems with a uniform appearance. The colour and transparency of sapphire can both be changed by heating the stone or subjecting it to radiation. The somewhat milky stones from the Missouri River, Montana, USA, are clarified by heat treatment, as are many Sri Lankan stones.

From medieval times until the end of the 19th century, green sapphire was referred to as "Oriental peridot" and yellow sapphire as "Oriental topaz". In ancient literature, the term "sapphire" appears to have mostly referred to lapis lazuli, although blue corundum has itself been prized since at least 800BC. One of the oldest sapphires is St. Edward's sapphire, set in the finial cross of the British Imperial State Crown (see p.157), and believed to date from Edward the Confessor's coronet in AD1042. Blue sapphire was widely used in the jewellery of the medieval kings of Europe, and was considered the stone most fitting for ecclesiastical rings.

## SOURCES AND USES OF SAPPHIRE

Corundum is most abundant in metamorphic rocks, and in silica-deficient igneous rocks such as nepheline syenites (see p.38). Large deposits are rare, however. Most sapphires and rubies are mined from alluvial deposits, where their higher density concentrates them when weathered from their original source. Myanmar, Sri Lanka, Nigeria, and Thailand are famous gemstone sources, but there are also deposits in Australia, Brazil, Kashmir, Cambodia, Kenya, Malawi, Colombia, and the USA. Some of the richest deposits of non-gemstone corundum occur in India, Russia, Zimbabwe, South Africa, and Nigeria.

Transparent sapphire is most commonly faceted, while star sapphire and other nontransparent varieties are cut *en cabochon*. Some sapphire is even carved or engraved despite its hardness. Non-gemstone corundum is used as an abrasive for grinding optical glass and for polishing metals, and has been made into sandpapers and grinding wheels; when naturally mixed with magnetite it forms emery. It is also used in high-temperature ceramics because of its high melting point (2,050°C/3,700°F). It has been synthesized in industrial quantities since the 1920s, and synthetic sapphire has almost entirely replaced the natural material in most applications.

**GEMS OF ALL COLOURS**
*Corundum comes in rainbow colours. All the colours are known as sapphire except red, which is known as ruby.*

## PROPERTIES

| | |
|---|---|
| **GROUP** | Oxides |
| **CRYSTAL SYSTEM** | Hexagonal – trigonal |
| **COMPOSITION** | $Al_2O_3$ |
| **COLOUR** | Occurs in most colours |
| **FORM/HABIT** | Pyramidal, prismatic barrel-shape |
| **HARDNESS** | 9 |
| **CLEAVAGE** | None |
| **FRACTURE** | Conchoidal to uneven |
| **LUSTRE** | Adamantine to vitreous |
| **STREAK** | Colourless |
| **SPECIFIC GRAVITY** | 4.0–4.1 |
| **TRANSPARENCY** | Transparent to translucent |
| **R.I.** | 1.76–1.77 |

**SINGLE CRYSTAL**

> "Kissing your hand may make you feel very very good but a diamond and a sapphire bracelet lasts forever"
>
> **ANITA LOOS**
> GENTLEMEN PREFER BLONDES

**TWIN CRYSTAL**

**CRYSTAL FORMS**
*These corundum crystals from Madagascar show superb crystal forms. The single crystal is 2.5cm (1in) long and has the classic corundum prismatic shape, with a triangular termination.*

**STAR CABOCHON**
*When sapphire containing rutile needles oriented along the growth directions of the crystal, is cut en cabochon, a star appears.*

**BRILLIANT**
*A brilliant cut is usually used for transparent sapphire to enhance its colour.*

**GROWING SAPPHIRES**
*This laboratory technician in the United States is watching sapphire crystals grow in a crystal grower. The crystals are used in sodium vapour lamps.*

# CORUNDUM: RUBY

ACCORDING TO THE 11TH-CENTURY French bishop Marbodius, "ruby is the solitary and glowing eye which dragons and wyverns carry in the middle of their foreheads". In less poetic modern terms, ruby is the red variety of the mineral corundum, an oxide of aluminium. It ranges in colour from deep cochineal to pale rose-red, sometimes with a tinge of purple; the most valued is a blood-red. Its coloration comes from traces of chromium – as small as 1 part in 5,000 – which replaces some of the aluminium in the structure. Ruby forms a continuous colour succession with pink sapphire, the colour deepening with an increase in chromium. Only stones of the darker hues are generally considered to be ruby. It was believed by ancient Hindu and Burmese miners that colourless or pale pink sapphires were rubies that had not completely ripened. Ruby has been mined from the gem gravels of Sri Lanka since the 8th century BC.

**RATNAPURA GEM MINES**
*Rubies and sapphires are mined from river gravels in Sri Lanka, where they have become concentrated following weathering and erosion.*

## KING OF PRECIOUS STONES

The name ruby comes from the Latin *ruber*, meaning "red". In Sanskrit, ruby is known as *ratnaraj*, "king of precious stones". Many rubies are heat treated to improve their clarity or colour (or both). Rubies tend to be small (stones more than 10 carats are unusual) as the presence of chromium has an inhibiting effect on crystal growth – hence the high value of large rubies. Ruby crystals tend to be hexagonal prisms with tapering or flat ends. They occur worldwide in igneous and metamorphic rocks, or as waterworn pebbles in alluvial deposits. Gem-quality ruby is a mineral of much more limited distribution. It is found in the following localities: in north-central Myanmar, where it occurs in bands of crystalline limestone; in gravels in Thailand with

**COWEE CREEK, USA**
*This locality in North Carolina is one of the US sources for gem-quality rubies.*

**HISTORIC RUBY CAMEO**
*Gems were mounted on this rock crystal ewer (or jug) in about 1660. The centrepiece is a ruby cameo of the head of Queen Elizabeth I of England.*

**ANCIENT RING**
*This ancient ruby ring is of European design.*

**STAR CABOCHON**
*When ruby containing rutile needles oriented along the growth directions of the crystal, is cut en cabochon, a star appears.*

**STEP-CUT SYNTHETIC**
*Synthetic ruby is flawless enough for cuts that display the interior of the stone to be used.*

**CUSHION MIX**
*Ruby often contains natural flaws that are hidden by using cuts, such as this one, with many small facets.*

*Striations on crystal*

*Rock groundmass*

*Prismatic crystal*

## PROPERTIES

| | |
|---|---|
| **GROUP** | Oxides |
| **CRYSTAL SYSTEM** | Hexagonal – trigonal |
| **COMPOSITION** | $Al_2O_3$ |
| **COLOUR** | Red |
| **FORM/HABIT** | Pyramidal, prismatic barrel-shape |
| **HARDNESS** | 9 |
| **CLEAVAGE** | None |
| **FRACTURE** | Conchoidal to uneven |
| **LUSTRE** | Adamantine to vitreous |
| **STREAK** | Colourless |
| **SPECIFIC GRAVITY** | 4.0–4.1 |
| **TRANSPARENCY** | Transparent to translucent |
| **R.I.** | 1.76–1.77 |

sapphires and spinels; and in the gem gravels of Sri Lanka. Other localities that produce minor amounts of gem material are Afghanistan, Madagascar, Vietnam, Cambodia, and North Carolina and Montana, USA. Bright red, opaque ruby crystals in green zoisite come from Tanzania. Star rubies occur in addition to star sapphires (see pp.152–53).

The distinction between ruby and other transparent red minerals has been made only relatively recently, and so the term "ruby" has often been misapplied in the past. A famous example is the Black Prince's Ruby, now in the Imperial State Crown of England and part of the Crown Jewels since 1367. It was not until the 19th century that it was discovered to be a spinel (see p.157). "Ruby" is a term applied to fine garnets as well: "Cape rubies", "Australian rubies", and "Arizona rubies" are all garnet. Red tourmaline, rubellite, has been called "Siberian ruby", and "balas ruby" is red spinel. Rubies have been synthesized since 1902.

**ANTIQUE RUBY RING**
*In this ring a square ruby has been set at right angles to its square setting.*

**RUBY AND DIAMOND BRACELET**
*In this glittering bracelet, rubies alternate with tiny diamonds to make a striking and bold arrangement.*

### THE LORE OF RUBY

To the Burmese, a fine ruby was a talisman of good fortune, bestowing invincibility. In classical antiquity rubies were reputed to banish sorrow, restrain lust, and resist poison. In later times, as a blood-coloured stone, ruby was used in the preparation of medicines to staunch bleeding. Russian popular tradition maintains that it is good for the heart, brain, vitality, and for clearing the blood. However, in all these cases, given the confusion that has surrounded the term "ruby", it is not certain which mineral was being used. Ruby also took on an important symbolism for Islam: after his expulsion from paradise and arrival in Mecca, Adam was shown a ruby canopy under which lay a glowing stone – a meteorite. He was instructed to build the Kabah, now the main Muslim shrine, over it.

**RUBY ELEPHANT**
*This 6.5cm (2$^1$/$_2$in) high elephant is carved from ruby in a zoisite matrix.*

**RUBY CRYSTALS**
*Ruskin's ruby (far left) was donated to the Natural History Museum, London, by John Ruskin in 1887. It is about 162 carats. The central crystal, embedded in calcite, is from the Mogok region in Myanmar. The crystal below is part of the Smithsonian Collection.*

RUSKIN'S RUBY CRYSTAL

MYANMAR CRYSTAL

SMITHSONIAN CRYSTAL

*Vitreous lustre*

**RUBY IN MATRIX**
*These prismatic Kashmir rubies are still embedded in the rock matrix in which they formed. They have a tapering prismatic form, with the classic flat terminations capped by a triangular face. Although Kashmir produces large quantities of ruby, relatively little is of gem quality.*

### THE DE LONG RUBY

One of the world's great star rubies, the 100.3-carat De Long Star Ruby was discovered in Burma (now Myanmar) in the 1930s, and donated to the American Museum of Natural History in 1938. On 29 October 1964, it was stolen, along with a number of other famous stones, in an audacious burglary, later termed the "Great Jewel Robbery", inspired by a similar robbery seen in a film by the three thieves. Two of them lowered themselves into the museum through an open window, and literally raked the stones out of their display cases. After making an easy getaway, their careless boasting soon led to their arrest, and most of the stones were recovered. The De Long Star Ruby had already passed into other hands, and after extensive negotiations, the payment of a ransom of $25,000 led to its return. Today, security at the museum is considerably improved.

**DE LONG STAR CABOCHON**
*The De Long Star Ruby was the prize of an audacious museum robbery that took place in 1964.*

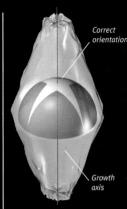

*Correct orientation*

*Growth axis*

# STAR STONES AND CAT'S EYES

THE APPARENT PRESENCE INSIDE A GEMSTONE OF A SINGLE BAND OF LIGHT, OR A NUMBER OF INTERSECTING LIGHT BANDS, CAN GIVE IT AN ETHEREAL, EVEN MAGICAL QUALITY. FOREMOST AMONG SUCH GEMS, BOTH IN BEAUTY AND IN VALUE, ARE STAR STONES AND CAT'S EYES.

**ORIENTING THE GEM**
*For a star stone to appear, a gem must be cut with a precise orientation to the rutile inclusions, which always align perpendicular to the length of the crystal. There is only one possible orientation that will produce a star stone in the finished gem.*

S tar stones and cat's eyes are not the names of specific mineral gem varieties. They are certain types of gemstone that exhibit a property called chatoyancy. Chatoyancy is the reflection of light from microscopic inclusions of other minerals within the gemstone. It is best observed when light from a single source, such as the Sun, is directed at the stone; under multiple light sources, the single or intersecting bands of light appear ill-defined or may disappear altogether. A single band is called a cat's-eye effect; intersecting stars are known as asterism.

A common cause of chatoyancy is the presence of the mineral rutile (titanium oxide). Corundum crystals, for example, grow during the metamorphism of clay-rich limestone. As the corundum (pure aluminium oxide) crystals grow, they incorporate atoms of titanium. As each crystal cools, the titanium atoms bond with oxygen atoms to form minute needle-like crystals of rutile (titanium oxide) within the corundum crystal. The rutile crystals

are forced to align, forming sets of parallel crystals. In star sapphires, microscopic inclusions of rutile orient themselves in three bands lying horizontally at 120 degrees to each other. Viewed from above, the bands intersect as a star. Star sapphires (and other star stones) are cut *en cabochon* so that the dome of the cabochon is oriented to coincide with the point of intersection with the rutile crystals (see left). If oriented in any other direction, no star appears.

Gems that produce star stones include sapphire and ruby, quartz (which produces six- or twelve-rayed stars), garnet (four- , six-, or eight-rayed stars), spinel (both four- and six-rayed stars), and rarely, kornerupine (four-rayed stars). In every case, the gemstone must be cut in such a way that the optical effect is made visible, and the intersecting sets of light-reflecting, tiny, hollow tubes or fibrous crystals appear as bright bands. The most valuable star stones have well-centred stars with straight arms of equal intensity, as well as excellent background colour.

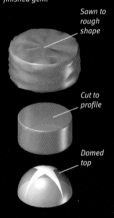

*Sawn to rough shape*

*Cut to profile*

*Domed top*

**CUTTING THE GEM**
*Once the gem is oriented, it is sawn into the rough shape required. It is then rounded in profile, and finally the top is cut into a smooth dome.*

## CAT'S EYES

The simplest form of chatoyancy is the cat's eye, in which the needle-like inclusions are oriented in only one direction, producing a single band of light – reminiscent of the narrow, vertical pupil in the eye of a cat. When a cat's-eye stone is to be cut, the cabochon is oriented along the direction of the inclusions, perpendicular to the length of the crystal. Often, a jeweller making a brooch from a cat's eye will centre the single line of light that appears so that it bisects the stone. In many smaller pieces the band of inclusions may be allowed to appear to the side of the gem.

*Gem is cut with the line of inclusions set off-centre*

**CAT'S-EYE CROSS**
*This pendant is composed of 12 chrysoberyl chatoyant gems set in the form of a cross. All of the gems are cut en cabochon to highlight the cat's-eye effect. For dramatic effect, the central, largest gem is set so the cat's eye is horizontal, while the surrounding 11 gems are set so the cat's- eye effect appears vertically.*

*Cat's-eye effect*

### STAR OF INDIA
In addition to the rutile bands in this 563-carat star sapphire, rutile crystals throughout give it a milky appearance.

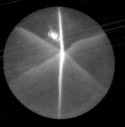

### ROSE QUARTZ
The relative depths of the three bands of inclusions may be seen inside this beautiful rose quartz.

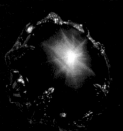

### BLACK SAPPHIRE
In a diamond and platinum ring setting, this 70.8-carat black sapphire from Sri Lanka shows a rare 12-rayed star.

### STAR OF ASIA
Weighing 330 carats, this star sapphire originated in Burma. It is famous for its size, colour, and sharply defined star.

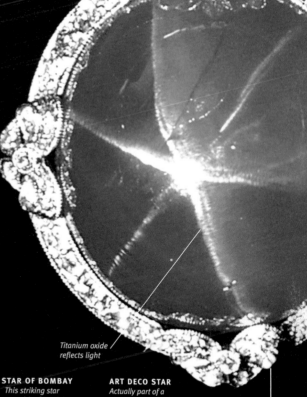

Titanium oxide reflects light

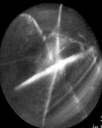

### ROSSER REEVES STAR RUBY
This 138.7-carat gem from Sri Lanka is probably the world's finest star ruby. In an earlier incarnation, it weighed 140 carats, but it was recut in the mid-20th century to centre the star.

### STAR OF BOMBAY
This striking star sapphire from Sri Lanka weighs 182 carats. Movie star Douglas Fairbanks gave it to his wife, Mary Pickford, who bequeathed it to the Smithsonian Institution in Washington, DC, USA.

### ART DECO STAR
Actually part of a platinum necklace in the Smithsonian's National Gem Collection, this 60-carat sky-blue star sapphire from Sri Lanka is mounted in a diamond-encrusted setting produced in the Art Deco style. Unlike the Star of India (above left), also in the Collection, the bands of microscopic inclusions are not uniformly even in their intensity.

High transparency

### YELLOW CHRYSOBERYL
This smooth cat's-eye gem is cut in a double cabochon (hollowed on both sides). Chrysoberyl is composed of beryllium aluminium oxide with iron and chromium, and ranges in colour from yellow to brown.

### RED-BROWN CHRYSOBERYL
The reddish-brown of this dark chrysoberyl double cabochon makes all the more dramatic the stripe of light reflecting from its microscopic inclusions. For jewellers, the term cat's eye refers to a chrysoberyl gem unless it is qualified by the name of another mineral.

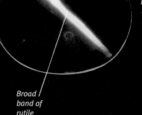

Broad band of rutile

Iron causes brown colour

HAWK'S EYE QUARTZ

PALE QUARTZ

### ENSTATITE CAT'S EYE
A magnesium silicate stone, enstatite ranges in colour from white, grey, to light brown or pale green. Heating can produce a deeper green. One bronze-coloured variety, called "bronzite", often contains inclusions of hematite or goethite that give a cat's-eye effect.

### CAT'S-EYE QUARTZ CABOCHONS
The three different varieties of quartz shown here all contain inclusions of crocidolite (blue asbestos), which create a cat's-eye effect when the gems are cut en cabochon. Chatoyant quartz is found in Sri Lanka, India, Brazil, Australia, and the USA.

TIGER'S EYE QUARTZ

# ICE

ALTHOUGH LARGELY ABSENT at
lower latitudes, ice is still the most abundant
mineral exposed at the Earth's surface. Yet, in spite
of being so common, ice is rarely seen in museums and
virtually never in private collections. Water is not
classified as a mineral because it has no crystalline form.

The name "ice" originates from the Middle English
word *iis*, and also from the Dutch *ijs* and the German
*eis*. As snow, it forms crystals that seldom exceed 7mm
(1/4in), although as massive aggregates in glaciers,
individual crystals may reach 45cm (18in). Other forms
include dendritic frost, frost as skeletal, hopper-shaped
prisms, and rounded polycrystalline bodies with
concentric structures (hailstones and icicles). Crystals are
generally colourless, but larger, massive bodies can be
light blue. Ice's common white colour is due to gaseous
inclusions of air. There are at least nine polymorphs –
different crystalline forms – of ice, each forming under
slightly different pressure and temperature conditions.

The hardness of ice varies with its crystal structure,
purity, and temperature. Ice exists only at temperatures
below 0°C (32°F). At –44°C (–47°F), which is not unusual
in arctic or high-alpine conditions, ice has a hardness of
4 on the Mohs scale, which is the same as fluorite (see
pp.172–73). At –77°C (–106°F), a temperature still within
reach in arctic or high-alpine conditions, it has the
hardness of feldspar – about 6. At this temperature, it
is hard enough to erode stone when windblown. If ice

**DIFFERENT FORMS OF ICE**
*Ice occurs in rock form as
glaciers (above), permafrost,
and icebergs in the Arctic and
Antarctic, and seasonally as
snow or as ice covering lakes
or rivers.*

**HOME-MADE ICE**
*Home-made or synthetic ice
is not a mineral, just as
synthetic gemstones are not
regarded as minerals.*

**SNOWFLAKE**
*Each snowflake is made up
of platy, hexagonal crystals
of ice.*

Airborne snow

Snow
(85–90% air)

Granular ice
(30–85% air)

Firn
(20–30% air)

Blue ice
(less than 20%
air, as bubbles)

**ICE-SHEET FORMATION**
*Ice sheets form through the gradual
accumulation, compaction, and
recrystallization of fallen snow.*

exists on any of the planets or moons in
the outer Solar System, it could be even
harder than this.

The classification applied to rocks
can also be used for ice, dividing it into
three types: igneous, when it crystallizes
from liquid water; sedimentary, when it
falls as snow; and metamorphic, when it
is in glaciers under tremendous pressure,
deforming and recrystallizing. Large
crystals of hoar ice, 10cm (4in) long, are
often found in caves, old mines, and
glacial crevasses where they have
crystallized directly from the air. Natural
ice has declined in economic value, but
as a medium for playing winter
sports, it is only marginally
supplanted by the
artificial form.

## PRESERVING FOOD

Once the preservative properties of chilling were discovered,
the harvesting and storing of ice for use as a refrigerant
became a major winter occupation in parts of Europe and
North America. The Chinese are said to have stored ice in caves
as early as the 8th century BC. The first ice-house in Britain was
built in the early 17th century, and by the 18th century an ice
house was a usual part of a large estate. Some were built as
pits, others at ground level. They were usually insulated with
straw both in the walls of the ice-house and around the blocks.
Some had a capacity of hundreds of tonnes. An ice crop could
last for up to three years, but
many did not last the entire summer, so it was an
unreliable method of long-term food preservation. With
the advent of modern refrigeration and the production
of artificial ice, natural ice is no longer an important
economic product as a refrigerant.

**PRIVATE ICE-HOUSE**
*This ice-house on the
Holkham Estate, Norfolk,
was typical of the 18th-
century English ice-house.*

**HARVESTING ICE**
*In 1900, workers cut sheets of
ice from the surface of Canneaut
Lake, Pennsylvania. The ice was
then floated to an ice-house.*

## ICE-CORE RESEARCH

**THIN SECTION**
*This section comes from an ice-core sample taken from the ice sheet in Greenland.*

Cores of ice recovered from holes bored deep into glaciers and ice sheets have provided important information about past atmospheric conditions, including temperature, pollutants, and the presence of dust particles (which can bring about climate change). An ice sheet preserves the entire history of its accumulation, in some cases extending back more than 300,000 years. Actual samples of ancient atmospheres are trapped in air bubbles within the ice, as are records of snow accumulation, air temperature, and fallout from volcanic, terrestrial, marine, cosmic, and man-made sources.

**SNOW CORING**
*These scientists are examining a newly extracted snow core, useful in avalanche research and for studying ice-sheet formation.*

### PROPERTIES

| | |
|---|---|
| **GROUP** | Oxides |
| **CRYSTAL SYSTEM** | Hexagonal |
| **COMPOSITION** | $H_2O$ |
| **COLOUR** | Colourless |
| **FORM/HABIT** | Platy, prismatic, dendritic, massive |
| **HARDNESS** | Varies |
| **CLEAVAGE** | Perfect, difficult |
| **FRACTURE** | Conchoidal, brittle |
| **LUSTRE** | Vitreous |
| **STREAK** | White |
| **SPECIFIC GRAVITY** | 1.0 |
| **TRANSPARENCY** | Transparent to translucent |

"In skating over **thin ice**, our safety is in our speed."

**RALPH WALDO EMERSON,**
"PRUDENCE", 1941

**ICE HOTEL**
*This chapel of ice is part of the Ice hotel in Jukkasjarvi, Sweden. At the start of every winter, the hotel is constructed from 20,000 tonnes of snow and 3,000 tonnes of ice. During the winter months, tourists can stay in this hotel and others in Sweden, Greenland, Alaska, and Canada.*

**INTERIOR OF THE PEDERSON GLACIER**
*Deep within a glacier, such as this one in Alaska, ice often looks blue. This is because ice absorbs other colours in the spectrum, but reflects blue.*

**IGLOO**
*Igloos are built from fine-grained, compact snow cut into blocks. They provide temporary winter homes or dwellings on hunting grounds for Inuit from Greenland and Canada.*

# CUPRITE

**CORNWALL**
*Cuprite is found in this part of southern England.*

A MAJOR ORE OF COPPER, cuprite is named from the Latin *cuprum*, meaning "copper". It can turn superficially dark grey on exposure to light. Cuprite typically has cubic crystals. In the variety called chalcotrichite or plush copper ore, the crystals are fibrous and found in loosely matted aggregates. It can also be massive and earthy. Cuprite is a secondary mineral, formed by the oxidation of copper sulphide veins. Fine specimens come from Namibia, Australia, Russia, France, and the USA.

## PROPERTIES

| | |
|---|---|
| **GROUP** | Oxides |
| **CRYSTAL SYSTEM** | Cubic |
| **COMPOSITION** | $Cu_2O$ |
| **COLOUR** | Shades of red to nearly black |
| **FORM/HABIT** | Octahedral |
| **HARDNESS** | $3^{1}/_{2}$–4 |
| **CLEAVAGE** | Distinct |
| **FRACTURE** | Uneven, brittle |
| **LUSTRE** | Adamantine, submetallic |
| **STREAK** | Brownish-red, shining |
| **SPECIFIC GRAVITY** | 6.1 |
| **TRANSPARENCY** | Transparent to almost opaque |

**CUPRITE CRYSTALS**
*Cuprite crystals are cubic, octahedral, or rarely dodecahedral.*

Adamantine lustre on crystal faces

**STEP CUT**
*Cuprite is extremely rare in faceted stones, and is cut for collectors only, here in a rectangular step cut.*

# ZINCITE

**STERLING HILL**
*Zincite is found with zinc in this mine in New Jersey, USA.*

RED OXIDE OF ZINC is another name for zincite. Natural crystals are rare; it is usually massive and granular. When crystals do occur, they are hollow pyramidal. Some so-called natural crystals of zincite in the collector's market are, in fact, smelter products. Zincite is found mainly as an accessory mineral in zinc ore deposits Crystals are found only in secondary veins or fractures. The key localities are Franklin and Sterling Hill, New Jersey, USA, where it is an ore of zinc. Tsumeb, Namibia, is another important source.

**MASSIVE ZINCITE**
*This massive specimen of zincite is from Sterling Hill, New Jersey, USA.*

Massive habit

Deep red zincite

## PROPERTIES

| | |
|---|---|
| **GROUP** | Oxides |
| **CRYSTAL SYSTEM** | Trigonal |
| **COMPOSITION** | $ZnO$ |
| **COLOUR** | Orange-yellow to deep red |
| **FORM/HABIT** | Massive |
| **HARDNESS** | 4 |
| **CLEAVAGE** | Perfect |
| **FRACTURE** | Conchoidal |
| **LUSTRE** | Resinous |
| **STREAK** | Orange-yellow |
| **SPECIFIC GRAVITY** | 5.7 |
| **TRANSPARENCY** | Almost opaque |

# MAGNETITE

**HARZ MOUNTAINS**
*This part of Germany is a key magnetite locality.*

MAGNETITE IS HIGHLY MAGNETIC: it will attract iron filings and deflect a compass needle. An iron oxide in the spinel group of minerals, it usually forms octahedral crystals, although it is sometimes found in highly modified dodecahedrons. Magnetite can also be granular, occurring as disseminated grains and as concentrations in black sand. It is similar in appearance to hematite, but hematite is non-magnetic and has a red streak. Magnetite is one of the most widespread of the oxide minerals, and is found in a wide range of geological environments. It is a high-temperature accessory mineral in igneous and metamorphic rocks and in sulphide veins. A major ore of iron, it forms large ore bodies in Norway; Kiruna, Sweden; and New York, USA. Excellent crystals are found at Binntal, Switzerland; Traversella, Italy; Zillertal, Austria; Värmland, Sweden; Durango, Mexico; and in New York and Vermont, USA. The most strongly magnetic material is found in the Ural Mountains of Russia and Mount Elba, Italy.

Magnetite is said to be named for the Greek shepherd boy Magnes, who noticed that the iron ferrule of his staff and the nails of his shoes clung to a magnetite-bearing rock.

**CRYSTAL**
*This magnetite crystal shows the mineral's classic octahedral habit.*

Iron filings

**MASSIVE MAGNETITE**
*This specimen of magnetite is covered with iron filings, an illustration of the strength of its intrinsic magnetism.*

Magnetic field

## PROPERTIES

| | |
|---|---|
| **GROUP** | Oxides |
| **CRYSTAL SYSTEM** | Cubic |
| **COMPOSITION** | $Fe_3O_4$ |
| **COLOUR** | Black to brownish-black |
| **FORM/HABIT** | Octahedral |
| **HARDNESS** | $5^{1}/_{2}$–6 |
| **CLEAVAGE** | None |
| **FRACTURE** | Conchoidal to uneven |
| **LUSTRE** | Metallic to semi-metallic |
| **STREAK** | Black |
| **SPECIFIC GRAVITY** | 5.2 |
| **TRANSPARENCY** | Opaque |

# SPINEL

SPINEL IS BOTH a mineral name and the name of a group of mostly rock-forming minerals, all of which are metal oxides, and all of which have the same crystal structure. Members of the group include gahnite (see p.162), franklinite (see p.158), and chromite (see p.158). Spinel minerals are usually found as glassy, hard octahedrons, grains, or masses. The spinel group contains the mineral spinel. The red variety is often called "ruby spinel" because of its blood-red colour, although it is also found in green, blue, brown, and black. It is frequently cut as a gemstone because, as a member of the aluminium spinels, it is harder and more transparent than the iron or chromium spinels. Its name comes from the Latin *spinella*, meaning "little thorn", a reference to the sharp points on the octahedral crystals. Its red colour is due to chromium.

Spinel is found in mafic igneous rocks, aluminium-rich metamorphic rocks, and contact metamorphosed limestones. Classic occurrences are the gem gravels of Myanmar, Sri Lanka, and Madagascar. Gem-quality stones are also found in Afghanistan, Pakistan, Brazil, Australia, Sweden, Italy, Turkey, Russia, and the USA. The earliest known gem spinel dates from 100BC and was discovered in a Buddhist tomb near Kabul, Afghanistan. Red spinel was also known to the Romans at the same time. Blue spinels from the Roman period (51BC–AD400) have been found in England.

## PROPERTIES

| | |
|---|---|
| **GROUP** | Oxides |
| **CRYSTAL SYSTEM** | Cubic |
| **COMPOSITION** | $MgAl_2O_4$ |
| **COLOUR** | Red, yellow, orange-red, blue, green, brown, black |
| **FORM/HABIT** | Octahedral |
| **HARDNESS** | $7\frac{1}{2}$–8 |
| **CLEAVAGE** | None |
| **FRACTURE** | Conchoidal to uneven |
| **LUSTRE** | Vitreous |
| **STREAK** | White |
| **SPECIFIC GRAVITY** | 3.6 |
| **TRANSPARENCY** | Transparent to translucent |
| **R.I.** | 1.71–1.73 |

## SPINEL "RUBIES"

Many historic "rubies" were actually spinel. Two are part of the British Crown Jewels: the Timur Ruby and the Black Prince's Ruby. The Timur Ruby is inscribed with the names and dates of six of its owners – the earliest of whom is Shah Jahgagir, 1612, and the latest Ahmad Shah Durrani, 1754. The superb spinel known as the Black Prince's Ruby was supposedly given to Edward, the Black Prince (son and heir-apparent of Edward III) by Pedro the Cruel, king of Castille, after the victory of Najera in 1367. It was later mounted in the British Imperial State Crown. Henry V wore and nearly lost it at the Battle of Agincourt in 1415. It has been passed down through the English, and later British, royal families.

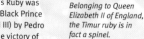

**TIMUR RUBY NECKLACE**
*Belonging to Queen Elizabeth II of England, the Timur ruby is in fact a spinel.*

— Sapphire in finial cross

— Spinel

**BLACK PRINCE'S RUBY**
*Mounted above the Cullinan II diamond, this "ruby" is in reality a 170-carat spinel from Badakhshan, Afghanistan.*

— Diamond

**CUSHION MIX**
*The transparency of spinel makes possible brilliant, fiery gems. Here brilliant-cut crown facets are combined with step-cut pavilion facets.*

**OCTAGONAL MIX**
*Although spinel is often thought of as a red gemstone, other spinel colours also produce superb gems.*

**BRILLIANT OVAL**
*This faceted brilliant oval shows the red coloration commonly associated with spinel and sometimes known as ruby spinel.*

**STAR CABOCHON**
*Star spinels are rare but, when found, show a four- or six-rayed star. Here a six-rayed star is brought out by the cabochon cut.*

Spinel crystal —

Gahnite crystal —

**SPINEL CRYSTALS**
*The blue spinel crystals in calcite (far left) are from Lake Baikal, Russia. Brown spinel-group crystals (centre) are known as gahnite, a zinc-rich variety of spinel. The ruby-red spinel crystals (left) have formed together into a crystal aggregate.*

**BLUE CRYSTALS IN CALCITE**

**BROWN CRYSTALS IN QUARTZ**

**AGGREGATE CRYSTALS**

Octahedral crystal —

Vitreous lustre —

Stream-rounded crystal —

**SPINEL CRYSTALS**
*These crystals are typical of those recovered from the gem gravels of Myanmar, Sri Lanka, and Madagascar.*

# FRANKLINITE

**STERLING HILL**
*Franklinite is mined at Sterling Hill near Franklin, New Jersey, USA.*

NAMED FOR THE LOCALITY of its discovery and major source, Franklin, New Jersey, franklinite is a zinc manganese iron oxide, and a member of the spinel group of minerals. It mainly crystallizes in octahedrons, commonly with rounded edges, and also as massive and granular aggregates. Occuring in zinc deposits in metamorphosed limestones and dolomites, franklinite is usually accompanied by a range of minerals such as willemite, garnet, and rhodonite. It has also been found in Germany, Sweden, and Romania.

**FRANKLIN**
*Franklin, New Jersey was named after the 18th-century American statesman, Benjamin Franklin.*

## PROPERTIES

| | |
|---|---|
| **GROUP** | Oxides |
| **CRYSTAL SYSTEM** | Cubic |
| **COMPOSITION** | (Zn,Mn,Fe)(Fe,Mn)$_2$O$_4$ |
| **COLOUR** | Iron-black |
| **FORM/HABIT** | Octahedral |
| **HARDNESS** | 5$^1$/$_2$–6$^1$/$_2$ |
| **CLEAVAGE** | None |
| **FRACTURE** | Conchoidal to uneven |
| **LUSTRE** | Metallic |
| **STREAK** | Reddish-black to iron-black |
| **SPECIFIC GRAVITY** | 5.2 |
| **TRANSPARENCY** | Opaque |

## FRANKLIN MINE

One of the world's most prolific mineral deposits in terms of number of species is located at the Franklin Mine, New Jersey, USA. The deposit has produced over 300 different mineral species, some unique to that mine. Of these 300, it is the type locality (the locality where a mineral is first recognized) for 60. Franklinite was abundant enough there to constitute an ore of zinc. The New Jersey Zinc Company, formed at the mine, produced ready-mixed zinc-oxide paint, the first to replace poisonous lead-based paint.

### VISITING THE MINE

Mining stopped at the Franklin Mine in 1954, but tours are available of some of the old workings, and specimens can still be collected on some of the remaining spoil heaps.

*Calcite groundmass*

**FLUORESCENT MINERAL**
*In this photograph of the interior of the mine taken under ultraviolet light, franklinite can be seen fluorescing.*

**OCTAHEDRAL CRYSTALS**
*Franklinite's metallic lustre and octahedral crystals are clearly seen in this specimen.*

*Octahedral crystal of franklinite*

# CHROMITE

**BENBOW MINE**
*Chromite is mined here in the Stillwater Complex, Montana, USA.*

A MEMBER OF THE SPINEL GROUP, chromite is an iron chromium oxide. Crystals of chromite are uncommon, and are usually either massive as lenses and tabular bodies, or disseminated as granules and streaks. It is dark brown to black and can contain some magnesium and aluminium. Chromite is most commonly found as an accessory mineral in ultramafic igneous rocks, or concentrated in sediments derived from them. In a few ultramafic bodies, it occurs as sedimentary–like layers of almost pure chromite. These rocks, known as chromitites, are the most important ores of chromium. Chromite also occurs in metamorphosed ultramafic rocks such as serpentinites. Layered deposits of chromite are found in the Bushveld Complex of South Africa, the Great Dyke of Zimbabwe, and the Stillwater Complex of Montana, USA. Good crystals occur in the Shetland Islands of Scotland and on the island of New Caledonia.

## USES OF CHROMIUM

Chromite is the only significant source of chromium metal. Chromium is alloyed with iron in the steel industry to manufacture high-speed tool steels and stainless steels. When they are alloyed with other metals such as molybdenum, tungsten, niobium, or titanium, chrome steels have superb high-temperature properties, and are used in furnace parts, burner nozzles, and kiln linings. Chromite itself is used as a refractory material for lining furnaces.

**LAYERS**
*Chromitite is found in layers, which can run continuously for many miles.*

**CHROME BUMPER**
*The large chrome rear bumper is a striking feature of this 1959 pink Cadillac Eldorado convertible.*

## PROPERTIES

| | |
|---|---|
| **GROUP** | Oxides |
| **CRYSTAL SYSTEM** | Cubic |
| **COMPOSITION** | FeCr$_2$O$_4$ |
| **COLOUR** | Dark brown, black |
| **FORM/HABIT** | Granular, massive |
| **HARDNESS** | 5$^1$/$_2$ |
| **CLEAVAGE** | None |
| **FRACTURE** | Uneven |
| **LUSTRE** | Metallic |
| **STREAK** | Brown |
| **SPECIFIC GRAVITY** | 4.7 |
| **TRANSPARENCY** | Opaque |

*Nodular chromite*

*Weathered crystal*

**NODULES**
*The metallic lustre of chromite is visible on broken surfaces of chromite nodules.*

*Groundmass of serpentinite*

# CHRYSOBERYL

ALTHOUGH CRYSTALS OF CHRYSOBERYL are not uncommon, the gemstone variety, alexandrite, is one of the rarest and most expensive gems. A beryllium aluminium oxide, chrysoberyl is hard and durable, inferior in hardness only to corundum and diamond. It generally occurs in granites or granitic pegmatites, although alexandrites are usually found in mica schists. Because of chrysoberyl's durability, crystals that weather out of the parent rock are often found in streams and gravel beds. Chrysoberyl crystals are commonly twinned.

The largest faceted chrysoberyl, from Russia, weighs 66 carats (about 14g/ ½oz), while alexandrites weighing more than 10 carats are rare. Aside from alexandrite, which appears green in daylight but red under tungsten

light, chrysoberyl occurs in other colours, from green, greenish-yellow, and yellow, to brown. Pale yellow Brazilian stones were used by the Spanish and Portuguese in the 17th and 18th centuries to make superb jewellery. Chrysoberyl has been known for thousands of years in Asia, where it was said to protect against the "evil eye". Chrysoberyl is found in Myanmar, Zimbabwe, Tanzania, Madagascar, and in Colorado and Connecticut, USA. The original deposit of alexandrite in the Urals of Russia is mainly worked out, but some is still mined in Brazil and Sri Lanka. Synthetic alexandrite has been made for a number of years. Chromium is the colour-producing trace element in alexandrite, replacing some of the aluminium in the structure.

**MINAS GERAIS, BRAZIL**
*Brazil's famous mining state is a source for chrysoberyl as well as many other minerals.*

**VICTORIAN BROOCH**
*Set in gold filigree, these stones are faceted yellowish-green chrysoberyls.*

**CHRYSOBERYL ORNAMENT**
*The chrysoberyls in this Portuguese piece from about 1760 are set in silver and probably come from Brazil.*

**COLOUR CHANGE**
*Alexandrite is known for its dramatic colour change from brilliant green in daylight to cherry-red under tungsten light.*

**TUNGSTEN LIGHT**

**DAYLIGHT**

**TSAR ALEXANDER II**
*Alexandrite was named for Tsar Alexander II of Russia. It was found, supposedly, on his birthday in 1830.*

**TWINNED CRYSTALS**
*These large, wedge-shaped crystals resting on a tabular crystal are typical of chrysoberyl.*

Multiple twinned crystals

## PROPERTIES

| | |
|---|---|
| **GROUP** | Oxides |
| **CRYSTAL SYSTEM** | Orthorhombic |
| **COMPOSITION** | $BeAl_2O_4$ |
| **COLOUR** | Green, yellow |
| **FORM/HABIT** | Tabular, stout prismatic |
| **HARDNESS** | $8\frac{1}{2}$ |
| **CLEAVAGE** | Distinct |
| **FRACTURE** | Uneven to conchoidal |
| **LUSTRE** | Vitreous |
| **STREAK** | Colourless |
| **SPECIFIC GRAVITY** | 3.7 |
| **TRANSPARENCY** | Transparent to translucent |
| **R.I.** | 1.74–1.75 |

**ALEXANDRITE CRYSTALS**
*This specimen of alexandrite with mica was mined in Siberia, Russia.*

Striations on crystal face

Vitreous lustre

Twinned crystals

## CAT'S EYE

A cloudy, opalescent, and chatoyant variety of chrysoberyl is sometimes called cymophane or cat's eye. When cut *en cabochon* some with parallel, acicular inclusions exhibit a cat's-eye effect: when the stone is properly oriented the reflected light is focused by the inclusions into a bright band on the surface. Cut cat's-eye stones rarely exceed 100 carats, and the finest are in the same price range as fine sapphires. Hindus once believed that a cat's eye guarded its owner's health and provided assurance against poverty.

**HONEY-YELLOW CAT'S EYE**
*The most desirable form of cat's eye is honey-yellow or greenish-yellow. The 11 stones set in this cross and the cabochon alongside are honey-yellow.*

**CABOCHON**
*This greenish-yellow chrysoberyl polished cabochon shows a faint cat's eye.*

**CUSHION MIXED**
*This transparent faceted stone ranges from dark green to brown in colour.*

**MIXED CUT**
*Photographed under incandescent light, this alexandrite stone appears golden brown.*

# FLUORESCENT MINERALS

VIEWED UNDER ULTRAVIOLET LIGHT, SOME CRYSTALS
EMIT VISIBLE LIGHT IN REMARKABLE COLOURS,
LENDING THEM AN EERIE, PSYCHEDELIC
APPEARANCE. FIRST NOTED IN FLUORITE, THE
PHENOMENON IS CALLED FLUORESCENCE.

Fluorescence is the property of some minerals to emit visible light of various colours when exposed to ultraviolet light (also known as "black light"). This particular kind of luminescence was first identified in 1824 by German mineralogist Friedrich Mohs (1773–1839), who also developed the Mohs scale of hardness. He observed that some fluorites, when viewed under ultraviolet light, appeared a completely different colour than when viewed in daylight.

Ultraviolet light can be produced in both short and long waves, and some minerals fluoresce only in one or the other; still others fluoresce in both. There is an unpredictability to fluorescence: some specimens of a mineral will manifest it, while other specimens will not. For example, willemite from the Franklin district of New Jersey, USA, may show brilliant fluorescent colours, while willemite from other localities may show none whatsoever. More surprisingly, specimens that are apparently identical and come from the same locality may or may not fluoresce. Also, the colour and intensity of the fluorescence can vary significantly, depending on the wavelengths of ultraviolet light used.

## WHY MINERALS FLUORESCE

Fluorescence occurs when certain atoms within the structure of a mineral move into a different, higher-energy state when exposed to ultraviolet light. These atoms absorb energy from the ultraviolet light and become "excited". They stay in this excited state for no more than tiny fractions of a second, then, as they move back into their "unexcited" state, termed their "ground" state, the energy they absorbed is released as fluorescent light. The atoms that cause fluorescence are often trace elements (see pp.92). Trace elements are prone to "excitation" because they are already in an electronically unstable situation as a result of not fitting perfectly into the crystal structure that contains them.

Another phenomenon related to the "excitation" of a mineral's atoms is phosphorescence. This is a continued emission of light as an afterglow following the removal of the "exciting" light source. Willemite, for example, can phosphoresce as well as fluoresce.

Fluorescence is exploited in mineral science in several ways: in some minerals, it serves as a means of identification – ultraviolet is used to find minerals that fluoresce; and, in the extraction of ore minerals that fluoresce, ultraviolet verifies whether extraction processes are effective. Fluorescence is also used in the microscopic examination of small-scale structures in crystals.

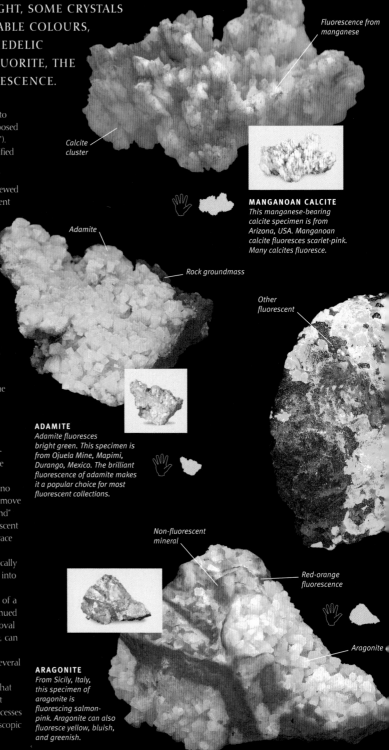

*Fluorescence from manganese*

*Calcite cluster*

**MANGANOAN CALCITE**
*This manganese-bearing calcite specimen is from Arizona, USA. Manganoan calcite fluoresces scarlet-pink. Many calcites fluoresce.*

*Adamite*

*Rock groundmass*

*Other fluorescent*

**ADAMITE**
*Adamite fluoresces bright green. This specimen is from Ojuela Mine, Mapimi, Durango, Mexico. The brilliant fluorescence of adamite makes it a popular choice for most fluorescent collections.*

*Non-fluorescent mineral*

*Red-orange fluorescence*

*Aragonite*

**ARAGONITE**
*From Sicily, Italy, this specimen of aragonite is fluorescing salmon-pink. Aragonite can also fluoresce yellow, bluish, and greenish.*

**FLUORITE ON QUARTZ**
Fluorite fluoresces violet-blue. This specimen is from Redburn Mine, Weardale, Durham, England. Blue fluorescence in fluorite is often caused by traces of erbium.

Rock groundmass

Scapolite

Fluorite

Quartz

**SCAPOLITE**
This scapolite specimen is from Quebec, Canada. It fluoresces yellow. Some scapolites fluoresce, others do not, depending on individual specimens.

Willemite

Calcite

Columnar calcite

Concentrated fluorescent material

**CALCITE**
The columnar crystals of this calcite specimen fluoresce white. Trace elements in calcite can cause other fluorescent colours.

**WILLEMITE AND CALCITE**
This specimen is from Franklin, New Jersey, USA. The willemite is fluorescing green and the calcite is fluorescing red.

Non-fluorescent mineral

Sodalite

Gypsum

**GYPSUM**
From Montmartre, Paris, France, this gypsum fluoresces a rich yellow. Fluorescence is rare in gypsum, although it can be caused by foreign molecules.

**SODALITE**
This Indian sodalite, from Kishangarh, Madhya Pradesh, is fluorescing orange. Sodalite specimens may also fluoresce orange-red fading to yellow-white.

Rock groundmass

# CASSITERITE

A TIN OXIDE, cassiterite is named for the Greek *kassiteros*, "tin". It can form heavily striated prisms and pyramids, and twinned crystals are quite common. It is also massive, and occurs as a botryoidal fibrous variety (wood tin) or water-worn pebbles (stream tin). Cassiterite forms in hydrothermal veins associated with granitic rocks, with tungsten minerals such as wolframite, and with molybdenite, tourmaline, and topaz. Durable and relatively dense, it becomes concentrated in sediments when eroded from its primary rocks. Many of the most productive tin ores are placer deposits. Fine crystals come from Portugal, Italy, France, the Czech Republic, Brazil, and Myanmar.

### PROPERTIES

| | |
|---|---|
| **GROUP** | Oxides |
| **CRYSTAL SYSTEM** | Tetragonal |
| **COMPOSITION** | $SnO_2$ |
| **COLOUR** | Medium to dark brown |
| **FORM/HABIT** | Short prismatic |
| **HARDNESS** | 6–7 |
| **CLEAVAGE** | Indistinct |
| **FRACTURE** | Subconchoidal to uneven |
| **LUSTRE** | Adamantine to metallic |
| **STREAK** | White, greyish, brownish |
| **SPECIFIC GRAVITY** | 7.0 |
| **TRANSPARENCY** | Transparent to opaque |
| **R.I.** | 2.0–2.1 |

**OVAL BRILLIANT**
*The characteristic golden-orange colour of this gem is transparent with a resinous lustre.*

**CRYSTALS**
*Originating from Pingwu, Sichuan, China, this specimen on a muscovite groundmass includes twinned cassiterite crystals.*

Muscovite

## MINING AND USING TIN

There are few tin minerals; cassiterite is almost the sole source of the metal. Massive alluvial (river) deposits occur in Malaysia, Thailand, and Indonesia. It is also worked from primary deposits underground in Bolivia. Tin lodes have been worked in south-west England, including Cornwall, since pre-Roman times, although there is virtually no production today. Tin from Cornwall was traded across the Mediterranean world by the Phoenicians, and then alloyed with copper from Cyprus and elsewhere to make bronze.

**TIN INGOT**

**CORNISH TIN MINE**
*An abandoned pump house marks the site of a pre-Roman tin mine on the Cornish coast.*

Varlamoffite
crystals

Short prismatic
crystal

Rock
groundmass

**VARLAMOFFITE CASSITERITE**

**CASSITERITE VARIETIES**
*The varlamoffite specimen was found at Gunheath Pit, Cornwall. The fibrous, botryoidal variety is known as wood tin cassiterite.*

**WOOD TIN CASSITERITE**

---

# GAHNITE

**SALIDA, USA**
*This area of Colorado produces fine gahnite crystals.*

A ZINC ALUMINIUM OXIDE, gahnite is a member of the spinel group, forming octahedral crystals. It is also found as irregular grains and masses. Gahnite crystals are usually dark green or blue to black and striated. It was named in 1807 for the Swedish chemist J.G. Gahn. Gahnite is found in schists and gneisses, in granite pegmatites, and in contact metamorphosed limestones. Good crystals come from the USA, Sweden, Finland, Australia, Brazil, and Mexico.

### PROPERTIES

| | |
|---|---|
| **GROUP** | Oxides |
| **CRYSTAL SYSTEM** | Cubic |
| **COMPOSITION** | $ZnAl_2O_4$ |
| **COLOUR** | Dark green or blue |
| **FORM/HABIT** | Octahedral |
| **HARDNESS** | 7½–8 |
| **CLEAVAGE** | Indistinct |
| **FRACTURE** | Conchoidal |
| **LUSTRE** | Vitreous |
| **STREAK** | Greyish |
| **SPECIFIC GRAVITY** | 4.6 |
| **TRANSPARENCY** | Translucent to nearly opaque |

Octahedral
crystal

Rock
groundmass

**BLUE GAHNITE**
*This specimen of octahedral crystals is from Franklin, New Jersey. Other localities in the USA are Colorado and Maine.*

---

# PYROLUSITE

**LAKE SUPERIOR**
*Pyrolusite is found in this part of Michigan, USA.*

A COMMON MANGANESE mineral, pyrolusite is the primary ore of manganese. It rarely forms crystals, and is usually found as light grey to black massive aggregates, metallic coatings, crusts, fibres, and nodules. Pyrolusite forms under highly oxidizing conditions as an alteration product of other manganese minerals such as rhodochrosite. It has also been found in bogs, lakes, shallow marine environments, and on the ocean floor. Rare crystals are found in Canada and the Czech Republic.

### PROPERTIES

| | |
|---|---|
| **GROUP** | Oxides |
| **CRYSTAL SYSTEM** | Tetragonal |
| **COMPOSITION** | $MnO_2$ |
| **COLOUR** | Steel-grey to black |
| **FORM/HABIT** | Massive |
| **HARDNESS** | 6–6½ |
| **CLEAVAGE** | Perfect |
| **FRACTURE** | Uneven, brittle, splintery |
| **LUSTRE** | Metallic to earthy |
| **STREAK** | Black or bluish-black |
| **SPECIFIC GRAVITY** | 4.4–5.1 |
| **TRANSPARENCY** | Opaque |

**MASSIVE HABIT**
*This dark grey specimen of massive pyrolusite has an uneven fracture.*

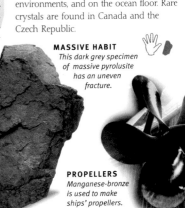

**PROPELLERS**
*Manganese-bronze is used to make ships' propellers.*

Dull lustre

# ANATASE

**SWISS ALPS**
*Crystals found in the Swiss Alps are up to 2cm (³⁄₄in) in diameter.*

**ANATASE CRYSTAL**

FORMERLY KNOWN AS OCTAHEDRITE, anatase, along with rutile and brookite, is a polymorph of titanium oxide. It is found as brilliant crystals, which can be brown, indigo-blue, green, grey, and black. Anatase forms in veins in metamorphic rocks such as schists and gneisses, and in pegmatites. It is found in sediments and can be concentrated in placer deposits. Notable vein deposits occur widely in the Alps. Other localities for fine crystals are Brazil, Norway, and Colorado, USA.

**BLACK CRYSTALS**
*This schist from Le Bourg-d'Oisans, Isère, France is scattered with tiny black anatase crystals.*

### PROPERTIES

| | |
|---|---|
| **GROUP** | Oxides |
| **CRYSTAL SYSTEM** | Tetragonal |
| **COMPOSITION** | $TiO_2$ |
| **COLOUR** | Various shades of brown, or black, indigo-blue |
| **FORM/HABIT** | Pyramidal |
| **HARDNESS** | 5¹⁄₂–6 |
| **CLEAVAGE** | Perfect |
| **FRACTURE** | Subconchoidal |
| **LUSTRE** | Adamantine to metallic |
| **STREAK** | White to pale yellow |
| **SPECIFIC GRAVITY** | 3.9 |
| **TRANSPARENCY** | Transparent to nearly opaque |

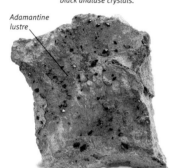

*Adamantine lustre*

# BROOKITE

**MINAS GERAIS**
*Brookite occurs in Minas Gerais, a famous mining area of Brazil.*

NAMED IN 1825 FOR English crystallographer H.J. Brooke, brookite typically occurs as brown metallic crystals. The orthorhombic polymorph of titanium oxide, it is found in veins in gneisses and schists and, more rarely, in zones of contact metamorphism. It generally occurs with rutile, anatase, and albite. Because it is relatively dense, it gets concentrated in placer deposits. Fine crystals are found in the same Alpine deposits as anatase; they are also found in Brazil, Norway, Wales, and Colorado, USA.

### PROPERTIES

| | |
|---|---|
| **GROUP** | Oxides |
| **CRYSTAL SYSTEM** | Orthorhombic |
| **COMPOSITION** | $TiO_2$ |
| **COLOUR** | Various shades of brown |
| **FORM/HABIT** | Tabular, elongated |
| **HARDNESS** | 5¹⁄₂–6 |
| **CLEAVAGE** | Indistinct |
| **FRACTURE** | Subconchoidal to uneven |
| **LUSTRE** | Metallic to adamantine |
| **STREAK** | White, greyish, yellowish |
| **SPECIFIC GRAVITY** | 4.1 |
| **TRANSPARENCY** | Opaque to transparent |

**TABULAR CRYSTAL**
*This transparent tabular brookite crystal rests on a mass of albite crystals.*

*Striated brookite crystal face*

*Albite*

# RUTILE

**CAROLINA**
*North Carolina is one of several American states that produce rutile crystals.*

TAKING ITS NAME FROM the Latin *rutilis*, "red" or "glowing", rutile is a form of titanium oxide. It ranges from reddish-brown to red, although some specimens are nearly black, and the rutile in quartz is usually pale golden-yellow. Crystals are generally prismatic, but they can also be slender and acicular. Prism faces often have vertical striations. Multiple twinning is common, and may form lattice-like structures. Other twinning also occurs. Rutile is a frequently occurring minor constituent of granites, pegmatites, gneisses, and schists. Good crystals are found in Sweden, Italy, France, Austria, Brazil, and in Pennsylvania, Arkansas, Georgia, North Carolina, California, and Virginia, USA.

Rutile commonly forms microscopic, oriented inclusions in other minerals, and produces the asterism shown by some rose quartz, rubies, and sapphires. Quartz containing long, translucent rutile needles is known as rutilated quartz, and has been used as an ornamental stone since ancient times.

### PROPERTIES

| | |
|---|---|
| **GROUP** | Oxides |
| **CRYSTAL SYSTEM** | Tetragonal |
| **COMPOSITION** | $TiO_2$ |
| **COLOUR** | Reddish-brown to red |
| **FORM/HABIT** | Slender prismatic |
| **HARDNESS** | 6–6¹⁄₂ |
| **CLEAVAGE** | Good |
| **FRACTURE** | Conchoidal to uneven |
| **LUSTRE** | Adamantine to sub-metallic |
| **STREAK** | Pale brown to yellowish |
| **SPECIFIC GRAVITY** | 4.2 |
| **TRANSPARENCY** | Transparent to opaque |
| **R.I.** | 2.62–2.90 |

*Metallic lustre*

*Vertical striations*

## USES OF TITANIUM

Titanium is one of the most important metals in modern times. It has high strength, low density, and excellent corrosion resistance. It is used in aircraft, spacecraft, missiles, and ships. As it is nonreactive with organic tissues, it is used to make artificial joints for hip replacements, and other prosthetic devices. Another important use of titanium oxide is as a white pigment.

**HARRIER JUMP JET**
*The use of titanium alloys in jet engines and airframes has made aircraft like the Harrier possible.*

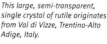

**SINGLE CRYSTAL**
*This large, semi-transparent, single crystal of rutile originates from Val di Vizze, Trentino-Alto Adige, Italy.*

**CLEAR QUARTZ**
*Needle-like rutile inclusions intersect at an angle of 60° in this polished, clear quartz cabochon.*

**ROSE QUARTZ**
*Microscopic needles of rutile create three intersecting lines of light in this rose quartz star stone.*

# URANINITE

**SAXONY**
*The German state of Saxony is a good source of fine uraninite crystals.*

A MAJOR ORE OF URANIUM, uraninite was named in 1792 for its composition – it is uranium oxide. It is highly radioactive and must be handled and stored with care. Uraninite occurs commonly in massive (known as pitchblende), botryoidal, or granular habits, and less commonly as octahedral or cubic crystals. It is black to brownish-black, dark grey, or greenish. Uraninite is found as crystals in granitic and syenitic pegmatites along with other uranium-bearing minerals and rare-earth minerals. It also forms in high-temperature hydrothermal veins with cassiterite and arsenopyrite, and in medium-temperature hydrothermal veins as pitchblende. In addition, it occurs as small grains in sandstones and conglomerates, where it has often weathered into secondary uranium minerals. Fine crystals are found in Córdoba, Spain; Saxony, Germany; Chihuahua, Mexico; and Topsham, Maine, USA.

## THE POWER OF URANIUM

Uraninite is the principal source of uranium. This is a dense, hard, silvery-white metal that is ductile and malleable. Until nuclear fission was discovered, uranium was used in very small amounts in the colouring of ceramics and as a specialized catalyst. Uranium metal is recovered from uraninite by first converting it to uranium tetrafluoride, and then by further reducing the uranium tetrafluoride with magnesium to release metallic uranium. The risks of the use of uranium – mainly for generating power and in nuclear weapons – are controversial and under intense debate. Many claim that it is unsafe, no matter how it is stored, used, and disposed of, and may remain so for hundreds or even thousands of years.

**NUCLEAR REACTOR**
*Uranium is essential for the operation of nuclear reactors for generating electricity.*

*Botryoidal habit*

*Dull to submetallic lustre*

### PROPERTIES

| | |
|---|---|
| **GROUP** | Oxides |
| **CRYSTAL SYSTEM** | Cubic |
| **COMPOSITION** | $UO_2$ |
| **COLOUR** | Black to brownish-black, dark grey, greenish |
| **FORM/HABIT** | Octahedral |
| **HARDNESS** | 5–6 |
| **CLEAVAGE** | None |
| **FRACTURE** | Uneven to subconchoidal |
| **LUSTRE** | Submetallic, pitchy, dull |
| **STREAK** | Brownish-black |
| **SPECIFIC GRAVITY** | 6.5–11.0 |
| **TRANSPARENCY** | Opaque |

**BOTRYOIDAL HABIT**
*This specimen of uraninite has a botryoidal habit, which is common in this mineral.*

**MARIE CURIE**
*In exploring the properties of pitchblende, Marie Curie (1867–1934) made numerous discoveries about radiation.*

---

# FERGUSONITE

*(hand scale icon)*

**JAPAN**
*Fergusonite deposits occur in Japan.*

THERE ARE SEVERAL MINERALS called fergusonite, named for the Scottish mineralogist Robert Ferguson. The most common is fergusonite-(Y), which is rich in yttrium. Its crystals are prismatic to pyramidal, and it is black to brownish-black in colour. Fergusonite-(Y) is found in granitic pegmatites associated with other rare-earth minerals. Localities include Hakatamura, Japan, and Ytterby, Sweden.

**SWEDISH FERGUSONITE**
*This specimen of fergusonite crystals in feldspar is from Ytterby, Sweden.*

### PROPERTIES

| | |
|---|---|
| **GROUP** | Oxides |
| **CRYSTAL SYSTEM** | Tetragonal |
| **COMPOSITION** | $YNbO_4$ |
| **COLOUR** | Black to brownish-black |
| **FORM/HABIT** | Prismatic to pyramidal |
| **HARDNESS** | $5^1/_2$–$6^1/_2$ |
| **CLEAVAGE** | Poor |
| **FRACTURE** | Subconchoidal, brittle |
| **LUSTRE** | Vitreous to submetallic |
| **STREAK** | Brown, yellow-brown, greenish-grey |
| **SPECIFIC GRAVITY** | 4.2–5.7 |
| **TRANSPARENCY** | Opaque |

*Feldspar*

*Fergusonite*

---

# SAMARSKITE

**URAL MOUNTAINS**
*Russia's Urals have deposits of samarskite.*

NAMED IN 1847 for Russian mining engineer Vasili Samarski-Bykhovets, samarskite is a complex oxide of yttrium, iron, niobium, tantalum, and uranium. Its radioactive crystals are prismatic, with a rectangular cross-section. They are found in granitic pegmatites, often associated with columbite, monazite, and garnet. Localities include its place of discovery at Miass, Russia; and Divino de Uba, Brazil.

### PROPERTIES

| | |
|---|---|
| **GROUP** | Oxides |
| **CRYSTAL SYSTEM** | Orthorhombic |
| **COMPOSITION** | $(Y,Fe,U)(Nb,Ta)O_4$ |
| **COLOUR** | Black |
| **FORM/HABIT** | Prismatic |
| **HARDNESS** | 5–6 |
| **CLEAVAGE** | Indistinct |
| **FRACTURE** | Conchoidal, brittle |
| **LUSTRE** | Vitreous to resinous |
| **STREAK** | Dark reddish-brown to black |
| **SPECIFIC GRAVITY** | 5.7 |
| **TRANSPARENCY** | Translucent to opaque |

**IRIDESCENT SHEEN**
*This specimen of samarskite exhibits an iridescent sheen on some surfaces.*

*Indistinct cleavage*

# ROMANÈCHITE

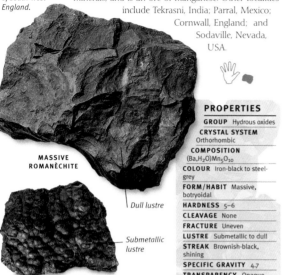

**CORNWALL**
*Romanèchite is mined in this area of south-west England.*

ONE OF THE MINERALS formerly known as psilomelane, romanèchite is named after its type locality, Romanèche-Thorins, France. Commonly massive or botryoidal, romanèchite can also be earthy. It is formed as an alteration product of other manganese minerals, and is an ore of manganese. Other localities include Tekrasni, India; Parral, Mexico; Cornwall, England; and Sodaville, Nevada, USA.

**MASSIVE ROMANÈCHITE**

*Dull lustre*

**BOTRYOIDAL ROMANÈCHITE**

*Submetallic lustre*

### PROPERTIES

| | |
|---|---|
| **GROUP** | Hydrous oxides |
| **CRYSTAL SYSTEM** | Orthorhombic |
| **COMPOSITION** | $(Ba,H_2O)Mn_5O_{10}$ |
| **COLOUR** | Iron-black to steel-grey |
| **FORM/HABIT** | Massive, botryoidal |
| **HARDNESS** | 5–6 |
| **CLEAVAGE** | None |
| **FRACTURE** | Uneven |
| **LUSTRE** | Submetallic to dull |
| **STREAK** | Brownish-black, shining |
| **SPECIFIC GRAVITY** | 4.7 |
| **TRANSPARENCY** | Opaque |

# BRAUNITE

**MINAS GERAIS**
*This area of Brazil is a good source of braunite.*

BRAUNITE IS CLASSIFIED as both an oxide and a silicate. It was named in 1831 for K.W. Braun, a German mineralogist. Crystals are pyramidal in habit, with striated faces; braunite can also be granular and massive. It is a major ore of manganese. Localities include Autlàn de Navarro, Mexico; Nombre de Dios, Panama; Ilmenau, Germany; Ouro Prêto, Brazil; and Nordmark, Sweden.

### PROPERTIES

| | |
|---|---|
| **GROUP** | Oxides |
| **CRYSTAL SYSTEM** | Tetragonal |
| **COMPOSITION** | $Mn_7SiO_{12}$ |
| **COLOUR** | Brownish-black to steel-grey |
| **FORM/HABIT** | Pyramidal |
| **HARDNESS** | 6–6½ |
| **CLEAVAGE** | Perfect |
| **FRACTURE** | Uneven to subconchoidal, brittle |
| **LUSTRE** | Submetallic |
| **STREAK** | Brownish-black to steel-grey |
| **SPECIFIC GRAVITY** | 4.8 |
| **TRANSPARENCY** | Translucent to opaque |

**RAILWAY LINES**
*Manganese is added to the steel used to make railway lines, which reduces its brittleness.*

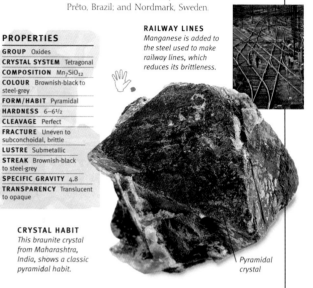

**CRYSTAL HABIT**
*This braunite crystal from Maharashtra, India, shows a classic pyramidal habit.*

*Pyramidal crystal*

# COLUMBITE

**MADAGASCAR**
*Large crystals of columbite occur in Madagascar.*

THREE DISTINCT COLUMBITE minerals are recognized: the more common ferrocolumbite and manganocolumbite, and the rarer magnesium-rich magnocolumbite. Columbite is an iron, manganese niobium tantalum oxide. When tantalum becomes significantly dominant over niobium in the structure, it becomes the mineral tantalite. Columbite and tantalite are found together in granite, granitic pegmatites, and placer deposits. Ferrocolumbite is an important ore of both tantalum and niobium. Localities include Greenland; Germany, Italy, France, Sweden, and Greece; and the pegmatite regions of Zimbabwe, South Africa, and Uganda.

### PROPERTIES

| | |
|---|---|
| **GROUP** | Oxides |
| **CRYSTAL SYSTEM** | Orthorhombic |
| **COMPOSITION** | $(Fe,Mn)(Nb,Ta)_2O_6$ |
| **COLOUR** | Iron-black to brown-black |
| **FORM/HABIT** | Short prismatic |
| **HARDNESS** | 6–6½ |
| **CLEAVAGE** | Distinct |
| **FRACTURE** | Subconchoidal to uneven |
| **LUSTRE** | Metallic |
| **STREAK** | Dark red to black |
| **SPECIFIC GRAVITY** | 5.2–6.8 |
| **TRANSPARENCY** | Translucent to opaque |

*Tabular crystal*

*Metallic lustre*

**FERROCOLUMBITE**
*This opaque, tabular crystal of ferrocolumbite exhibits a metallic to resinous lustre.*

# PYROCHLORE

**COLORADO**
*This mineral-rich state in the USA is a source of pyrochlore.*

PYROCHLORE TAKES ITS NAME from the Greek for "fire" and "green" – some specimens turn green after heating. It is a complex niobium sodium calcium oxide, an important niobium ore. It forms octahedral crystals, often twinned, and irregular masses. It is found in felsic igneous rocks and associated pegmatites, and in contact metamorphic zones. Localities include Miass, Russia; Oka, Quebec, Canada; and Conakry, Guinea.

**RELATED MINERALS**
*Microlite (below) contains tantalum in place of the niobium in pyrochlore (right).*

*Dull lustre*

### PROPERTIES

| | |
|---|---|
| **GROUP** | Oxides |
| **CRYSTAL SYSTEM** | Cubic |
| **COMPOSITION** | $(Na,Ca)_2Nb_2(O,OH,F)_7$ |
| **COLOUR** | Brown to black |
| **FORM/HABIT** | Octahedral |
| **HARDNESS** | 5–5½ |
| **CLEAVAGE** | Distinct |
| **FRACTURE** | Subconchoidal to uneven |
| **LUSTRE** | Vitreous to resinous |
| **STREAK** | Light brown, yellowish-brown |
| **SPECIFIC GRAVITY** | 4.5 |
| **TRANSPARENCY** | Translucent to opaque |

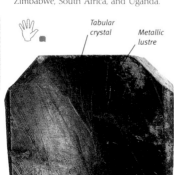

*Twinned octahedra*

# HYDROXIDES

THE HYDROXIDES FORM when metallic elements combine with the hydroxyl radical (OH). All hydroxides form at low temperatures and are found predominantly as weathering products of other minerals. Hydroxide minerals tend to be less dense than the oxides and also softer. They are often important ore minerals. For example, bauxite, the ore of aluminium, is principally composed of the aluminium hydroxides diaspore, böhmite, and gibbsite. Goethite, an iron hydroxide, is an iron ore.

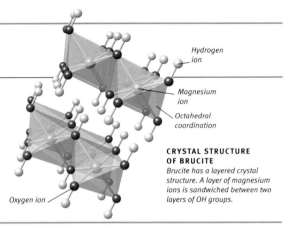

Hydrogen ion

Magnesium ion

Octahedral coordination

**CRYSTAL STRUCTURE OF BRUCITE**
*Brucite has a layered crystal structure. A layer of magnesium ions is sandwiched between two layers of OH groups.*

Oxygen ion

## GIBBSITE

*Massive habit*

**URALS**
*Gibbsite is found in these Russian mountains.*

Pearly lustre

**MASSIVE GIBBSITE**
*This specimen is massive but gibbsite can also be earthy or compact.*

GIBBSITE WAS NAMED in 1822 for George Gibbs, whose minerals formed the core of the Yale University collection. Its crystals are tabular, and may appear hexagonal. Twinning is very common. Often white, it can also be greyish, greenish, or yellowish, depending on impurities. Gibbsite is principally an alteration product of aluminous minerals whose silica has been leached out, and forms primarily in tropical and subtropical environments. It can also form in low-temperature hydrothermal veins. It is one of the main constituents of bauxite (see p.167).

**PROPERTIES**

| | |
|---|---|
| **GROUP** | Hydroxides |
| **CRYSTAL SYSTEM** | Monoclinic |
| **COMPOSITION** | $Al(OH)_3$ |
| **COLOUR** | White |
| **FORM/HABIT** | Tabular |
| **HARDNESS** | $2^{1}/_2$–$3^{1}/_2$ |
| **CLEAVAGE** | Perfect |
| **FRACTURE** | Uneven |
| **LUSTRE** | Pearly to vitreous |
| **STREAK** | Undetermined |
| **SPECIFIC GRAVITY** | 2.3–2.4 |
| **TRANSPARENCY** | Transparent |

## DIASPORE

**STERLING HILL, USA**
*This mine in New Jersey is a source of diaspore.*

A HYDROUS ALUMINIUM OXIDE, diaspore takes its name from the Greek *diaspora*, meaning "scattering", a reference to the way diaspore crackles when strongly heated. Its crystals are thin and platy, elongated, tabular, prismatic, or acicular. It can be massive and is also found as disseminated grains. In colour it is white, greyish-white, colourless, greenish-grey, light brown, yellowish, lilac, or pink. It can be strongly pleochroic (see p.92), showing violet-blue in one direction, asparagus-green in another, and reddish-plum in a third. Diaspore forms in metamorphic rocks, such as schists and marbles, where it is often associated with corundum, manganite, and spinel. It is also found in hydrothermally altered rocks, and in sediments, often as a major constituent of bauxite. A relatively widespread mineral, diaspore occurs in significant amounts in Shokozan Mine, Honshu, Japan; Jordanów, Poland; the Ural Mountains of Russia; and at Franklin in North Carolina, Chester in Maine, and Laws, California, USA.

**CRYSTAL STRUCTURE OF DIASPORE**
*In diaspore aluminium ions are in octahedral coordination with OH groups, forming strips of octahedra.*

Hydrogen ion

**DARK RED DIASPORE**
*In this specimen a mass of thin, platy crystals rests in a groundmass of emery. Manganese has substituted in part for the aluminium, giving it a dark red colour.*

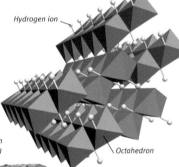

Octahedron

**FACETED GEM**
*Gem-quality crystals come from Mugla, Turkey, and are either faceted, as here, or cut en cabochon.*

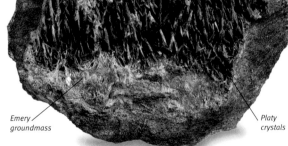

Emery groundmass

Platy crystals

**PROPERTIES**

| | |
|---|---|
| **GROUP** | Hydroxides |
| **CRYSTAL SYSTEM** | Orthorhombic |
| **COMPOSITION** | $AlO(OH)$ |
| **COLOUR** | White, grey, yellow, lilac, or pink |
| **FORM/HABIT** | Thin, platy |
| **HARDNESS** | $6^{1}/_2$–7 |
| **CLEAVAGE** | Perfect, imperfect |
| **FRACTURE** | Conchoidal, brittle |
| **LUSTRE** | Vitreous |
| **STREAK** | White |
| **SPECIFIC GRAVITY** | 3.4 |
| **TRANSPARENCY** | Transparent to translucent |
| **R.I.** | 1.68–1.75 |

# BAUXITE

BAUXITE IS NOT A MINERAL, but a product of rock weathering that contains several constituent minerals. It occasionally forms recognizable hand specimens. It is composed of several hydrated aluminium oxides, principally gibbsite $Al(OH)_3$, boehmite $AlO(OH)$, and diaspore, also $AlO(OH)$ (see opposite). Clay minerals, hematite, magnetite, goethite, siderite, and quartz are common impurities, and most deposits contain lesser amounts of rutile, anatase, and zircon. It may require the use of a range of sophisticated techniques, including thin-section analysis and combined X-ray diffraction, to ascertain the components of any individual specimen. This is of major importance because almost all of the aluminium produced to date has come from bauxite (see p.174 for details of the release of aluminium metal from bauxite through the use of molten cryolite). By far the largest producer is Queensland, Australia, followed by Jamaica, Brazil, and Guinea. These four localities account for more than half of the world's bauxite production. Other significant producers are France, Italy, Greece, China, the USA, and Ghana.

Bauxite is formed as a weathering product of many different

## PROPERTIES

| | |
|---|---|
| **GROUP** | Hydroxides |
| **CRYSTAL SYSTEM** | Amorphous mixture |
| **COMPOSITION** | Mixture of hydrous aluminium oxides |
| **COLOUR** | White, yellowish, red, reddish-brown |
| **FORM/HABIT** | Amorphous |
| **HARDNESS** | 1–3 |
| **CLEAVAGE** | None |
| **FRACTURE** | Uneven |
| **LUSTRE** | Earthy |
| **STREAK** | Usually white |
| **SPECIFIC GRAVITY** | 2.3–2.7 |
| **TRANSPARENCY** | Opaque |

## ALUMINIUM ORE

Not all bauxites are suitable for mining. Ores must have an aluminium-oxide content of 30 per cent or more to be economic. On average, 4 tonnes of high-grade ore yields 1 tonne of aluminium. Of all bauxite mined, 90 per cent is smelted into aluminium, with the remaining 10 per cent used to make, among other things, abrasives and refractories. High-grade bauxite mined in Russia's Urals is mainly diaspore.

**PROCESSING ALUMINIUM**
*Ingots of newly smelted aluminium await shipment for processing into sheets of foil.*

**BAUXITE ORE**
*Bauxite ore is being transported on this conveyor belt at the Kaiser Refinery, Jamaica.*

rocks. It often forms from laterites, soils that develop in warm, tropical climates and that tend to be leached of all soluble material. Bauxites are currently forming in tropical regions in Australia, Brazil, western Africa, and elsewhere. They vary greatly in physical appearance, depending on their composition and impurities, and according to the origin and geologic history of their deposits. They range in colour from yellowish-white to grey, or from pink to dark red or brown if high in iron oxides. Some deposits are soft, easily crushed, and structureless; some are hard, dense, and pea-like; others are porous but structurally strong, or are stratified, or largely retain the form of their parent rock.

*Aluminium oxide groundmass*

*Spherical pisolith*

**PISOLITIC BAUXITE**
*This bauxite specimen shows a classic pisolitic habit, meaning that it is made up of numerous grains up to pea-size known as pisoliths.*

**BAUXITE MINE**
*Most of the world's bauxite lies at or near the surface and thus is easily mined. This mine is in Jamaica.*

**OIL FIELD**
*Bauxite is used in the petroleum industry as a proppant – to assist with the flow of oil from the reservoir to the bore hole – in the recovery of crude oil.*

# BRUCITE

NAMED FOR American mineralogist Archibald Bruce in 1824, brucite is magnesium hydroxide. Its crystals can be tabular or aggregates of plates. It can also be massive, foliated, fibrous, or, more rarely, granular. Usually white, it can be pale green, grey, or blue. Manganese may substitute to some degree for magnesium, producing yellow to red coloration. Brucite is pyroelectric, and easily soluble in acids. It is found in metamorphic rock such as schist, and in low-temperature hydrothermal veins associated with calcite, aragonite, magnesite, and talc. Fine, often transparent, crystals come from the Ural Mountains of Russia, and from Pennsylvania, USA. It is used as a primary source of medical magnesia.

**KILN LINING**
*Brucite has such a high melting point that it is used to line kilns, such as the potter's kiln being used here at Hawaii Volcanoes National Park, Hawaii.*

*Vitreous lustre*

*Single fibres*

*Fibrous habit*

**FIBROUS BRUCITE**
*This fibrous mass of brucite is from Timmins, Ontario, Canada. Brucite is also found in Austria, England, Russia, Sweden, and Turkey, as well as California, New York, Pennsylvania, and New Jersey, USA.*

## PROPERTIES

| | |
|---|---|
| **GROUP** | Hydroxides |
| **CRYSTAL SYSTEM** | Hexagonal/trigonal |
| **COMPOSITION** | $Mg(OH)_2$ |
| **COLOUR** | White, pale green, grey or blue |
| **FORM/HABIT** | Broad tabular |
| **HARDNESS** | $2^1/_2$ |
| **CLEAVAGE** | Perfect |
| **FRACTURE** | Uneven; sectile |
| **LUSTRE** | Waxy to vitreous/pearly |
| **STREAK** | White |
| **SPECIFIC GRAVITY** | 2.4 |
| **TRANSPARENCY** | Transparent |

# MANGANITE

MANGANITE IS A WIDESPREAD and important ore of manganese – after which it was named in 1827. Most crystals are pseudo-orthorhombic prisms. It is also found in crystal bundles, fibrous masses, and massive aggregates. Multiple twinning is common. It occurs in low-temperature hydrothermal deposits associated with baryte, calcite, and siderite; in replacement deposits with goethite; and in shallow marine deposits, lakes, and bogs. It alters to romanèchite. Well-crystallized specimens come from Germany; South Africa; and Virginia, USA.

**MANGANITE CRYSTALS**
*This specimen is a mass of pseudo-orthorhombic prisms, showing the typical deep striations on the crystal faces.*

*Uneven fracture*

*Striated crystal*

*Opaque crystals*

*Submetallic lustre*

## PROPERTIES

| | |
|---|---|
| **GROUP** | Hydroxides |
| **CRYSTAL SYSTEM** | Monoclinic |
| **COMPOSITION** | MnO(OH) |
| **COLOUR** | Steel-grey to iron-black |
| **FORM/HABIT** | Prismatic, striated |
| **HARDNESS** | 4 |
| **CLEAVAGE** | Perfect, good |
| **FRACTURE** | Uneven |
| **LUSTRE** | Submetallic |
| **STREAK** | Reddish-brown to black |
| **SPECIFIC GRAVITY** | 4.3 |
| **TRANSPARENCY** | Opaque |

# GOETHITE

**PIKE'S PEAK, USA**
*Notable goethite specimens come from this part of Colorado.*

**PIKE'S PEAK, USA**
*Notable goethite specimens come from this part of Colorado.*

THIS SECONDARY MINERAL can take various forms and colours. It can occur as prismatic and vertically striated crystals, velvety, radiating fibrous aggregates, flattened tablets or scales, reniform or botryoidal masses, or in stalactitic, or massive forms. Goethite is usually black, but can be brownish, yellowish, or reddish, depending on impurities. It is an iron oxide hydroxide, although manganese can substitute for up to 5 per cent of the iron. It is formed as a weathering product in the oxidation zones of veins of iron minerals such as magnetite, pyrite, and siderite. It may occur with these minerals in the weathered capping of iron ore deposits, called the gossan, or "iron hat". It can also form as a biogenic precipitate known as bog iron ore. Fine crystallized specimens come from Cornwall, England; Chaillac, France; and Lake Onega and other localities in the Ural Mountains of Russia. It is found as pseudomorphs after pyrite and siderite at Diamantina, Brazil and Pelican Point, Utah, USA.

## PROPERTIES

| | |
|---|---|
| **GROUP** | Hydroxides |
| **CRYSTAL SYSTEM** | Orthorhombic |
| **COMPOSITION** | FeO(OH) |
| **COLOUR** | Orangeish to blackish-brown |
| **FORM/HABIT** | Prismatic, elongated |
| **HARDNESS** | $5-5^1/_2$ |
| **CLEAVAGE** | Perfect |
| **FRACTURE** | Uneven |
| **LUSTRE** | Adamantine to metallic |
| **STREAK** | Brownish-yellow to ochre-red |
| **SPECIFIC GRAVITY** | 4.3 |
| **TRANSPARENCY** | Translucent to opaque |

*Botryoidal mass*

*Radiating crystals*

*Quartz groundmass*

**GOETHE**
*Goethite was named in 1806 for Johann Wolfgang von Goethe, the German poet and author, who was an enthusiastic mineralogist.*

**VELVET SURFACE**
*This botryoidal mass of radiating goethite crystals has a velvety appearance to its surface. The specimen is from Merehead Quarry, Somerset, England.*

---

# LIMONITE

**PIKE'S PEAK, USA**
*This area of Colorado is one of the localities for this mineral.*

LIMONITE IS THE NAME given to unidentified iron oxides and, very commonly and more specifically, to unidentified hydroxides. It is a mixture in the same manner as bauxite (see p.167). In practice, most so-called limonite is goethite, and the name has been used widely as a synonym of goethite. It can form concretions or be stalactitic, mammillary, or earthy. It does not form crystals. Its name comes from the Greek *leimon*, "meadow", referring to the marshy localities where the variety known as bog iron is found. Limonite is a secondary product formed from the oxidation of other iron minerals. It also occurs by precipitation in the sea and fresh water, and in bogs. It is often found as pseudomorphs after pyrite and other iron minerals. In ancient times it was used as an ore of iron.

## PROPERTIES

| | |
|---|---|
| **GROUP** | Hydroxides |
| **CRYSTAL SYSTEM** | Mixture |
| **COMPOSITION** | $2Fe_2O_3 \cdot H_2O$, approx |
| **COLOUR** | Various shades of brown, yellow |
| **FORM/HABIT** | Massive, oolitic, stalactitic |
| **HARDNESS** | $5-5^1/_2$ |
| **CLEAVAGE** | None |
| **FRACTURE** | Uneven |
| **LUSTRE** | Earthy, sometimes submetallic or dull |
| **STREAK** | Yellowish-brown |
| **SPECIFIC GRAVITY** | 3.6–4.0 |
| **TRANSPARENCY** | Opaque |

**BRIGHT MINERAL**
*The brilliant coloration of this specimen shows why limonite is chosen as a pigment.*

## USE IN PIGMENTS

The use of limonite as a pigment goes back to ancient Egypt. Its various colours yield ochres (yellow-browns), siennas (orange-browns), and umbers (browns). More recently, the Flemish painter Van Dyck (1599–1641), best known for his portraits of the aristocracy, was particularly noted for his delicate use of ochre and sienna.

**RAW UMBER AND YELLOW OCHRE**
*Limonite has provided the pigment in umbers and ochres for millennia.*

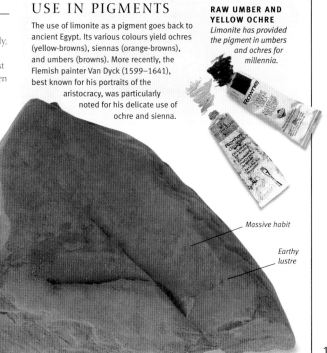

*Massive habit*

*Earthy lustre*

# HALIDES

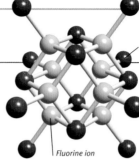

Calcium ion

Fluorine ion

**FLUORITE CRYSTAL STRUCTURE**
*In the crystal structure of fluorite every calcium ion is coordinated with eight fluorine ions at the corners of a cube.*

HALIDES ARE MINERALS consisting of metals combined with one of the common halogen elements: fluorine, chlorine, bromine, or iodine. Compositionally and structurally there are three broad categories of halide minerals: the simple halides, the halide complexes, and the oxyhydroxy–halides. A simple halide is formed when a metal combines with one of the halogens. Halite (sodium and chlorine) and fluorite (calcium and fluorine) are examples. Many of the simple halides are soluble in water. In the

halide complexes, the halide is usually bound to aluminium, creating a molecule that behaves as a single unit, which is in turn bound to a metal. For example, in cryolite aluminium and fluorine are bound to sodium. Halide complexes are relatively rare, and generally insoluble in water. The last group, the oxyhydroxy–halides, are very rare. They are often formed by the action of halide–bearing waters upon the oxidation products of previously existing sulphide minerals. Atacamite is an example of these.

## HALITE

**SALT LAKE**
*White crystals of salt can be seen encrusting this block of stone protruding from this salt lake.*

BETTER KNOWN AS COMMON SALT, halite is sodium chloride. Its name is derived from the Greek *hals*, meaning "salt". Its crystals are usually cubes, but sometimes it forms "hopper" crystals – crystals where the outer edges of the cube faces have grown more rapidly than their centres, leaving cavernous faces. It is commonly found in massive and bedded aggregates as rock salt (see p.58). It is colourless, white, orange, blue, or purple. The orange colour is derived from inclusions of hematite; the blue and purple colours derive from defects in the crystal structure. Halite is widespread in large saline evaporite deposits, where it has formed by the evaporation of sea water. It is often accompanied by anhydrite, gypsum, and sylvite. It is widely distributed worldwide, with large-scale commercial deposits in Stassfurt, Germany; Salzburg, Austria; Wittelsheim, France; Salar de Uyuni, Bolivia; the salt domes along the Gulf of Mexico, USA; and the world's largest rock-salt mine at Retsof, New York.

### USES OF SALT

Besides its culinary use, salt is a preservative and source of sodium carbonate (soda ash), used in the manufacture of soap and glass, and sodium bicarbonate (baking soda). It is also used as a flux for the melting of metals, as a glaze for porcelain enamels, and as a source of chlorine for hydrochloric acid and other chlorine compounds, particularly PVC.

**TABLE SALT**
*Salt is one of the most vital minerals for human and animal health.*

**SHINTO**
*In the Japanese Shinto religion, salt is set in small heaps at the entrance to houses and on the rims of wells to cleanse them.*

| PROPERTIES | |
|---|---|
| **GROUP** Halides | |
| **CRYSTAL SYSTEM** Cubic | |
| **COMPOSITION** NaCl | |
| **COLOUR** Colourless to white | |
| **FORM/HABIT** Cubes | |
| **HARDNESS** 2½ | |
| **CLEAVAGE** Perfect cubic | |
| **FRACTURE** Conchoidal | |
| **LUSTRE** Vitreous | |
| **STREAK** White | |
| **SPECIFIC GRAVITY** 2.1–2.6 | |
| **TRANSPARENCY** Transparent to translucent | |

Vitreous lustre

**SALT MINE**
*Underground mining of salt has been carried out since prehistoric times in Europe.*

Rock groundmass

Cubic crystals

**HALITE CRYSTALS**
*Cubic crystals of halite cover a rock groundmass in this specimen from Inowroclaw, Poland.*

# SYLVITE

**PERMIAN BASIN, USA**
This area of Texas contains major sylvite deposits.

FOUND AS GLASSY CUBES or as granular, crystalline masses, sylvite is potassium chloride. Its name is derived from its old medicinal name, *sal digestivus Sylvii*; it is also known as sylvine. It is usually colourless to white or greyish, but can be tinged blue, yellow, purple, or red. A bitter taste helps to distinguish it from halite. Sylvite forms in evaporite deposits along with halite and gypsum, although it is rarer than halite. It is found in evaporite deposits in the vicinity of Stassfurt, Germany; Saskatchewan, Canada; and in the Permian Basin deposits of Texas and New Mexico, USA. It also forms in volcanic fumaroles and occurs as encrustations on Mount Vesuvius, Italy, where it was first discovered in 1823. Sylvite is mined to make artificial potassium compounds, such as potash fertilizer (see p.175).

**FERTILIZING CROPS**
Sylvite is an important raw material in the production of potash fertilizer.

## PROPERTIES

| | |
|---|---|
| GROUP | Halides |
| CRYSTAL SYSTEM | Cubic |
| COMPOSITION | KCl |
| COLOUR | Colourless to white |
| FORM/HABIT | Cubes |
| HARDNESS | 2½ |
| CLEAVAGE | Perfect cubic |
| FRACTURE | Uneven |
| LUSTRE | Vitreous |
| STREAK | White |
| SPECIFIC GRAVITY | 2.0 |
| TRANSPARENCY | Transparent to translucent |

Interlocking cubic crystals

Transparent at crystal margins

**SYLVITE CRYSTALS**
The pinkish interlocking cubic crystals in this specimen are typical of sylvite.

Vitreous lustre

---

# CHLORARGYRITE

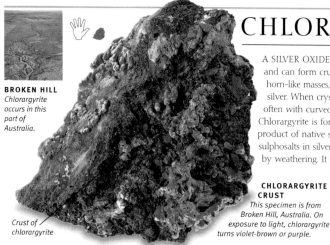

**BROKEN HILL**
Chlorargyrite occurs in this part of Australia.

**CHLORARGYRITE CRUST**
This specimen is from Broken Hill, Australia. On exposure to light, chlorargyrite turns violet-brown or purple.

Crust of chlorargyrite

A SILVER OXIDE, chlorargyrite is usually massive and can form crusts and coatings, sometimes in horn-like masses, which give it another name, horn silver. When crystals form, they are usually cubes, often with curved and poorly developed faces. Chlorargyrite is formed as a secondary alteration product of native silver, silver sulphides, and sulphosalts in silver deposits that have been oxidized by weathering. It is often associated with cerussite, limonite, and malachite. In several silver-mining localities, it was an important ore of silver. These localities include: Saxony, Germany; Bohemia, the Czech Republic; Potosí, Bolivia; Tombstone, Arizona, Leadville, Colorado, Frisco, Utah, and the Bullfrog district, Nevada, USA.

## PROPERTIES

| | |
|---|---|
| GROUP | Halides |
| CRYSTAL SYSTEM | Cubic |
| COMPOSITION | AgCl |
| COLOUR | Pearl-grey, greenish, white, colourless |
| FORM/HABIT | Usually massive |
| HARDNESS | 1–2 |
| CLEAVAGE | None |
| FRACTURE | Subconchoidal |
| LUSTRE | Dull to adamantine |
| STREAK | White |
| SPECIFIC GRAVITY | 5.6 |
| TRANSPARENCY | Transparent to nearly opaque |

---

# CALOMEL

**TERLINGUA, TEXAS, USA**
This is a prime source of calomel crystals.

ITS NAME DERIVING from the Greek for "beautiful" and "honey", alluding to its sweet taste (although it is in fact toxic), calomel is mercury chloride. It is very soft, heavy, and plastic, with crystals that are tabular, prismatic, or pyramidal, often with complex twinning. It is also found as crusts, and can be massive and earthy. It fluoresces brick-red. Calomel occurs as a secondary mineral in the oxidized zone of mercury-bearing deposits, together with native mercury, cinnabar, calcite, and limonite. Crystals are found at Alava, Serbia, and at Terlingua, Texas, and in Howard and Maricopa counties, Arizona, USA. It is also found at Landsberg, Germany; Huahuaxtla, Mexico; and Almadén, Spain. Calomel was once a popular purgative, in use since the 16th century – until its toxicity was discovered. It is still used in some insecticides and fungicides, and as an ore of mercury.

## PROPERTIES

| | |
|---|---|
| GROUP | Halides |
| CRYSTAL SYSTEM | Tetragonal |
| COMPOSITION | HgCl |
| COLOUR | White, colourless, yellow-grey, or greyish |
| FORM/HABIT | Tabular |
| HARDNESS | 1–2 |
| CLEAVAGE | Distinct |
| FRACTURE | Conchoidal |
| LUSTRE | Adamantine |
| STREAK | Pale yellow-white |
| SPECIFIC GRAVITY | 6.5 |
| TRANSPARENCY | Transparent to nearly opaque |

**CALOMEL ENCRUSTATIONS**
A thin crust of calomel coats this specimen.

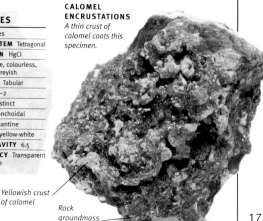

Yellowish crust of calomel

Rock groundmass

**SWISS ALPS**
*The Alps on the borders of France and Switzerland are the foremost locality for rare pink octahedral fluorite crystals.*

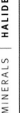

**RECTANGULAR STEP**
*This rectangular step-cut has been faceted from purple fluorite.*

**YELLOW EMERALD**
*This specimen is a yellow, emerald-cut fluorite. Faceted fluorites are made only for collectors.*

**PINK EMERALD**
*This emerald cut has been carefully faceted from a specimen of delicate pink fluorite.*

**GREEN BRILLIANT**
*Fluorite's modest refractive index means that faceted stones can lack sparkle.*

# FLUORITE

OCCURRING IN WELL-FORMED CRYSTALS of many different colours, fluorite is calcium fluoride. Violet, green, and yellow are the most common colours, although it is colourless and transparent when free of trace elements – as much as 20 per cent yttrium or cerium may replace the calcium in the composition of fluorite. Colours can occur in distinct zones within the same crystal, following the contour of the crystal faces. Fluorite crystals are widely found in cubes; octahedrons, which are often twinned, are much less common. Fluorite can also be massive, granular, or compact.

The phenomenon of fluorescence – the emission of visible light on exposure to ultraviolet light (see pp.160–61) – was first observed in fluorite and takes its name from it. The name fluorite comes from the Latin *fluere*, meaning "to flow", a reference to the ease with which it melts when used as a flux in the smelting and refining of metals. An old name for fluorite is fluorspar; this is now used only in industry to refer to the bulk stone.

## USES OF FLUORITE

Fluorite finds many industrial uses: as a flux in the manufacture of steel, in the production of hydrofluoric acid, and as a catalyst in the manufacture of high-octane fuels. It used in the manufacture of artificial cryolite (see p.174) for the refining of aluminium, lead, and antimony, in the formation of opalescent glass (see panel, below), and in iron and steel enamelware. Clear, optical-quality fluorite, with its low refractive index and low dispersion, is used for apochromatic lenses for microscopes, which eliminate distortion of colour. Fluorite is also a source of fluorine, which is

**PURPLE FLUORITE CRYSTALS**

**FLUORITE WITH GALENA**

**YELLOW FLUORITE CRYSTALS**

**GREEN FLUORITE**

**BLUE JOHN**

**COLOURS OF FLUORITE**
*Fluorite has the widest colour range of any mineral. It can be colourless, white, or various shades of purple, green, blue, yellow, or rarely, pink and red. Coloration is controlled by various trace elements substituting in its structure. Several colours can occur in the same specimen.*

## OPALESCENT GLASS

Fluorite plays an important part in the manufacture of glazed tiles and opalescent glass, a semi-opaque pressed glass, tinged with blue and having a milky iridescence. The colour is produced by the slower cooling of the molten glass in those parts which are thick, causing some crystallization inside the glass. American artist John LaFarge first developed and patented opalescent glass in 1879 but it was Louis Comfort Tiffany who created the most striking and colourful Art Nouveau opalescent glass designs. During the 1920s and 1930s, many French companies made beautiful opalescent Art Deco vases, bottles and lampshades. Among the best known were those made by René Lalique.

**PLAFONNIERE**
*This shallow, circular, opalescent glass bowl by Mark Sabino features a design of veil-tailed fishes against blue. It is fixed to the ceiling by three metal chains.*

**OPALESCENT GLASS BOTTLE**
*Dating from 1911, the milky white sections of this Chinese opalescent glass bottle are painted with fish swimming among fronds.*

**PINK FLUORITE**
*This superb octahedral specimen of rare pink fluorite is perched on a cluster of smoky quartz crystals. It is from the Alps, the world's largest source of pink fluorite.*

### PROPERTIES

| | |
|---|---|
| **GROUP** | Halides |
| **CRYSTAL SYSTEM** | Cubic |
| **COMPOSITION** | CaF$_2$ |
| **COLOUR** | Occurs in most colours |
| **FORM/HABIT** | Cubic, octahedral |
| **HARDNESS** | 4 |
| **CLEAVAGE** | Perfect octahedral |
| **FRACTURE** | Flat conchoidal |
| **LUSTRE** | Vitreous |
| **STREAK** | White |
| **SPECIFIC GRAVITY** | 3.0–3.3 |
| **TRANSPARENCY** | Transparent to translucent |
| **R.I.** | 1.43 |

used for the fluoridation of water and in Teflon coating, where the fluorine helps to provide the "non-stick" surface on Teflon cooking pans.

Fluorite's wide range of intense colours, the frequent occurrence of more than one colour in a single stone, and the zoning or patchy distribution of colour would make it a popular gemstone were it not for its softness and its fragility due to its perfect octahedral cleavage. Stones are sometimes faceted for collectors. The ancient Egyptians used fluorite in statues and to carve scarabs, and the Chinese have used it in carvings for more than 300 years. There has even been an attempt to pass off Chinese fluorite carvings as jade, but their softness provides a ready means of identification. In the 18th century, powdered fluorite was added to water to relieve the symptoms associated with kidney disease.

### OCCURRENCE OF FLUORITE

Fluorite occurs as a vein mineral, often associated with lead and silver ores; it also occurs in pegmatite cavities, sedimentary rocks, and in hot-springs areas. It is found in numerous localities worldwide, but more important ones include Canada, the USA, South Africa, Mexico, China, Mongolia, Thailand, Peru, Poland, Hungary, the Czech Republic, Norway, England, Spain, and Germany.

**BLUE JOHN RING**
*This 20th-century silver ring is set with a cabochon of Blue John from Castleton, England.*

**BLUE JOHN MINES, DERBYSHIRE**
*This open vein at the Odin Mine, Castleton, Derbyshire, is part of the Blue John Mines in the hillside of Treak cliff, which were worked by the Romans.*

# BANDED FLUORITE

The banded, purple and white or yellow variety of fluorite is popularly known as Blue John. The name may be a corruption of the French bleu jaune, "blue yellow", colours that are often interbanded in Blue John. A major source of this colour-zoned fluorite is Castleton in Derbyshire, England, where it is found in at least 14 differently patterned veins. It was first mined there during Roman times and used to make vessels. The Romans particularly prized fluorite vessels because of the special flavour given to the wine drunk from them. This flavour was actually the result of the resin used to help hold the crystals together during manufacture.

### REDISCOVERY OF THE MINES

At the beginning of the 18th century, two miners – John Kirk and Joseph Hall – followed a vein of Blue John and stumbled into the old Roman mines. They and others went on to work the mines for the banded fluorite, which craftsmen, including Matthew Boulton (1728–1809), an early pioneer of steam power, used to fashion vases, ornaments, and jewellery. The mines are still worked today for the same purposes. Because of the brittleness of the mineral, it is often bonded with resins to increase its durability.

**ROMAN CUP**
*This one-handled cup was carved in the 1st century AD from a single piece of fluorite, which is richly veined with purple, green, yellow, and white. It is decorated with a low-relief panel of vine tendrils.*

Fluorite octahedron

Smoky quartz

# CRYOLITE

FEW PEOPLE HAVE HEARD of cryolite, yet it is arguably one of the most important minerals of our age. Modern aircraft could not fly without it, and modern engineering of all kinds would be stunted in its absence. Cryolite is an essential ingredient in aluminium production. Although aluminium is the most abundant metal in the Earth's crust, and exceeded only by silicon and oxygen in weight, prior to 1886 it scarcely existed as a separate metal (see panel, below).

Cryolite takes its modern name from its ancient Greek name *kronos*, meaning "ice stone", alluding to its translucent, ice-like appearance. It is a colourless to white halide

**MODERN AIRCRAFT**
*Light yet strong enough to bear great pressure, aluminium is ideal for aircraft bodies.*

**PIKE'S PEAK, COLORADO, USA**
*Small quantities of massive cryolite can still be found at a deposit at Pike's Peak, El Paso County, Colorado.*

mineral, sodium aluminum fluoride. Rarely, it can be brown, yellow, reddish-brown, or black. It forms mainly in certain granitic pegmatites and granites. Cryolite is usually found in coarse, cleavable masses. Distinct crystals are rare, but when found they are pseudocubic and often deeply striated. The refractive index of cryolite is so near to that of water that when a cryolite specimen is placed in water, it virtually disappears. This test distinguishes it from similar-looking minerals.

The largest deposit of cryolite, at Ivigtut, Greenland, is now exhausted. Lesser amounts are found in Spain, Russia, Colorado, USA, and small yellow crystals occur at Mont St.-Hilaire, Quebec, Canada. In addition to its use in aluminium processing, it is also used in the glass and enamel industries, other metallurgical applications, and in the manufacture of insecticides. Synthetic cryolite made from fluorite (see p.172) has largely replaced natural cryolite in industry.

*Greasy lustre*

*Pseudocubic outline*

## SMELTING ALUMINIUM

In 1886 both Charles M. Hall of the United States and Paul Louis-Toussaint Héroult of France discovered almost simultaneously that if alumina, aluminium oxide (such as occurs in bauxite, see p.167), is dissolved in molten cryolite and electrolyzed with direct current, molten aluminium metal is released. Their process, the Hall–Héroult process, still remains the basis for today's aluminium industry. Once released, the aluminium often requires alloying with other minerals to increase its strength,

**HALL**
*C.M. Hall (1863–1914) found backers for his smelting process in Pittsburgh.*

durability, and other mechanical properties. Aircraft aluminium, for example, is alloyed with a small amount of copper and manganese to increase its strength, making an alloy called duralumin.

**HÉROULT**
*P.L.T. Héroult (1863–1914) also invented the electric arc furnace for steel in 1900.*

### PROPERTIES

| | |
|---|---|
| **GROUP** | Halides |
| **CRYSTAL SYSTEM** | Monoclinic |
| **COMPOSITION** | $Na_3AlF_6$ |
| **COLOUR** | Colourless to snow-white |
| **FORM/HABIT** | Pseudocubic |
| **HARDNESS** | 2½ |
| **CLEAVAGE** | None |
| **FRACTURE** | Uneven |
| **LUSTRE** | Vitreous to greasy |
| **STREAK** | White |
| **SPECIFIC GRAVITY** | 3.0 |
| **TRANSPARENCY** | Transparent to translucent |

**MASSIVE CRYOLITE**
*This specimen of massive cryolite from Ivigtut, Greenland, exhibits a typically greasy lustre and the pseudocubic outline of a cleaved mass.*

**GLAZE**
*Cryolite is used in the production of speciality glazes for glass and ceramics, such as the ceramic leaf shown here.*

**MOLTEN ALUMINIUM**
*In the Hall–Héroult process, bauxite is fed into molten cryolite in an aluminium smelting*

# CARNALLITE

FIRST DISCOVERED IN GERMANY, carnallite was named for Rudolph von Carnall, a Prussian mining engineer. It is hydrated potassium and magnesium chloride, and is generally massive to granular. Crystals are rare because they absorb water from the air and dissolve. When found, they are thick tabular, pseudohexagonal, or pyramidal. Carnallite is usually white or colourless, but it may appear reddish or yellowish depending on the presence of hematite or goethite impurities. It is relatively easy to distinguish from similar-looking minerals because of its low density. Carnallite forms in the upper layers of marine evaporite salt deposits, where it occurs with other potassium and magnesium evaporite minerals. It is an important source of potash for fertilizers, and is Russia's most important source of magnesium. The mineral is found in abundance in the northern German salt deposits; in the Permian Basin of Texas and New Mexico, USA; in the Barcelona and Lleida provinces of Spain; and in Silesia and Galicia, Poland.

**PERMIAN BASIN, USA**
*Carnallite occurs in abundance in the Permian Basin of Texas.*

## USES OF POTASH

Carnallite is one of several minerals from which potash fertilizer is made, the most important being sylvite (see p.171). In agriculture, maintaining the proper balance between potash and nitrogen is important. Many diseases, rots, and mildews affecting both commercial and garden-grown crops are due to the overapplication of nitrogen, and respond to applications of potash. Potash is potassium oxide. A hydroxide, caustic potash, is also produced. Caustic potash is used in the production of speciality soaps and indigo dyes.

**POTASH PLANT**
*These evaporating ponds are part of a plant that produces potash for fertilizers.*

### PROPERTIES

| | |
|---|---|
| **GROUP** | Halides |
| **CRYSTAL SYSTEM** | Orthorhombic |
| **COMPOSITION** | $KMgCl_3 \cdot 6H_2O$ |
| **COLOUR** | Milky-white, often reddish |
| **FORM/HABIT** | Massive to granular |
| **HARDNESS** | $2\frac{1}{2}$ |
| **CLEAVAGE** | None |
| **FRACTURE** | Conchoidal |
| **LUSTRE** | Greasy |
| **STREAK** | White |
| **SPECIFIC GRAVITY** | 1.6 |
| **TRANSPARENCY** | Translucent to opaque |

*Granular surface*

*Massive habit*

**GRANULAR CARNALLITE**
*This granular mass of carnallite is coloured red by inclusions of hematite.*

# ATACAMITE

**ATACAMA DESERT, CHILE**
*Crystals of atacamite up to 2.5cm (1in) long are found in the Atacama Desert.*

NAMED AFTER THE ATACAMA DESERT of Chile, atacamite is a secondary mineral formed, principally under arid conditions, by the oxidation of other copper minerals. Exceptionally, atacamite has condensed directly from vapours on the slopes of Mount Vesuvius, Italy. It is a copper chloride hydroxide. Crystals are slender prismatic, tabular, or pseudo-octahedral. It is also found in crystalline, massive, granular, and fibrous aggregates. Localities include the Atacama region of Chile; Boleo, Mexico; and Tintic, Utah and Majuba Hill Mine, Nevada, USA. Atacamite is a major corrosion product on both ancient and modern bronzes and other copper alloys. It is also a minor ore of copper in some localities.

**STATUE OF LIBERTY**
*The copper-alloy Statue of Liberty in New York, USA, is coloured green by a surface covering of atacamite.*

### PROPERTIES

| | |
|---|---|
| **GROUP** | Halides |
| **CRYSTAL SYSTEM** | Orthorhombic |
| **COMPOSITION** | $Cu_2Cl(OH)_3$ |
| **COLOUR** | Bright green to black-green |
| **FORM/HABIT** | Slender, prismatic |
| **HARDNESS** | $3–3\frac{1}{2}$ |
| **CLEAVAGE** | Perfect, fair |
| **FRACTURE** | Conchoidal, brittle |
| **LUSTRE** | Adamantine to vitreous |
| **STREAK** | Apple-green |
| **SPECIFIC GRAVITY** | 3.8 |
| **TRANSPARENCY** | Transparent to translucent |

**ATACAMITE CRYSTALS**
*In this specimen, a mass of dark green, tabular crystals of atacamite is in a rock cavity surrounded by associated quartz and malachite.*

*Bright green malachite, an associated mineral*

*Dark green, tabular crystals of atacamite*

# CARBONATES

CARBONATES ARE AN IMPORTANT group of minerals in the Earth's crust. The carbonate minerals calcite and dolomite are found in sediments, such as chalk and limestone, in sea shells and coral reefs, and in metamorphic rocks, such as marble. They also occur in low–temperature hydrothermal veins and evaporate deposits. All carbonate minerals contain the carbonate group, $CO_3$, as the basic compositional and structural unit. This group has a carbon atom centrally located in an equilateral triangle of oxygen atoms, which gives rise to the trigonal symmetry of many carbonate minerals. This basic unit is joined by one or more cations to form a carbonate mineral. For example, calcite is a compound of a $CO_3$ group and a calcium ion. Carbonates tend to be soft.

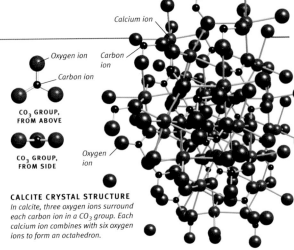

Calcium ion
Oxygen ion
Carbon ion
Carbon ion

**CO₃ GROUP, FROM ABOVE**

**CO₃ GROUP, FROM SIDE**

Oxygen ion

**CALCITE CRYSTAL STRUCTURE**
*In calcite, three oxygen ions surround each carbon ion in a $CO_3$ group. Each calcium ion combines with six oxygen ions to form an octahedron.*

# TRONA

**MINING AT TRONA, USA**
*Trona, California, is named for the large deposit of the mineral trona discovered there.*

TAKING ITS NAME FROM TRON, an abbreviation of the Arabic *natrun*, meaning "salt", trona is sodium bicarbonate hydrate. It is generally massive, and is also found in powdery surface layers on the walls of mines or on the surface of soils in desert regions. Crystals are rare, but when found they are elongated prismatic, tabular, or fibrous. It is colourless to grey or yellowish-white, and has a strongly alkaline taste.

Trona is an evaporate mineral, formed in saline lake deposits or as an evaporation product. It also occurs as an efflorescence on the soil surface in arid regions. Trona is usually associated with halite, gypsum, borax, dolomite, glauberite, and sylvite.

It occurs widely in saline desert environments, especially near Mit Rahina (formerly the ancient city of Memphis) in the Lower Nile Valley of Egypt; at Lake Goodenough, British Columbia, Canada; Bilma, Libya; Lake Chad, Chad; Lake Texcoco, Mexico; in the alkali deserts of Mongolia and Tibet; and in the USA at Searles Lake and Borax, California, Fallon, Nevada, and what is probably the world's largest deposit, at Sweetwater, Wyoming. Trona is a commercial source of sodium in localities where there is a rich deposit.

## PANAMINT VALLEY, CALIFORNIA

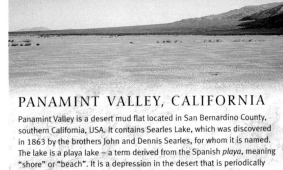

Panamint Valley is a desert mud flat located in San Bernardino County, southern California, USA. It contains Searles Lake, which was discovered in 1863 by the brothers John and Dennis Searles, for whom it is named. The lake is a playa lake – a term derived from the Spanish *playa*, meaning "shore" or "beach". It is a depression in the desert that is periodically covered by water, which evaporates, causing the deposition of salt and other minerals. It formed as part of a drainage network in the Pleistocene Epoch (1.8 to 0.01 mya). Trona is among the minerals deposited there, along with a deposit of borax, which the brothers began mining in 1873.

Layered structure

**MASSIVE TRONA**
*This specimen of massive trona shows the layered structure that is characteristic of its mode of deposition.*

Massive habit

### PROPERTIES

**GROUP** Carbonates

**CRYSTAL SYSTEM** Monoclinic

**COMPOSITION** $Na_3(HCO_3)(CO_3)\cdot2H_2O$

**COLOUR** Colourless to grey, yellow-white

**FORM/HABIT** Massive

**HARDNESS** $2^{1}/_2$–3

**CLEAVAGE** Perfect

**FRACTURE** Uneven to subconchoidal

**LUSTRE** Vitreous, glistening

**STREAK** White

**SPECIFIC GRAVITY** 2.1

**TRANSPARENCY** Transparent to translucent

**ESKIFJORD, ICELAND**
*The deposits of optical calcite at Eskifjord give calcite its popular name of Iceland spar.*

# CALCITE

ALTHOUGH OVER 70 carbonate mineral species are known, three of them, calcite, dolomite, and siderite, account for most of the carbonate material in the Earth's crust. Calcite is the most common form of calcium carbonate and is known for the great variety and beautiful development of its crystals. Its crystals occur most often as scalenohedra and are commonly twinned, sometimes forming heart-shaped butterfly twins. Crystals showing rhombohedral terminations are also common, and calcite readily cleaves into rhombohedra. Those with steep rhombohedral or scalenohedral terminations are known as dogtooth spar; those with shallow rhombohedral terminations are called nailhead spar; and highly transparent crystals are sometimes called optical spar, a reference to their use in polarizing filters.

Calcite perfectly demonstrates the optical property of double refraction: light passing through is split into two components, giving a double image of any object viewed through it. In its pure form it is colourless, pale-coloured, or white, but it is found in virtually all colours, including blue and black.

Although it forms spectacular crystals, most calcite is massive, occurring either as limestone or marble. It is also found as fibres, nodules, stalactites, and as an earthy aggregate. It is found in many geological environments. In hydrothermal deposits, the habits of its crystals are good indicators of depositional temperature and other conditions. The world's most famous source of optical calcite was at Eskifjord, Iceland, in cavernous basalt. One crystal from this locality measured 7 x 2.5m (23 x 8ft). Calcite in general is so widespread that crystallized specimens are found in virtually every country.

## ANCIENT CARVINGS

A famous ancient source of calcite was Hattsub, Egypt, where white or yellow calcite was quarried to make, among other things, buildings, vases, and inlaid eyes in statues. Some of the objects in Tutankhamun's tomb were probably carved from this source. The Egyptians referred to the material as "alabaster", but virtually all Egyptian alabaster was calcite. Its modern name has its origin in the Latin *chalx*, meaning "burnt lime".

**ALABASTER SPHINX**
*This sphinx is thought to represent Amenhotep II, and dates from the 18th or 19th dynasty.*

**SAND CALCITE**

**NAILHEAD SPAR**

**BUTTERFLY TWIN**

*Rhombohedral termination*

*Vitrous lustre*

**TREE CALCITE**

**CALCITE CRYSTALS**
*These specimens show the hugely varied habits as well as some of the many colours of calcite crystals.*

**PART OF SCALENOHEDRON**

*Perfect cleavage*

### PROPERTIES

| | |
|---|---|
| **GROUP** | Carbonates |
| **CRYSTAL SYSTEM** | Hexagonal/trigonal |
| **COMPOSITION** | $CaCO_3$ |
| **COLOUR** | Colourless, white |
| **FORM/HABIT** | Scalenohedral, rhombohedral |
| **HARDNESS** | 3 |
| **CLEAVAGE** | Perfect rhombohedral |
| **FRACTURE** | Subconchoidal, brittle |
| **LUSTRE** | Vitreous |
| **STREAK** | White |
| **SPECIFIC GRAVITY** | 2.7 |
| **TRANSPARENCY** | Transparent to translucent |
| **R.I.** | 1.48–1.66 |

**DOGTOOTH SPAR**
*This fine group of dogtooth spar crystals comes from St. Andreasberg, Lower Saxony, Germany.*

**CALCITE STRAWS**
*Hollow tubes of crystalline calcite have formed in a cave to create this impressive spray of extremely long, slender stalactites that dwarf the person standing in the cave.*

**CHIHUAHUA**
*Notable deposits of smithsonite are found in the Chihuahua area of Mexico.*

# SMITHSONITE

ALTHOUGH SMITHSONITE rarely forms crystals, spectacular examples – more than 2cm (1in) long – come from Tsumeb, Namibia. A zinc carbonate, smithsonite is most commonly found as botryoidal or stalactitic masses, or as honeycombed aggregates called "dry-bone" ore. Rhombohedral crystals of smithsonite generally have curved faces. It is a fairly common mineral found in the oxidation zones of most zinc ore deposits, and in adjacent calcareous rocks. There are significant deposits in Italy, Germany, Mexico, Zambia, and Australia, and fine specimens come from Colorado and New Mexico, USA. Smithsonite is occasionally cut as ornaments, and *en cabochon* as gems. Frequently mined as an ore of zinc, it may have provided the zinc component of brass in ancient metallurgy.

*Rounded masses showing botryoidal habit*

*Pearly lustre*

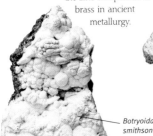

*Botryoidal smithsonite*

**WHITE SMITHSONITE**

**CABOCHON**
*Pieces of smithsonite that are of a sufficient thickness can be cut and polished as cabochon gems.*

*Green smithsonite*

**BLUE SMITHSONITE**
*This botryoidal mass of translucent blue smithsonite shades into green towards the bottom of the specimen.*

**JAMES SMITHSON**
*Smithsonite is named for the English founder of the Smithsonian Institution, USA.*

## PROPERTIES

| | |
|---|---|
| **GROUP** | Carbonates |
| **CRYSTAL SYSTEM** | Hexagonal/trigonal |
| **COMPOSITION** | $ZnCO_3$ |
| **COLOUR** | White, blue, green, yellow, brown, pink, colourless |
| **FORM/HABIT** | Botryoidal, rhombohedral, scalenohedral |
| **HARDNESS** | $4-4\frac{1}{2}$ |
| **CLEAVAGE** | Perfect rhombohedral |
| **FRACTURE** | Uneven to conchoidal |
| **LUSTRE** | Vitreous to pearly |
| **STREAK** | White |
| **SPECIFIC GRAVITY** | 4.4 |
| **TRANSPARENCY** | Translucent to opaque |
| **R.I.** | 1.62–1.85 |

# SIDERITE

**CORNWALL**
*Siderite is one of many minerals found in this part of south-west England.*

AN IRON CARBONATE, siderite takes its name from the Greek *sideros*, meaning "iron". Formerly known as chalybite, siderite has a calcite structure and can form rhombohedral crystals, often with curved surfaces, or scalenohedral, tabular, or prismatic crystals. It is more commonly massive or granular, and sometimes botryoidal or globular. Siderite is widespread, located in sedimentary, igneous, and metamorphic rocks. In sedimentary rocks it occurs in concretions and in thin beds with shale, clay, and coal seams. Well-formed crystals are found in hydrothermal veins and in some granitic and syenitic pegmatites. Single crystals up to 15cm (6in) are found in Quebec, Canada, and at the Morro Velho Mine in Brazil. Fine specimens have been found in Devon and Cornwall, England.

## PROPERTIES

| | |
|---|---|
| **GROUP** | Carbonates |
| **CRYSTAL SYSTEM** | Hexagonal/trigonal |
| **COMPOSITION** | $FeCO_3$ |
| **COLOUR** | Yellowish- to dark brown |
| **FORM/HABIT** | Rhombohedral |
| **HARDNESS** | $3\frac{1}{2}-4$ |
| **CLEAVAGE** | Perfect rhombohedral |
| **FRACTURE** | Uneven or subconchoidal |
| **LUSTRE** | Vitreous to pearly |
| **STREAK** | White |
| **SPECIFIC GRAVITY** | 3.9 |
| **TRANSPARENCY** | Translucent |

**MASSIVE SIDERITE**
*Numerous acicular quartz crystals coat this specimen of massive siderite.*

# IRON SMELTING

Siderite is a local, but not a major, source of iron. The smelting of iron from ores began in about 2000BC in Anatolia and Persia, but it did not take place in China until as late as 600BC. Iron proved a superior metal to bronze in almost every way, and its use led ultimately to the Industrial Revolution. Precisely where siderite as an ore fitted into these developments is yet to be determined.

**RHOMBOHEDRAL SIDERITE**
*This mass of rhombohedral siderite crystals, many twinned, rests on a quartz groundmass.*

*Twinned crystals*

*Quartz groundmass*

**BOTRYOIDAL SIDERITE**

**RUSTY IRON TONGS**
*This pair of hand-forged iron tongs, found in Norfolk, England, dates from the 1st century BC.*

# MAGNESITE

DISTINCT CRYSTALS OF MAGNESITE are rare, but when found are rhombohedral or prismatic. Magnesite is generally massive, lamellar, fibrous, or granular. A carbonate of magnesium, magnesite takes its name from its composition. Most commonly white or light grey, it can be yellow or brownish when iron substitutes for some of the magnesium. Magnesite forms principally as an alteration product in magnesium-rich rocks, such as peridotites, and through the action of magnesium- containing solutions upon calcite. It also occurs as a primary mineral in limestones and talc or chlorite schists and is found in some meteorites. Localities include Australia and Brazil. It fizzes in warm hydrochloric acid.

**BAHIA**
*The Brumado Mine in Bahia, Brazil contains distinct crystals of magnesite.*

### PROPERTIES

| | |
|---|---|
| **ROCK TYPE** | Carbonates |
| **CRYSTAL SYSTEM** | Hexagonal/trigonal |
| **COMPOSITION** | $MgCO_3$ |
| **COLOUR** | White, light grey, yellowish , brownish |
| **FORM/HABIT** | Massive |
| **HARDNESS** | 4 |
| **CLEAVAGE** | Perfect rhombohedral |
| **FRACTURE** | Conchoidal, brittle |
| **LUSTRE** | Vitreous |
| **STREAK** | White |
| **SPECIFIC GRAVITY** | 3.0 |
| **TRANSPARENCY** | Transparent to translucent |

**RHOMBOHEDRAL CLEAVAGE**
*This cleavage mass of magnesite is with associated phlogopite and serpentine.*

Phlogopite    Serpentine

*Perfect rhombohedral cleavage*

## INDUSTRIAL USES OF MAGNESIUM

A source of magnesium, magnesite is used as a refractory material, as a catalyst and filler in the production of synthetic rubber, and in the preparation of magnesium chemicals and fertilizers. Magnesium is alloyed with aluminium, zinc, or manganese to give it structural strength for use in aircraft, spacecraft, road vehicles, and household appliances (see p.181).

**FURNACE LINING**
*Magnesite is virtually impossible to melt, making it ideal for lining furnaces.*

---

# RHODOCHROSITE

THE CLASSIC COLOUR of rhodochrosite is rose-pink, although it can also be brown or grey. It was named in 1800 from the Greek *rhodokhros*, "of rosy colour". A manganese carbonate, most samples have some calcium and iron substituting for manganese; some also contain magnesium. It is found in hydrothermal ore veins formed at moderate temperatures, in high-temperature metamorphic deposits, and as a secondary mineral in sedimentary manganese deposits. At Butte, Montana, USA, it is so abundant that it is mined as an ore of manganese. Other noted localities are in Romania, Gabon, Mexico, Russia, Japan, and the Sweet Home Mine, Colorado, USA.

**HOKKAIDO**
*Rhodochrosite crystals are found in this area of Japan.*

### PROPERTIES

| | |
|---|---|
| **GROUP** | Carbonates |
| **CRYSTAL SYSTEM** | Hexagonal/trigonal |
| **COMPOSITION** | $MnCO_3$ |
| **COLOUR** | Rose-pink, brown or grey |
| **FORM/HABIT** | Rhombohedral |
| **HARDNESS** | 3½–4 |
| **CLEAVAGE** | Perfect rhombohedral |
| **FRACTURE** | Uneven |
| **LUSTRE** | Vitreous to pearly |
| **STREAK** | White |
| **SPECIFIC GRAVITY** | 3.6 |
| **TRANSPARENCY** | Transparent to translucent |
| **R.I.** | 1.6–1.8 |

## ORNAMENTAL USE

Gem-quality crystals occur at several mines in Colorado, USA, and at Hotazel in South Africa. These are sometimes cut for collectors, but the more common, fine-grained, banded, stalactitic rock is the one normally used for decoration. Rhodochrosite is both soft and very fragile: faceted stones are rare and demand a great deal of skill from the cutter.

**EARRINGS**
*Lustrous rhodochrosite beads adorn these earrings.*

**DECORATIVE DUCKS**
*These ducks are carved from rhodochrosite and calcite.*

**RHODOCHROSITE CRYSTALS**
*This superb group of transparent, gem-quality rhodochrosite crystals is partly coated with quartz.*

**STALACTITIC GROWTH**
*Particularly fine stalactitic growths, sometimes called "Inca rose", come from Catarnarca, Argentina.*

Quartz

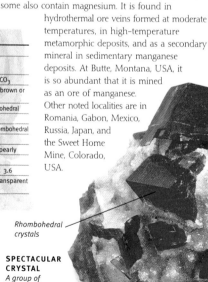

*Rhombohedral crystals*

**SPECTACULAR CRYSTAL**
*A group of rhodochrosite rhombohedrons perch on radiating quartz crystals.*

**TEARDROP**
*This rhodochrosite is faceted in a pendaloque, or teardrop, cut.*

**BRILLIANT**
*Transparent rhodochrosite is sometimes faceted into gems for collectors.*

# ARAGONITE

CHEMICALLY IDENTICAL to the more common mineral calcite, but forming under more limited geological conditions, aragonite is calcium carbonate. It is relatively unstable and can alter to calcite. Aragonite crystals are orthorhombic; those of calcite are trigonal. When aragonite is found in single crystals, they are short to long prismatic, but crystals are commonly twinned, with some multiple twins taking on a hexagonal appearance. It can also be columnar, stalactitic, radiating, or fibrous. Although it sometimes appears similar to calcite, aragonite is easily distinguished by the absence of rhombohedral cleavage. It can be white, colourless, grey, yellowish, green, blue, reddish, violet, or brown.

Aragonite is found in the oxidized zone of ore deposits and elsewhere, formed at low temperatures near the surface of the Earth. It is also found in caves as stalactites, around hot springs, and in mineral veins. When it occurs in coral-like aggregates in iron-ore deposits in

**CARD CASE AND SHELL**
*Pearl, many shells, and the tests (hard outer coverings) of numerous marine invertebrates are aragonite.*

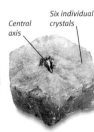

association with siderite, its iron carbonate counterpart, it is called *flos-ferri*, meaning "flowers of iron". Banded, stalactitic material is sometimes polished as an ornamental stone, and transparent crystals are rarely faceted for collectors. Aragonite is also produced by biological processes yet to be fully understood. Facet-quality crystals come from the Czech Republic, superb cave growths from Mexico, *flos-ferri* from Austria and Greece, and fine crystals from Morocco, England, France, Germany, Italy, Hungary, Japan, and numerous localities in the USA.

**ARAGON, SPAIN**
*Aragonite is named for its original locality, the Aragon region of Spain.*

## PROPERTIES

| | |
|---|---|
| **GROUP** | Carbonates |
| **CRYSTAL SYSTEM** | Orthorhombic |
| **COMPOSITION** | $CaCO_3$ |
| **COLOUR** | Colourless, white, grey, yellowish, reddish, green |
| **FORM/HABIT** | Prismatic, acicular |
| **HARDNESS** | $3\frac{1}{2}$–4 |
| **CLEAVAGE** | Distinct |
| **FRACTURE** | Subconchoidal, brittle |
| **LUSTRE** | Vitreous inclining to resinous |
| **STREAK** | White |
| **SPECIFIC GRAVITY** | 2.9 |
| **TRANSPARENCY** | Transparent to translucent |

**POLISHED STONE**
*Layered aragonite is sometimes polished and sold as "onyx".*

*Radiating habit*

*Semi-transparent crystal*

*Prismatic crystals*

**ARAGONITE**
*This radiating group of prismatic, semitransparent, pseudohexagonal aragonite crystals comes from Morocco. Such groups are commonly sold as aragonite "sputniks".*

**INLAID SHEATH**
*The mother-of-pearl in the inlay of this dagger sheath is principally aragonite.*

**VARIETIES OF ARAGONITE**
*The huge range of aragonite habits can be seen from these specimens.*

*Rock groundmass*

*Prismatic crystals*

**LONG PRISMATIC ARAGONITE**

*Tree-like structure*

**FLOS-FERRI ARAGONITE**

*Radiating habit*

**RADIATING ARAGONITE**

*Central axis*

*Six individual crystals*

**CYCLIC TWIN**

# WITHERITE

**ALSTON MOOR, CUMBRIA**
*Witherite occurs in this part of northern England.*

WITHERITE WAS NAMED IN 1790 for English mineralogist William Withering. Its crystals always form multiple twins, often yielding pseudohexagonal, paired pyramids. They can also be short to long prismatic or tabular. Witherite is white, colourless, or tinged yellow, brown, or green. It forms in low-temperature hydrothermal veins. Well-formed crystals occur at Cave-in-Rock, Illinois and El Portal, California, USA. Witherite is found in commercial quantities in England, France, Turkmenistan, and Japan.

## PROPERTIES

| | |
|---|---|
| **GROUP** | Carbonates |
| **CRYSTAL SYSTEM** | Orthorhombic |
| **COMPOSITION** | $BaCO_3$ |
| **COLOUR** | White, colourless, yellow, brown, or green |
| **FORM/HABIT** | Pseudohexagonal |
| **HARDNESS** | $3-3\frac{1}{2}$ |
| **CLEAVAGE** | Distinct, imperfect |
| **FRACTURE** | Uneven, brittle |
| **LUSTRE** | Vitreous |
| **STREAK** | White |
| **SPECIFIC GRAVITY** | 4.3 |
| **TRANSPARENCY** | Transparent to translucent |

**WITHERITE CRYSTALS**
*This specimen of witherite also contains galena.*

*Twinned crystals*

*Galena*

# STRONTIANITE

**STRONTIAN**
*Strontianite was discovered in this area of Scotland.*

STRONTIANITE'S CRYSTALS are acicular, short columnar, spear-shaped, or in radiating masses. It can also be massive and granular. It is usually colourless to grey, but can be pale green, yellow, and yellow-brown to reddish. Strontianite forms as a low-temperature hydrothermal mineral, and is also found in geodes and concretions. Localities include Germany, Canada, India, and California, USA. Strontianite is the principal source of strontium, and is used in sugar refining, and to produce the red colour in fireworks.

## PROPERTIES

| | |
|---|---|
| **GROUP** | Carbonates |
| **CRYSTAL SYSTEM** | Orthorhombic |
| **COMPOSITION** | $SrCO_3$ |
| **COLOUR** | Colourless, grey, green, yellow, or reddish |
| **FORM/HABIT** | Acicular, columnar |
| **HARDNESS** | $3\frac{1}{2}-4$ |
| **CLEAVAGE** | Good |
| **FRACTURE** | Uneven, brittle |
| **LUSTRE** | Vitreous |
| **STREAK** | White |
| **SPECIFIC GRAVITY** | 3.7 |
| **TRANSPARENCY** | Transparent to translucent |

*Acicular crystal habit*

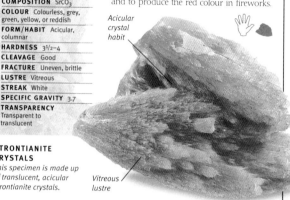

**STRONTIANITE CRYSTALS**
*This specimen is made up of translucent, acicular strontianite crystals.*

*Vitreous lustre*

# DOLOMITE

**SAXONY**
*This part of Germany has extensive deposits of dolomite.*

NAMED AFTER FRENCH mineralogist D. de Dolomieu (1750–1801), dolomite is calcium magnesium carbonate. Crystals are commonly rhombohedral or tabular, and often have curved faces. Dolomite can also be coarse to fine granular, massive, or (rarely) fibrous. Dolomite is important as a rock-forming mineral in carbonate rocks, and is the principal component of the rock of the same name (see p.54). It also occurs as a secondary replacement of limestone by the action of magnesium-bearing solutions, in marbles, talc schists, and other magnesium-rich metamorphic rocks, and in hydrothermal veins associated with lead, zinc, or copper ores. Dolomite is widespread; crystals come from Algeria, Brazil, Switzerland, Namibia, Mexico, and several US localities.

## PROPERTIES

| | |
|---|---|
| **GROUP** | Carbonates |
| **CRYSTAL SYSTEM** | Hexagonal/trigonal |
| **COMPOSITION** | $CaMg(CO_3)_2$ |
| **COLOUR** | Colourless, white, or cream |
| **FORM/HABIT** | Rhombohedral |
| **HARDNESS** | $3\frac{1}{2}-4$ |
| **CLEAVAGE** | Perfect rhombohedral |
| **FRACTURE** | Subconchoidal |
| **LUSTRE** | Vitreous |
| **STREAK** | White |
| **SPECIFIC GRAVITY** | 2.8–2.9 |
| **TRANSPARENCY** | Transparent to translucent |

*Curved faces*

*Saddle-shaped crystals*

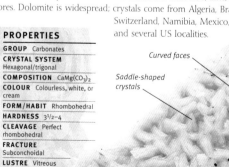

## USES OF MAGNESIUM

Dolomite is a minor source of magnesium. Magnesium and magnesium alloys are widely used in industry where a combination of lightness and strength are required. Alloyed with manganese, aluminium, or titanium, magnesium parts are used in aircraft, spacecraft, machinery, cars, portable tools, and household appliances. Magnesium is also an important mineral in the human diet, and research has shown that a high magnesium intake aids high blood pressure and cardiovascular conditions.

**FLASHBULB**
*Fine strands of magnesium are ignited in flashbulbs.*

**RACING YACHT**
*Magnesium alloys are used in the hulls of racing yachts.*

*Vitreous lustre*

**DOLOMITE CRUST**
*The tabular crystals of pink dolomite form a crust on this specimen from Morocco. Its pale pink colouring is due to increasing substitution of manganese. Substitution of iron causes a pale brown colouring.*

# CERUSSITE

KNOWN SINCE ANTIQUITY and named for the Latin *cerussa*, a white lead pigment, cerussite is lead carbonate. Its crystal habits are highly varied, but include tabular, prismatic, dipyramidal, pseudohexagonal, and acicular. Thin tabular crystals are less common. It is generally colourless or white to grey, but may be blue to green with copper impurities. Cerussite is a secondary mineral, which occurs in the oxidation zone of lead veins, and is formed by the action of carbonated water on other lead minerals, particularly galena and anglesite. It is a widespread mineral, with excellent specimens coming from Tsumeb, Namibia; Broken Hill, NSW, Australia; Monteponi Mine, Sardinia; Oruro, Bolivia; Murcia, Spain; and at Phoenixville, Pennsylvania, and numerous mines in Arizona and California in the USA.

*Deep striations*

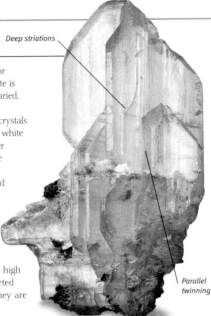

After galena, cerussite is the commonest ore of lead. It has a high refractive index, making the faceted stones especially brilliant, but they are too soft to be worn.

*Parallel twinning*

**BRILLIANT**
*Although very soft and fragile, transparent cerussite is occasionally faceted for collectors.*

**CERUSSITE CRYSTALS**
*This group of transparent, parallel-twinned, colourless cerussite crystals shows good tabular development.*

## PROPERTIES

| | |
|---|---|
| **GROUP** | Carbonates |
| **CRYSTAL SYSTEM** | Orthorhombic |
| **COMPOSITION** | $PbCO_3$ |
| **COLOUR** | White, grey, blue to green |
| **FORM/HABIT** | Tabular, prismatic |
| **HARDNESS** | $3–3\frac{1}{2}$ |
| **CLEAVAGE** | Distinct |
| **FRACTURE** | Conchoidal, brittle |
| **LUSTRE** | Adamantine to vitreous |
| **STREAK** | Colourless |
| **SPECIFIC GRAVITY** | 6.5 |
| **TRANSPARENCY** | Transparent to translucent |
| **R.I.** | 1.8–2.1 |

# AZURITE

**GLOBE, ARIZONA, USA**
*The Globe Copper Mines are a great source of azurite.*

TAKING ITS NAME from the same Persian word as lapis lazuli - *lazhuward*, meaning "blue" – azurite is a deep blue copper carbonate hydroxide. It is azure-blue to very dark blue, and forms tabular or prismatic crystals with a wide variety of faces - over 45 may be commonly seen, as well as over 100 rarer faces. It can also be massive, stalactitic, and botryoidal.

Azurite is a secondary mineral, formed in the oxidized portion of copper deposits through the action of carbonated waters on other copper minerals. It is often found with malachite. Finely crystallized specimens come from Utah, Arizona, and New Mexico, USA, and from France, Mexico, Chile, Australia, Russia, Morocco, and especially from Tsumeb, Namibia.

It was mined in ancient Egypt, from Sinai and the eastern desert, and used as a source of metallic copper, as a pigment, and probably in the production of blue glaze. Although used primarily as an ore of copper, it was also used as a pigment in 15th to 17th century European art. Massive azurite used for ornamental purposes is sometimes called chessylite, after Chessy, France.

*Azurite pigment*

**DEEP BLUE**
*Azurite pigment was used in this 15th-century French painting in an illuminated manuscript.*

*Deep azure colour*

*Bladed crystals*

## PROPERTIES

| | |
|---|---|
| **GROUP** | Carbonates |
| **CRYSTAL SYSTEM** | Monoclinic |
| **COMPOSITION** | $Cu_3(CO_3)_2(OH)_2$ |
| **COLOUR** | Azure- to dark-blue |
| **FORM/HABIT** | Tabular, prismatic |
| **HARDNESS** | $3\frac{1}{2}–4$ |
| **CLEAVAGE** | Perfect |
| **FRACTURE** | Conchoidal, brittle |
| **LUSTRE** | Vitreous to dull to earthy |
| **STREAK** | Blue |
| **SPECIFIC GRAVITY** | 3.8 |
| **TRANSPARENCY** | Transparent to translucent |
| **R.I.** | 1.73–1.84 |

**CONCRETION**
*Spherical concretions of radiating azurite crystals, such as this, can measure more than 2.5cm (1in) in diameter.*

**CABOCHON**
*Azurite can be cut en cabochon. The top surface of this specimen has been faceted.*

*Vitreous lustre*

**AZURITE CRYSTAL**

**AZURITE-MALACHITE CABOCHON**

**DIFFERENT AZURITE HABITS**
*A dark blue bladed azurite crystal (far left) contrasts with the vivid blue of an azurite and green malachite cabochon.*

**AZURITE CRYSTALS**
*This group of bladed azurite crystals comes from Chessy, near Lyon, France, a traditional source of azurite to be ground for use as a pigment, though little remains now.*

# ANKERITE

**AUSTRIA**
*Erzberg, Austria, is a source of well-formed ankerite crystals.*

ESSENTIALLY CALCIUM CARBONATE with varying amounts of iron, magnesium, and manganese in the structure, ankerite was named in 1825 for Austrian mineralogist M.J. Anker. It forms rhombohedral crystals similar to those of dolomite, and can also be massive or coarsely granular. Usually pale buff coloured, it can also be colourless, white, grey, and brownish. Much ankerite is fluorescent. Ankerite forms as a secondary mineral from the action of magnesium–bearing fluids on limestone or dolomite. Specimens can be found in Namibia; South Africa; Austria; Japan; Canada; western USA; and elsewhere.

**PROPERTIES**

| | |
|---|---|
| **GROUP** Carbonates | |
| **CRYSTAL SYSTEM** Hexagonal/trigonal | |
| **COMPOSITION** Ca(Fe,Mg,Mn,)(CO$_3$)$_2$ | |
| **COLOUR** Colourless to pale buff | |
| **FORM/HABIT** Rhombohedral | |
| **HARDNESS** 3$^1$/$_2$–4 | |
| **CLEAVAGE** Perfect | |
| **FRACTURE** Subconchoidal | |
| **LUSTRE** Vitreous to pearly | |
| **STREAK** White | |
| **SPECIFIC GRAVITY** 2.9 | |
| **TRANSPARENCY** Translucent | |

*Perfect cleavage*  *Rhombohedral crystal*

**ANKERITE CRYSTALS**
*These rhombohedral ankerite crystals have a characteristic brown colour.*

# BARYTOCALCITE

**ALSTON MOOR, ENGLAND**
*Barytocalcite is found in this part of Cumbria.*

BARYTOCALCITE derives its name from its components: barium and calcite (calcium carbonate). It forms short to long prismatic crystals, usually striated, and can also be massive. It is colourless to white, greyish, greenish, or yellowish. Barytocalcite forms in hydrothermal veins, especially where hydrothermal solutions have invaded limestone. It is found with baryte and calcite, and can be a minor ore of barium. Barytocalcite effervesces in hydrochloric acid. Crystals up to 5cm (2in) long are found in Cumbria, England. Other localities include Langban, Sweden, and the Krasnoyarsk region of Siberia, Russia.

**PROPERTIES**

| | |
|---|---|
| **GROUP** Carbonates | |
| **CRYSTAL SYSTEM** Monoclinic | |
| **COMPOSITION** BaCa(CO$_3$)$_2$ | |
| **COLOUR** White, greyish, greenish, or yellowish | |
| **FORM/HABIT** Prismatic | |
| **HARDNESS** 4 | |
| **CLEAVAGE** Perfect, imperfect | |
| **FRACTURE** Uneven, brittle | |
| **LUSTRE** Vitreous to resinous | |
| **STREAK** White | |
| **SPECIFIC GRAVITY** 3.7 | |
| **TRANSPARENCY** Transparent to translucent | |

**PRISMATIC CRYSTALS**
*These prismatic crystals are on a limestone groundmass.*

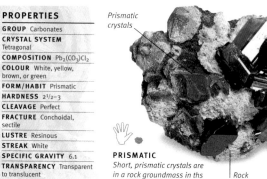

*Prismatic crystals*  *Limestone groundmass*

# AURICHALCITE

**CHIHUAHUA, MEXICO**
*Copper Canyon has deposits of aurichalcite.*

ITS NAME DERIVING from the Latin for "golden copper", aurichalcite is a zinc copper carbonate hydroxide. It forms acicular, sky-blue, greenish-blue, or pale green crystals, often found as radiating, tufted masses or as velvet-like encrustations. Occasionally it is massive, columnar, or lamellar. Crystals rarely exceed several millimetres long. A secondary mineral, it forms in the oxidized zones of copper and zinc deposits. Localities include the western USA; Nagato, Japan; Chihuahua, Mexico; Chessy, France; and Tsumeb, Namibia.

**PROPERTIES**

| | |
|---|---|
| **GROUP** Carbonates | |
| **CRYSTAL SYSTEM** Monoclinic | |
| **COMPOSITION** (Zn,Cu)$_5$(CO$_3$)$_2$(OH)$_6$ | |
| **COLOUR** Sky-blue, green-blue, or pale green | |
| **FORM/HABIT** Acicular | |
| **HARDNESS** 1–2 | |
| **CLEAVAGE** Perfect, brittle | |
| **FRACTURE** Uneven | |
| **LUSTRE** Silky to pearly | |
| **STREAK** Pale blue-green | |
| **SPECIFIC GRAVITY** 4.2 | |
| **TRANSPARENCY** Transparent to translucent | |

*Limonite groundmass*

**TUFTED CRYSTALS**
*These masses of small, tufted aurichalcite crystals are on a groundmass of limonite.*

*Aurichalcite*

# PHOSGENITE

**BROKEN HILL, AUSTRALIA**
*This area is a source of phosgenite.*

PHOSGENITE WAS NAMED in 1841 for the colourless, poisonous gas phosgen (COCl$_2$), because it contains some of the elements that are also found in phosgen. Its crystals are short prismatic, or less commonly, thick tabular. It can be massive or granular as well. Its colour varies from white to yellowish, brown, or greenish. It occurs as a product of secondary alteration of other lead minerals under surface conditions. Localities include Sardinia, Italy; Tsumeb, Namibia; Dundas, Tasmania; Broken Hill, NSW, Australia; Derbyshire, England; and, in the USA, Inyo County, California, and Custer County, Colorado.

**PROPERTIES**

| | |
|---|---|
| **GROUP** Carbonates | |
| **CRYSTAL SYSTEM** Tetragonal | |
| **COMPOSITION** Pb$_2$(CO$_3$)Cl$_2$ | |
| **COLOUR** White, yellow, brown, or green | |
| **FORM/HABIT** Prismatic | |
| **HARDNESS** 2$^1$/$_2$–3 | |
| **CLEAVAGE** Perfect | |
| **FRACTURE** Conchoidal, sectile | |
| **LUSTRE** Resinous | |
| **STREAK** White | |
| **SPECIFIC GRAVITY** 6.1 | |
| **TRANSPARENCY** Transparent to translucent | |

*Prismatic crystals*

**PRISMATIC**
*Short, prismatic crystals are in a rock groundmass in this specimen from Sardinia, Italy.*

*Rock groundmass*

183

# MALACHITE

POSSIBLY THE EARLIEST ORE OF COPPER, having been mined in the Sinai and eastern deserts of ancient Egypt from as early as 3000BC, malachite is a green copper carbonate hydroxide. It was used as an eye paint, a pigment for wall painting, and in glazes and the colouring of glass. Single crystals are uncommon; when found, they are short to long prismatic. Usually found as botryoidal or encrusting masses, often with a radiating fibrous structure and banded in various shades of green, malachite is also found as delicate fibrous aggregates and as concentrically banded stalactites. Its name comes from the Greek for "mallow", in reference to its leaf-green colour.

Malachite occurs in the altered zones of copper deposits, where it is usually accompanied by lesser amounts of azurite (see p.182). The brilliant green of malachite is usually visible on the surface of copper deposits, and is often the first indication a prospector has of this metal's presence.

**COPPER MINE, ZAMBIA**
*Malachite and chrysocolla have formed as an alteration product of copper and are visible here on the surface of the Kansanshi open-pit copper mine, Zambia.*

**BOHEMIAN BOTTLE**
*This mid-19th century Bohemian perfume bottle is made from green malachite glass and contains a gilt spherical stopper.*

**CARVED MALACHITE NECKLACE**
*This necklace is fitted with ten panels of malachite, each carved with allegorical figures. It is part of a necklace, bracelet, earring, and pin set made by Francis Peck, Ltd., Nassau.*

It is a minor ore of copper, but its principal use is as an ornamental material and gemstone. In ancient China, it was highly prized and called *shilu*, taking its name from its source near Shilu, Guandong Province. In the 19th century, huge deposits in the Ural Mountains of Russia supplied large amounts of malachite to Europe, with some single masses weighing up to 51 tonnes (see pp.186–87). It was worn in Italy as protection against the "evil eye".

The finest crystals come from Kolwezi, Democratic Republic of Congo, and can approach 10cm (4in) in length. The Congo also provides much of the world's supply of ornamental material. Material suitable for cutting also comes from South Australia; Morocco; Bisbee, Arizona, USA; and Lyon in France.

## ANCIENT USES

Malachite must rank as one of the most important minerals in mankind's history. It was discovered in around 4000BC that heating malachite in a hot fire reduced it to native copper. This was probably the first smelting of metal from an ore. This led in turn to the search for other metallic ores, and to the development of metallurgy. For the first time, specialist skills and a new organization of work, in which some people devoted all their time to the mining, smelting, and forging of metal, were developed – all as a direct result of the properties of malachite. The mineral was in use in Egypt as a cosmetic and a pigment from at least 3000BC. Its use as an eye paint may have been an attempt to ward off eye infections – it is now known that the green eye paint containing malachite had medicinal properties. In ancient Greece, it was used in amulets for children.

**EGYPTIAN WALL PAINTING**
*Malachite was used as the green pigment in this Egyptian wall painting from the 18th dynasty (15th century BC). The painting is from the tomb of Nakht at Luxor and depicts Nakht and his family hunting water fowl.*

**MALACHITE PIGMENT**

## PROPERTIES

**GROUP** Carbonates
**CRYSTAL SYSTEM** Monoclinic
**COMPOSITION** $Cu_2CO_3(OH)_2$
**COLOUR** Bright green
**FORM/HABIT** Massive, botryoidal
**HARDNESS** $3^{1}/_{2}$–4
**CLEAVAGE** Perfect
**FRACTURE** Subconchoidal to uneven, brittle
**LUSTRE** Adamantine to silky
**STREAK** Pale green
**SPECIFIC GRAVITY** 3.9–4.0
**TRANSPARENCY** Translucent

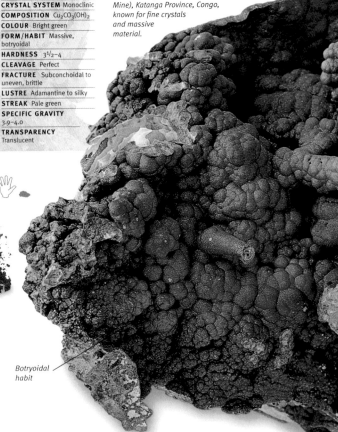

**BOTRYOIDAL MALACHITE**
*This botryoidal malachite on chrysocolla is from Etoile du Congo Mine (Star of the Congo Mine), Katanga Province, Congo, known for fine crystals and massive material.*

*Botryoidal habit*

## MALACHITE HABITS

Malachite comes in a variety of habits, all of which provide beautiful and interesting patterns when cut and polished.

Radiating crystals

**STALACTITIC**

Growth layering

**STALACTITIC CROSS SECTION**

Radiating crystals

**FIBROUS**

Malachite encrustation

Azurite

**MALACHITE WITH AZURITE**

---

"A field of ripe cabbages with their prevailing hue of malachite green"

**WALT WHITMAN**, "STRAW-COLOR'D AND OTHER PSYCHES", 1892

### AZTEC KNIFE

The handle of this Aztec knife from the 15th century is covered with a mosaic of malachite, turquoise, and shell. The blade is made of chalcedony.

Chrysocolla

### KOREAN PAINTING

Malachite pigment is used to dramatic effect in this painting on hemp cloth of Dhratarastra, Guardian King of the East. Probably from Kyongsang Province, Korea, the painting is from the Choson Dynasty and dates from the late 18th century.

### POLISHED EGG

The light and dark layers of malachite create a striking pattern in this polished egg, which has been cut to reveal the concentric banding.

### PILLOW

This polished piece of malachite is cut across the botryoidal structure, to show an intricate pattern of swirls and bands.

### VASE

This Chinese ritual food vessel, dating from the Sung Dynasty (10th to 12th centuries AD) is decorated with inlaid malachite and silver.

Rock groundmass

185

# RUSSIAN MALACHITE

MINED IN THE URAL MOUNTAINS FROM EARLY IN THE 19TH CENTURY, RUSSIAN MALACHITE HAS BEEN USED DECORATIVELY TO FORM HUGE COLUMNS WITHIN BUILDINGS, USEFUL PIECES OF FURNITURE, AND DELICATE CARVED ARTEFACTS.

**PAINTED CEILING**
*The ceiling of St. Isaac's Cathedral was painted by Karl Bryullov, president of the Academy of Fine Arts. The title of his work is* Virgin in Glory.

**THE MALACHITE ROOM**
*In this 1865 watercolour painting of the Malachite Room by Konstantin A. Ukhtomsky, the green malachite columns and surfaces are highlighted by the white walls. During the Russian Revolution, between June and October 1917, the Provisional Government met in this room.*

Two of the world's largest gemstone deposits were found in the Ural Mountains of Russia in the early 19th century. Malachite was found in huge quantities in mines at Yekaterinburg, situated 1,650km (1,025 miles) east of Moscow, and at the Demidoff Mine at Nizhni-Tagil (160km/100 miles to the north of Yekaterinburg), where a single, banded mass contained more than 1,100 tonnes of fine material. It has been used to make objects as large as the enormous malachite pillars of St. Isaac's Cathedral in St. Petersburg, as well as intricately worked objects by Carl Fabergé. Many artefacts created using malachite from these deposits can be seen in the Hermitage Museum in Russia, and the Uffizi and St. Peter's in Italy.

St. Isaac's Cathedral on the banks of the River Neva in St. Petersburg is the fourth largest domed cathedral in the world. Begun in 1818, the building took 40 years to complete, and its interior contains 16 varieties of marble, granite, malachite, and lapis lazuli. Designed by French architect Auguste de Montferrand, it incorporates eight massive malachite pillars in the mosaic-covered wall decorated with icons that separates the altar from the rest of the church.

The Malachite Room is one of the most spectacular rooms of the Winter Palace at St. Petersburg. It was rebuilt in 1837 as a drawing room for the wife of Tsar Nicholas I, Alexandra Fyodorovna, to the design of architect and painter Alexander Bryullov. It features pilasters, columns, and mantelpieces in malachite mosaic set against white walls decorated with figures representing day, night, and poetry. Gilt doors and crimson hangings complete the sumptuous interior. The room is further ornamented with malachite vases and other artefacts made of malachite during the early 19th century, which were produced in the workshop of Peter Gambs from sketches by Auguste de Montferrand. In the late 19th century, small cupboards were added, decorated with mosaic panels produced by the Peterhof Lapidary Works. Virtually all the malachite from which the decoration of the room was produced was derived from the Russian deposits in the Ural Mountains.

**MALACHITE JEWEL BOX**
*Malachite remains a favourite carving stone today. This malachite jewel box was crafted in 1989 by E. Zhiriakova and is entitled* Mistress of Copper Mountain.

**INTERIOR OF ST. ISAAC'S CATHEDRAL**
*Eight giant green columns made of Russian malachite and two of lapis lazuli grace the iconostasis of this gold-domed, Russian Orthodox cathedral. They stand out dramatically against the white marble and gold leaf that cover the wall.*

# PHOSPHATES, ARSENATES, AND VANADATES

PHOSPHATE, ARSENATE, AND VANADATE minerals are grouped together because of the similarity of their crystal structures. The phosphates are the most numerous of the three groups with more than 200 known species, including the bright blue lazulite and turquoise.

## PHOSPHATES

Phosphate minerals contain phosphorus and oxygen combined in a 1:4 ratio, written as $PO_4$. This acts as a single unit that combines with another element or elements to form phosphate minerals. There are three groups of phosphates. Primary phosphates usually crystallize from aqueous fluids derived from the late stages of crystallization, and are common in granitic pegmatites. Secondary phosphates form when primary phosphates are altered in the presence of water. Fine–grained rock phosphates form from phosphorus–bearing organic material, primarily underwater.

*Phosphorus*

*Oxygen*

## ARSENATES

The arsenate minerals have a basic structural unit of arsenic and oxygen, written AsO4. This unit combines with another element or elements to form arsenate minerals. Most arsenates are rare, and many are brilliantly coloured. Their oxidation products can be a useful sign for prospectors.

*Oxygen*

*Arsenic*

## VANADATES

The vanadates mostly contain the same type of structural tetrahedra as the phosphates and arsenates, written $VO_4$. Vanadate structures are complex, and their restricted conditions of crystallization make them relatively rare. Apart from carnotite, which is an important source of uranium, most vanadates have no economic importance. However, they are prized by mineral collectors for their brilliant colours.

*Oxygen*

*Vanadium*

# MONAZITE

**EIFEL**
*The German district of Eifel contains large monazite deposits.*

A PHOSPHATE MINERAL, monazite has three different species, all sharing the same crystal structure, and each designated by the chemical symbol of the element it contains. Thus, monazite-(Ce) is cerium phosphate; monazite-(La) is lanthanum phosphate; and monazite-(Nd) is neodymium phosphate. The most widespread is monazite-(Ce), which forms prismatic, flattened, or elongated crystals, which are occasionally large and coarse, and commonly twinned. Its colour is yellowish- or reddish-brown to brown, greenish, or nearly white. Monazite

is a common accessory mineral in granites and gneisses, and detrital monazite in the form of monazite sands can accumulate in commercial quantities. It is mined on the west coast of Sri Lanka; in the Ural Mountains of Russia; the tin deposits of Malaysia; and the Mbabane district of Swaziland. Crystals come from Iceland, Norway, Brazil, Germany, South Africa, and from Colorado, California, New Mexico, and North Carolina, USA.

*Striated face*

**RED MONAZITE**

## PROPERTIES

| | |
|---|---|
| **GROUP** | Phosphates |
| **CRYSTAL SYSTEM** | Monoclinic |
| **COMPOSITION** | (Ce,La,Th,Nd)$PO_4$ |
| **COLOUR** | Yellowish-brown to brown, greenish, or nearly white |
| **FORM/HABIT** | Prismatic |
| **HARDNESS** | 5 |
| **CLEAVAGE** | Perfect, good, poor |
| **FRACTURE** | Conchoidal to uneven, brittle |
| **LUSTRE** | Resinous, waxy, or vitreous |
| **STREAK** | White |
| **SPECIFIC GRAVITY** | 4.6–5.4 |
| **TRANSPARENCY** | Translucent |

*Termination*

*Prismatic crystal*

*Prism face*

**MONAZITE CRYSTAL**
*This twinned and striated crystal of monazite is from Arendal, Aust-Agder, Norway, an important monazite locality.*

## USES OF ELEMENTS IN MONAZITE

Monazite-(Ce) is a major source of cerium. Cerium oxide is an important polishing compound, and is widely used for polishing gemstones and glass products, especially lenses for cameras and other optical instruments. Lanthanum is used in catalysts in the process of refining oil. Neodymium is used for doping glass to give it a violet-purple colour.

**FLUORESCENT LAMP**
*Lanthanum is used as a phosphor in fluorescent lamps.*

# XENOTIME

XENOTIME IS YTTRIUM PHOSPHATE. It takes its name from the Greek for "vain honour", because the yttrium in xenotime was mistakenly believed to be a new element when it was described in 1832. It is generally found as glassy, short to long prismatic crystals, or as rosette-shaped crystal aggregates. Its colour can be yellowish- to reddish-brown, fleshy red, greyish-white, or pale to wine-yellow. Large proportions of erbium commonly replace yttrium in the structure. Xenotime occurs as an accessory mineral in igneous rocks and in associated pegmatites, where it can form large crystals. It also occurs in quartzose and micaceous gneisses, and commonly in detrital material. It is often found with zircon, anatase, rutile, sillimanite, columbite, monazite, and ilmenite. It is widely distributed, with notable occurrences in Japan, Sweden, Norway, Germany, Brazil, Madagascar, and, in the USA, in New York, North Carolina, and Colorado.

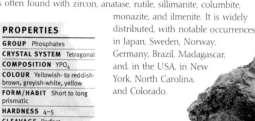

**HIGH-SPEED LASER**
*Garnet containing yttrium is used to make lasers, such as the one shown here. Mirrors change the direction of the beam, which is coloured by the use of dyes and filters.*

**TV PIXELS**
*Red yttrium phosphors are used in the production of colour TV screens.*

*Pyramidal crystal*

*Aggregate of rough crystals*

## PROPERTIES

| | |
|---|---|
| **GROUP** | Phosphates |
| **CRYSTAL SYSTEM** | Tetragonal |
| **COMPOSITION** | $YPO_4$ |
| **COLOUR** | Yellowish- to reddish-brown, greyish-white, yellow |
| **FORM/HABIT** | Short to long prismatic |
| **HARDNESS** | 4–5 |
| **CLEAVAGE** | Perfect |
| **FRACTURE** | Uneven to splintery |
| **LUSTRE** | Vitreous to resinous |
| **STREAK** | Pale brown or reddish |
| **SPECIFIC GRAVITY** | 4.4–5.1 |
| **TRANSPARENCY** | Transparent to opaque |

**XENOTIME CRYSTALS**
*These pyramidal xenotime crystals are on an aggregate of rough xenotime crystals.*

---

# CARNOTITE

NAMED IN 1899 FOR FRENCH chemist and mining engineer Marie-Adolphe Carno, carnotite is a potassium uranyl vanadate hydrate. Generally it is found as powdery or microcrystalline masses, as tiny, disseminated grains, or as crusts. When crystals form, they are platy, rhombohedral, or lath-like. It is bright to lemon-yellow, or greenish-yellow. It is radioactive and easily soluble in acids. Carnotite is a secondary mineral, formed by the alteration of primary uranium-vanadium minerals. It occurs chiefly with tyuyamunite (its calcium analogue) in sandstone, either disseminated or in concentrations around fossil wood or other fossilized vegetable matter. The largest area of carnotite deposits is in the south-western USA, particularly in the Colorado Plateau area and the adjoining states of Utah, Arizona, and New Mexico. Other commercial deposits occur in Uzbekistan, Congo, and Australia.

## USES OF URANIUM

Carnotite is an important source of uranium – pure carnotite contains about 53 per cent uranium, 12 per cent vanadium, and trace amounts of radium. It was mined for vanadium and radium prior to World War II, and afterwards for uranium, although it is still an important source for vanadium and radium. Uranium is a vital material for the generation of nuclear energy, and is used in atomic weapons. Another use of uranium is in the production of transuranium elements – those with an atomic number greater than 92. Twenty have so far been created, yielding new insights into the fundamental structure.

**ATOMIC EXPLOSION**
*The emotive image of the atomic mushroom cloud will forever be synonymous with uranium.*

## PROPERTIES

| | |
|---|---|
| **GROUP** | Vanadates |
| **CRYSTAL SYSTEM** | Monoclinic |
| **COMPOSITION** | $K_2(UO_2)_2(VO_4)_2 \cdot 3H_2O$ |
| **COLOUR** | Yellow |
| **FORM/HABIT** | Powdery microcrystalline |
| **HARDNESS** | 2 |
| **CLEAVAGE** | Perfect |
| **FRACTURE** | Uneven |
| **LUSTRE** | Pearly to dull |
| **STREAK** | Yellow |
| **SPECIFIC GRAVITY** | 4.7 |
| **TRANSPARENCY** | Semi-transparent to opaque |

*Crust of carnotite*

*Rock groundmass*

**CARNOTITE CRUST**
*A crust of vivid yellow, powdery carnotite coats this fragment of sandstone.*

# AUTUNITE

A POPULAR COLLECTORS' MINERAL, in part because it fluoresces under ultraviolet light, autunite contains uranium and is radioactive, so must be stored away carefully and handled as little as possible to minimize exposure to radiation. Autunite forms crystals, scaly foliated aggregates, and crusts with crystals standing on edge giving a serrated appearance. As the mineral is mildly heated, water loss transforms tetragonal autunite into orthorhombic meta-autunite. Most museum specimens have at least partly converted to meta-autunite. A moist atmosphere helps to prevent dehydration. It is best stored in a well-ventilated place out of the house. If it is sealed in a container, radon builds up, so ventilation is needed upon opening.

Named for the place of its discovery, Autun, France, autunite is a secondary mineral, formed in the oxidation zone of uranium ore bodies as an alteration product of uraninite and other uranium-bearing minerals. It also occurs in hydrothermal veins and in pegmatites. It is found in England, Germany, Japan, India, Australia, Brazil, and the USA.

## PROPERTIES

**GROUP** Phosphates
**CRYSTAL SYSTEM** Tetragonal
**COMPOSITION** $Ca(UO_2)_2(PO_4)_2 \cdot 10-12H_2O$
**COLOUR** Lemon-yellow to pale green
**FORM/HABIT** Tabular
**HARDNESS** $2-2\frac{1}{2}$
**CLEAVAGE** Perfect basal
**FRACTURE** Uneven
**LUSTRE** Vitreous to pearly
**STREAK** Pale yellow
**SPECIFIC GRAVITY** 3.1–3.2
**TRANSPARENCY** Translucent to transparent

*Vitreous lustre*

*Tabular, twinned crystal aggregate*

**TABULAR CRYSTALS**
*Thin, tabular, twinned crystals characteristic of autunite are clearly visible in this lemon-yellow specimen.*

*Perfect basal cleavage*

## RADIOACTIVE DECAY

Radioactivity is the spontaneous decay of an atom's nucleus to form a new nucleus. As it decays, the atom gives off neutrons or protons, along with energy in the form of radiation. Uranium is the element that makes autunite radioactive. It has three isotopes, all of which are radioactive. The most common isotope, Uranium-238, which eventually decays to lead, has a half-life of about 4.5 billion years. This means that over 4.5 billion years, there is a 50 per cent chance that any single U-238 atom will decay.

**B.B. BOLTWOOD**
*B.B. Boltwood devised the method of dating rocks by measuring radioactive decay.*

---

# VIVIANITE

COLOURLESS WHEN freshly exposed, vivianite becomes pale blue to greenish-blue on oxidation. It was named in 1817 for its discoverer, the English mineralogist J.G. Vivian. Vivianite occurs as elongated, prismatic, or bladed tabular crystals, or can be massive or fibrous. It is a widespread secondary mineral, forming in the weathered zones of iron ore and phosphate deposits. It can also be found as powdery concretions in clays, in recent soils and sediments, and as an alteration coating on bone. Large crystals can be found in pegmatites such as those in Maine, New Hampshire, and Georgia, USA, and in Brazil. Other deposits occur in Japan, Germany, Bolivia, Canada, Russia, and Serbia.

## PROPERTIES

**GROUP** Phosphates
**CRYSTAL SYSTEM** Monoclinic
**COMPOSITION** $Fe_3(PO_4)_2 \cdot 8H_2O$
**COLOUR** Colourless to green or blue
**FORM/HABIT** Elongated to prismatic
**HARDNESS** $1\frac{1}{2}-2$
**CLEAVAGE** Perfect
**FRACTURE** Uneven
**LUSTRE** Vitreous to earthy
**STREAK** Bluish-white
**SPECIFIC GRAVITY** 2.7
**TRANSPARENCY** Transparent to translucent

*Clusters of elongated, prismatic crystals*

*Vitreous lustre*

**CRYSTAL CLUSTERS**
*After long exposure to light, vivianite becomes bluish-black as here.*

## A NATURAL BLUE PIGMENT

Before the development of modern synthetics, paint pigments were exclusively natural in origin. Durable blue has always been a prized colour, and vivianite provides that. Vivianite pigment becomes bluer with exposure, whereas azurite becomes greener. Like most natural pigments, it has now been replaced by a synthetic version.

**PRUSSIAN BLUE**
*Powdered vivianite was used to make Prussian blue, which can be seen in this 14th century painting on the choir screens of Cologne cathedral.*

# TORBERNITE

**CORNWALL**
*Mines in Cornwall, south-west England, yield excellent torbernite.*

NAMED IN 1793 FOR Swedish mineralogist Torbern Olaf Bergmann, torbernite is one of the principal uranium-bearing minerals and a minor ore of uranium. It forms thin to thick tabular crystals, commonly square in outline, as well as foliated, mica-like masses. A copper uranyl phosphate hydrate, torbernite is a secondary mineral formed as an alteration product of uraninite or other uranium-bearing minerals. It is associated with autunite, and is also radioactive, needing to be stored and handled with great care. Fine specimens occur in Cornwall, England, and the Democratic Republic of Congo.

**CRYSTALS**
*Tabular torbernite crystals rest on an iron-rich groundmass.*

Rock groundmass

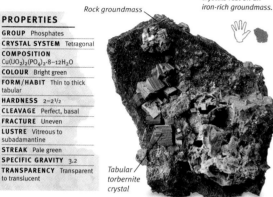

## PROPERTIES

**GROUP** Phosphates
**CRYSTAL SYSTEM** Tetragonal
**COMPOSITION**
$Cu(UO_2)_2(PO_4)_2 \cdot 8-12H_2O$
**COLOUR** Bright green
**FORM/HABIT** Thin to thick tabular
**HARDNESS** 2–2½
**CLEAVAGE** Perfect, basal
**FRACTURE** Uneven
**LUSTRE** Vitreous to subadamantine
**STREAK** Pale green
**SPECIFIC GRAVITY** 3.2
**TRANSPARENCY** Transparent to translucent

*Tabular torbernite crystal*

# ERYTHRITE

**ONTARIO**
*Some of the best erythrite comes from Ontario, Canada.*

COMMONLY CALLED COBALT BLOOM, erythrite is a cobalt arsenate hydrate that forms in the oxidized zone of cobalt and nickel deposits. Due to its bright colour, it is used to indicate the presence of cobalt-nickel-silver ores. Erythrite occurs as deeply striated, prismatic to acicular crystals, commonly radiating, globular tufts of crystals, or as coatings. Fine specimens come from Ontario, Canada, and Morocco. It is also found in south-western USA, Mexico, France, the Czech Republic, Germany, and Australia.

## PROPERTIES

**GROUP** Arsenates
**CRYSTAL SYSTEM** Monoclinic
**COMPOSITION**
$CO_3(AsO_4)_2 \cdot 8H_2O$
**COLOUR** Purple-pink
**FORM/HABIT** Prismatic to acicular
**HARDNESS** 1½–2½
**CLEAVAGE** Perfect
**FRACTURE** Uneven, sectile
**LUSTRE** Adamantine to vitreous, pearly
**STREAK** Pale red
**SPECIFIC GRAVITY** 3.1
**TRANSPARENCY** Transparent to translucent

**ACICULAR CRYSTALS**
*These brightly coloured, acicular crystals of erythrite are from Bou Azzer, Morocco.*

Rock groundmass

*Prismatic crystals*

# VARISCITE

**BISBEE, USA**
*Variscite occurs in this mining district in Arizona.*

VARISCITE WAS NAMED FOR VARISCIA, the old name for the German district of Voightland, where it was first discovered. It rarely forms crystals, and is predominantly found as cryptocrystalline or fine-grained masses, in veins, crusts, or nodules. It forms in cavities in near-surface deposits, produced by the action of phosphate-rich waters on aluminous rocks. It commonly occurs in association with apatite and wavellite, and with chalcedony and various hydrous oxides of iron. Located in Austria, the Czech Republic, Australia, Venezuela, and in North Carolina, Utah, and Arizona, USA, variscite is valued as a semiprecious gemstone and an ornamental material. It is porous, and, when worn next to the skin, tends to absorb body oils, which discolour it.

**CONCRETIONARY VARISCITE**
*Variscite is often found in nodules and concretions like the sliced specimen shown here. It is sometimes mistaken for turquoise.*

Waxy lustre

**PENDANT**
*This variscite pendant on a sterling silver chain was designed by Jan McClellan. Next to the variscite are three garnets mounted in silver.*

**CABOCHON**
*Variscite is sometimes polished and transformed into an inexpensive cabochon gem.*

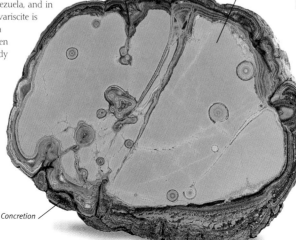

**CRYSTAL FORMATION**
*Variscite rarely forms crystals. When it does so, they are often visible only under the electron microscope, as shown here.*

*Concretion*

## PROPERTIES

**GROUP** Phosphates
**CRYSTAL SYSTEM** Orthorhombic
**COMPOSITION** $AlPO_4 \cdot 2H_2O$
**COLOUR** Pale to apple-green
**FORM/HABIT** Cryptocrystalline aggregates
**HARDNESS** 4½
**CLEAVAGE** Good but rarely visible
**FRACTURE** Splintery in massive types
**LUSTRE** Vitreous to waxy
**STREAK** White
**SPECIFIC GRAVITY** 2.6
**TRANSPARENCY** Opaque
**R.I.** 1.6–1.7

191

# SCORODITE

**ZACATECAS**
*A good source of scorodite is Zacatecas, Mexico.*

SCORODITE TAKES ITS NAME from the Greek for "garlic-like", alluding to its odour when heated. It forms aggregates of crystals, crusts, or earthy masses. Scorodite is widespread, found in hydrothermal veins and hot spring deposits, as well as in the oxidized zones of arsenic-rich ore bodies.

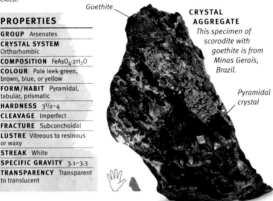

Goethite

**CRYSTAL AGGREGATE**
*This specimen of scorodite with goethite is from Minas Gerais, Brazil.*

Pyramidal crystal

## PROPERTIES

| | |
|---|---|
| **GROUP** Arsenates | |
| **CRYSTAL SYSTEM** Orthorhombic | |
| **COMPOSITION** $FeAsO_4 \cdot 2H_2O$ | |
| **COLOUR** Pale leek-green, brown, blue, or yellow | |
| **FORM/HABIT** Pyramidal, tabular, prismatic | |
| **HARDNESS** $3\frac{1}{2}$–4 | |
| **CLEAVAGE** Imperfect | |
| **FRACTURE** Subconchoidal | |
| **LUSTRE** Vitreous to resinous or waxy | |
| **STREAK** White | |
| **SPECIFIC GRAVITY** 3.1–3.3 | |
| **TRANSPARENCY** Transparent to translucent | |

# CLINOCLASE

**URALS**
*Clinoclase is found in the Ural Mountains in Russia.*

CLINOCLASE TAKES ITS NAME from the Greek "to incline", and "to break", referring to its oblique basal cleavage. Its crystals can be elongated or tabular, and occur as single, isolated crystals, or it can be found in aggregates forming rosettes, or as crusts or coatings with a fibrous structure. Clinoclase forms as a secondary mineral in the oxidized zones of deposits containing copper sulphides. Specimens are found in Australia, England, France, Namibia, and Utah and Nevada, USA.

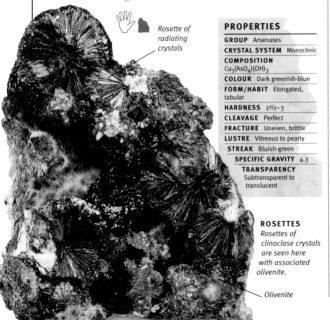

Rosette of radiating crystals

## PROPERTIES

| | |
|---|---|
| **GROUP** Arsenates | |
| **CRYSTAL SYSTEM** Monoclinic | |
| **COMPOSITION** $Cu_3(AsO_4)(OH)_3$ | |
| **COLOUR** Dark greenish-blue | |
| **FORM/HABIT** Elongated, tabular | |
| **HARDNESS** $2\frac{1}{2}$–3 | |
| **CLEAVAGE** Perfect | |
| **FRACTURE** Uneven, brittle | |
| **LUSTRE** Vitreous to pearly | |
| **STREAK** Bluish-green | |
| **SPECIFIC GRAVITY** 4.3 | |
| **TRANSPARENCY** Subtransparent to translucent | |

**ROSETTES**
*Rosettes of clinoclase crystals are seen here with associated olivenite.*

Olivenite

# HERDERITE

**MINAS GERAIS**
*Minas Gerais, Brazil, has violet hydroxylherderite crystals up to 15cm (6in) long.*

FORMING PRINCIPALLY IN granitic pegmatites, in association with quartz, albite, topaz, and tourmaline, herderite is a calcium beryllium phosphate mineral. Its crystals can be euhedral, stout prismatic, or thick tabular, and it also forms botryoidal or spheroidal radiating aggregates of crystals. Some herderite specimens fluoresce deep blue when viewed under ultraviolet light. Localities include Germany, Finland, Russia, and the USA.

Prismatic crystals

## PROPERTIES

| | |
|---|---|
| **GROUP** Phosphates | |
| **CRYSTAL SYSTEM** Monoclinic | |
| **COMPOSITION** $CaBePO_4(F,OH)$ | |
| **COLOUR** Colourless to pale yellow or greenish-white | |
| **CRYSTAL HABIT** Prismatic, tabular | |
| **HARDNESS** 5–$5\frac{1}{2}$ | |
| **CLEAVAGE** Irregular | |
| **FRACTURE** Subconchoidal | |
| **LUSTRE** Vitreous | |
| **STREAK** White | |
| **SPECIFIC GRAVITY** 3.0 | |
| **TRANSPARENCY** Transparent to translucent | |

Vitreous lustre

**PRISMATIC CRYSTALS**
*Stout prismatic herderite crystals cluster on a rock groundmass.*

# TRIPLITE

**URALS**
*The Urals in Russia are a source of triplite.*

THE THREE CLEAVAGES oriented at right angles to each other give triplite its name, from the Greek for "threefold". Its crystals tend to form massive aggregates and are rough and poorly developed. Although it is a manganese phosphate, in most samples, the manganese is partially replaced by iron. Manganese-rich crystals are reddish-brown to salmon-pink. Triplite is a primary mineral in pegmatites along with apatite and other phosphate minerals. Localities include Cornwall, England, the Urals, Russia, Namibia, Sweden, the Czech Republic, Argentina, and Afghanistan.

## PROPERTIES

| | |
|---|---|
| **GROUP** Phosphates | |
| **CRYSTAL SYSTEM** Monoclinic | |
| **COMPOSITION** $(Mn,Fe,Mg)_2PO_4(F,OH)$ | |
| **COLOUR** Dark brown to chestnut-brown | |
| **FORM/HABIT** Rough masses of crystals | |
| **HARDNESS** 5–$5\frac{1}{2}$ | |
| **CLEAVAGE** Good in three directions | |
| **FRACTURE** Uneven to subconchoidal | |
| **LUSTRE** Vitreous to resinous | |
| **STREAK** White to brown | |
| **SPECIFIC GRAVITY** 3.5–3.9 | |
| **TRANSPARENCY** Translucent | |

**MASSIVE AGGREGATE**
*This sample of massive triplite is from Megilligar Rocks in Cornwall, England.*

Massive habit

Quartz

# BRAZILIANITE

**MINAS GERAIS**
*Brazilianite was discovered in Conselheira Pena in Minas Gerais, south-east Brazil in 1945.*

A RELATIVELY HARD but fragile and brittle phosphate mineral, brazilianite is a sodium aluminium phosphate hydroxide. It has short prismatic or elongated crystals with commonly striated prism faces. It is also found with a globular or radiating fibrous structure. Most brazilianite is chartreuse-yellow to pale yellow. Brazilianite forms in granitic pegmatites, associated with tourmaline and apatite. It is named for Brazil, where it was discovered, and where gem-quality crystals, measuring up to 15cm (6in) long, are found. Brazilianite has been found in other localities, including Maine and New Hampshire, USA.

## PROPERTIES

| | |
|---|---|
| **GROUP** | Phosphates |
| **CRYSTAL SYSTEM** | Monoclinic |
| **COMPOSITION** | $NaAl_3(PO_4)_2(OH)_4$ |
| **COLOUR** | Yellow |
| **FORM/HABIT** | Equant to short prismatic |
| **HARDNESS** | $5^{1/2}$ |
| **CLEAVAGE** | Good in one direction |
| **FRACTURE** | Conchoidal |
| **LUSTRE** | Vitreous |
| **STREAK** | Colourless |
| **SPECIFIC GRAVITY** | 3.0 |
| **TRANSPARENCY** | Transparent |
| **R.I.** | 1.60–1.62 |

## CUTTING BRAZILIANITE

Brazilianite would be a popular gem were it not for two factors: it is fragile and brittle. When it is cut, the lapidary takes great care not to knock the stone against anything, and when it is stuck to its holder for cutting, it is done with easily removable adhesive. It is cut only for collectors. Another factor making it impractical as a popular gemstone is its relative scarcity – very little gem-grade material is found each year.

**EMERALD CUT**

**RECTANGULAR**

**TRIANGULAR STEP**

**LIMITED CUTS**
*Because of its colour and fragility, brazilianite is usually faceted in step cuts.*

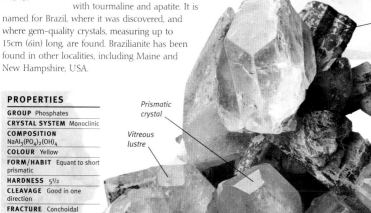

Prismatic crystal

Vitreous lustre

Apatite crystal

**BRAZILIAN CRYSTALS**
*These well-formed, prismatic brazilianite crystals, shown with accessory apatite, are from Minas Gerais, Brazil.*

Transparent crystal

Short prismatic crystal

**PRISMATIC CRYSTAL**
*This greenish, prismatic brazilianite crystal is from the Smithsonian collection.*

# AMBLYGONITE

**MYANMAR**
*Amblygonite deposits are found in Myanmar.*

USED AS A SOURCE OF lithium and phosphorus, and, to a lesser extent, as a gem mineral, amblygonite is a lithium phosphate mineral. The crystals tend to be short and prismatic, often with rough faces. More commonly, the mineral is found in very large, white, translucent masses. Gem amblygonite tends to be yellow, greenish-yellow, or lilac. Huge amblygonite crystals occur in granitic pegmatites in Zimbabwe and in South Dakota and Maine, USA.

## PROPERTIES

| | |
|---|---|
| **GROUP** | Phosphates |
| **CRYSTAL SYSTEM** | Triclinic |
| **COMPOSITION** | $(Li,Na)AlPO_4(F,OH)$ |
| **COLOUR** | White, yellow, or lilac |
| **FORM/HABIT** | Prismatic |
| **HARDNESS** | $5^{1/2}$–6 |
| **CLEAVAGE** | Perfect |
| **FRACTURE** | Uneven to subconchoidal |
| **LUSTRE** | Vitreous to greasy or pearly |
| **STREAK** | White |
| **SPECIFIC GRAVITY** | 3.0 |
| **TRANSPARENCY** | Transparent to translucent |
| **R.I.** | 1.57–1.60 |

**GEM-QUALITY CRYSTAL**
*This creamy-white fragment comes from a gem-quality amblygonite crystal.*

**WHITE MASS**
*This amblygonite from China is with orange wavellite and quartz.*

Mass of amblygonite

Translucent

Orange wavellite

## MEDICAL USES OF LITHIUM

Lithium, in the form of lithium carbonate, has proven highly effective in treating manic depression. It calms the manic episode, and lifts the depressed state. It also serves to prevent or drastically reduce manic-depressive mood swings and their severity, and it can also help suppress the psychotic behaviour shown by some manic-depressives.

**LITHIUM TABLETS**
*Medical lithium carbonate is taken in tablet form.*

**DARK YELLOW**
*This oval brilliant cut makes an attractive gem, although it is too soft to wear.*

**PALE YELLOW**
*Cut for a collector, this pale yellow stone is a perfect oval brilliant.*

# OLIVENITE

**DEVON**
*Devon is an English source of olivenite.*

OLIVENITE'S CRYSTALS ARE SHORT to long prismatic, acicular, or tabular. Other habits are globular, fibrous, reniform, massive, granular, and earthy. Aside from its olive-green colour, olivenite can also be brownish-green, straw-yellow, or greyish-white. It is thought to form a solid-solution series with libethenite (see right). Olivenite forms as a secondary mineral in the oxidation zone of copper-bearing ore deposits, often accompanied by malachite, goethite, calcite, dioptase, and azurite. Localities include Namibia, Australia, Germany, Greece, Chile, and the western USA.

## PROPERTIES

| | |
|---|---|
| **GROUP** | Arsenates |
| **CRYSTAL SYSTEM** | Orthorhombic |
| **COMPOSITION** | $Cu_2AsO_4(OH)$ |
| **COLOUR** | Olive-green, brown-green, yellow, or grey-white |
| **FORM/HABIT** | Prismatic |
| **HARDNESS** | 3 |
| **CLEAVAGE** | Indistinct |
| **FRACTURE** | Conchoidal to uneven |
| **LUSTRE** | Adamantine |
| **STREAK** | Green to brown |
| **SPECIFIC GRAVITY** | 3.9–4.5 |
| **TRANSPARENCY** | Translucent to opaque |

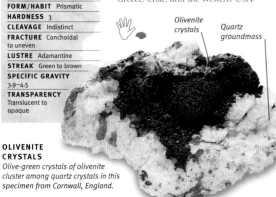

*Olivenite crystals*

*Quartz groundmass*

**OLIVENITE CRYSTALS**
*Olive-green crystals of olivenite cluster among quartz crystals in this specimen from Cornwall, England.*

# LIBETHENITE

**BROKEN HILL**
*This Australian locality provides libethenite.*

A COPPER PHOSPHATE HYDROXIDE, libethenite is thought to form a solid-solution series with olivenite, in which phosphorus substitutes for the arsenic in olivenite. Its crystals are short prismatic or slightly elongated and vertically grooved or striated. Libethenite is a secondary mineral formed in the oxidized zone of copper deposits. It is often accompanied by malachite and azurite. Localities include the Ural Mountains, Russia; Cornwall, England; and Arizona, USA.

## PROPERTIES

| | |
|---|---|
| **GROUP** | Phosphates |
| **CRYSTAL SYSTEM** | Orthorhombic |
| **COMPOSITION** | $Cu_2PO_4(OH)$ |
| **COLOUR** | Olive-green |
| **FORM/HABIT** | Short prismatic |
| **HARDNESS** | 4 |
| **CLEAVAGE** | Indistinct |
| **FRACTURE** | Conchoidal to uneven |
| **LUSTRE** | Vitreous |
| **STREAK** | Olive-green |
| **SPECIFIC GRAVITY** | 4.0 |
| **TRANSPARENCY** | Translucent |

*Libethenite crystal*

**LIBETHENITE IN GROUNDMASS**
*These libethenite crystals are from Lubietová (formerly Libethen), Slovakia, the locality that gave the mineral its name.*

*Rock groundmass*

# ADAMITE

**TSUMEB, NAMIBIA**
*This area of Namibia is a source of adamite.*

NAMED FOR FRENCH MINERALOGIST G.J. Adam, who discovered the mineral in Chile, adamite is a zinc arsenate hydroxide. It forms elongated, tabular, or blocky crystals that are often rounded almost to the point of appearing spherical. Other habits include rosettes and spherical masses of radiating crystals. It is often brightly coloured by traces of other elements: copper substitutes for zinc to yield yellow or green crystals, depending on its concentration; and cobalt substituting for zinc results in pink or violet crystals. It is rarely colourless or white. While it has no commercial uses, its bright and lustrous crystals are highly sought after by mineral collectors.

Adamite forms as a secondary mineral in the oxidized zones of zinc and arsenic deposits, often associated with azurite, smithsonite, mimetite, hemimorphite, scorodite, olivenite, and limonite. Well-crystallized specimens come from Chañarcillo, Chile; Tsumeb, Namibia; Mapimí, Mexico; Reichenbach, Germany; Mount Valerio, Italy; Hyères, France; and Franklin, New Jersey, and Mohawk, California in the USA.

## PROPERTIES

| | |
|---|---|
| **GROUP** | Arsenates |
| **CRYSTAL SYSTEM** | Orthorhombic |
| **COMPOSITION** | $Zn_2AsO_4(OH)$ |
| **COLOUR** | Yellow, green, pink or violet |
| **FORM/HABIT** | Tabular, prismatic |
| **HARDNESS** | $3^1/_2$ |
| **CLEAVAGE** | Good |
| **FRACTURE** | Subconchoidal to uneven, brittle |
| **LUSTRE** | Vitreous |
| **STREAK** | White |
| **SPECIFIC GRAVITY** | 4.4 |
| **TRANSPARENCY** | Transparent to translucent |

## FLUORESCING MINERAL

Much adamite is highly fluorescent, and spectacular specimens are often included in fluorescent displays (see pp.160–61). Like many fluorescent minerals, its fluorescence, or lack of it, depends on the locality from which the specimen comes. Some specimens fluoresce but others show no fluorescence at all.

**FLUORESCING ADAMITE**
*Green fluorescence is typical of adamite.*

**ADAMITE CRYSTALS**
*These rounded, whitish adamite crystals are on a rock groundmass.*

*Rounded crystal*

*Vitreous lustre*

# APATITE

**KOLA PENINSULA, RUSSIA**
*This mineral-rich area of Russia is a notable locality for apatite. One of the world's largest deposits of this important mineral is found here.*

APATITE IS A GENERAL TERM for any one of a series of phosphate minerals. All of the apatites are structurally identical calcium phosphates. They differ in their composition: fluorapatite contains fluorine; hydroxylapatite contains a hydroxyl (OH); and chlorapatite contains chlorine. Fluorapatite is the most important mineral commercially, being the world's major source of phosphorus; chlorapatite is relatively rare. Carbonate fluorapatite occurs in fossil bones and teeth and associated sediments. Carbonate hydroxylapatite is the principal component of contemporary human bones and teeth.

Apatite is commonly found as well-formed and transparent, coloured, glassy crystals, and in masses, or nodules. The name is derived from the Greek apate, meaning "deceit", and referring to its similarity to crystals of other minerals such as aquamarine, amethyst, and olivine. It can be intensely coloured, occurring in green, blue, violet-blue, purple, colourless, white, yellow, flesh, or rose-red forms. Crystals are short to long prismatic, thick tabular, or prismatic with complex forms.

Apatite is an accessory mineral in a wide range of igneous rocks, including pegmatites and high-temperature hydrothermal veins. Most gem-quality material is associated with pegmatites. It is also found in bedded marine deposits, which are frequently mined as a source of phosphorus. Fine apatite crystals occur in many localities. The USA, Mexico, Namibia, and Russia all have notable deposits, and crystals weighing up to 200kg (485lb) have been found in Canada. Faceted stones are bright with strong colours, but tend to be soft.

**YELLOW APATITE**
*Brilliant yellow crystals from Mexico proved so popular with collectors and jewellery makers that the deposit has been exhausted.*

Prismatic crystal

Muscovite

**CHLORAPATITE**
*A double-terminated crystal of chlorapatite rests on albite in this specimen from Buskerud, Norway. Chlorapatite is one of the rarest of the apatite minerals.*

Arsenopyrite

**RECTANGULAR STEP**
*Beautifully coloured gem apatite, such as this step-cut stone, is usually found in pegmatites.*

**CUSHION MIXED**
*The wide variety of intense colours in which apatite occurs would make it a major gem if it were not so soft a material.*

**OCTAGONAL STEP**
*Owing to the brittleness of this blue apatite gem, one of the facet's edges has become chipped, spoiling its appearance.*

**CABOCHON**
*Fibrous blue apatite from Myanmar and Sri Lanka is cut en cabochon to show a cat's eye. Chatoyant stones are also found in Brazil.*

## PROPERTIES

**GROUP** Phosphates
**CRYSTAL SYSTEM** Hexagonal or monoclinic
**COMPOSITION** $Ca_5(PO_4)_3(F,OH,Cl)$
**COLOUR** Green, blue, violet, purple, colourless, yellow, or rose
**FORM/HABIT** Short to long prismatic, tabular
**HARDNESS** 5
**CLEAVAGE** Indistinct, variable
**FRACTURE** Conchoidal to uneven
**LUSTRE** Vitreous, waxy
**STREAK** White
**SPECIFIC GRAVITY** 3.1–3.2
**TRANSPARENCY** Transparent to translucent
**R.I.** 1.63 – 1.64

**USE IN MATCHES**
*Apatite is an important source of phosphorus, used in matches.*

**APATITE CRYSTALS**
*These spectacular apatite crystals are with muscovite and a small amount of arsenopyrite. The specimen comes from Panasqueira Mine, Beira Baixa, Portugal.*

195

# TURQUOISE

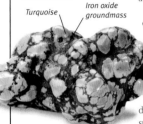

**BISBEE, ARIZONA**
*The copper mines around Bisbee, Arizona, USA, are noted for their rich, relatively dark turquoise.*

TURQUOISE WAS ONE OF THE FIRST gemstones to be mined: turquoise beads dating from about 5000BC have been found in Mesopotamia (present-day Iraq). Turquoise was extracted by the Egyptians from sources in the Sinai Peninsula before the 4th century BC, and records from the reign of the Pharaoh Semerkhet (c.2923-2915BC) detail extensive mining operations that employed thousands of labourers. The magnificent turquoise-adorned breastplate of Pharaoh Sesostris II (1844–1837BC) is now at the Metropolitan Museum of Art in New York.

Turquoise varies in colour from sky-blue to green, depending on the amount of iron and copper it contains. Crystals are rare; it usually occurs in massive or microcrystalline forms, as encrustations or nodules, or in veins. When crystals are found they seldom exceed 2mm in length, and occur as short prisms. Malachite occurs principally in arid environments as a secondary mineral, probably derived from the decomposition of apatite (see p.195) and copper sulphides, and deposited from circulating waters.

*Turquoise — Iron oxide groundmass*

**BLUE NUGGET**

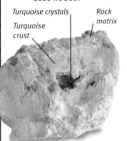

*Turquoise crystals — Rock matrix*
*Turquoise crust*

**CRUST AND CRYSTALS**

**TURQUOISE FORMS**
*Turquoise comes in different shades of blue and green. It usually occurs as nuggets, but it also forms crusts and crystals.*

## TURQUOISE GEMS AND CARVINGS
Some gem material is very porous, so it may be impregnated with wax or resin to maintain its appearance, or enhance its colour – so-called "stabilized" turquoise. For most gem uses, turquoise is cut *en cabochon*, but natural turquoise is very brittle, and cabochons are frequently backed with epoxy resin to strengthen them. The delicate veining caused by impurities in turquoise from some localities is desired by some. Turquoise is also carved or engraved, and irregular pieces are often set in mosaics with jasper, obsidian, and mother-of-pearl. Natural aggregate of turquoise with limonite or other substances – called turquoise matrix – is sometimes cut and polished.

*Word for turquoise*
*King Sanakht*

**INSCRIPTION**
*This Egyptian inscription from around 2680BC commemorates an expedition to turquoise mines.*

**EGYPTIAN SINAI**
*The Sinai area of Egypt was one of the first localities where the systematic mining of turquoise took place.*

## PERSIAN TURQUOISE

**IRANIAN PURSE**
*Cabochons of turquoise are set in concentric rings on the surface of this silver purse, which is designed for carrying the Koran.*

In many respects turquoise is regarded by the Iranians as jade is by the Chinese. It has been highly prized in Iran (ancient Persia) since antiquity. The local turquoise source was the mines of Neyshabur, in the Khorasan region of Iran. It was used to embellish thrones, sword hilts, horse trappings, daggers, bowls, cups, and ornamental objects of all kinds. Large stones were embellished with goldwork or sometimes with passages from the Koran.

Persian turquoise tends to be harder and of a more even colour than North American turquoise. It is always sky-blue, never green. It was traded far afield, and was in use in the Indus Valley civilization of India as long ago as the second millennium BC. Prior to World War I, turquoise was Iran's principal export, with production coming from nearly 100 mines across the country.

**PERSIAN ORNAMENT**
*This antique turquoise ornament is engraved and inlaid with gold.*

**GREEN TURQUOISE**
*This nugget of hard, green turquoise is from a deposit near Silver City, New Mexico, USA.*

| PROPERTIES | |
|---|---|
| **GROUP** | Phosphates |
| **CRYSTAL SYSTEM** | Triclinic |
| **COMPOSITION** | $CuAl_6(PO_4)_4(OH)_8 \cdot 4H_2O$ |
| **COLOUR** | Blue, green |
| **FORM/HABIT** | Massive |
| **HARDNESS** | 5–6 |
| **CLEAVAGE** | Good |
| **FRACTURE** | Conchoidal |
| **LUSTRE** | Waxy to dull |
| **STREAK** | White to green |
| **SPECIFIC GRAVITY** | 2.6–2.8 |
| **TRANSPARENCY** | Usually opaque |

Turquoise from several sources was first transported to Europe through Turkey, probably accounting for its name, which is French for "Turkish". Sky-blue turquoise from Iran has been mined for centuries, and is regarded as the most desirable. In Tibet, where turquoise is very popular, a greener variety is preferred. Turquoise also occurs in northern Africa, Australia, Siberia, England, Belgium, France, Poland, Ethiopia, Mexico, Chile, and China.

## LORE OF TURQUOISE

In Persia (present-day Iran), good luck was believed to come to someone who saw the reflection of a New Moon on a turquoise. Indeed, turquoise was the national gemstone of Persia, adorning everything from thrones to horse trappings. Turquoise seals decorated with pearls and rubies were the emblems of high office. More widely, turquoise has been thought to warn the wearer of danger or illness by changing colour. This is not altogether a false notion, because turquoise is a porous mineral, so when it is worn next to the skin it absorbs body oils, and can consequently change colour.

*Matrix*

*Waxy lustre*

*Green turquoise*

# NATIVE AMERICANS AND TURQUOISE

Native Americans have worked the numerous deposits in the south-western United States since AD1000. The Pueblo Indians used turquoise for beads and made beautiful necklaces and pendants from shells covered with turquoise mosaics. In 1853 silver working was introduced to the Navajo for the first time by Mexican smiths. The Navajo soon began reworking Spanish American designs in traditional Native American styles. In 1872 the Zuni learned silversmithing from the Navajo. Zuni work is characterized by finely set, small turquoise inlays; Navajo work is distinguished by die-stamped designs and large turquoise cabochons.

### FALCON BROOCH
*This finely crafted Zuni brooch is inlaid with jet, turquoise, shell, and mother-of-pearl.*

### ASSOCIATIONS
Some Native Americans regard turquoise as "male" or "female", depending on its colour. "Male" blue is associated with the sky – Father Sky – and "female" green with earth – Mother Earth. The highly secret cult-room of the Pueblo Indian High-priest of the Rains is said to contain an altar comprising two small crystal and turquoise columns and a heart-shaped stone, the heart of the world. According to Pueblo belief, a piece of turquoise attached to a gun or bow assured the firer a perfect aim.

### TURQUOISE FETISHES
*Native American fetishes are charms believed to ensure the success of a hunt.*

### NAVAJO TURQUOISE JEWELLER
*This Navajo jeweller from New Mexico is wearing striking examples of his turquoise work.*

### POLISHED STONE
*This turquoise cabochon shows the dark lanes of matrix called "spider web".*

### CARVED ELEPHANT
*Turquoise is a favourite of Chinese stone carvers, who produced this charming turquoise elephant.*

### COPPER MINE
*Turquoise is found as an accessory mineral of copper. This copper strip mine near Tucson, Arizona, USA, was a rich source of blue turquoise before it was closed in 1974.*

# MESOAMERICAN TURQUOISE

THE ANCIENT CIVILIZATIONS OF MESOAMERICA
(MEXICO AND CENTRAL AMERICA) TREASURED
TURQUOISE AS A MOSAIC DECORATION FOR THEIR
SOPHISTICATED ARTEFACTS, FROM MIRRORS AND KNIFE
HANDLES TO SACRED STATUES AND FUNERARY MASKS.

Turquoise was first used in Mexico and Central America during the Classic Period of Mesoamerican civilization, between about 200 and 900AD. The source of the gemstone was the Mount Chalchihuitl Mine near Cerillos, in what is now New Mexico, USA. There, a turquoise deposit was dug out with primitive hand tools, leaving a pit 60m (200ft) across and up to 40m (130ft) deep. It was probably mined by native Pueblo Indians, and then passed along trade routes to the ancient trading centre of Casas Grandes, now in Chichihuitlahua State, Mexico – a distance of several thousand miles.

The use of turquoise spanned more than 1,000 years of Mesoamerican history. The people of the city-state of Teotihuacan, which dominated Mesoamerica from around the 4th to the 7th century AD, were among the first to use turquoise in ancient Mesoamerica. It was still of major importance by the time of the Aztecs, who emerged as a civilization in the 13th century, and ruled a large empire in what is now central and southern Mexico in the 15th and early 16th centuries. For example, the Aztec fire-god, the supreme solar deity, was called *Xiuhtecuhtli*, meaning Lord of the Turquoise, and was depicted costumed in turquoise, with a turquoise crown and breast pendant, and a shield covered in turquoise mosaics.

## TURQUOISE MOSAICS

Turquoise was used with great technical skill in Mesoamerican mosaics. Usually, small, irregularly shaped pieces were highly polished, and then cut to fit tightly together to suit the design. Aztec turquoise mosaics can consist of as many as 14,000 individual pieces. The mosaic was typically held in place by a vegetal pitch or gum, or by a kind of cement.

Turquoise mosaics were used to encrust shields, helmets, knife handles, collars, medallions, ear plugs, mirrors, animal figures, cult statues, and, most strikingly, masks. The latter range from the trapezoid Teotihuacan masks, many of which have been found in ancient burial places, to Aztec funerary masks in which turquoise, gold, and shell were overlaid on a human skull.

**TEOTIHUACAN RUINS**
*Built in AD50 on the north-east side of the Valley of Mexico, the city of Teotihuacan was laid out in a grid pattern with the huge Pyramid of the Sun forming an awe-inspiring backdrop.*

**TURQUOISE SERPENT**
*Shaped in the form of a serpent, this 15th-century Aztec pin is inlaid with turquoise. The turquoise serpent was a powerful symbol in Aztec mythology. The Aztecs believed that when the god of the Sun awoke he drove the Moon and stars from the sky, armed with his mystic weapon, the "turquoise serpent".*

**TEOTIHUACAN MASK**
*Made of turquoise and shell, this mosaic sculpture mask is typical of the Teotihuacan style, being trapezoid in shape and having a straight forehead, almond-shaped eyes, a strong nose, and a sensuous half-open mouth*

# VANADINITE

THIS RELATIVELY RARE MINERAL is a lead vanadate chloride. Its crystals are usually long prisms, but can be acicular to hair-like, and, on occasion, form hollow prisms. Small amounts of calcium, zinc, and copper may substitute for lead, and arsenic can completely substitute for vanadium in the crystal structure to form the mineral mimetite (see below). Vanadinite is soluble in both hydrochloric and nitric acids. It forms as a secondary mineral in oxidized ore deposits containing lead, often associated with galena, baryte, wulfenite, and limonite. It is a major source of vanadium and a minor source of lead. Superb specimens come from Minas Gerais, Brazil; Chihuahua, Mexico; Mibladen, Morocco; Leadhills, Scotland; Tsumeb, Namibia; and from many mines in the USA.

**BROKEN HILL**
*Vanadinite occurs in this part of New South Wales, Australia.*

*Short, prismatic crystals*

*Romanèchite groundmass*

**VANADINITE CRYSTALS ON ROMANÈCHITE**

*Adamantine lustre*

*Rock groundmass*

**SIR HENRY ROSCOE**
*The English chemist Sir Henry Roscoe first isolated vanadium from vanadinite in 1867.*

## USES OF VANADIUM

Most of the vanadium that is extracted goes into making vanadium steels. The vanadium refines the grain of the steel, and forms carbides with the carbon present. Vanadium steel is particularly strong and hard, with considerable shock resistance. Vanadium compounds are used as catalysts in the manufacture of sulphuric acid, nylon, and in the oxidation of some organic substances.

| PROPERTIES | |
|---|---|
| **GROUP** | Phosphates |
| **CRYSTAL SYSTEM** | Hexagonal |
| **COMPOSITION** | $Pb_5(VO_4)_3Cl$ |
| **COLOUR** | Orange-red, yellow |
| **FORM/HABIT** | Hexagonal prisms |
| **HARDNESS** | 3 |
| **CLEAVAGE** | None |
| **FRACTURE** | Uneven, brittle |
| **LUSTRE** | Adamantine |
| **STREAK** | Whitish-yellow |
| **SPECIFIC GRAVITY** | 6.9 |
| **TRANSPARENCY** | Transparent to translucent |

**RED VANADINITE**
*Vanadinite crystals are often brilliantly coloured in shades of red and yellow. This specimen is typical, with smooth-faced, prismatic crystals.*

*Prismatic vanadinite crystals*

# MIMETITE

THIS MINERAL FORMS HEAVY, barrel-shaped crystals or rounded masses, and is also found in botryoidal, granular, tabular, and acicular aggregates. The arsenic-rich end member of a solid-solution series with pyromorphite (see opposite), mimetite is named after the Greek *mimetes*, "imitator", referring to its resemblance to pyromorphite. Its colour range includes shades of yellow, orange, brown, and sometimes green; it can be colourless. Mimetite is a secondary mineral, which forms in the oxidized zone of lead deposits and other localities where lead and arsenic occur together. Excellent specimens come from Chihuahua, Mexico; Saxony, Germany; Attica, Greece; Broken Hill, Australia; and in Bisbee and Tombstone, Arizona, USA. One single crystal from Tsumeb in Namibia measured 6.4cm (2½in) by 2.5cm (1in).

**CUMBRIA**
*The rocks of Cumbria in the north of England are a source of mimetite.*

**CAMPYLITE**
*This specimen, from Caldbeck Fells, Cumbria, England, contains a variety of mimetite known as campylite. It also includes baryte on nodules of manganese.*

*Baryte*

*Resinous lustre*

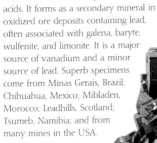

*Rounded masses of mimetite*

*Black coronadite matrix*

| PROPERTIES | |
|---|---|
| **GROUP** | Arsenates |
| **CRYSTAL SYSTEM** | Hexagonal |
| **COMPOSITION** | $Pb_5(AsO_4)_3Cl$ |
| **COLOUR** | Pale yellow to yellowish-brown, orange, green |
| **FORM/HABIT** | Barrel-shaped crystals and rounded masses |
| **HARDNESS** | 3½–4 |
| **CLEAVAGE** | Poor |
| **FRACTURE** | Conchoidal to uneven, brittle |
| **LUSTRE** | Resinous |
| **STREAK** | White |
| **SPECIFIC GRAVITY** | 7·3 |
| **TRANSPARENCY** | Subtransparent |

# LAZULITE

**BISBEE**
*Lazulite occurs in several localities in the USA, including Bisbee in Arizona.*

LAZULITE TAKES ITS NAME from the old German *lazurstein*, "blue stone". Crystals are pyramidal, with tabular prism faces; it can also be massive and granular. It is azure-blue, sky-blue, or bluish-white to blue-green in colour. It can be similar in appearance to lapis lazuli (lazurite). Lazulite occurs in aluminous metamorphic rocks, quartz veins, and granite pegmatites. Localities for fine specimens are Sweden; Austria; Switzerland; Mount Fitton, Yukon, Canada; and, in the USA, the White Mountains, California; Newport, New Hampshire; Chowders Mountain, North Carolina; and Graves Mountain, Georgia.

**LAZULITE IN QUARTZ**
*This specimen from Lincoln County, Georgia, USA, has well-formed lazulite crystals in a groundmass of quartz.*

*Dipyramidal crystals*

*Quartz groundmass*

## PROPERTIES

| | |
|---|---|
| **GROUP** | Phosphates |
| **CRYSTAL SYSTEM** | Monoclinic |
| **COMPOSITION** | $(Mg,Fe)Al_2(PO_4)_2(OH)_2$ |
| **COLOUR** | Various shades of blue |
| **FORM/HABIT** | Pyramidal |
| **HARDNESS** | 5–6 |
| **CLEAVAGE** | Indistinct |
| **FRACTURE** | Uneven to splintery |
| **LUSTRE** | Vitreous |
| **STREAK** | White |
| **SPECIFIC GRAVITY** | 3.1 |
| **TRANSPARENCY** | Transparent to translucent |
| **R.I.** | 1.61–1.64 |

**POLISHED**
*Granular lazulite is cut en cabochon. Beads and decorative items are sometimes made from it.*

**SINGLE CRYSTAL**
*This single crystal of lazulite displays a dipyramidal habit.*

# PYROMORPHITE

**DYFED, WALES**
*Superb examples of pyromorphite are found in this part of the UK.*

PYROMORPHITE GETS ITS NAME from the Greek *pyr*, meaning "fire", and *morphe*, "form", alluding to its property of taking up a crystalline form on cooling after it has melted to a globual. A lead phosphate chloride, pyromorphite forms a solid-solution series with mimetite (see opposite) in which phosphorus and arsenic replace one another. Crystals may be simple hexagonal prisms or rounded and barrel-shaped, spindle-shaped, or cavernous. Pyromorphite can also be globular, reniform, or granular. It is dark green to yellow-green, shades of brown, a waxy yellow, or yellow-orange. A minor ore of lead, it occurs as a secondary mineral in the oxidized zone of lead deposits with cerussite, smithsonite, vanadinite, galena, and limonite. Excellent specimens come from Dyfed, Wales; Leadhills, Scotland; Cumbria, England; Bad Ems, Germany; Broken Hill, Australia; Ussel, France; Castilla-León, Spain; Chihuahua, Mexico; and, in the USA, Phoenixville, Pennsylvania; Leadville, Colorado; Bisbee, Arizona; and Coeur d'Alene, Idaho.

**LIME-GREEN PYROMORPHITE**

**YELLOW-GREEN PYROMORPHITE**

*Prismatic crystals*

*Barrel-shaped crystals*

## PROPERTIES

| | |
|---|---|
| **GROUP** | Phosphates |
| **CRYSTAL SYSTEM** | Hexagonal |
| **COMPOSITION** | $Pb_5(PO_4)_3Cl$ |
| **COLOUR** | Green, yellow, orange or brown |
| **FORM/HABIT** | Prismatic |
| **HARDNESS** | $3^1/2$–4 |
| **CLEAVAGE** | Poor |
| **FRACTURE** | Uneven to subconchoidal, brittle |
| **LUSTRE** | Resinous |
| **STREAK** | White |
| **SPECIFIC GRAVITY** | 7.0 |
| **TRANSPARENCY** | Subtransparent to translucent |

*Resinous lustre*

**BARREL-SHAPED CRYSTALS**
*This mass of pyromorphite on a limonite groundmass shows typical barrel-shaped crystals.*

*Limonite groundmass*

# EUCHROITE

**MISSOULA**
*Euchroite is found in the copper mining area of Missoula, Montana, USA.*

A RELATIVELY RARE MINERAL, euchroite was named in 1823 from the Greek for "beautiful colour". It is a copper arsenate hydrate, and forms short prismatic crystals. Its colour ranges from brilliant emerald-green to leek-green. It was originally found in Slovakia. It also occurs in Cramer Creek, near Missoula, Montana, USA, where it is found lining crevices in mica schist, and there are important deposits in Bulgaria and Greece.

*Vitreous lustre*

*Euchroite crystals*

### PROPERTIES

| | |
|---|---|
| **GROUP** | Arsenates |
| **CRYSTAL SYSTEM** | Orthorhombic |
| **COMPOSITION** | $Cu_2(AsO_4)(OH)\cdot3H_2O$ |
| **COLOUR** | Emerald-green |
| **FORM/HABIT** | Prismatic |
| **HARDNESS** | $3\frac{1}{2}$–4 |
| **CLEAVAGE** | Imperfect |
| **FRACTURE** | Uneven to subconchoidal |
| **LUSTRE** | Vitreous |
| **STREAK** | Green |
| **SPECIFIC GRAVITY** | 3.4 |
| **TRANSPARENCY** | Transparent to translucent |

**PRISMATIC CRYSTALS**
*These short, prismatic crystals on a rock groundmass are from Lubietová, Slovakia.*

# CHILDRENITE

**SAXONY**
*Mineral-rich Saxony in Germany has deposits of childrenite.*

NAMED FOR THE English mineralogist J.G. Children in 1823, childrenite is an iron aluminium phosphate hydrate. Its crystals are pyramidal to short prismatic, thick tabular, or platy. It is sometimes found in radiating, botryoidal masses, crusts with a fibrous structure, or as massive aggregates. Prism faces are usually striated. As increasing amounts of manganese substitute for iron, it grades into eosphorite. Childrenite is found in Maine and South Dakota, USA, England, Germany, and Brazil.

### PROPERTIES

| | |
|---|---|
| **GROUP** | Arsenates |
| **CRYSTAL SYSTEM** | Orthorhombic |
| **COMPOSITION** | $FeAl(PO_4)(OH)_2\cdot H_2O$ |
| **COLOUR** | Brown to yellowish-brown, white, or brownish-black |
| **FORM/HABIT** | Pyramidal to short prismatic |
| **HARDNESS** | 5 |
| **CLEAVAGE** | Poor |
| **FRACTURE** | Subconchoidal to uneven |
| **LUSTRE** | Vitreous |
| **STREAK** | White to yellowish |
| **SPECIFIC GRAVITY** | 3.2 |
| **TRANSPARENCY** | Translucent |

*Prismatic childrenite crystals*

**CRYSTALS**
*This sample is from George and Charlotte Mine, Devon, England.*

*Quartz*

# PHARMACOSIDERITE

**CORNWALL**
*Pharmacosiderite was first discovered in Cornwall, south-west England.*

PHARMACOSIDERITE IS A POTASSIUM iron arsenate hydrate, and takes its name from the Greek *pharmakon*, meaning "poison" or "drug", an allusion to its arsenic content, and *sideros*, "iron". Its crystals are commonly diagonally striated cubes, and not many exceed a few millimetres. It is also rarely granular and earthy, and is quite sectile. Pharmacosiderite occurs in hydrothermal deposits, or as a secondary alteration product of arsenic-bearing minerals, such as arsenopyrite. It was first discovered in Cornwall, England; it is also found in Germany, Greece, Italy, Algeria, Namibia, Brazil, and in Utah, New Jersey, and Arizona, USA.

### PROPERTIES

| | |
|---|---|
| **GROUP** | Arsenates |
| **CRYSTAL SYSTEM** | Cubic |
| **COMPOSITION** | $KFe_4(AsO_4)_3(OH)_4\cdot6$–$7(H_2O)$ |
| **COLOUR** | Variable greens, yellowish-brown, brown |
| **FORM/HABIT** | Cubic |
| **HARDNESS** | $2\frac{1}{2}$ |
| **CLEAVAGE** | Imperfect to good |
| **FRACTURE** | Uneven |
| **LUSTRE** | Adamantine to greasy |
| **STREAK** | Pale green to brown, pale yellow |
| **SPECIFIC GRAVITY** | 2.8 |
| **TRANSPARENCY** | Transparent to translucent |

*Cubic pharmacosiderite crystals*

*Quartz and sulphide matrix*

*Adamantine lustre*

# ARSENIC POISONING

On 5 May 1821 Napoleon died in the drawing room at Longwood on the island of St. Helena after more than four years of illness. A lock of hair examined shortly after his death was found to contain arsenic. Recently, fragments of wallpaper purporting to be from the walls of the drawing room and bedroom were analysed. Both were dyed with a green pigment that contained arsenite. In the humidity the wallpaper would have become damp, causing the arsenite to become a poisonous vapour form of arsenic.

**DEATH OF NAPOLEON**
*The green patterning in the wallpaper is visible in this painting by Carl von Steuben (1788–1856).*

*Scrap of wallpaper from Napoleon's bedroom*

**CUBIC CRYSTALS**
*This mass of green pharmacosiderite crystals was found in Wheal Gorland, Cornwall, England. Pharmacosiderite also occurs in olive or brown colours.*

# CHALCOPHYLLITE

**TSUMEB**
*Fine deposits of chalcophyllite are found in Tsumeb, Namibia.*

A VIVID BLUE-GREEN in colour, chalcophyllite takes its name from the Greek for "copper" and for "leaf", which is a reference to its copper content and its common foliated habit. Crystals of chalcophyllite are tabular and hexagonal. Chalcophyllite is a copper aluminium arsenate sulphate hydrate. It is a widespread secondary mineral that occurs in the alteration zone of copper deposits. Although similar in appearance to other copper minerals, it is soluble in both nitric acid and ammonia. Specimens come from south-west England, France, Germany, Austria, Russia, Namibia, Chile, and from Utah, Arizona, and Nevada, USA.

## PROPERTIES

**GROUP** Arsenates
**CRYSTAL SYSTEM** Hexagonal/trigonal
**COMPOSITION** $Cu_{18}Al_2(AsO_4)_2(SO_4)_3(OH)_{27}\cdot33H_2O$
**COLOUR** Vivid blue-green
**FORM/HABIT** Tabular
**HARDNESS** 2
**CLEAVAGE** Perfect basal
**FRACTURE** Uneven to subconchoidal
**LUSTRE** Pearly to vitreous
**STREAK** Pale green
**SPECIFIC GRAVITY** 2.7
**TRANSPARENCY** Transparent to translucent

## COPPER IN ART

The casting of native copper began around 4000BC, and the smelting of copper from copper-bearing rock started sometime before 3800BC. Although copper was first used for making weapons and implements, its ductility and malleability soon made it a material of choice for artisans. The artisans hammered the copper and cast it into statuettes and heads, which are still admired today for their artistry as well as for the skill of their craftsmen.

**STATUE OF LAMMA**
*Dating from 1800–1600BC, this statue of the deity was found at Ur.*

**HEAD OF SARGON I**
*This copper head of the King of Akkad was made in Mesopotamia between 2334 and 2279BC.*

**BULL'S HEAD**
*Dating from 2500–2400BC, this copper bull's head was crafted for Sumerian royalty.*

*Tabular chalcophyllite crystals*

*Rock groundmass*

**TABULAR CRYSTALS**
*A mass of vivid blue, tabular chalcophyllite crystals rests on a rock groundmass.*

# DUFRENITE

**CORNWALL**
*The best specimens of dufrenite are found in Cornwall, south-west England.*

USUALLY FOUND IN botryoidal masses or crusts, dufrenite is a hydrous phosphate of iron. Crystals, which are relatively uncommon, are found in sheaf-like aggregates with rounded ends. The mineral is named in 1833 for French mineralogist P.A. Dufrenoy. It occurs as a secondary mineral in the weathered zone of metallic veins and in iron ore deposits. It is found in Cornwall, England; Hureau, France; Bushmanland, South Africa; and New Hampshire and Alabama, USA.

## PROPERTIES

**GROUP** Phosphates
**CRYSTAL SYSTEM** Monoclinic
**COMPOSITION** $CaFeFe_5(PO_4)_4(OH)_6\cdot2H_2O$
**COLOUR** Dark olive-green to black
**FORM/HABIT** Massive
**HARDNESS** $3^{1}/_{2}-4^{1}/_{2}$
**CLEAVAGE** Perfect in two directions
**FRACTURE** Indeterminate, brittle
**LUSTRE** Vitreous to silky to dull
**STREAK** Yellow-green
**SPECIFIC GRAVITY** 3.3
**TRANSPARENCY** Sub-translucent to opaque

**BOTRYOIDAL MASS**
*This specimen of dufrenite on goethite is from Hirschberg, Germany.*

*Dufrenite masses*

*Goethite*

# WAVELLITE

**BOHEMIA**
*Good wavellite specimens occur in this part of the Czech Republic.*

AN ALUMINIUM PHOSPHATE hydroxide hydrate, wavellite typically occurs as translucent, greenish, globular aggregates of radiating crystals, as crusts, or occasionally as stalactitic deposits. It is a secondary mineral that forms in crevices in low-grade, aluminous, metamorphic rocks, limonite and phosphate rock deposits, and, rarely, hydrothermal veins. It is found in the Czech Republic, France, Bolivia, Germany, Zaire, Australia, England, Ireland, and in the USA.

## PROPERTIES

**GROUP** Phosphates
**CRYSTAL SYSTEM** Orthorhombic
**COMPOSITION** $Al_3(PO_4)_2(OH,F)_3\cdot5H_2O$
**COLOUR** Green or white
**FORM/HABIT** Radiating aggregates
**HARDNESS** $3^{1}/_{2}-4$
**CLEAVAGE** Good
**FRACTURE** Subconchoidal to uneven
**LUSTRE** Vitreous to resinous
**STREAK** White
**SPECIFIC GRAVITY** 2.4
**TRANSPARENCY** Translucent

*Radiating wavellite crystals*

**RADIATING CRYSTALS**
*Acicular crystals of wavellite form radiating aggregates on this rock groundmass.*

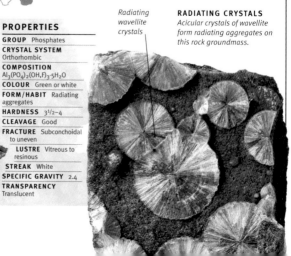

203

# BORATES AND NITRATES

## BORATES

Borate minerals are compounds containing boron and oxygen. The majority of borate minerals are rare, but a few, such as borax, ulexite, colemanite, and kernite, form large, commercially mined deposits. Structurally, boron and oxygen may form a triangle ($BO_3$) or a tetrahedron ($BO_4$), each with a central boron atom.

Borates appear in two geological environments. In the first, borate–bearing solutions that result from volcanic activity flow into a closed basin, where evaporation takes place. Over time, stratified deposits of borate minerals accumulate. Basin deposits usually occur in desert regions, such as the Mojave Desert and Death Valley in California; since they are soluble in water, they are better preserved in dry environments. Borax, colemanite, and kernite are all found in evaporate deposits. In the second environment, borate minerals are formed as a result of alteration of the rocks by heat and pressure at relatively high temperatures. These borate minerals usually consist

**CRYSTAL STRUCTURE OF BORACITE**
*In boracite, densely packed boron tetrahedra combine with the metal ion magnesium.*

*Magnesium ion*
*Magnesium tetrahedron*
*Boron tetrahedron*

of densely packed $BO_3$ triangles combined with metal ions such as magnesium, manganese, aluminium, or iron. Boracite and sussexite form in this type of environment.

## NITRATES

Nitrates are a small group of nitrogen– and oxygen–bearing compounds, which, because of their solubility and instability, are practically confined to arid regions and soils. The only major deposit of nitrates is in the Atacama Desert of northern Chile, where they occur under the loose soil as beds of a greyish, cemented mixture of nitrates, sulphates, halides, and sand. They contain an $NO_3$ group in a triangular arrangement.

## BORAX

BORAX TAKES ITS NAME from the Arabic *buraq*, meaning "white". A sodium borate hydrate, its crystals are short prismatic to tabular, although in commercial deposits it is predominantly massive. Its colourless crystals dehydrate in air to become the chalky mineral tincalconite. Borax fuses easily to become a colourless glass. In mineralogy, molten beads of borax are used to test other minerals for composition: a powdered mineral sample is fused with the bead to see what colour it turns and therefore what chemicals the mineral contains.

Borax is an evaporite mineral formed in dry desert lake beds, and accompanied by halite, other borates, and various evaporite sulphates and carbonates. Borax has been mined since ancient times from saline lakes in Kashmir and Tibet. Today it is the principal source of boron compounds widely used in industry. About half of the world's supply comes from the borax crusts and brine of Searles Lake in southern California, USA. Other sources include: Boron, California, USA; Turkey; Salar Cauchari, Salta Province, Argentina; Ladakh and Kashmir in the Indian Himalayas; and Inder, Kazakhstan.

**BORAX CRUST**
*Searles Lake, California, is the prime source of borax.*

**BORAX CRYSTALS**
*This group of prismatic borax crystals coated with an opaque layer of tincalconite is situated on a rock groundmass.*

*Prismatic crystals*

*Coating of white tincalconite*

*Rock groundmass*

## USES OF BORON

Boron and its compounds are widely used in industry. It is an important component of glass and pottery glazes. In metallurgy, it is used as a solvent for metal oxide slags in steel-making and metal-casting, and as a flux in welding and soldering. It is also widely used as a fertilizer additive, a water softener, a disinfectant, and a soap additive.

**MOUTHWASH**
*Boron is one of the ingredients in proprietary mouthwashes.*

### PROPERTIES

| | |
|---|---|
| **GROUP** | Borates |
| **CRYSTAL SYSTEM** | Monoclinic |
| **COMPOSITION** | $Na_2B_4O_5(OH)_4 \cdot 8H_2O$ |
| **COLOUR** | Colourless |
| **FORM/HABIT** | Short prismatic |
| **HARDNESS** | $2–2\frac{1}{2}$ |
| **CLEAVAGE** | Perfect, imperfect |
| **FRACTURE** | Conchoidal |
| **LUSTRE** | Vitreous to earthy |
| **STREAK** | White |
| **SPECIFIC GRAVITY** | 1.7 |
| **TRANSPARENCY** | Transparent to translucent |

# ULEXITE

ULEXITE, LIKE BORAX, is an important economic mineral. It was first discovered at Iquique, Tarapace, Chile, formerly part of Peru. A hydrated sodium calcium borate hydrate, ulexite is commonly found in nodular, rounded, or lens-like crystal aggregates (often resembling cotton wool balls). In these forms, it is white and has a silky or satiny lustre. It less commonly forms parallel, fibrous crystals, which are colourless and have a vitreous lustre; this form is known as "television stone" (see panel, right). The German chemist George Ludwig Ulex (1811-1883) determined the composition of ulexite in 1850, and it was named after him. It has also been called boronatrocalcite and natroborocalcite.

Ulexite is most commonly found in playa lakes and other evaporite basins in desert regions, where it is derived from boron-rich magmatic fluids. It occurs in the dry plains of Chile, Argentina, and Kazakhstan, and in the huge evaporate deposits of the Kramer District of Death Valley, California, USA. It is commonly in association with anhydrite, colemanite, and glauberite.

Ulexite's industrial uses are similar to those for borax (see panel, opposite).

**SOAP**
*Boron compounds derived from borax and, to a lesser extent, ulexite are important components of many soaps.*

## TELEVISION STONE

An unusual property of one form of ulexite is its ability to "transmit" images. This form of ulexite develops as tightly bundled, acicular crystals that effectively function like optic fibres. A piece crystallized in this way, and with each end flattened and polished perpendicular to the long direction of the crystal bundle will, when placed on print or other images, show that image on the opposite end, as in the photograph below.

**TRANSMITTING PRINT**
*The dark line underneath this ulexite specimen appears to be on its surface. It has been "transmitted" from the back to the front.*

**DEATH VALLEY**
*Death Valley, California, is characterized by the playa lakes where evaporite deposits form.*

**GLAZED POTS**
*Boron compounds derived from ulexite and borax are used extensively in pottery glazes. They are also effective fluxes for melting metal and welding.*

## PROPERTIES

| | |
|---|---|
| **GROUP** | Borates |
| **CRYSTAL SYSTEM** | Monoclinic |
| **COMPOSITION** | $NaCaB_5O_6(OH)_6 \cdot H_2O$ |
| **COLOUR** | Colourless, white |
| **FORM/HABIT** | Nodular, acicular |
| **HARDNESS** | 2½ |
| **CLEAVAGE** | Perfect |
| **FRACTURE** | Uneven |
| **LUSTRE** | Vitreous to silky |
| **STREAK** | White |
| **SPECIFIC GRAVITY** | 2.0 |
| **TRANSPARENCY** | Transparent to translucent |

Parallel, acicular crystals

**ULEXITE SLICE**
*This ulexite specimen from the boron deposits of California, USA, has been sliced and polished to show its fibre-optic effect. This also shows its fibrous structure, composed of parallel, acicular crystals.*

Transparent specimen

Surface polished, showing fibre-optic effect

**FERTILIZER**
*Borate is added to many agricultural fertilizers because boron is a vital trace element for plant growth.*

# HOWLITE

*Significant howlite deposits are found in Death Valley, California, USA.*

NAMED IN 1868 for Canadian chemist Henry How, who discovered it, howlite is a calcium borosilicate hydroxide. Crystals are tabular but rare; generally it forms nodular masses. It is usually white, often with veins of other minerals running through the nodules. When dyed, it resembles turquoise, although it is easily distinguished from turquoise by its inferior hardness (turquoise is 5–6). In addition, howlite is easily fusible. It occurs with other boron minerals, such as kernite (below right) and borax (see p.204), and is found in quantity in California, USA, especially the Kramer district of Death Valley and San Bernadino County. It is also found in Nova Scotia and Newfoundland, Canada; Magdalena, Mexico; Saxony, Germany; the southern Urals of Russia; and Suserluk, Turkey.

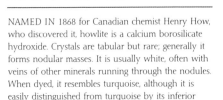

**STAINED HOWLITE**
*Tumble-polished and dyed or stained howlite looks similar to turquoise, and is sometimes sold as such.*

Lighter colour

**TURQUOISE**
*Turquoise can be told apart from dyed howlite by its greater hardness and depth of colour.*

Greater depth of colour

Howlite

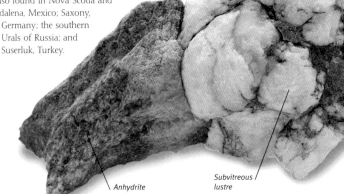

### PROPERTIES

| | |
|---|---|
| GROUP | Borates |
| CRYSTAL SYSTEM | Monoclinic |
| COMPOSITION | $Ca_2B_5SiO_9(OH)_5$ |
| COLOUR | White |
| FORM/HABIT | Nodular masses |
| HARDNESS | $3^{1}/_2$ |
| CLEAVAGE | None |
| FRACTURE | Conchoidal to uneven |
| LUSTRE | Subvitreous |
| STREAK | White |
| SPECIFIC GRAVITY | 2.6 |
| TRANSPARENCY | Translucent to opaque |

Anhydrite

Subvitreous lustre

**NODULAR HOWLITE**
*This howlite nodule, from the Fraser Quarry, Hants County, Nova Scotia, Canada, still retains part of the anhydrite in which it was formed.*

# COLEMANITE

**CALIFORNIA, USA**
*The mineral-rich state of California includes deposits of colemanite.*

THIS BORATE MINERAL was named in 1884 for William Coleman, the owner of the mine in California, USA, where it was discovered. Typically it occurs as crystals up to several centimetres across, and it can be massive, granular, or compact. It is pyroelectric. Colemanite is found in playas and other evaporite deposits where it replaces other borate minerals, such as borax, that were originally deposited in huge inland lakes. These deposits may be several metres thick. Colemanite in commercial deposits is usually massive, but individual crystals up to 20cm (8in) long have been found. The chief localities are California, USA; Kazakhstan; Turkey; and Argentina. Colemanite is an important source of commercial borates and boric acid. It was the main source of boron (see p.207) until the 1930s.

**COMPLEX CRYSTALS**
*Colemanite commonly occurs as colourless, brilliant, complex crystals, as in this specimen.*

## HEAT-RESISTANT GLASS

Heat-resistant glass has been developed by the inclusion of boron compounds in place of some of the silica of normal glass. These compounds are called borosilicates. They make glass resistant to heat, chemicals, and electricity. Borosilicate glass is used in car headlights, laboratory glassware, industrial equipment, thermometers, and ovenware (including glass sold under the well-known trademark Pyrex).

**FIREMAN'S VISOR**
*Borosilicates are used in a wide range of heat-resistant items.*

Prismatic crystals

Translucent crystal

### PROPERTIES

| | |
|---|---|
| GROUP | Borates |
| CRYSTAL SYSTEM | Monoclinic |
| COMPOSITION | $CaB_3O_4(OH)_3 \cdot H_2O$ |
| COLOUR | Colourless, white |
| FORM/HABIT | Short prismatic |
| HARDNESS | $4–4^{1}/_2$ |
| CLEAVAGE | Perfect, distinct |
| FRACTURE | Uneven to subconchoidal |
| LUSTRE | Vitreous to adamantine |
| STREAK | White |
| SPECIFIC GRAVITY | 2.4 |
| TRANSPARENCY | Transparent to translucent |

Vitreous lustre

# BORACITE

**SAXONY, GERMANY**
*Boracite crystals were first found in Lüneburg, Lower Saxony.*

THIS GLASSY BORATE MINERAL is, as its name suggests, a minor source of the element boron. Its crystals are pseudocubic, and also occur in massive aggregates. Generally white to grey, it shades towards light green with increasing iron content. Strongly pyroelectric and piezoelectric, it also exhibits double refraction. Boracite is found as crystals embedded in sedimentary deposits of anhydrite, gypsum, and halite. Localities include the Khorat Plateau of Thailand; Stassfurt, Hannover, and Lüneburg, Germany; Cleveland, England; Lunéville, France; Inowroclaw, Poland; and Missouri and Louisiana in the USA.

**BORACITE CRYSTALS**
*These boracite crystals from Saxony, Germany, appear to be dodecahedrons, a classic example of pseudocubic habit.*

*Green colour from iron*

*Pseudocubic habit*

*Vitreous lustre*

## PROPERTIES

| | |
|---|---|
| **GROUP** | Borates |
| **CRYSTAL SYSTEM** | Orthorhombic |
| **COMPOSITION** | $(Mg,Fe)_3B_7O_{13}Cl$ |
| **COLOUR** | White to grey, green |
| **FORM/HABIT** | Pseudocubic |
| **HARDNESS** | $7–7\frac{1}{2}$ |
| **CLEAVAGE** | None |
| **FRACTURE** | Conchoidal to uneven |
| **LUSTRE** | Vitreous |
| **STREAK** | White |
| **SPECIFIC GRAVITY** | 3.0 |
| **TRANSPARENCY** | Subtransparent to translucent |

# KERNITE

**BORON, USA**
*This area of California has extensive borate deposits, including kernite.*

KERNITE IS A SODIUM BORATE HYDRATE, named for the place where it was discovered, Kern County, California, USA. Crystals are relatively uncommon, but when found can be very large, often 60–90cm (2–3ft) thick. The largest discovered so far measured 240cm by 90cm (8ft by 3ft). Crystals are colourless and transparent but are usually covered by a surface film of opaque white tincalconite, a dehydration product of borate minerals. Kernite is associated with other borate minerals as veins and irregular masses formed in evaporite deposits, and crystals embedded in shale. The Boron area of Kern County produces a large portion of the boron used in the USA, and kernite is one of its principal mineral sources. The deposits at Inyo, California; Catamarca and Salta Provinces, Argentina; and Kirka, Turkey, are other major world sources.

**FIBROUS MASS**
*This specimen of kernite is from the type locality in Kern County. It shows characteristic perfect cleavage and vitreous lustre.*

## PROPERTIES

| | |
|---|---|
| **GROUP** | Borates |
| **CRYSTAL SYSTEM** | Monoclinic |
| **COMPOSITION** | $Na_2B_4O_6(OH)_2 \cdot 3H_2O$ |
| **COLOUR** | Colourless, white |
| **FORM/HABIT** | Fibrous mass |
| **HARDNESS** | $2\frac{1}{2}$ |
| **CLEAVAGE** | Perfect |
| **FRACTURE** | Splintery |
| **LUSTRE** | Vitreous, silky, dull |
| **STREAK** | White |
| **SPECIFIC GRAVITY** | 1.9 |
| **TRANSPARENCY** | Transparent |

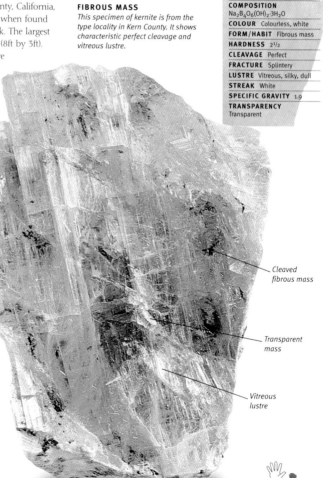

*Cleaved fibrous mass*

*Transparent mass*

*Vitreous lustre*

## REINFORCING AGENT

Boron, derived from minerals such as kernite and colemanite, is a black, dense metal in its pure form. Boron carbide is one of the hardest synthetic substances known, being nearly as hard as diamond. It is used as an abrasive, as a reinforcing agent for military armour, and to control the rate of fission in nuclear reactors. Because of its hardness, it is also used in sandblasting nozzles and pump seals, and it is added to increase the hardness of steel. In semiconductors, small amounts of boron are also added to silicon and germanium to modify their electrical conductivity.

**ARMOURED TANK**
*Boron compounds are added to steel to harden it. Super-hardened steel is used in military armour.*

# SULPHATES, CHROMATES, TUNGSTATES, AND MOLYBDATES

SULPHATES, CHROMATES, tungstates, and molybdates are grouped together because they behave the same structurally and chemically. Around 200 sulphates are known; gypsum and baryte are very common, but most are rare. The chromates, tungstates, and molybdates are all relatively rare, but when found in concentration are important ores of the metals they contain.

## CRYSTAL STRUCTURE

The sulphate minerals have a crystal structure in which four oxygen atoms are located at the corners of a tetrahedron with a sulphur atom in the centre. The sulphate tetrahedron behaves chemically as a single negatively charged

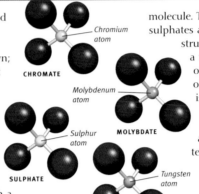

Chromium atom

CHROMATE

Molybdenum atom

MOLYBDATE

Sulphur atom

SULPHATE

Tungsten atom

TUNGSTATE

molecule. The chemical formulae of the sulphates all contain an $SO_4$ group. The basic structural unit of the chromates is also a tetrahedron formed from four oxygen atoms, each at one corner of the tetrahedron. But in the centre is a chromium atom instead of a sulphur atom. The chemical formulae for chromate minerals all contain a $CrO_4$ group. The same type of tetrahedron forms in the molybdates and tungstates, but with an atom of molybdenum (Mo) or tungsten (W) in the centre, respectively. Their chemical formulae all contain a $WO_4$ or a $MoO_4$ group.

---

## ANGLESITE

ANGLESITE HAS A VARIETY of crystal habits: thin to thick tabular, prismatic, pseudorhombohedral, or pyramidal. It is also commonly massive, granular, or compact. Its crystals resemble those of baryte and celestine, with which it shares a similar structure. It is colourless to white, greyish, yellow, green, or blue and often fluoresces yellow under ultraviolet light. Anglesite is lead sulphate, an alteration product of galena, and is sometimes found in concentric layers with a core of unaltered galena. It is common in the oxidation zone of lead deposits. It readily alters to cerussite, lead carbonate. Exceptionally large crystals, up to 80cm (32in) long, have come from Touissit, Morocco. Other fine specimens come from Tsumeb, Namibia; Broken Hill, NSW, Australia; Derbyshire, England; Siegen, Germany; Monteponi Mine, Sardinia, Italy; and Coeur d'Alene, Idaho, USA.

Rock groundmass

Prismatic crystal

**ANGLESEY**
*This island in Wales is the type locality for anglesite.*

Galena

### ANGLESITE CRYSTALS
*A large, striated prismatic crystal of anglesite is on a rock groundmass along with several smaller crystals and accessory mineral, galena.*

Pyramidal crystal

Rock groundmass

**ANGLESITE CRYSTAL**

### PROPERTIES

| | |
|---|---|
| **GROUP** | Sulphates |
| **CRYSTAL SYSTEM** | Orthorhombic |
| **COMPOSITION** | $PbSO_4$ |
| **COLOUR** | Colourless to white, yellow, green, or blue |
| **FORM/HABIT** | Thin to thick tabular |
| **HARDNESS** | $2^1/_2$–3 |
| **CLEAVAGE** | Good, distinct |
| **FRACTURE** | Conchoidal, brittle |
| **LUSTRE** | Adamantine to resinous, vitreous |
| **STREAK** | Colourless |
| **SPECIFIC GRAVITY** | 6.4 |
| **TRANSPARENCY** | Transparent to opaque |

## ROMAN LEAD MINING IN BRITAIN

The lure of metal wealth was one of the factors prompting the Roman invasion of Britain in 43BC. Lead was mined extensively in a number of localities, with some deposits carrying significant portions of silver. Anglesite was one of the ores the Romans sought. Lead was used extensively in the Roman world, especially for water pipes. Our modern word "plumbing" comes from the Latin for lead, *plumbum*, and lead was still used for piping until about 75 years ago.

**WATER PIPE**
*This Roman lead water pipe still lies where it was installed nearly two millennia ago.*

**DISUSED LEAD MINE**
*This disused lead mine in Lathkill Dale, Derbyshire, was worked profitably from the 13th century until 1851.*

# GLAUBERITE

A SODIUM CALCIUM SULPHATE, glauberite was named in 1802 for its similarity to another chemical, glauber's salt, which in turn was named for the German alchemist Johann Glauber. Crystals can be prismatic, tabular, and dipyramidal, all with combinations of forms. Usually grey or yellowish, it can sometimes be colourless or reddish. It has a slightly saline taste, and fuses to a white enamel. Glauberite forms under a variety of conditions, although it is primarily an evaporite, forming in both marine and salt-lake environments. It is also found in cavities in basaltic igneous rocks and in volcanic fumaroles. Glauberite was first found in crystals in the halite deposit in Villarrubia, Toledo, Spain. It also occurs in crystals in the Stassfurt potash region of Germany. Other localities include the Salt Range, Pakistan; Gypsumville, Canada; Lorraine, France; and Arizona and New Jersey, USA.

**SEARLES LAKE**
*This Californian lake is a source of evaporites such as glauberite.*

**GLAUBERITE CRYSTALS**
*This group of dipyramidal glauberite crystals is from Ciempozuelos, Madrid, Spain.*

*Dipyramidal crystals*

*Vitreous lustre*

## ALCHEMY

Alchemy is the search for a method of changing base metals into gold. It was a popular if mysterious pursuit, especially in the 16th and 17th centuries, when Johann Glauber was practising it. Alchemy was a predecessor of chemistry, and its practitioners made discoveries that laid the foundations of modern chemical knowledge.

**ALCHEMIST'S WORKSHOP**
*This 1570 painting by Jan van der Sraet shows some of the pseudoscientific methods of alchemists.*

### PROPERTIES

**GROUP** Sulphates

**CRYSTAL SYSTEM** Monoclinic

**COMPOSITION** $Na_2Ca(SO_4)_2$

**COLOUR** Grey, yellowish, colourless, or reddish

**FORM/HABIT** Prismatic, tabular

**HARDNESS** $2\frac{1}{2}$–3

**CLEAVAGE** Perfect, indistinct

**FRACTURE** Conchoidal

**LUSTRE** Vitreous to waxy

**STREAK** White

**SPECIFIC GRAVITY** 2.8

**TRANSPARENCY** Transparent to translucent

---

# ANHYDRITE

**CHIHUAHUA**
*Fine anhydrite crystals are found here.*

AN IMPORTANT ROCK-FORMING mineral, anhydrate, along with gypsum, is a form of calcium sulphate. Its name comes from the Greek, *anhydrous*, meaning "without water". Gypsum, which contains water in its structure, is a much more common mineral. Anhydrite alters to gypsum in humid conditions. It is usually massive, granular, or coarsely crystalline. Individual crystals are uncommon, but when found are blocky or thick tabular. It is colourless to white, but also may be brownish, reddish, or greyish, or pale shades of pink, blue, and violet. Anhydrite is one of the major minerals in evaporite deposits, and occurs most often in salt deposits associated with gypsum and halite. Massive deposits of anhydrite occur in the cap rock of the Texas–Louisiana salt domes in the USA. It is found in Nova Scotia, Canada, and in the rock salt deposits of the French Pyrénées. Fine blue crystals are found in Chihuahua, Mexico; crystals up to 10cm (4in) long come from Swiss deposits; and fine white crystals come from Tuscany, Italy. Anhydrite is used in fertilizers, and as a drying agent in plasters, cement, paints, and varnishes.

### PROPERTIES

**GROUP** Sulphates

**CRYSTAL SYSTEM** Orthorhombic

**COMPOSITION** $CaSO_4$

**COLOUR** Colourless to white, pink, blue, violet, brownish, reddish, or greyish

**FORM/HABIT** Massive

**HARDNESS** $3\frac{1}{2}$

**CLEAVAGE** Perfect, good

**FRACTURE** Uneven to splintery

**LUSTRE** Vitreous to pearly

**STREAK** White

**SPECIFIC GRAVITY** 3.0

**TRANSPARENCY** Transparent to translucent

**GERMAN ANHYDRITE**
*This reddish specimen of anhydrite from Germany shows perfect cleavage.*

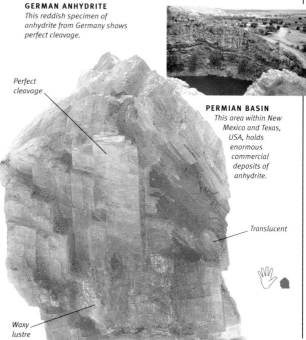

*Perfect cleavage*

*Translucent*

*Waxy lustre*

**PERMIAN BASIN**
*This area within New Mexico and Texas, USA, holds enormous commercial deposits of anhydrite.*

**OCTAGONAL MIXED**
*Transparent baryte is sometimes faceted for collectors, although cutting it is difficult even for master cutters.*

# BARYTE

THE NAME BARYTE originates from the Greek *barys*, meaning "heavy", a reference to its high specific gravity. For the same reason, it has also been called heavy spar. It is the most common barium mineral, and the element's principal ore.

Baryte crystals are sometimes tinged yellow, blue, or brown, and a rare golden baryte comes from Colorado, USA. It is frequently well crystallized, and tabular or prismatic crystals are common, as are cockscomb (or crested) aggregates. Baryte can also be fibrous, massive, stalactitic, or concretionary. It also occurs as aggregate rosettes of crystals known as "desert roses". Transparent blue baryte crystals can resemble aquamarine, but their softness, heaviness, and crystal shape make them easily distinguishable. Were the stone not so soft and fragile, it would make a beautiful gem, and it is cut for collectors on rare occasions.

Baryte is a common accessory mineral in lead and zinc veins, where it occurs with galena, sphalerite, fluorite, and calcite. It is also found in sedimentary rocks such as limestone, in clay deposits formed by the weathering of limestone, in marine deposits, and in cavities in igneous rock. It is abundant in Spain, Germany, Canada, France, India, Romania, and other localities in the USA. Baryte is an important economic mineral for oil and gas production, because it is used as drilling mud in oil- and gas-well drilling. It is also used as a filler in paper- and cloth-making, as an inert body in coloured paints, and as a white pigment. Barium sulphate is used in medicine as a "barium meal" (see below).

**STALAGMITE SECTION**
*Baryte is sometimes found forming stalagmites. In cross section, these contain structures that look like the rings of a tree.*

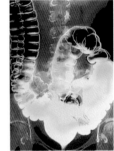

**BARIUM MEAL**
*Barium sulphate is swallowed to highlight problems in the gastrointestinal tract on X-rays.*

## PROPERTIES

| | |
|---|---|
| **GROUP** | Sulphates |
| **CRYSTAL SYSTEM** | Orthorhombic |
| **COMPOSITION** | $BaSO_4$ |
| **COLOUR** | Colourless, white, grey, bluish, greenish, beige |
| **FORM/HABIT** | Tabular to prismatic |
| **HARDNESS** | $3–3\frac{1}{2}$ |
| **CLEAVAGE** | Perfect |
| **FRACTURE** | Uneven |
| **LUSTRE** | Vitreous, resinous, pearly |
| **STREAK** | White |
| **SPECIFIC GRAVITY** | 4.5 |
| **TRANSPARENCY** | Transparent to translucent |
| **R.I.** | 1.63–1.65 |

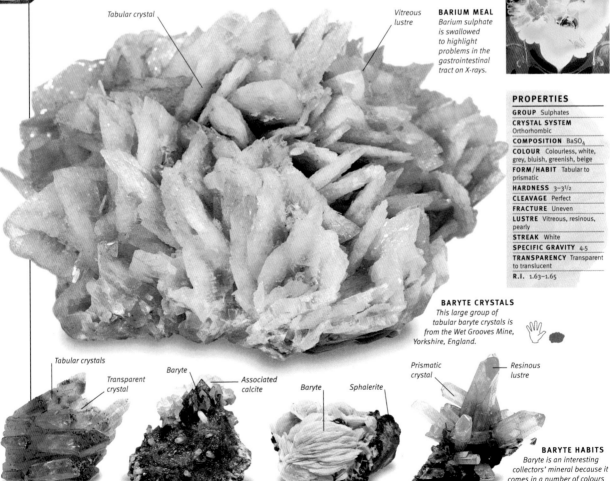

Tabular crystal

Vitreous lustre

**BARYTE CRYSTALS**
*This large group of tabular baryte crystals is from the Wet Grooves Mine, Yorkshire, England.*

Tabular crystals

Transparent crystal

Baryte

Associated calcite

Baryte

Sphalerite

Prismatic crystal

Resinous lustre

**BARYTE HABITS**
*Baryte is an interesting collectors' mineral because it comes in a number of colours and habits. The specimens to the left are just a few examples.*

**CRYSTALLINE**

**BARYTE WITH CALCITE**

**COCKSCOMB**

**PRISMATIC CRYSTALS**

# CELESTINE

OFTEN FORMING BEAUTIFUL transparent light to medium-blue tabular crystals, celestine can be colourless, white, light red, green, blue, or brown. It can also be massive, fibrous, granular, or nodular. It takes its name from the Latin *coelestis*, meaning "heavenly", an allusion to its colour. Celestine is strontium sulphate, but barium substitutes freely for strontium in the structure, and calcium may also be present. Celestine forms in sedimentary rock such as limestones, dolomites, and sandstones. It can also be precipitated directly from sea water, and commonly occurs in evaporite deposits as a minor constituent. Occasionally it forms in hydrothermal deposits.

Collectors' specimens come from Madagascar, Mexico, Italy, Canada, and the USA. Lesser amounts come from England, Slovakia, Austria, and Germany. Beds of massive celestine 3–6m (10–20ft) thick occur in California, USA. Well-formed crystals commonly reach 10cm (4in) in length, and crystals more than 75cm (30in) long are known.

**MADAGASCAR**
*The deposits of Madagascar are a prime locality for celestine crystals, including those mined and faceted for collectors.*

**SINGLE CRYSTAL**
*This is a superb example of a light blue, single prismatic crystal of celestine.*

## PROPERTIES

| | |
|---|---|
| **GROUP** | Sulphates |
| **CRYSTAL SYSTEM** | Orthorhombic |
| **COMPOSITION** | $SrSO_4$ |
| **COLOUR** | Colourless, white, red, green, blue, or brown |
| **FORM/HABIT** | Tabular |
| **HARDNESS** | $3–3\frac{1}{2}$ |
| **CLEAVAGE** | Perfect |
| **FRACTURE** | Uneven |
| **LUSTRE** | Vitreous, pearly on cleavage |
| **STREAK** | White |
| **SPECIFIC GRAVITY** | 4.0 |
| **TRANSPARENCY** | Transparent to translucent |
| **R.I.** | 1.62–1.63 |

**FIREWORKS**
*Along with strontianite, celestine is the principal source of strontium, which is used to create the red colour in fireworks, signal flares, and tracer bullets.*

Sulphur groundmass

**MIXED**
*Transparent celestine is sometimes cut for collectors. It is too soft and fragile to wear and is difficult to cut.*

Prismatic crystals

**CRYSTALS IN MATRIX**
*This mass of colourless, prismatic crystals of celestine is on a sulphur groundmass.*

Vitreous lustre

Tabular crystal

**CELESTINE CRYSTALS**
*This specimen of superbly crystallized, blue celestine crystals is from Madagascar. The largest crystal is over 3.5cm (1½in) in length.*

Blue coloration

Granular celestine

Small celestine crystals

# GYPSUM

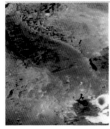

**"SUN-SPANGLED" BEAD**
*This polished bead contains cracks that may have been caused by heat treatment and produce a "sun-spangled" effect.*

### PROPERTIES

| | |
|---|---|
| **GROUP** | Sulphates |
| **CRYSTAL SYSTEM** | Monoclinic |
| **COMPOSITION** | $CaSO_4 \cdot 2H_2O$ |
| **COLOUR** | Colourless, white, light brown, yellow, pink |
| **FORM/HABIT** | Prismatic to tabular |
| **HARDNESS** | 2 |
| **CLEAVAGE** | Perfect |
| **FRACTURE** | Splintery |
| **LUSTRE** | Subvitreous to pearly |
| **STREAK** | White |
| **SPECIFIC GRAVITY** | 2.3 |
| **TRANSPARENCY** | Transparent to translucent |
| **R.I.** | 1.52–1.53 |

ONE OF NATURE'S MOST spectacular – and largest – mineral deposits lies in the desert country of south-central New Mexico, USA. White Sands is 580 square km (225 square miles) of dazzling white gypsum sand that has been blown into dunes up to 18m (60ft) high. In another spectacular deposit at the Cave of Swords, Naica, Chihuahua, Mexico, numerous transparent, sword-like selenite gypsum crystals reach lengths of 2m (6ft) or more. Selenite, from the Greek word *selene*, meaning "the moon", is the name for transparent crystals of gypsum, but is often incorrectly used for gypsum in general. Gypsum takes its name from the Greek *gypsos*, meaning "chalk", "plaster", or "cement".

## CHARACTERISTICS OF GYPSUM

Gypsum is a widespread calcium sulphate hydrate that is found in a number of forms, and is of great economic importance. It is colourless or white, but impurities tint it light brown, grey, yellow, green, or orange. It often occurs in well-developed crystals. Single crystals can be blocky with a slanted parallelogram outline, tabular, bladed, or in a long, thin shape like a "ram's horn". Twinned crystals are common, and frequently form characteristic "swallowtails" or "fishtails". Gypsum is also found in a parallel, fibrous variety with a silky lustre, called satin spar. The massive, fine-grained variety is called alabaster. Rosette-shaped crystals are called desert roses, and are more common and less dense than the baryte "sand crystals" (see p.210) that they resemble.

Gypsum occurs in extensive beds formed by the evaporation of ocean brine, along with other minerals similarly formed – in particular, anhydrite (see p.209)

and halite (see p.170). It has low solubility and so is the first mineral to separate from evaporating sea water. Once it has been deposited, it is commonly blown into dunes. Gypsum also occurs as an alteration product of sulphides in ore deposits; as disseminated crystals and rosette-shaped aggregates in sedimentary deposits, including sands and clays; and as deposits around volcanic fumaroles.

**FISH**
*This fish has been carved from local alabaster in Loveland, Colorado, USA.*

## ALABASTER

Alabaster, the fine-grained, massive form of gypsum, has been used for centuries for statuary, carvings, and other ornamental purposes. Florence, Milan, and Livorno in Italy and Berlin in Germany are important modern centres of the alabaster trade. Alabaster is normally snow-white and translucent but can be artificially dyed, or it may be heat treated to make it similar in appearance to marble. In ancient times, the term was applied to other minerals of similar appearance; for example, the "alabaster" containers recovered from Tutankhamun's tomb were, in fact, made of calcite (onyx).

**COUPLE**
*This simple modern carving has been made from pale, translucent alabaster.*

**WHITE SANDS**
*The huge expanse of the deposit of gypsum sand at White Sands, New Mexico, USA, can be seen below and from space (left).*

**SELENITE CRYSTAL**
*This single, prismatic selenite crystal comes from the Cave of Swords in Mexico.*

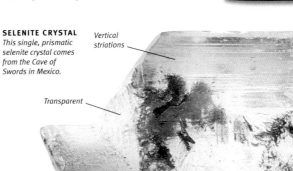

Vertical striations

Transparent

Attachment point

Gypsum occurs widely throughout the world, but the USA, Canada, Australia, Spain, France, Italy, and England are among the leading commercial producers.

## USES OF GYPSUM

An important economic mineral since the time of ancient Egyptian civilization, gypsum was mined to the west of Alexandria, near Suez, in Al Fayyum, and near the Red Sea coast. It was used for mortar and plaster in the Giza necropolis, and it was used elsewhere to plaster walls, cover bodies, make statues and masks, as an adhesive, and as a filler in pigments. The Romans discovered that heating gypsum to 300˚C (600˚F) made a plaster that sets hard when mixed with water. This plaster was used for building, and is still widely used today.

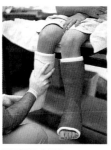

**PLASTER CAST**
*About 75 per cent of the gypsum mined worldwide is used to make plaster of Paris.*

### SELENITE GYPSUM FORMS
*Selenite gypsum comes in a variety of crystalline forms, including single monoclinic trapezoids, fibrous satin spar, characteristic "fishtail" twins, and in radiating clusters called desert roses.*

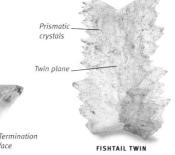

Prismatic crystals

Twin plane

**FISHTAIL TWIN**

Prism face

Termination face

**SINGLE GYPSUM CRYSTAL**

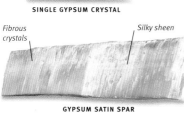

Fibrous crystals

Silky sheen

**GYPSUM SATIN SPAR**

Bladed crystals

Radiating form

**DESERT ROSE**

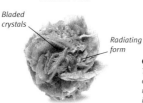

**CAVE OF SWORDS**
*The selenite crystals in the Cave of Swords, Mexico (above and right), reach more than 2m (6ft) in length. The cave is named for these enormous selenite crystals.*

Termination face

Pearly lustre

# MELANTERITE

**NEVADA, USA**
*This state and Arizona are two of many localities for melanterite in the USA.*

DRAWING ITS NAME from the Greek *melas*, meaning "sulphate of iron", melanterite is a hydrous iron sulphate. It rarely forms crystals, and is generally found in stalactitic or concretionary masses. When crystals occur, they are short prisms or pseudo-octahedrons. Most samples are colourless to white, but become green to blue with increasing substitution of copper for iron. Melanterite is water soluble. It is a secondary mineral formed by the oxidation of pyrite, marcasite, and other iron sulphides, and is frequently deposited on the timbers of old mine workings. It also occurs in the altered zones of pyrite-bearing rocks, especially in arid climates, and in coal and lignite deposits, where it is an alteration product of marcasite. It is found in Russia, Italy, Germany, the Czech Republic, and at various localities in the USA.

## PROPERTIES

| | |
|---|---|
| **GROUP** Sulphates | |
| **CRYSTAL SYSTEM** Monoclinic | |
| **COMPOSITION** $FeSO_4 \cdot 7H_2O$ | |
| **COLOUR** White, greenish, bluish | |
| **FORM/HABIT** Stalactitic, concretionary | |
| **HARDNESS** 2 | |
| **CLEAVAGE** Perfect | |
| **FRACTURE** Conchoidal, brittle | |
| **LUSTRE** Vitreous | |
| **STREAK** White | |
| **SPECIFIC GRAVITY** 1.9 | |
| **TRANSPARENCY** Translucent | |

## INDUSTRIAL USES

Melanterite, which is a sulphide of iron, has been used for hundreds of years to make ink. During the Middle Ages, oak was steeped in water with ferrous sulphate to produce a dark purplish ink. Today, iron sulphate is used in water purification as a coagulant – a substance added to water to coagulate and thus drop out minute particles that otherwise do not settle. It is also used as a fertilizer.

**WATER PURIFICATION**
*Melanterite helps to purify water by separating out tiny particles.*

**MELANTERITE NODULE**
*This melanterite nodule shows typical massive form and some minor crystallization.*

*Massive habit*

*Blue indicates presence of copper*

# CHALCANTHITE

**BUTTE, USA**
*In this part of Montana, chalcanthite is an ore of copper.*

A WIDESPREAD, naturally occurring hydrated copper sulphate, chalcanthite takes its name from the Greek *khalkos*, "copper" and *anthos* "flower". It used to be known as blue vitriol. Natural crystals are relatively rare; it usually occurs in veinlets, and as massive and stalactitic aggregates. Commonly peacock-blue, in some cases it tends to greenish. Chalcanthite forms through the oxidation of chalcopyrite and other copper sulphates, occurring in the oxidized zone of copper deposits. It is often found forming crusts and stalactites on the timbers and walls of mine workings where it has crystallized from mine waters. In arid areas, such as Chile, it can be an important ore of copper.

## PROPERTIES

| | |
|---|---|
| **GROUP** Sulphates | |
| **CRYSTAL SYSTEM** Triclinic | |
| **COMPOSITION** $CuSO_4 \cdot 5H_2O$ | |
| **COLOUR** Blue | |
| **FORM/HABIT** Stalactitic, encrustations | |
| **HARDNESS** $2\frac{1}{2}$ | |
| **CLEAVAGE** Not distinct | |
| **FRACTURE** Conchoidal | |
| **LUSTRE** Vitreous | |
| **STREAK** Colourless | |
| **SPECIFIC GRAVITY** 2.3 | |
| **TRANSPARENCY** Transparent | |

## MINING COPPER

In its earliest form, copper mining was carried out by workers using hand picks. Today, most copper mining takes place in open-cast (open-pit) mines. These huge craters scooped out of solid rock can be more than 1.5km (1 mile) across. Pits such as those at Chuquicamata, Chile, are so deep that a truck loaded with ore leaving the bottom of the pit can take over an hour to drive to the surface.

**CHUQUICAMATA MINE**
*At more than 3km (2 miles) across, Chuquicamata, Chile, is the world's largest open-cast copper mine.*

*Kaolinite*

*Crystalline chalcanthite*

*Granular chalcanthite*

*Groundmass of rock*

**TWO FORMS**
*This specimen on a rock groundmass with patches of kaolinite exhibits massive and crystalline forms.*

# BROCHANTITE

THIS SECONDARY MINERAL usually forms needle-like or prismatic crystals, seldom more than a few millimetres long. It is also found in druse crusts and fine-grained masses. Twinning is common in crystals. A copper sulphate hydroxide, brochantite forms in the oxidation zone of copper deposits, and is usually associated with azurite, malachite, and other copper minerals. Splendid specimens come from Namibia, and Bisbee, Arizona, USA, where prismatic crystals may exceed 10mm (¹/₂in). It is also found in Algeria, Morocco, Chile, Greece, Germany, Italy, England, Spain, and Australia. In Arizona and Chile it is so abundant that it is an ore of copper.

**TSUMEB**
*Very high-quality deposits of brochantite are found in Tsumeb, Namibia.*

**NEEDLE-LIKE CRYSTALS**
*This mass of acicular brochantite crystals rests on a groundmass of iron oxides. The specimen is from Chile.*

Iron oxide groundmass

Mass of acicular brochantite crystals

## BROCHANT DE VILLIERS

Brochantite was named for the French geologist and mineralogist A.J.M. Brochant de Villiers in 1824. Born in 1772, he was the first pupil admitted to the École des Mines, and he later became its Professor of Geology and Mines. He was also Inspector-General of Mines.

**EMINENT MINERALOGIST**
*Brochant de Villiers oversaw the construction of the geological map of France.*

### PROPERTIES

| | |
|---|---|
| **GROUP** | Sulphates |
| **CRYSTAL SYSTEM** | Monoclinic |
| **COMPOSITION** | $Cu_4SO_4(OH)_6$ |
| **COLOUR** | Emerald-green |
| **FORM/HABIT** | Short prismatic to acicular |
| **HARDNESS** | 3¹/₂–4 |
| **CLEAVAGE** | Perfect |
| **FRACTURE** | Uneven to subconchoidal |
| **LUSTRE** | Vitreous |
| **STREAK** | Pale green |
| **SPECIFIC GRAVITY** | 4.0 |
| **TRANSPARENCY** | Translucent |

---

# EPSOMITE

FIRST FOUND AROUND mineral springs near the town of Epsom in Surrey, epsomite is an important source of magnesium salts, which is marketed as a laxative called Epsom salts. Magnesium sulphate occurs in solution in sea water, saline lake water, and spring water. When the water evaporates, epsomite precipitates, forming lake deposits. It is also found as crusts in coal, in the weathered portions of magnesium-rich rocks, in the oxidized zones of sulphide ore deposits, and in limestone caves, such as those in Kentucky, USA. Zinc and nickel can substitute for magnesium in the structure.

**EPSOM**
*Epsomite was found in spring waters in the town of Epsom in Surrey, England.*

**EPSOM SALTS**
*This widely used medication is derived from epsomite.*

**EPSOMITE FIBRES**
*The fibrous habit of this specimen is clearly visible.*

Fibrous strands

Vitreous to silky lustre

### PROPERTIES

| | |
|---|---|
| **GROUP** | Sulphates |
| **CRYSTAL SYSTEM** | Orthorhombic |
| **COMPOSITION** | $MgSO_4\cdot7H_2O$ |
| **COLOUR** | Colourless or white |
| **FORM/HABIT** | Fibrous crusts |
| **HARDNESS** | 2–2¹/₂ |
| **CLEAVAGE** | Perfect |
| **FRACTURE** | Conchoidal |
| **LUSTRE** | Vitreous to silky |
| **STREAK** | White |
| **SPECIFIC GRAVITY** | 1.7 |
| **TRANSPARENCY** | Translucent |

---

# ALUNITE

ALUNITE IS ALSO CALLED "alum stone". Crystals are rare; it is usually massive or granular, occurring in pockets or seams in altered volcanic rocks, such as rhyolites, trachytes, and andesites, where it forms through the rocks' chemical reaction with sulphurous vapours. Earthy masses of alunite are commonly mixed with quartz and kaolinite clay. Large deposits are found in the Ukraine, Spain, and New South Wales, Australia.

**DEATH VALLEY**
*Deposits of alunite are found in Death Valley in California, USA.*

### PROPERTIES

| | |
|---|---|
| **GROUP** | Sulphates |
| **CRYSTAL SYSTEM** | Hexagonal/trigonal |
| **COMPOSITION** | $KAl_3(SO_4)_2(OH)_6$ |
| **COLOUR** | White, yellowish |
| **FORM/HABIT** | Massive |
| **HARDNESS** | 3¹/₂–4 |
| **CLEAVAGE** | Distinct |
| **FRACTURE** | Conchoidal to splintery |
| **LUSTRE** | Dull to vitreous to pearly |
| **STREAK** | White |
| **SPECIFIC GRAVITY** | 2.8 |
| **TRANSPARENCY** | Translucent |

**POTASH ALUM**
*Alunite is a source of the compound potash alum, which is used as a fixative for dyes.*

**MASSIVE ALUNITE**
*This specimen of massive alunite exhibits a compact habit.*

Pearly lustre

215

# CYANOTRICHITE

THIS SULPHUR MINERAL takes its name from the Greek words for "blue" and "hair". Its name well describes its usual occurrence: plush-like encrustations of tiny, acicular, radiating crystals that are pale to dark blue in colour. Cyanotrichite forms in the oxidized zones of copper deposits along with other secondary copper minerals. Localities include France, Greece, Romania, Russia, and Arizona and Utah, USA.

**URALS**
*These Russian mountains are a source of cyanotrichite.*

**CYANOTRICHITE CRYSTALS**
*Cyanotrichite's intense blue shows clearly in this specimen from Cap Garonne, Var, France.*

Rock groundmass

Radiating acicular crystals

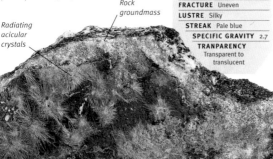

## PROPERTIES

| | |
|---|---|
| **GROUP** | Sulphates |
| **CRYSTAL SYSTEM** | Orthorhombic |
| **COMPOSITION** | $Cu_4Al_2SO_4(OH)_{12}\cdot2H_2O$ |
| **COLOUR** | Pale to dark blue |
| **FORM/HABIT** | Acicular, fibrous |
| **HARDNESS** | 2–3 |
| **CLEAVAGE** | None |
| **FRACTURE** | Uneven |
| **LUSTRE** | Silky |
| **STREAK** | Pale blue |
| **SPECIFIC GRAVITY** | 2.7 |
| **TRANSPARENCY** | Transparent to translucent |

# CROCOITE

CROCOITE IS LEAD CHROMATE, taking its name from the Greek for "saffron", after its colour. Crystals are prismatic, commonly slender and elongated, and sometimes cavernous or hollow. It can also be granular or massive. Crocoite forms as a secondary mineral in the oxidized zone of lead deposits. Superb specimens come from Tasmania; other localities include Russia, Germany, and California, USA.

**TASMANIA**
*Fine crocoite occurs on this Australian island.*

**ORANGE CROCOITE**
*This brilliant orange specimen exhibits attractive, elongated prismatic crystals.*

**RED CROCOITE**
*These well-formed prismatic crocoite crystals are from the classic locality of Dundas, Tasmania.*

Adamantine lustre

Prismatic crystals

## PROPERTIES

| | |
|---|---|
| **GROUP** | Chromates |
| **CRYSTAL SYSTEM** | Monoclinic |
| **COMPOSITION** | $PbCrO_4$ |
| **COLOUR** | Orange, red |
| **FORM/HABIT** | Prismatic |
| **HARDNESS** | $2^1/_2$–3 |
| **CLEAVAGE** | Distinct in one direction |
| **FRACTURE** | Conchoidal to uneven, brittle |
| **LUSTRE** | Vitreous |
| **STREAK** | Orange-yellow |
| **SPECIFIC GRAVITY** | 6.0 |
| **TRANSPARENCY** | Transparent to translucent |

# HÜBNERITE

A MANGANESE IRON TUNGSTATE, hübnerite is found as prismatic, tabular, or flattened crystals with striated faces, and is commonly twinned. It occurs principally in granitic pegmatites and hydrothermal veins with quartz, cassiterite, topaz, and lepidolite. Good specimens come from Peru and Colorado, USA. Formerly called wolframite, it is a major tungsten ore.

**SAN JUAN, USA**
*This area of Colorado produces superb hübnerite crystals.*

**HÜBNERITE CRYSTALS**
*Here translucent hübnerite crystals and tarnished tetrahedrite are on quartz.*

Prismatic hübnerite crystal

Adamantine lustre

Quartz groundmass

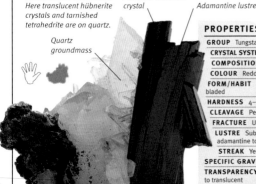

## PROPERTIES

| | |
|---|---|
| **GROUP** | Tungstates |
| **CRYSTAL SYSTEM** | Monoclinic |
| **COMPOSITION** | $MnWO_4$ |
| **COLOUR** | Reddish-brown |
| **FORM/HABIT** | Prismatic, bladed |
| **HARDNESS** | $4–4^1/_2$ |
| **CLEAVAGE** | Perfect |
| **FRACTURE** | Uneven |
| **LUSTRE** | Submetallic/ adamantine to resinous |
| **STREAK** | Yellow to brown |
| **SPECIFIC GRAVITY** | 7.3 |
| **TRANSPARENCY** | Transparent to translucent |

# FERBERITE

FERBERITE IS AN IRON TUNGSTATE which, with hübnerite (see left), constitutes the mineral formerly known as wolframite. Its black crystals are commonly elongated or flattened with a wedge-shaped appearance. Twinning is also common. It is found in granitic pegmatites and high-temperature hydrothermal veins, often with cassiterite, hematite, or arsenopyrite. Excellent specimens come from Japan, Portugal, Romania, South Korea, and Rwanda. It is an ore of tungsten.

**MYANMAR**
*Ferberite crystals can be found in this country.*

Submetallic lustre

**FERBERITE CRYSTAL**
*This crystal is from Cínovec, Czech Republic.*

## PROPERTIES

| | |
|---|---|
| **GROUP** | Tungstates |
| **CRYSTAL SYSTEM** | Monoclinic |
| **COMPOSITION** | $FeWO_4$ |
| **COLOUR** | Black |
| **FORM/HABIT** | Bladed, prismatic |
| **HARDNESS** | $4–4^1/_2$ |
| **CLEAVAGE** | Perfect in one direction |
| **FRACTURE** | Uneven, brittle |
| **LUSTRE** | Submetallic |
| **STREAK** | Black to brown |
| **SPECIFIC GRAVITY** | 7.5 |
| **TRANSPARENCY** | Opaque |

**TV TUBES**
*Tungsten is used in the making of television vacuum tubes.*

# SCHEELITE

**BOHEMIA**
*A mineral-rich part of the Czech Republic, Bohemia is a source of scheelite.*

NAMED IN 1821 for the Swedish chemist C.W. Scheele (1742–86), scheelite is calcium tungstate – a major source of tungsten. Crystals are generally bipyramidal, and twinned, but also form in granular or massive aggregates. A small amount of molybdenum is commonly present. Most scheelite fluoresces, the colour ranging from blue to white. It commonly occurs in contact with metamorphic deposits, high-temperature hydrothermal veins, and less commonly in granitic pegmatites. Opaque crystals weighing up to 7kg (15lb) come from Arizona, USA, and superb, often transparent crystals from Austria, Italy, Brazil, Rwanda, and Colorado, USA.

## PROPERTIES

| | |
|---|---|
| **GROUP** | Tungstates |
| **CRYSTAL SYSTEM** | Tetragonal |
| **COMPOSITION** | $CaWO_4$ |
| **COLOUR** | White, yellow, brown, or green |
| **FORM/HABIT** | Bipyramidal |
| **HARDNESS** | $4\frac{1}{2}$–5 |
| **CLEAVAGE** | Distinct |
| **FRACTURE** | Uneven to subconchoidal |
| **LUSTRE** | Vitreous to greasy |
| **STREAK** | White |
| **SPECIFIC GRAVITY** | 6.1 |
| **TRANSPARENCY** | Transparent to translucent |

**PROSPECTING**
*This prospector north of Abha, Saudi Arabia, uses a panning device called a "goldhound" to pan for scheelite.*

*Bipyramidal scheelite crystals*

*Magnetite groundmass*

## USES OF TUNGSTEN

Tungsten is one of the most crucial metals used in industry. Tungsten-steel is used for rocket nozzles and other very high-temperature applications. Tungsten carbide is used in drill bits, dies, and tools for shearing metal. Pure tungsten is used in electric-light filaments. Cobalt-chromium-tungsten alloys are used for the hard-facing of wear-resistant valves, bearings, propeller shafts, and cutting tools.

**ROCKET NOZZLE**
*The nozzle of the Saturn V rocket that launched the Apollo 11 in 1969 was made of tungsten-steel.*

**SCHEELITE WITH MAGNETITE**
*This group of orange-yellow scheelite crystals with magnetite clearly shows a tetragonal bipyramidal crystal habit.*

---

# WULFENITE

**TOMBSTONE**
*Arizona, USA, is a key locality for fine wulfenite crystals.*

THE SECOND MOST COMMON molybdenum mineral after molybdenite, wulfenite "lead molybdate" is a minor source of molybdenum. It ordinarily forms as thin, square plates, or square, bevelled, tabular crystals, but can also be massive or granular. Its colour can be yellow, orange, red, grey, or brown. Tungsten substitutes for the molybdenum to varying degrees, although in most specimens it is present only in trace amounts. Wulfenite was named in 1845 for F.X. Wülfen (1728–1805), an Austro-Hungarian mineralogist. It is a secondary mineral, formed in the oxidized zone of lead and molybdenum deposits, and occurs with minerals such as cerussite, vanadinite, and pyromorphite. It is relatively widespread, and is often found in superb crystals, occasionally up to 10cm (4in) on an edge. Excellent crystals come from Mexico, Slovenia, Zambia, China, and Arizona, USA.

## PROPERTIES

| | |
|---|---|
| **GROUP** | Molybdates |
| **CRYSTAL SYSTEM** | Tetragonal |
| **COMPOSITION** | $PbMoO_4$ |
| **COLOUR** | Yellow, orange, red |
| **FORM/HABIT** | Square tabular, prismatic |
| **HARDNESS** | $2\frac{1}{2}$–3 |
| **CLEAVAGE** | Distinct |
| **FRACTURE** | Subconchoidal to uneven |
| **LUSTRE** | Subadamantine to greasy |
| **STREAK** | White |
| **SPECIFIC GRAVITY** | 6.5–7.0 |
| **TRANSPARENCY** | Transparent to translucent |

**SQUARE CRYSTALS**
*The thin, tabular, square crystals visible in this specimen are typical of wulfenite.*

*Square, platy crystal*

*Iron oxide groundmass*

*Greasy lustre*

*Tabular crystal*

*Greasy lustre*

*Tabular wulfenite crystal*

**RED CLOUD WULFENITE**
*The Red Cloud Mine in southern Arizona, USA, has produced spectacular wulfenite specimens.*

**YELLOW WULFENITE**
*The wulfenite crystals on this groundmass composed principally of iron oxides show classic square, platy development.*

217

# SILICATES

THE MOST IMPORTANT chemical group of minerals, silicates make up about 25 per cent of all known minerals and 40 per cent of the most common ones. As well as being a major mineral component of the Earth, they are also significant constituents of lunar samples and meteorites. Quartz, feldspar, and mica, the predominant components of granite, are silicates. All silicates contain silicon and oxygen. Silicon is a light, shiny metal that looks like pencil lead; oxygen is a colourless, odourless gas. In the silicates, these two elements combine to form structural tetrahedra, each with a silicon atom in the centre and oxygen atoms at the corners. Silicate tetrahedra may exist as discrete, independent units and connect with only other silicate tetrahedra (as in quartz), or with other chemical elements such as iron, magnesium, aluminium, and so on. Tetrahedra may also share their oxygen atoms at corners, edges, or, more rarely, faces, creating variations in structure as a result. The different linkages also create voids of varying sizes, which are

**QUARTZ OUTCROP**
*After silica, quartz is the most abundant mineral in the Earth's crust. The bulk of quartz is massive, as in this outcrop in the Shining Rock Wilderness, North Carolina, USA.*

occupied by the ions of various metals according to the size and atomic coordination of the void. Silicates are divided into six main groups according to the structural configurations that result from the different ways in which tetrahedra and other elements are linked. The inosilicate group is sub-divided into single- and double-chain inosilicates. Within these main groups there are further sub-divisions, such as the feldspars and zeolites.

**THE SIX SILICATE GROUPS**
*These illustrations show the basic arrangement of the silica tetrahedra in the six silicate groups. Both single- and double-chain inosilicates are shown.*

No shared corners

Two shared corners to form rings

Two shared corners to form chains

Three shared corners

All four corners shared

**NESOSILICATE**
One shared corner

**SOROSILICATE**

**CYCLOSILICATE**

**INOSILICATE (SINGLE-CHAIN)**
Two or three corners shared to form double chains

**INOSILICATE (DOUBLE-CHAIN)**

**PHYLLOSILICATE**

**TECTOSILICATE**

# TECTOSILICATES

THE TECTOSILICATE STRUCTURE is a three-dimensional network of silica tetrahedra, with each tetrahedral unit sharing all of its four oxygen atoms. Its name comes from the Greek *tecton*, meaning "builder". Sometimes referred to as framework silicates, the tectosilicates are the largest single group of silicates, and include several rock-forming minerals. An important factor in the size of the group is the ability of aluminium to substitute for silicon in the tetrahedron. Aluminium follows only oxygen and silicon in volume in the Earth's crust. Its ability to form tetrahedral groups of a similar size to silicate tetrahedra allows these aluminium groups to be easily incorporated in tectosilicate structures. Aluminium can also occupy other sites

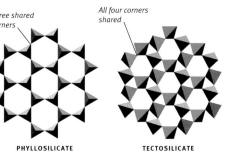

$SiO_4$ tetrahedron

**THREE-FOLD HELIX**

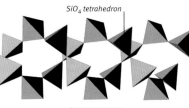

$SiO_4$ tetrahedron

**ATOMIC STRUCTURE OF QUARTZ**
*The silica tetrahedra that form the mineral quartz are arranged in the form of either a three-fold or a six-fold helix. This gives quartz a trigonal symmetry (see p.100).*

**SIX-FOLD HELIX**

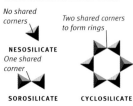

$SiO_4$ tetrahedron

**TRIGONAL SYMMETRY**

within the structure, where it can substitute for elements such as magnesium, ferrous iron, and titanium. The result is a large number of minerals that are produced in great abundance in the crust. They all share an $SiO_4$ unit in their chemical formulae.

# ROCK CRYSTAL

ROCK CRYSTAL IS A COLOURLESS, transparent variety of quartz. Vessels of all kinds and spheres have been carved from large crystals since ancient times, and the name rock crystal emerged in the late Middle Ages to differentiate it from the newly perfected colourless glass, to which the name crystal or crystal glass had become attached. Water-clear crystals were known to the ancient Greeks as *krystalos* – hence the name crystal.

Quartz is silicon dioxide, the third most common mineral in the Earth's crust after ice and feldspar. The name quartz comes from Old German and first appears in the writings of Georgius Agricola in 1530, although its origin is uncertain. Quartz comes essentially in two forms: crystalline or fully crystalline; and cryptocrystalline, formed of microscopic crystalline particles (see p.102). Crystalline quartz is usually colourless and transparent (rock crystal) or white and translucent (milky quartz, see p.220) but it can occur also in many coloured varieties (see pp.220–23). Crystallized impurities occur in some crystalline quartz varieties, such as the hair-like inclusions of rutile (see p.163), needles (rutilated

**MINAS GERAIS, BRAZIL**
Fine examples of rock crystals are among many minerals found in this Brazilian state.

**CRYSTAL LIP PIECE**
This gold and rock crystal lip piece was awarded by the Aztec Emperor Moctezuma to his bodyguard in the 16th century.

**DUCK BOWL**
Carved in the shape of a duck, this rock crystal bowl was made in Athens c.1200BC.

quartz, see p.225), or green, moss-like clumps of chlorite. Quartz occurs in nearly all silica-rich metamorphic, sedimentary, and igneous rocks. The optical properties of rock crystal led to its extensive use in lenses and prisms, and as an inexpensive gemstone. Initially oscillators for electronic components were made from natural rock crystal, but synthetic rock crystal has replaced it. Today, much rock crystal is mined for New Age crystal healing.

**OSCILLATOR**
Quartz is used as oscillators in electronics.

**POLISHED BEAD**
Quartz is easy to shape and, even with primitive equipment, lapidaries can produce beautiful items like this fluted bead.

**CUSHION**
Complex faceting techniques are often easy to use on rock crystal, producing glittering yet inexpensive results.

**ROUND BRILLIANT**
Rhinestones were originally cut from pebbles of quartz that were found in the Rhine river, Germany.

## PROPERTIES

**GROUP** Silicates – tectosilicates
**CRYSTAL SYSTEM** Hexagonal/trigonal
**COMPOSITION** $SiO_2$
**COLOUR** Colourless
**FORM/HABIT** Prismatic
**HARDNESS** 7
**CLEAVAGE** None
**FRACTURE** Conchoidal
**LUSTRE** Vitreous
**STREAK** White
**SPECIFIC GRAVITY** 2.7
**TRANSPARENCY** Transparent
**R.I.** 1.54–1.55

**PRISMATIC CRYSTALS**
This group of long prismatic crystals is from Dauphiné, France, which is a classic locality that gives its name to a type of twinned quartz crystal.

*Rhombohedral termination*

*Agate core*

*Drusy coating*

**CRYSTALS ON AGATE**

*Crystal cluster*

**CRYSTAL CLUSTER**

*Twin plane*

*Two crystals*

**TWINNED CRYSTAL**

*Striations on prism face*

*Prism face*

**FORMS OF CRYSTAL**
Rock crystal is found in a number of different coatings, clusters, and twinned forms.

## ROCK CRYSTAL LORE

Rock crystal is commonly regarded in shamanistic practice as a "light-stone", an instrument of clairvoyance representing a level intermediate between the visible and the invisible. Rock crystal was traditionally used by Australian Aborigines and the Prairie Indians of North America as a talisman and to produce visions. The Navajo believed it to be rock crystal that first caused the Sun to cast its light upon the world. In ancient times it was believed that rock crystal was ice that had frozen too hard to melt.

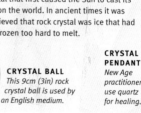

**CRYSTAL BALL**
This 9cm (3in) rock crystal ball is used by an English medium.

**CRYSTAL PENDANT**
New Age practitioners use quartz for healing.

# SMOKY QUARTZ

SMOKY QUARTZ IS THE LIGHT BROWN to nearly black variety of crystalline quartz. Black crystals are properly called morion. There is no clear colour boundary between colourless and smoky quartz. It is relatively abundant, and as such it is worth less than either amethyst or citrine. In some countries, faceted smoky quartz is sold as smoky topaz quartz, in an effort to associate it with the more valuable topaz and thus increase its saleability and its price. Smoky quartz on sale is often made by irradiating rock crystal (typically that from Arkansas, USA) to turn it dark brown.

Very dark, natural smoky quartz may be heated to give it a lighter, more attractive hue, and may turn it yellow so that it can be sold as the more valuable citrine (see opposite). Large crystals of smoky quartz are commonly found in pegmatites. Unusually fine specimens come from the Swiss Alps and, in the USA, Pike's Peak, Colorado, and Alexander and Lincoln Counties, North Carolina.

**PIKE'S PEAK**
*This area of Colorado, USA, produces superb crystals of smoky quartz.*

**INTAGLIO SEAL**
*This portrait of an ancient Roman has been carved in smoky quartz and set in a polished oval of obsidian.*

*Vitreous lustre*

## PROPERTIES

**GROUP** Silicates – tectosilicates
**CRYSTAL SYSTEM** Hexagonal/trigonal
**COMPOSITION** $SiO_2$
**COLOUR** Light brown to black
**FORM/HABIT** Prismatic
**HARDNESS** 7
**CLEAVAGE** None
**FRACTURE** Conchoidal
**LUSTRE** Vitreous
**STREAK** White
**SPECIFIC GRAVITY** 2.7
**TRANSPARENCY** Translucent to nearly opaque
**R.I.** 1.54–1.55

**SMOKY QUARTZ CRYSTALS**
*This spectacular double-terminated crystal of smoky quartz is on a groundmass of milky quartz.*

*Pyramidal termination*

*Termination face*

*Polished face*

**DIFFERENT HABITS**
*Smoky quartz is often found as single, prismatic crystals, and in transparent, gem-quality, water-rounded cobbles.*

**SINGLE CRYSTAL**  **CAIRNGORM COBBLE**

*Prism face*

*Milky quartz groundmass*

**BRILLIANT**
*In some countries, faceted smoky quartz is sold as smoky topaz quartz.*

**MIXED**
*Smoky quartz from the Cairngorms, Scotland, is known as cairngorm.*

# MILKY QUARTZ

THE WHITE TO GREYISH-WHITE, translucent to nearly opaque variety of silicon dioxide, milky quartz forms hexagonal prismatic crystals that are identical to rock crystal in all respects apart from colour. Both milky and transparent areas can occur within the same crystal. Crystals of milky quartz and rock crystal often occur in the same deposit. The milkiness has been attributed to the presence of minute gas bubbles trapped inside the crystal. It is by far the most common variety of quartz. White veins of massive milky quartz may be seen penetrating a wide number of rock types, and in quartz mining, milky crystals may exceed those of rock crystal in quantity.

Large quantities of milky crystals are recovered in major quartz mining areas such as Hot Springs, Arkansas, USA, and Minas Gerais, Brazil.

**MINAS GERAIS**
*Milky quartz is abundant in this part of Brazil.*

## PROPERTIES

**GROUP** Silicates – tectosilicates
**CRYSTAL SYSTEM** Hexagonal/trigonal
**COMPOSITION** $SiO_2$
**COLOUR** White to greyish-white
**FORM/HABIT** Prismatic
**HARDNESS** 7
**CLEAVAGE** None
**FRACTURE** Conchoidal
**LUSTRE** Vitreous
**STREAK** White
**SPECIFIC GRAVITY** 2.7
**TRANSPARENCY** Translucent to nearly opaque
**R.I.** 1.5

**MILKY QUARTZ**
*This single, prismatic crystal from Cornwall, England, rises from a group of smaller crystals.*

*Vitreous lustre*

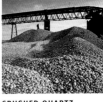

**CRUSHED QUARTZ**
*Milky quartz is often crushed for use as a ground-cover or ornamental stone.*

*Vitreous lustre*

*Termination face*

**DOUBLE-TERMINATED CRYSTAL**

**OVAL CUSHION**
*Milky quartz can be transparent enough to facet.*

# ROSE QUARTZ

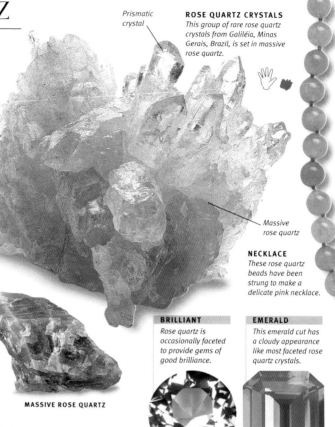

Prismatic crystal

**ROSE QUARTZ CRYSTALS**
This group of rare rose quartz crystals from Galiléia, Minas Gerais, Brazil, is set in massive rose quartz.

**MADAGASCAR**
This island is one of the world's leading producers of rose quartz.

THE TRANSLUCENT or transparent pink variety of crystalline quartz, known as rose quartz, is rarely found in crystals, and when these occur they seldom exceed 1cm (½in) in length. It is far more commonly found as a massive aggregate. Its coloration has been attributed to traces of titanium, while its milky appearance may be due to tiny, needle-like inclusions of rutile. When rose quartz from certain localities – such as Madagascar – is cut *en cabochon* with proper orientation, it shows asterism (a star-shaped figure) like that found in sapphires, although the effect is neither as sharp nor as intense. Rose quartz is generally found in pegmatites, sometimes in masses weighing hundreds of kilograms. Significant occurrences are in Madagascar, Brazil, Sweden, Namibia, Russia, Spain, Scotland, and in California and Maine, USA.

Rose quartz has been carved since ancient times, and in modern times it has been used by "crystal healers", who attribute the properties of unconditional love and emotional healing to it if it is placed against the skin.

## PROPERTIES

**GROUP** Silicates – tectosilicates

**CRYSTAL SYSTEM** Hexagonal/trigonal

**COMPOSITION** $SiO_2$

**COLOUR** Pink, rose

**FORM/HABIT** Massive

**HARDNESS** 7

**CLEAVAGE** None

**FRACTURE** Conchoidal

**LUSTRE** Vitreous

**STREAK** White

**SPECIFIC GRAVITY** 2.7

**TRANSPARENCY** Translucent to nearly opaque

**R.I.** 1.54–1.55

Massive rose quartz

**NECKLACE**
These rose quartz beads have been strung to make a delicate pink necklace.

**BRILLIANT**
Rose quartz is occasionally faceted to provide gems of good brilliance.

**EMERALD**
This emerald cut has a cloudy appearance like most faceted rose quartz crystals.

**MASSIVE ROSE QUARTZ**

---

# CITRINE

**ISLE OF ARRAN**
Citrine is found on this island off the west of Scotland.

CITRINE IS YELLOW TO BROWNISH quartz (silicon dioxide) and resembles yellow topaz. It is coloured by hydrous iron oxide, and is found in the same hexagonal crystals as the other varieties of crystalline quartz. Natural citrine is much less common than amethyst or smoky quartz, both of which can be heat treated to turn their colour into that of citrine. Most citrine that is available is in fact heat-treated amethyst, although heat-treated smoky quartz comes from some locations. As with smoky quartz, citrine is often marketed under names that confuse it with topaz, in order to inflate its price. It is easily distinguished from topaz by its inferior hardness.

Citrine occurs principally in localities that produce amethyst, and it is sometimes found as a zone of citrine in amethyst, when it is known as ametrine. Gem-quality citrine is found on the Isle of Arran, Scotland; in the Ural Mountains of Russia; near Hyderabad, India; in Dauphiné, France; in Minas Gerais, Brazil; in the Salamanca Province of Spain; and in North Carolina, USA.

**BROOCH**
A faceted citrine forms the centre of this gold brooch.

**HEAT-TREATED AMETHYST**

**CITRINE CRYSTAL**
This gem-quality single crystal of citrine shows some slight pitting of its faces and rounding of its edges due to erosion in a stream.

Prism face

Rhombohedral face

Vitreous lustre

## PROPERTIES

**GROUP** Silicates – tectosilicates

**CRYSTAL SYSTEM** Hexagonal/trigonal

**COMPOSITION** $SiO_2$

**COLOUR** Yellow, yellow-brown

**FORM/HABIT** Prismatic

**HARDNESS** 7

**CLEAVAGE** None

**FRACTURE** Conchoidal

**LUSTRE** Vitreous

**STREAK** Colourless

**SPECIFIC GRAVITY** 2.7

**TRANSPARENCY** Translucent to nearly opaque

**R.I.** 1.54–1.55

**BRILLIANT**
This faceted example of citrine shows a rich, yellow-brown colour.

**OVAL MIXED**
Some citrine, such as this faceted example, is pale yellow in colour.

**COLORADO, USA**
*This mineral-rich state is one of several North American localities for amethyst.*

**POLISHED PEBBLE**
*Polished and tumbled amethyst stones have an attractive mottled appearance despite their poor quality.*

**CRYSTAL SLICE**
*This slice shows how the colouring material of amethyst is attracted to certain crystal faces and not to others (known as preferential absorption).*

**OVAL MIXED**
*Faceted as an oval mixed cut, this polished amethyst stone is of a typical purplish-violet colour.*

**RECTANGULAR STEP**
*This polished rectangular step cut features a band of amethyst in the centre bordered on each side with clear quartz.*

**PRISMATIC CRYSTAL**
*This tall prismatic crystal from La Garita, Colorado, USA, is typical of those worn uncut as pendants. It is shown actual size, and has a slight red tinge to its purple colour due to traces of manganese.*

# AMETHYST

A VARIETY OF VITREOUS QUARTZ with purple, violet, or red-purple coloration, amethyst derives its name from the ancient Greek *amethustos*, meaning literally "not drunk" as it was believed to guard against drunkenness (see panel, opposite). Traditionally associated with purity and piety, amethyst has also always been favoured by royalty as purple is considered a regal hue.

## LOCATION OF AMETHYST

Found in most countries where granitic rocks are exposed, amethyst occurs in alluvial deposits and geodes. Its coloration is principally due to traces of iron, and it is sometimes colour-zoned due to twinning or preferential absorption on the rhombohedral faces. Major commercial sources of amethyst are Brazil, where it occurs in geodes that are frequently human-sized; Uruguay, Siberia, and North America. Crystals from Brazilian and Uruguayan deposits are most often found as radiating masses, with individual crystals appearing as pyramids. Lower grade Brazilian and Uruguayan amethyst is frequently turned into citrine (see p.221) by heat treatment, which changes its colour. Where both amethyst and citrine occur naturally in the same stone, the name ametrine is sometimes used.

## TRADITIONAL GEMSTONE

Amethyst has a long history as a gemstone. In the ancient civilizations of Mesopotamia, amethyst was highly valued and was used to create cylinder seals, engraved with a religious design and the owner's name. The engraving was transferred to legal documents in the form of clay tablets by rolling the cylinder over the tablets. The ancient Egyptians also valued amethyst, using it in much of their jewellery. Egyptian amethyst came principally from Nubia, once a province of Egypt.

In modern times, amethyst is both faceted and cut *en cabochon*, and it has widespread use as a carving material. Its most valued shades are a deep, rich purple, and a deep purple with a reddish tinge. Uncut, prismatic crystals, such as those mined in Querétaro, Mexico, are frequently worn as jewellery.

**VICTORIAN EARRINGS**
*This delicate pair of drop earrings features rose-cut amethyst crystals and pearls.*

**CONTEMPORARY NECKLACE**
*Fragments of polished amethyst crystals are linked together on three individual silver chains to create a pretty massed effect.*

*Translucent crystals*

**HEAT-TREATED AMETHYST**
*When heated, some amethyst crystals change colour to yellow-brown. Many crystals sold as citrine (see p.221) are actually heat-treated amethyst.*

*Major rhombohedral face*

## PROPERTIES

**GROUP** Silicates ~ tectosilicates
**CRYSTAL SYSTEM** Hexagonal/trigonal
**COMPOSITION** $SiO_2$
**COLOUR** Violet
**FORM/HABIT** Normally pyramidal terminations, sometimes prismatic
**HARDNESS** 7
**CLEAVAGE** None
**FRACTURE** Conchoidal
**LUSTRE** Vitreous
**STREAK** White
**SPECIFIC GRAVITY** 2.7
**TRANSPARENCY** Opaque to translucent
**R.I.** 1.54–1.55

## CYLINDER SEAL
*This Babylonian amethyst cylinder seal dates from the 18th century BC. It bears the inscription: "Ishtar Lamassi, daughter of Lushtammar, slave girl of Ninisina".*

### AMETHYST TIE PIN
*An octagonal step-cut amethyst crystal adorns this handsome late-19th-century gold tie pin.*

"Amethyst dissipates **evil thoughts** and quickens the **intelligence**."

**LEONARDO DA VINCI**

# GEMSTONE OF VIRTUE

The ancient Greeks believed that drinking wine from a cup of amethyst would make them immune to intoxication. This belief is founded upon the origin of amethyst described in Greek mythology.

According to this myth, amethyst was created by Dionysus, god of fruitfulness and wine. Angered by the purity of a young woman named Amethyst, who preferred to pay homage to the goddess Diana, he ordered two voracious tigers to devour her. Diana came to her rescue, transforming Amethyst into white quartz. Overcome with remorse, Dionysus shed tears into his goblet of red wine. Some of the wine spilled, running onto the white stone, which absorbed its colour, creating the stone amethyst.

## DOUBLE CUP
*This hemispherical amethyst cup was cut from a 1.5kg (3¹/₂lb) crystal of Brazilian amethyst.*

## SPIRITUAL ASSOCIATIONS
In the early Christian church, amethyst was also believed to have sobering properties. The mineral was adopted as a symbol of the high spiritual state its bishops must attain, and an amethyst ring soon became part of a bishop's regalia. Even today the highest grade of amethyst is referred to by gem-cutters as "Bishop's Grade".

In India, amethyst is associated with the crown chakra, or energy centre, in the body. Meditation upon it is said to lead to mystical union with the universal consciousness. In Tibet, amethyst is considered sacred to Buddha and is often used to make rosaries.

### CLERIC'S RING
*This 18th-century Austrian clerical ring features a faceted step-cut amethyst surrounded by diamonds.*

### AARON'S BREASTPLATE
*An amethyst adorns the breastplate of Aaron, high priest of the Hebrews. It represents the prophet Math, who was filled with desire to please God.*

*Vitreous lustre*

*Minor rhombohedral face*

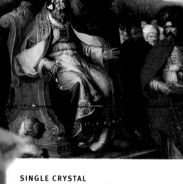

## SINGLE CRYSTAL
*This single amethyst crystal is about 10cm (4in) long but crystals exceeding 15cm (6in) in length are sometimes found.*

*Rich purple colouring*

## MASS OF CRYSTALS
*In this mass of pyramid-shaped amethyst crystals, both minor and major rhombohedral faces can be seen. The colour of pyramidal crystals darkens towards the tips.*

*Colourless quartz*

*Agate base*

223

# AVENTURINE

**MINAS GERAIS**
*Aventurine is found in this area of Brazil.*

AVENTURINE IS THE NAME given to a variety of quartz that has a spangled appearance due to sparkling internal reflections from uniformly oriented minute inclusions of other minerals. Green aventurine is coloured by green fuchsite mica; brown aventurine is coloured by pyrite; and reddish-brown aventurine is due to hematite. Other inclusions can colour the mineral orange, bluish-white, yellow, or bluish-green. It is always massive.

Aventurine is found in the Ural Mountains of Russia, in the state of Tamil Nadu, India; at numerous localities in Minas Gerais, Brazil; and in Rutland, Vermont, USA. Aventurine quartz is used for jewellery where it is cut *en cabochon*, for vases and bowls, and for other ornamental objects.

*Granular texture*

*Massive form*

## PROPERTIES

**GROUP** Silicates – tectosilicates
**CRYSTAL SYSTEM** Hexagonal/trigonal
**COMPOSITION** $SiO_2$
**COLOUR** Various
**FORM/HABIT** Massive
**HARDNESS** 7
**CLEAVAGE** None
**FRACTURE** Conchoidal
**LUSTRE** Vitreous
**STREAK** White
**SPECIFIC GRAVITY** 2.7
**TRANSPARENCY** Translucent

*No cleavage*

*Vitreous lustre*

**GRANULAR AVENTURINE**
*This specimen of green aventurine shows its granular texture.*

**CABOCHON**
*A cabochon of yellow-brown aventurine shows the sparkle of its inclusions.*

**CABOCHON**
*This green cabochon has been faceted with flat faces to show off its sparkle.*

# CAT'S-EYE QUARTZ

**SRI LANKA**
*The gem gravels of Sri Lanka are a rich source of cat's-eye quartz.*

THE TERM "CAT'S EYE" is applied to two different minerals: one a variety of chrysoberyl (see p.159), the other of quartz. When cut *en cabochon*, both show a single white line across the stone. Cat's-eye quartz is sometimes called Occidental cat's eye to differentiate it from the more valuable Oriental cat's eye – chrysoberyl. The two may be distinguished by their specific gravities: chrysoberyl is much denser. Cat's-eye quartz owes its chatoyancy (cat's-eye effect) and greyish-green or greenish colour to parallel fibres of asbestos in the quartz; a more reddish or golden colour derives from minute fibres of rutile. The main source of cat's-eye quartz is the gem gravels of Sri Lanka. It is also found in India and Australia, while inferior green stones are obtained from Bavaria, Germany.

*Water-worn surface*

*Parallel fibres*

## PROPERTIES

**GROUP** Silicates – tectosilicates
**CRYSTAL SYSTEM** Hexagonal/trigonal
**COMPOSITION** $SiO_2$
**COLOUR** Greenish
**FORM/HABIT** Massive
**HARDNESS** 7
**CLEAVAGE** None
**FRACTURE** Conchoidal
**LUSTRE** Vitreous
**STREAK** White
**SPECIFIC GRAVITY** 2.7
**TRANSPARENCY** Translucent

**FIBROUS CAT'S-EYE QUARTZ**
*In this specimen of unpolished cat's eye, no chatoyancy can be seen. The "eye" emerges only on cutting, requiring considerable expertise when selecting specimens.*

**GEM PANNING**
*In the gem fields of Sri Lanka and Myanmar, gems are usually recovered by panning stream gravels.*

*Vitreous lustre*

**CABOCHON**
*This cabochon of greyish-green cat's-eye quartz shows a strong "eye".*

**CABOCHON**
*A fine, translucent yellow-grey cabochon also shows the "eye" effect well.*

**CABOCHON**
*This round cabochon of cat's eye has a weak "eye", but shows fibrous structure.*

# TIGER'S EYE

**WITTENOOM GORGE**
*This Western Australian gorge has tiger's eye.*

TIGER'S-EYE QUARTZ, ALSO SPELLED tiger-eye, is a semiprecious variety of quartz exhibiting chatoyancy, a vertical luminescent band like that of a cat's eye. Unlike cat's-eye quartz, tiger's eye is formed when parallel veins of crocidolite (blue asbestos) fibres are first altered to iron oxides and then replaced by silica. As a result it is more opaque, and has a rich yellow to brown colour. When cut *en cabochon*, the gem has a fine lustre. The major source of tiger's eye is Griquatown West, in South Africa. It is also found at Wittenoom Gorge, Western Australia. Hawk's eye is similar to tiger's eye except that the crocidolite is replaced by quartz without altering to iron oxide, and so it retains the grey-blue of the original asbestos.

## PROPERTIES

| | |
|---|---|
| **GROUP** | Silicates – tectosilicates |
| **CRYSTAL SYSTEM** | Hexagonal/trigonal |
| **COMPOSITION** | $SiO_2$ |
| **COLOUR** | Yellow-brown |
| **FORM/HABIT** | Fibrous |
| **HARDNESS** | 7 |
| **CLEAVAGE** | None |
| **FRACTURE** | Conchoidal |
| **LUSTRE** | Vitreous |
| **STREAK** | White |
| **SPECIFIC GRAVITY** | 2.7 |
| **TRANSPARENCY** | Translucent |

**HAWK'S EYE**
*Although similar to tiger's eye, hawk's eye retains the grey-blue of the original asbestos fibres.*

*Crocidolite fibres*

**FIBROUS TIGER'S EYE**
*In this piece of tiger's eye, the fibres are clearly visible. The richness of its colour is brought out by polishing.*

Ironstone

Altered crocidolite fibres

No cleavage

**NECKLACE**
Beads of tiger's eye have been made into this richly coloured necklace.

**POLISHED SECTION**
*This section cut through a piece of tiger's eye has been polished to show off its fibrous structure.*

**SPHERE**
*This sphere made of tiger's eye shows off the golden sheen from the fibres.*

---

# RUTILATED QUARTZ

**BAHIA, BRAZIL**
*The prime locality for rutilated quartz crystals is Novo Horizonte, Bahia, Brazil.*

THIS VARIETY OF QUARTZ takes its name from needles of rutile (see p.163) that are enclosed within it. It is generally transparent, with any amount of needles from a few to so many that the stone may be nearly opaque. The needles can occur as sprays, or they can be randomly oriented. The rutile is generally golden in colour, but it can be reddish to deep red, appearing black without intense light. The quartz can be found in well-formed crystals or in stream-rounded fragments. It is also known as Venus' hairstone. Black tourmaline, green actinolite, and green chlorite needles are also found encased in quartz in a similar manner to rutile needles.

Tourmaline needles

Rutile needles

**SNUFF BOTTLE**
*This finely carved Chinese snuff bottle has been made from tourmalated quartz.*

## PROPERTIES

| | |
|---|---|
| **GROUP** | Silicates – tectosilicates |
| **CRYSTAL SYSTEM** | Hexagonal/trigonal |
| **COMPOSITION** | $SiO_2$ |
| **COLOUR** | Colourless with gold or red needles |
| **FORM/HABIT** | Prismatic |
| **HARDNESS** | 7 |
| **CLEAVAGE** | None |
| **FRACTURE** | Conchoidal |
| **LUSTRE** | Vitreous |
| **STREAK** | White |
| **SPECIFIC GRAVITY** | 2.7 |
| **TRANSPARENCY** | Transparent |

*Hexagonal quartz prisms*

**RUTILATED CRYSTALS**
*Here, two fine crystals of rutilated quartz are held in a matrix.*

*Reddish-brown rutile needles*

**RUTILE NEEDLES**
*The faces of this irregular piece of rutilated quartz have been polished to show the rutile needles.*

Spray of rutile

Golden rutile

**PENDANT**
*This cabochon of rutilated quartz, domed on top and faceted on the bottom, is set in a silver and gold pendant mount.*

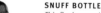

# PHANTOM QUARTZ

**MINAS GERAIS**
*This area of Brazil is one of many localities for phantom quartz.*

WHEN GAS BUBBLES or tiny crystals of other minerals accumulate on the termination of a quartz crystal during its growth, the colour of the crystal is subtly changed. This can happen at several stages, leaving a "shadow" or "phantom" of the termination at each point. Green phantoms usually result from chlorite, reddish-brown from various iron minerals, blue from riebeckite, and white from gas or liquid bubbles or as a result of etching. Phantom quartz is widespread. Blue phantoms come from São Paulo, Brazil; green phantoms from Hot Springs, Arkansas and Quartzsite, Arizona, USA; and reddish-brown phantoms from Minas Gerais, Brazil. White phantoms are found widely throughout Brazil, and also in Arkansas, USA.

*Polished face*

*"Phantoms"*

**PHANTOM QUARTZ**
*This quartz crystal has the faces polished to show the "phantoms" of earlier terminations within.*

### PROPERTIES

| | |
|---|---|
| **GROUP** | Silicates – tectosilicates |
| **CRYSTAL SYSTEM** | Hexagonal – trigonal |
| **COMPOSITION** | $SiO_2$ |
| **COLOUR** | Colourless, white, green, red-brown, or blue |
| **FORM/HABIT** | Prismatic |
| **HARDNESS** | 7 |
| **CLEAVAGE** | None |
| **FRACTURE** | Conchoidal |
| **LUSTRE** | Vitreous |
| **STREAK** | White |
| **SPECIFIC GRAVITY** | 2.7 |
| **TRANSPARENCY** | Transparent |

# CHALCEDONY

**QUARTZSITE, ARIZONA, USA**
*Chalcedony is found in this area of Arizona.*

A COMPACT VARIETY of microcrystalline quartz, chalcedony is composed of microscopic fibres. It can be mamillary, botryoidal, or stalactitic, and is found in veins, geodes, and concretions. It is white when pure, but much chalcedony contains trace elements or microscopic inclusions of other minerals, giving a range of colours. Many of these coloured chalcedonies have their own variety names (see pp.227–29). Chalcedony that shows distinct banding is called agate (see pp.230–31). Occurrences of all varieties of chalcedony are extremely numerous worldwide. Its name may derive from the ancient port of Khalkedon in Asia Minor (now Turkey), where there were deposits.

Chalcedony forms in cavities, cracks, and by replacement when low-temperature, silica-rich waters percolate through pre-existing rocks, in particular volcanic rocks. Chalcedony is relatively porous, and much chalcedony (in the form of agate, for example) on the commercial market has been dyed to enhance or artificially colour it.

**CAMEO**
*A white layer of chalcedony has been carved away to a yellow layer of agate below to create a relief in the shape of a woman's head with intricately carved hair.*

*Arborescent form*

*Waxy lustre*

**UNUSUAL HABIT**
*This chalcedony specimen has developed in an almost arborescent form, probably as a dripstone in a hollow.*

*Botryoidal habit*

**CHALCEDONY BOWL**
*This elegant Austrian bowl has been created from chalcedony and has intricate decorations on its rim, base, and handles.*

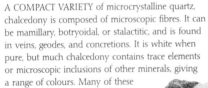

### PROPERTIES

| | |
|---|---|
| **GROUP** | Silicates – tectosilicates |
| **CRYSTAL SYSTEM** | Hexagonal – trigonal |
| **COMPOSITION** | $SiO_2$ |
| **COLOUR** | White |
| **FORM/HABIT** | Microcrystalline |
| **HARDNESS** | 7 |
| **CLEAVAGE** | None |
| **FRACTURE** | Uneven |
| **LUSTRE** | Waxy to dull |
| **STREAK** | White |
| **SPECIFIC GRAVITY** | 2.7 |
| **TRANSPARENCY** | Transparent to translucent |

**PINK CHALCEDONY**
*This form of botryoidal pink chalcedony from Quartzsite, Arizona, USA, is, along with some forms of selenite gypsum, referred to as a "desert rose".*

# CHRYSOPRASE

**QUEENSLAND**
*A relatively recent source of this mineral is Queensland, Australia.*

CHRYSOPRASE IS A TRANSLUCENT, apple-green variety of chalcedony. Its colour is derived from the presence of nickel. The colour of some chrysoprase may fade in sunlight, and lighter-coloured material may be confused with fine jade in cut stones. It was used by both the Greeks and Romans, and is still the most valued of the chalcedonies. Prase, a leek-green chalcedony, has a more sombre hue and is much rarer. Fine chrysoprase once came from mines in Poland and former Czechoslovakia, but since 1965 the best quality material has come from Queensland, Australia. Lesser amounts are found in the Ural Mountains of Russia, California, USA, and Brazil.

This mineral's name is derived from the Greek *chrysos* and *prase*, meaning "golden leek", and was probably applied originally to a lighter or more yellowish chrysoprase. In ancient times it was reputed to strengthen the eyesight and relieve internal pain.

*Waxy lustre*

*Massive habit*

**MASSIVE CHRYSOPRASE**
*Examples such as this massive specimen of apple-green chrysoprase have been used for ornamentation since prehistoric times.*

*Uneven fracture*

*Colour from chromium*

**PRASE CAMEO**
*This subtle cameo uses the dark tones of prase in a gold setting.*

## PROPERTIES

| | |
|---|---|
| **GROUP** | Silicates – tectosilicates |
| **CRYSTAL SYSTEM** | Hexagonal – trigonal |
| **COMPOSITION** | $SiO_2$ |
| **COLOUR** | Apple-green |
| **FORM/HABIT** | Massive or microcrystalline |
| **HARDNESS** | 7 |
| **CLEAVAGE** | None |
| **FRACTURE** | Uneven |
| **LUSTRE** | Waxy or dull |
| **STREAK** | White |
| **SPECIFIC GRAVITY** | 2.7 |
| **TRANSPARENCY** | Translucent |

# CARNELIAN

**PUNE, INDIA**
*India produces the best examples of carnelian.*

A BLOOD-RED to reddish-orange translucent variety of chalcedony, carnelian is also occasionally called cornelian. Its coloration is due to the presence of iron oxide, and it can be uniformly coloured or banded. Strongly banded material is known as carnelian agate. Scotland, Brazil, and Washington, USA, are among the localities that produce finer-quality carnelian. Freshly mined carnelian, especially Indian material, is often placed in the sun to change brown tints to red. Carnelian was once thought to still the blood and calm the temper. Conversely, it was also said to give the owner courage in battle, and help timid speakers to be eloquent.

**POLISHED STONE**
*This reddish-orange carnelian from India has been cut and polished.*

**MASSIVE CARNELIAN**
*This specimen of carnelian has been collected from the Narmada River in northern India.*

**EGYPTIAN RING**
*Carved from a single piece of carnelian, this Egyptian ring dates from 1400BC.*

**CARNELIAN VEIN**
*This vivid red vein of carnelian runs through a mass of rock crystal.*

**BROOCH**
*Converted from a pendant, this Arts and Crafts brooch features a carnelian cabochon as its centrepiece.*

*Carnelian vein*

*Massive habit*

*Vitreous lustre*

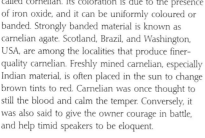

*Colour from iron oxide*

## PROPERTIES

| | |
|---|---|
| **GROUP** | Silicates – tectosilicates |
| **CRYSTAL SYSTEM** | Hexagonal – trigonal |
| **COMPOSITION** | $SiO_2$ |
| **COLOUR** | Red |
| **FORM/HABIT** | Massive |
| **HARDNESS** | 7 |
| **CLEAVAGE** | None |
| **FRACTURE** | Uneven |
| **LUSTRE** | Waxy to dull |
| **STREAK** | White |
| **SPECIFIC GRAVITY** | 2.7 |
| **TRANSPARENCY** | Translucent |

# ONYX

ONYX IS THE STRIPED, semiprecious variety of agate with white and black alternating bands. Relatively uncommon in nature, it can be produced artificially through dyeing pale, layered agate. The name comes from the Greek *onux*, meaning "nail" or "claw", referring to the mineral's colour, and it was used by the Romans for a variety of stones including alabaster, chalcedony, and what is now called onyx marble. The name onyx is properly applied only to the agate variety. Other varieties include carnelian onyx, with white and red bands, and sardonyx (see below), with white and brown bands. As with the other chalcedonies, onyx forms from the deposition of silica at low temperatures from silica-rich waters percolating through cracks and fissures in other rocks. Natural onyx comes from India and South America. Onyx is popular for carved cameos and intaglios because its layers can be cut to show a colour contrast.

**PUNE, INDIA**
*Fine natural onyx is found in this area of western India.*

## PROPERTIES

**GROUP** Silicates – tectosilicates
**CRYSTAL SYSTEM** Hexagonal/trigonal
**COMPOSITION** SiO$_2$
**COLOUR** Black and white
**FORM/HABIT** Cryptocrystalline
**HARDNESS** 7
**CLEAVAGE** None
**FRACTURE** Conchoidal
**LUSTRE** Vitreous
**STREAK** White
**SPECIFIC GRAVITY** 2.7
**TRANSPARENCY** Translucent to nearly opaque

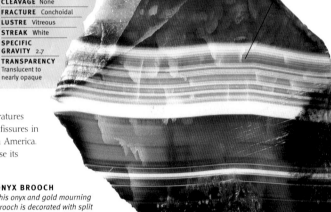

*Translucent slab*

**ALTERNATING BANDS**
*This polished slab of onyx displays its characteristic alternating bands of colour.*

*Parallel bands of black and white*

**ONYX SEAL**
*This striking Georgian copy of a Roman seal exploits the parallel layering of brown and white onyx.*

**ONYX CAMEO**
*The layers of contrasting colours in onyx make it an ideal material for carving cameos.*

**ONYX BROOCH**
*This onyx and gold mourning brooch is decorated with split pearls and has a locket back.*

# SARD

**SRI LANKA**
*Ratnapura in south-west Sri Lanka is a famous locality for sard.*

A TRANSLUCENT, light to dark brown chalcedony, sard takes its name from the Greek *Sardis*, the capital of ancient Lydia. Until the Middle Ages it shared the name "sardion" with carnelian. Bands of sard and white chalcedony are called sardonyx. Sard forms from the deposition of silica at low temperatures from silica-rich waters percolating through cracks and fissures in other rocks. One famous locality for sard is Ratnapura, Sri Lanka. Other sources are in India, Brazil, and Uruguay. It was used at Harappa, one of the oldest centres of the Indus civilization (c.2300–1500BC), by the Mycenaeans (1450–1100BC), and also by the Assyrians (1400–600BC).

**POLISHED SARDONYX**
*White and brown-red bands are characteristic of sardonyx.*

## SARDONYX JEWELLERY

Both sard and sardonyx have been used since ancient times for making cameos and intaglios. Wearing sard jewellery has always been considered fortunate. In the 4th century, Epiphanius, Bishop of Salamis, noted sard's medicinal value for wounds, and in the 11th century Bishop Marbodius remarked on it as a protector against incantation and sorcery.

**CAMEO**
*This classical cameo has been carved out of sardonyx.*

*Polished surface*

*Pitted, partly waterworn surface*

## PROPERTIES

**GROUP** Silicates – tectosilicates
**CRYSTAL SYSTEM** Hexagonal/trigonal
**COMPOSITION** SiO$_2$
**COLOUR** Light to dark brown
**FORM/HABIT** Cryptocrystalline
**HARDNESS** 7
**CLEAVAGE** None
**FRACTURE** Conchoidal
**LUSTRE** Vitreous
**STREAK** White
**SPECIFIC GRAVITY** 2.7
**TRANSPARENCY** Translucent to opaque
**R.I.** 1.5

**POLISHED INTERIOR**
*This piece of sard rough has been polished to show its internal brown coloration.*

**CABOCHON**
*Sard that is cut as a semi-precious gem is always cut en cabochon. This translucent stone has patches of brownish-red.*

**PEBBLE**
*This smooth, oval-shaped pebble of dark brown, semi-translucent sard has a slightly pitted waterworn surface.*

# BLOODSTONE

**BAHIA**
*Extensive deposits of bloodstone can still be found in this part of Brazil.*

ALSO KNOWN AS HELIOTROPE, bloodstone is a dark green variety of chalcedony coloured by traces of iron silicates and with patches of bright red jasper distributed throughout its mass. The name heliotrope is derived from the Greek *helio*, meaning "sun", and *trepein*, "turning". Both polished and unpolished stones show red spots on a dark green background, resembling drops of blood: hence the name bloodstone. The ancient source of bloodstone was the Kathiawar Peninsula of India, with modern sources in Brazil and Australia. Its occurrence is, as with the other chalcedonies, through deposition from low-temperature, silica-rich waters percolating through cracks and fissures in other rocks.

**BLOODSTONE BOWL**
*This silver gilt and bloodstone bowl and setting was made by the silver engraver Paul Storr in 1824.*

**SNUFFBOX**
*This rectangular Louis XV lacquer snuffbox is inlaid with bloodstone and gold.*

## PROPERTIES

| | |
|---|---|
| **GROUP** | Silicates – tectosilicates |
| **CRYSTAL SYSTEM** | Hexagonal/trigonal |
| **COMPOSITION** | $SiO_2$ |
| **COLOUR** | Dark green with red spots |
| **FORM/HABIT** | Cryptocrystalline |
| **HARDNESS** | 7 |
| **CLEAVAGE** | None |
| **FRACTURE** | Conchoidal |
| **LUSTRE** | Vitreous |
| **STREAK** | White |
| **SPECIFIC GRAVITY** | 2.7 |
| **TRANSPARENCY** | Translucent to opaque |
| **R.I.** | 1.53–1.54 |

## ANCIENT LORE

In the first century BC, bloodstone was – according to Greek natural philosopher Damigueron – a preserver of health, and offered protection against deception. During the Middle Ages in Europe it was used in sculptures representing flagellation and martyrdom. It was believed to prevent nosebleeds, remove anger and discord, and to be a remedy for haemorrhage and inflammatory diseases. It was said to be most effective if first dipped in cold water.

**BLOODSTONE RELIQUARY**
*This late 12th-century wooden French reliquary of St. Stephen is decorated with silver gilt and bloodstones.*

**POLISHED SLAB**
*Red spots and veins of iron silicates are scattered throughout this polished slab of dark green bloodstone.*

*Conchoidal fracture*

*Jasper spots*

**RED PATCHES**
*One face of this bloodstone has been polished to show the red spots of jasper to good effect.*

# JASPER

**ARIZONA, USA**
*The Petrified Forest area of Arizona preserves some fine jasper specimens.*

AN OPAQUE, FINE-GRAINED or dense variety of cryptocrystalline quartz, jasper is related to chert (see pp.70–71). Its name is from the Greek *iaspis*, which is probably of Semitic origin. Jasper is a chalcedony incorporating various amounts of other materials that give it both its opacity and colour. Brick-red to brownish-red jasper contains hematite; clay gives rise to a yellowish-white or grey, and goethite produces brown or yellow. Jasper is formed through deposition from low-temperature, silica-rich waters percolating through cracks and fissures in other rocks, incorporating a variety of materials in the process. It is found worldwide wherever crypto-crystalline quartz occurs. Beautiful examples of jasperized fossil wood are found in Arizona, USA (see pp.334–35).

## PROPERTIES

| | |
|---|---|
| **GROUP** | Silicates – tectosilicates |
| **CRYSTAL SYSTEM** | Hexagonal/trigonal |
| **COMPOSITION** | $SiO_2$ |
| **COLOUR** | Red, yellow, brown, numerous others |
| **FORM/HABIT** | Cryptocrystalline |
| **HARDNESS** | 7 |
| **CLEAVAGE** | None |
| **FRACTURE** | Conchoidal |
| **LUSTRE** | Vitreous |
| **STREAK** | White |
| **SPECIFIC GRAVITY** | 2.7 |
| **TRANSPARENCY** | Opaque |

**JASPER INTAGLIO**
*This early 3rd-century jasper intaglio from northern England features Silvanus-Cocidius, a Romano-Celtic hunter deity, with a hare and hound.*

## ANCIENT LORE

Jasper has been used for jewellery and ornamentation since Palaeolithic times. The Babylonians believed that jasper influenced women's diseases, and was a symbol of childbirth. The 11th-century Bishop of Rouen, Marbodius, stated that a jasper "placed on the belly of a woman in childbirth relieves her pangs". This comes from a time when all stones were believed to be alive and have a sex, jasper being a female stone.

**PHARAOH**
*This Egyptian pharaoh in the guise of a falcon (1555–1080BC), is made of jasper.*

*Colour variation from incorporated materials*

*White quartz vein*

**COLOUR VARIATION**
*Hematite colours this example of jasper brownish-red. Threads of white quartz veins criss-cross the specimen.*

229

# AGATE

A COMMON, SEMIPRECIOUS type of chalcedony, agate is the compact, microcrystalline variety of quartz. In general, its physical properties are those of quartz. Most agates form in cavities in ancient lavas or other extrusive igneous rocks. They are characterized for the most part by colour bands in a concentric form, and less often by moss-like inclusions (moss agate). The characteristic bands usually follow the outline of the cavity in which the mineral has formed. The band colours are determined by the differing impurities present, and occur in shades of white, yellow, grey, pale blue, brown, pink, red, or black. Much of the sliced agate offered on the market in particularly bright colours is dyed or stained to enhance the natural colour, an easy process because of its porous nature.

## TYPES OF BANDED AGATE

Other names often precede the word agate, and indicate either its visual characteristics or its place of origin. Fortification agate is a type of banded agate with angularly arranged bands that resemble an aerial view of an ancient fortress. There are several types of fortification agate. Brazilian agate is a fortification agate with banding in angled concentric circles like the fortifications of a castle. Botswana agate from Africa has beautiful dark and light banding. Mexican lace – sometimes called "crazy-lace" – is a multi-coloured fortification agate with highly convoluted layering. Blue lace agate from South Africa – one of the

most common kinds for sale today – is a delicate light blue with a fine interlayering of colourless agate. Fire agate has inclusions of reddish to brown hematite that give an internal iridescence to polished stones.

Banded agate is produced by a series of processes that take place in cavities in a solidified lava. As the lava cools, steam and other gases form bubbles in the liquid rock that are preserved as the lava hardens, forming cavities. Long after the rock has solidified, silica-bearing water solutions penetrate into a bubble and coagulate to

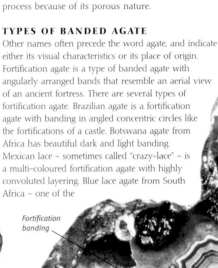

*"Moss"*

**MOSS AGATE**

*Fortification banding*

**FORTIFICATION AGATE**

**AGATE VARIETIES**
*Above are examples of two agate varieties. The fortification agate is a historic specimen from the Idar-Oberstein deposit (see panel, opposite).*

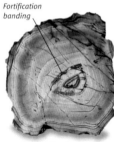

## PROPERTIES

| | |
|---|---|
| **GROUP** Silicates – tectosilicates | |
| **CRYSTAL SYSTEM** Hexagonal/trigonal | |
| **COMPOSITION** $SiO_2$ | |
| **COLOUR** Colourless, white, yellow, grey, brown, blue, or red | |
| **FORM/HABIT** Cryptocrystalline | |
| **HARDNESS** 7 | |
| **CLEAVAGE** None | |
| **FRACTURE** Conchoidal | |
| **LUSTRE** Vitreous to waxy | |
| **STREAK** White | |
| **SPECIFIC GRAVITY** 2.7 | |
| **TRANSPARENCY** Translucent to opaque | |

**AGATE DIE**
*This Roman die is made of agate. The dots have been carved into the surface of the die.*

*Yellowish colour probably due to iron oxide (limonite)*

*Fortification banding*

*Fortification banding*

**FORTIFICATION AGATE**
*This sliced specimen of fortification agate shows the parallel banding that gives it its name. The variations in colour are due to different mineral inclusions picked up by the agate as the bands formed.*

form a layer of silica gel. This sequence is repeated until the hollow is filled. Some of the layers will have picked up traces of iron or other soluble material to give the bands their distinct colouring. Finally, the whole mass crystallizes, with the water being lost but the bands remaining undisturbed.

Agates have been worked since prehistoric times, and were among the world's first lapidary materials. Today, it is usually cut as thin slabs, or polished as ornaments, brooches, or pendants. When cut for jewellery, it is cut *en cabochon*.

## OTHER TYPES OF AGATE

Moss agate does not have banding. It is white or grey with brown, black, or green moss- or tree-like (dendritic) inclusions, suggestive of vegetable growth. These dendrites are not organic, but are inclusions of other minerals, most often iron or manganese oxides or chlorite. Moss agate with brown inclusions is sometimes called mocha stone. Indian moss agate often has green, moss-like dendrites. Sweetwater agate from the Sweetwater River area of Wyoming, USA, is characterized by fine, black dendrites. Moss agate forms by a variety of processes by which other minerals are included in the agate, during or after its formation. Agatized wood is fossilized wood

**MOSS AGATE SLICE**
*This beautiful moss agate, with a strong dendritic form, has been cut and polished.*

that has had its organic matter replaced by agate, in some cases, cell-by-cell. The Petrified Forest in Arizona, USA, with hundreds of acres of agatized wood in a multitude of colours, is one of nature's most spectacular sights (see pp.334–35). Bodies of agate that fill round hollows in old volcanic rocks in parallel straight bands, rather than in curved fortifications, are called "thunder eggs".

Agate is found worldwide, but Brazil, Botswana, South Africa, Mexico, Egypt, China, and Scotland are prolific sources and produce interesting varieties. Oregon, Washington, Idaho, and Montana, are particularly abundant sources in the USA.

**MOSS AGATE PIN TRAY**
*A thick slice of moss agate has been carved to form a shallow bowl, or pin tray.*

**AGATE BRACELET**
*This elegant silver-mounted Scots agate bracelet was crafted in 1865.*

Red colour probably due to iron oxide (hematite)

**POLISHED AGATE**
*These two examples show more variety in patterning. The geode has an agate lining and an amethyst centre. The slice is Brazilian fortification agate with simple oval banding.*

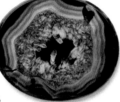

**POLISHED GEODE**

**SLICE OF BRAZILIAN AGATE**

## IDAR–OBERSTEIN

Idar-Oberstein is a city in the Rhineland of south-west Germany, straddling the river Nahe at its confluence with the Idarbach. Agates were abundant in the area for at least 700 years, and were collected and mined there until the 20th century. An agate-working industry was well established in the then separate towns of Idar and Oberstein by 1548, using the water power of the Nahe river to drive the cutting and polishing wheels made of local sandstone. The agate workings were eventually

**GEM-CUTTING SHOP**
*This photograph shows a cutter at work in a gem-cutting shop in the then separate town of Idar in about 1930.*

extended underground in huge galleries that can be visited today. From about 1900, the local supplies were exhausted, but were replaced with agates from Brazil and Uruguay. Today Idar-Oberstein is still a gemstone- and jewellery-making centre, and it has some notable museums of gems. It also hosts an annual gem and jewellery trade fair that showcases the latest trends in jewellery-making, particularly from this historic area.

**VISITING THE MINE TODAY**
*The abandoned mines, known as Steinkaulenberg, have been cleared and divided into a researchers' area, a collectors' area, and a visitors' area.*

**ABANDONED OPAL MINE**
*These shafts of old opal test holes are part of an underground opal mine in Australia. Australia produces fine, precious and black opal.*

# OPAL

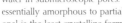

KNOWN SINCE ANTIQUITY, opal derives its name from the Roman word *opalus*, a Latinized version of the older Sanskrit *upala*, meaning "precious stone". Opal is hardened silica gel, and usually contains 5 to 10 per cent water in submicroscopic pores. Its structure varies from essentially amorphous to partially crystalline. Precious opal is the least crystalline form of the mineral, consisting of a regular arrangement of tiny, transparent, silica spheres with water in the intervening spaces.

**AUSTRALIAN OPAL MINE**
*This miner is drilling for opals at an opal mine in Coober Pedy, South Australia. Coober Pedy is an aboriginal phrase meaning "white man in a hole".*

## HOW OPAL FORMS

Opal is very widespread. In its pure form it is essentially colourless. The vast majority is common or "potch" opal in opaque, dull yellows and reds imparted by iron oxides, or black from manganese oxides and organic carbon. In potch opal, where silica spheres are present, they are of many different sizes. Opal is deposited at low temperatures from silica-bearing, circulating waters. It is found as nodules, stalactitic masses, veinlets, and encrustations in most kinds of rocks. It is especially abundant in areas of hot-spring activity and, as the siliceous skeletons of diatoms, radiolarians, and sponges, opal constitutes important parts of many sedimentary accumulations such as diatomaceous earth. It is commonly

**ZUNI EARRINGS**
*In these earrings by the Native American Zuni tribe, opal is flanked by mother-of-pearl and inlaid with jet.*

**OPAL AND PEARL CROSS**
*This gold pendant is set with five opal cabochons and two pearls. The opals have a fine colour play, showing red, green, and blue flashes.*

**FOSSIL BONE AND BOULDER OPAL**
*In the Australian opal fields, bones and shells are sometimes found replaced by precious opal (above). Boulder opal occurs when opal fills cavities in a host rock. In this specimen (right), ironstone has been split open to reveal the opal.*

**PRECIOUS OPAL**
*This specimen of precious opal is from Coober Pedy, Queensland, Australia. It shows the ironstone matrix and streaks of yellowish potch opal that are characteristic of opals from this field.*

**VICTORIAN RING**
*This 19th-century ring uses a single opal in an intricate gold setting.*

Ironstone matrix /

Potch opal /

## PROPERTIES

| | |
|---|---|
| **GROUP** | Silicates – tectosilicates |
| **CRYSTAL SYSTEM** | Amorphous |
| **COMPOSITION** | $SiO_2 \cdot nH_2O$ |
| **COLOUR** | Colourless, white, yellow, orange, rose-red, black, or dark blue |
| **FORM/HABIT** | Massive |
| **HARDNESS** | 5–6 |
| **CLEAVAGE** | None |
| **FRACTURE** | Conchoidal |
| **LUSTRE** | Vitreous |
| **STREAK** | White |
| **SPECIFIC GRAVITY** | 1.9–2.3 |
| **TRANSPARENCY** | Transparent to translucent |
| **R.I.** | 1.37–1.47 |

found as fossilized wood, where it preserves the wood's external appearance and cellular structure. Fossil bones and seashells have been discovered in Australia replaced by precious opal, and it also forms pseudomorphs after gypsum, calcite, feldspars, and other minerals.

Precious opals can form only in undisturbed space within another rock that is capable of holding a clean solution of silica from which water is slowly removed over a long period – perhaps thousands of years. The silica spheres slowly settle out of solution and arrange themselves into an orderly three-dimensional formation. Unless the spheres are regularly arranged and of the correct size, there is no colour play, which is caused by the diffraction of light through the spheres; opal is, in effect, a diffraction grating. The larger the spheres (within the limits imposed by the wavelengths of visible light), the greater the range of colour. All precious opal is probably relatively young in geological terms, as precious opal cannot withstand the heat and pressure of burial and metamorphism.

## SOURCES AND USES

Until late in the 19th century, the primary source of precious opal, including that used by the Romans, was in present-day Slovakia, where opal occurred in andesite. Today the chief producer of precious opal is Australia, where it was discovered in 1887, with deposits in South Australia, Queensland, and New South Wales. In Australia, opal is found in sedimentary rocks such as sandstone and ironstone. The Lightning Ridge field in New South Wales, Australia, produces the rare and prized black opal, with a very dark grey or blue to black body-colour and a superb colour play. Until the discovery of this field in 1903, black opal was virtually unknown, and it is still the rarest form of opal. Much smaller amounts of precious opal come from India, New Zealand, Honduras, and the western United States. The other form of gem opal is called fire opal. This transparent, intensely coloured opal in red, orange, or yellow also shows flashes of colour. The principal sources of fire opal are Mexico and Honduras, where it is found in trachyte.

Common opal in its various forms is widely mined for use as abrasives, insulation media, fillers, and ingredients in ceramic manufacture. Precious opals are usually finished *en cabochon* because their colour play shows best on smoothly rounded surfaces; fire opals are normally facet cut. Precious opal with a variety of intense colours, including red and violet, is generally the most expensive. Unlike most other gemstones, opal may crack or lose its colour if it dries. Thin slices of precious opal are often either glued to a thicker quartz base (forming a "doublet") or sandwiched between layers of quartz (forming a "triplet"), both giving the appearance of a denser, more valuable stone. Black stones, such as obsidian, are sometimes used as the backing for doublets or triplets, giving the appearance of rare black opal.

**FORMS OF OPAL**
These opal specimens show opal in both precious and common forms. The sil-sinter with a spherular habit is also known as fiorite, after its source: Santa Fiora, Tuscany, Italy. The black opal is from Lightning Ridge, Australia. The rose opal is potch, from France. The potch opal is from Cornwall, England.

BLACK OPAL

SIL-SINTER OPAL

ROSE OPAL

POTCH OPAL

Blue to yellow colour play

**PEACOCK BROOCH**
The centrepiece of this eye-catching brooch by Harry Winston, Inc. is a 32-carat black opal from Lightning Ridge, Australia. Sapphires, rubies, emeralds, and diamonds adorn the tail.

Vitreous lustre

Colour play

Conchoidal fracture

**HUNGARIAN**
This white opal is from an area of Slovakia, once part of Hungary. Opal from here is often called Hungarian.

**SPIDERWEB OPAL**
This opal gains its name from the web-like network of fine lines across its surface.

**WHITE OPAL**
This precious white opal cut into a cabochon shows flashes of several colours.

**BLACK OPAL**
Precious opal with a black body-colour is rare and expensive. Here it is carved en cabochon.

**FIRE OPAL**
This beautiful, rich orange octagonal-cut fire opal shows an unusual degree of transparency.

233

# FELDSPARS

AMONG THE MOST IMPORTANT silicate groups is the feldspars. A group of aluminosilicate minerals that contain calcium, sodium, or potassium, they are the most common minerals in the Earth's crust. Feldspars are the major component of most igneous rocks, and are found not only on Earth but also on the Moon and in meteorites. There are two groups of feldspars: the alkali feldspars and the plagioclase feldspars. The plagioclases can be differentiated from the alkali feldspars by the presence of polysynthetic twinning (see above right).

## ALKALI FELDSPARS

When formed at high temperatures, a solid–solution series (see p.91) exists within the alkali feldspars between the potassium feldspar orthoclase ($KAlSi_3O_8$) and the sodium feldspar albite ($NaAlSi_3O_8$). Unlike the plagioclases, which all crystallize in the triclinic crystal system, the alkali feldspars crystallize both in the triclinic and monoclinic systems. There are also monoclinic varieties in which barium replaces some or all of the potassium. These rather rare minerals are known as barium feldspars. Albite appears as the end member of both the alkali feldspars and the plagioclases, and is considered to be both an alkali and a plagioclase feldspar.

CARLSBAD    MANEBACH    BAVENO

**ORTHOCLASE TWINNING**
*Three very distinctive types of intergrown, twin crystals are formed by orthoclase.*

SINGLE    TWIN    POLYSYNTHETIC TWINNING

**ALBITE TWINNING**
*Polysynthetic (parallel and repeated) twinning is characteristic of plagioclase feldspars, such as albite.*

## PLAGIOCLASE FELDSPARS

The plagioclases are a major group of rock–forming feldspars that are a continuous series of solid solutions between the end members albite ($NaAlSi_3O_8$, or sodium aluminosilicate, abbreviated Ab), and anorthite ($CaAl_2Si_2O_8$, or calcium aluminosilicate, abbreviated An). Various members of intermediate composition are given names dependent on the percentages of albite or anorthite they contain. There is no other mineralogical difference between them. Selected feldspars are shown in the table below.

**PINK GRANITE**
*The pink colour of granite is due to the high percentage of alkali feldspar in its composition. The pink's intensity varies according to location. This weathered outcrop is part of the Pink Granite Coast near Ploumanac'h in northern Brittany, France.*

**BARIUM FELDSPARS**

MONOCLINIC

CELSIAN
$BaAl_2Si_2O_8$

HYALOPHANE
$(K,Na,Ba)(Al,Si)_4O_8$

**ALKALI FELDSPARS**

MONOCLINIC

ORTHOCLASE
$KAlSi_3O_8$

SANIDINE
$(K,Na)AlSi_3O_8$

TRICLINIC

MICROCLINE
$KAlSi_3O_8$

ANORTHOCLASE
$(Na,K)AlSi_3O_8$

ALBITE (Ab)
$NaAlSi_3O_8$

**PLAGIOCLASE FELDSPARS**

ANDESINE
$NaAlSi_3O_8 – CaAl_2Si_2O_8$

BYTOWNITE
$(NaSi,CaAl)AlSi_2O_8$

OLIGOCLASE
$(Na,Ca)(Al,Si)AlSi_3O_8$

LABRADORITE
$(Ca,Na)Al(Al,Si)Si_2O_8$

ANORTHITE
(An)
$CaAl_2Si_2O_8$

# ORTHOCLASE

THE POTASSIUM-BEARING end member of the alkali feldspars (see opposite), orthoclase is a major rock-forming mineral. It is a major component of granite – its pink crystals give common granite its characteristic pink colour. Orthoclase also occurs in colourless, white, cream, pale yellow, and brownish-red crystals, which are frequently twinned. It appears as well-formed, short prismatic crystals, and also in massive form. Pure orthoclase is relatively rare, as some sodium is usually present in the structure.

Orthoclase is abundant in igneous rocks rich in potassium or silica, in pegmatites, and in gneisses. It occurs worldwide. Geologists include orthoclase along with other potassium-rich feldspars under the general term K-feldspars, or K-spar for short. The name orthoclase is derived from the Greek word for "straight fracture" and alludes to the crystal's perpendicular cleavages. Adularia (see p.237) is a colourless or white variety of orthoclase, attributed to orthoclase formed in Alpine-type veins. The term valencianite was at one time used to describe adularia from the Valenciana Mine in Mexico. Moonstone is an opalescent variety of orthoclase and some other feldspars, which has a blue or white sheen known as schiller. This sheen is a result of the interlayering of orthoclase with albite (see p.239). Very thin layers yield a blue sheen, much prized in jewellery, while thicker layers give a white sheen.

**YELLOW CRYSTAL**

**MOONSTONE PEBBLE**

**MOONSTONE CRYSTALS IN ROCK**

**ORTHOCLASE FORMS**
*Colourless and yellow orthoclase crystals are the most common. Moonstone is usually white.*

## SACRED STONES

Moonstone derived its name from its opalescent moon-like appearance. It was sacred in India, where it was said that if lovers placed it in their mouths during the full moon, their futures would be revealed. In 11th-century Europe, moonstone was believed to bring about the reconciliation of lovers. Sunstone – a variety of oligoclase or labradorite – is associated, like the Sun, with health, physical energy, passion, and courage. Tiny inclusions of hematite are in fact responsible for the red to orange sun-like colour.

*Blue sheen* — **MOONSTONE**

*Hematite inclusions*

**SUNSTONE**

Orthoclase is important in ceramics, where it can be used either to make the artefact itself or as a glaze. Moonstone was used in jewellery by the Romans from AD100, and even earlier in the Orient. Today, its best sources are Myanmar (Burma), Sri Lanka, India, Brazil, Tanzania, and the USA.

**MOONSTONE NECKLACE**
*Moonstone gems are almost always cut en cabochon to accentuate the blue sheen, as in this contemporary necklace.*

## PROPERTIES

| | |
|---|---|
| **GROUP** | Silicates / tectosilicates |
| **CRYSTAL SYSTEM** | Monoclinic |
| **COMPOSITION** | $KAlSi_3O_8$ |
| **COLOUR** | Colourless, white, cream, yellow, pink, brown-red |
| **HABIT** | Short prismatic |
| **HARDNESS** | $6–6\frac{1}{2}$ |
| **CLEAVAGE** | Perfect |
| **FRACTURE** | Subconchoidal to uneven, brittle |
| **LUSTRE** | Vitreous |
| **STREAK** | White |
| **SPECIFIC GRAVITY** | 2.5–2.6 |
| **TRANSPARENCY** | Transparent to translucent |
| **R.I.** | 1.51 – 1.54 |

**ORTHOCLASE CRYSTALS**
*In this group of three orthoclase crystals, the crystal on the right demonstrates good prismatic habit. The largest crystal is twinned.*

*Crystal has unequal sides*

*Twinning*

**MOONSTONE BROOCH**
*This 1960s' Georg Jensen leaf-design brooch features moonstones with a blue sheen.*

**OVAL CUSHION**
*This orthoclase is cut in an oval cushion; the smoky, bluish sheen of schiller may be seen within it.*

**YELLOW STEP**
*A transparent yellow orthoclase from Madagascar is here faceted in a rectangular step cut.*

**PALE BROWN STEP**
*This step-cut, pale brown orthoclase is attractive but the mineral is too brittle for regular wear.*

**CABOCHON**
*This orthoclase moonstone is cut en cabochon to highlight the ethereal beauty of its schiller.*

**EARRINGS**
*Rectangular cabochon moonstones are set in these silver earrings.*

# HYALOPHANE

ONE OF THE LESS COMMON feldspar minerals, hyalophane is a potassium barium aluminosilicate. It takes its name from the Greek for "glass", and "to appear", a reference to its transparent, glassy crystals. Those crystals, which are frequently twinned, are prismatic and either similar in appearance to those of adularia or short prismatic like those of orthoclase. It also occurs as glassy masses. The mineral is colourless, white, pale yellow, or pale to deep pink. It is intermediate in composition between orthoclase and celsian. Hyalophane is found with manganese deposits in contact-metamorphic zones, often in association with rhodonite, spessartine, epidote, and analcime. Localities include Tochigi Prefecture, Japan; Jakobsberg, Sweden; Rhiw, Wales; Broken Hill, NSW, Australia; Binntal, Switzerland; Pahau River, New Zealand; and, in the USA, Franklin, New Jersey, and the Highwood Mountains of Montana. Zorgrazski Creek, near Busovaca in Bosnia, produces large and spectacular twinned crystals.

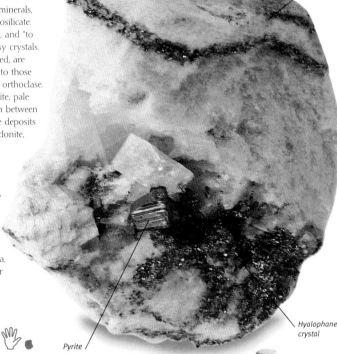

Dolomite marble

Hyalophane crystal

Pyrite

### PROPERTIES

**GROUP** Silicates – tectosilicates
**CRYSTAL SYSTEM** Monoclinic
**COMPOSITION** $(K,Ba)(Al,Si)_4O_8$
**COLOUR** Colourless, white, pale yellow, or pale pink
**FORM/HABIT** Prismatic
**HARDNESS** 6–6½
**CLEAVAGE** Perfect, good
**FRACTURE** Conchoidal, brittle
**LUSTRE** Vitreous
**STREAK** White
**SPECIFIC GRAVITY** 2.6–2.8
**TRANSPARENCY** Transparent to translucent

**HYALOPHANE CRYSTAL**
*This 6mm (¼in) long hyalophane crystal, with pyrite in dolomite marble, is from Binntal, Valais, Switzerland.*

# SANIDINE

FORMING COLOURLESS OR WHITE, glassy, transparent crystals, which can also be grey or other pale tints, sanidine is an alkali feldspar, a high-temperature form of potassium aluminosilicate. Its crystals are generally short prismatic in habit, and may reach 50cm (20in) in length. It is also found as granular or cleavable masses. Some sanidine falls under the general category of moonstone (see p.235), exhibiting the "schiller" effect – the internal reflections that create the "moonglow" appearance in cut stones.

Sanidine occurs in felsic volcanic rocks such as rhyolite and trachyte. It is also found in high-temperature and low-pressure metamorphic rocks. It is a very widespread mineral, with significant occurrences at Auvergne, France; the Alban Hills near Rome, Italy; La Cruz, Mexico; Mont St.-Hilaire, Canada; Eifel, Germany; and, in the USA, Grant County, New Mexico, Klamath County, Oregon, and the Thomas Range of Utah. Moonstone comes from India and Sri Lanka.

**CERAMICS**
*Sanidine is one of the feldspars ground for use in ceramics.*

Sanidine crystal

Trachyte groundmass

**MOONSTONE RING**
*This gold ring is set with a sanidine moonstone. Sanidine has white to light blue "schiller".*

**SANIDINE CRYSTAL**
*This single, well-formed prismatic crystal of sanidine rests in a groundmass of trachyte rock.*

### PROPERTIES

**GROUP** Silicates – tectosilicates
**CRYSTAL SYSTEM** Monoclinic
**COMPOSITION** $(K,Na)AlSi_3O_8$
**COLOUR** Colourless, white
**FORM/HABIT** Tabular
**HARDNESS** 6–6½
**CLEAVAGE** Perfect, good
**FRACTURE** Conchoidal to uneven
**LUSTRE** Vitreous
**STREAK** White
**SPECIFIC GRAVITY** 2.6
**TRANSPARENCY** Transparent to translucent

# CELSIAN

ONE OF THE LESS COMMON of the feldspar minerals, celsian is named for Anders Celsius (1701–1744), who invented the Celsius temperature scale. A barium aluminosilicate, it has short prismatic crystals that have a similar habit to adularia crystals. Crystals are commonly twinned and can be colourless, white, or yellow, often with glassy, lustrous faces. The mineral also occurs in cleavable masses.

Celsian is found in manganese- and barium-rich metamorphic rocks, often in association with diopside, witherite, and quartz. Localities include Jakobsberg, Sweden; Tochigi Prefecture, Japan; Rhiw, Wales; Broken Hill, NSW, Australia; Otjosondu, Namibia; and, in the USA, Fresno and Santa Cruz counties, California.

**ALASKA RANGE**
*These mountains in Alaska, USA, contain masses of celsian.*

**ANDERS CELSIUS**
*Celsian is named for Anders Celsius, a Swedish naturalist and astronomer.*

Granular habit

Uneven fracture

Vitreous lustre

**GRANULAR CELSIAN**
*This creamy white granular specimen of celsian is from Big Creek, Fresno County, California, USA.*

## PROPERTIES

| | |
|---|---|
| **GROUP** Silicates – tectosilicates | |
| **CRYSTAL SYSTEM** Monoclinic | |
| **COMPOSITION** $BaAl_2Si_2O_8$ | |
| **COLOUR** White, colourless, or yellow | |
| **FORM/HABIT** Short prismatic | |
| **HARDNESS** $6–6\frac{1}{2}$ | |
| **CLEAVAGE** Perfect, good | |
| **FRACTURE** Uneven, brittle | |
| **LUSTRE** Vitreous | |
| **STREAK** White | |
| **SPECIFIC GRAVITY** 3.1–3.5 | |
| **TRANSPARENCY** Transparent to translucent | |

# ADULARIA

THIS FELDSPAR MINERAL is generally considered to be a more highy ordered form of orthoclase. It is found in hydrothermal veins in mountainous areas, such as the Adular Mountains of Switzerland, after which it is named. It commonly forms colourless to white, cream, pale yellow to pink, or reddish-brown, glassy, prismatic, twinned crystals. It can also be massive, granular, or cryptocrystalline.

Adularia is found in low-temperature veins of felsic plutonic rocks, in hydrothermal veins, and in cavities in crystalline schists. Noteworthy localities include the Alps, where it occurs in schists, and at Betroka, Madagascar where large, transparent crystals are found.

**SWISS ALPS**
*The schists of the Alps provide excellent adularia crystals.*

**YELLOW CRYSTAL**
*Internal cracks or flaws are visible in this yellow-tinged adularia crystal.*

**ADULARIA CRYSTALS**
*This group of glassy white adularia crystals rests on a groundmass of green actinolite.*

Adularia crystal

Short prismatic habit

Twinned crystals

Vitreous lustre

Actinolite groundmass

## PROPERTIES

| | |
|---|---|
| **GROUP** Silicates – tectosilicates | |
| **CRYSTAL SYSTEM** Monoclinic/triclinic | |
| **COMPOSITION** $KAlSi_3O_8$ | |
| **COLOUR** Colourless, white, yellow, pink, or reddish-brown | |
| **FORM/HABIT** Short prismatic | |
| **HARDNESS** $6–6\frac{1}{2}$ | |
| **CLEAVAGE** Perfect, good | |
| **FRACTURE** Conchoidal to uneven | |
| **LUSTRE** Vitreous | |
| **STREAK** Colourless | |
| **SPECIFIC GRAVITY** 2.5–2.6 | |
| **TRANSPARENCY** Transparent to translucent | |

237

# MICROCLINE

**URAL MOUNTAINS**
*This Russian mountain range is a prime locality for microcline.*

USED IN CERAMICS and as a mild abrasive, microcline is one of the most common feldspar minerals. It is one of two forms of potassium aluminosilicate, the other being orthoclase (see p.235). Microcline forms short prismatic or tabular crystals, often of considerable size: single crystals from granite pegmatites can weigh several tonnes, and reach tens of metres in length. Crystals are often multiple twinned and have two sets of fine lines (exsolution lamellae) at right angles to each other, giving a "plaid" effect that distinguishes microcline from other feldspars. It can also have a massive habit. It can be colourless, white, cream to pale yellow, salmon-pink to red, and bright green to blue-green. Its name, from the Greek for "little" and "lean", is derived from the planes of its two cleavage directions, which form a slightly oblique angle with each other.

Microcline is the most common feldspar in deep-seated, felsic rocks such as granite, granodiorite, and syenite. It is found in granite pegmatites and in metamorphic rocks such as schists and gneisses. It can be a major component of detrital sedimentary rocks. It is a very widespread mineral; exceptional specimens are found in Baveno, Italy; Irkutsk, Russia; Mont St.-Hilaire, Canada; and in South Dakota and New Mexico, both in the USA.

Blue-green specimens of microcline are called amazonstone or amazonite, after the Amazon River. This variety is used in jewellery, generally cut into cabochons. Gem-quality amazonite is found in the Ilmen and Ural Mountains of Russia, the Pikes Peak district of Colorado, USA, and in Minas Gerais, Brazil.

**PROPERTIES**

| | |
|---|---|
| **GROUP** | Silicates – tectosilicates |
| **CRYSTAL SYSTEM** | Triclinic |
| **COMPOSITION** | $KAlSi_3O_8$ |
| **COLOUR** | White, pale yellow |
| **FORM/HABIT** | Short prismatic |
| **HARDNESS** | 6–6½ |
| **CLEAVAGE** | Perfect, good |
| **FRACTURE** | Conchoidal to uneven, brittle |
| **LUSTRE** | Vitreous, dull |
| **STREAK** | White |
| **SPECIFIC GRAVITY** | 2.6 |
| **TRANSPARENCY** | Transparent to translucent |
| **R.I.** | 1.5 |

**AMAZONSTONE**
*Also called amazonite, this blue-green variety of microcline is a semiprecious gemstone.*

Fine lines on surface

Vitreous lustre

**POLISHED**
*Prized for its pastel blue-green colour, translucent amazonstone is cut en cabochon.*

**AMAZONSTONE BRACELET**
*This broad, flexible bracelet of loop-and-ring sculptural design is set with oval amazonstones.*

Amazonstone cabochon

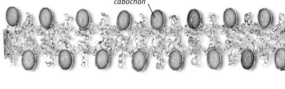

---

# ANORTHOCLASE

**COLORADO**
*Gem anorthoclase is found in Colorado, USA.*

THIS MEMBER OF THE feldspar group takes its name from the Greek for "not straight fracture" (orthoclase means "straight fracture"), referring to its oblique cleavage. It is one of several feldspars that is known as moonstone when cut *en cabochon*. Crystals are prismatic or tabular, and are often multiple twinned. Like microcline, it shows two sets of fine lines (exsolution lamellae) at right angles to each other, but in anorthoclase the lamellae are much finer. It can also be massive and granular. It is colourless, white, cream, pink, pale yellow, grey, and green. Anorthoclase is a sodium potassium aluminosilicate formed in sodium-rich igneous environments. It is commonly found with augite, apatite, and ilmenite. Noteworthy specimens come from Scotland, Kenya, New Zealand, Australia, Sicily, Norway, and Antarctica.

## MOONSTONE

Anorthoclase is one of several feldspars that show a white or silvery iridescence when cut *en cabochon*. They are classed as moonstone; other moonstones are orthoclase, sanidine, albite, and oligoclase. The iridescence, called "schiller", results from the minute interlayering of a different feldspar that develops by internal chemical separation during the process of crystallization.

**SETTING STONE**
*This rainbow moonstone is being mounted in a silver setting.*

**PROPERTIES**

| | |
|---|---|
| **GROUP** | Silicates – tectosilicates |
| **CRYSTAL SYSTEM** | Triclinic |
| **COMPOSITION** | $(Na,K)AlSi_3O_8$ |
| **COLOUR** | Colourless, white |
| **FORM/HABIT** | Prismatic |
| **HARDNESS** | 6–6½ |
| **CLEAVAGE** | Perfect, good |
| **FRACTURE** | Conchoidal to uneven, brittle |
| **LUSTRE** | Vitreous |
| **STREAK** | White |
| **SPECIFIC GRAVITY** | 2.6 |
| **TRANSPARENCY** | Transparent to translucent |
| **R.I.** | 1.5 |

**LARGE CRYSTAL**
*This specimen of a pink and grey prismatic anorthoclase crystal shows well-developed crystal faces.*

**MOONSTONE NECKLACE**
*This gold and silver necklace (c.1912) is mounted with moonstones, carnelians, and labradorite.*

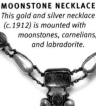

Single prismatic crystal

Vitreous lustre

# ALBITE

THE PRIMARY GEOLOGICAL importance of albite is as a rock-forming mineral. Crystals are tabular or platy, often twinned, glassy, and brittle. It can also be massive and granular. Albite takes its name from the Latin *albus*, "white", a reference to its usual colour, but it can also be colourless, yellowish, pink, or green. The end member of both the plagioclase and the alkali feldspars (see p.234), albite occurs widely in pegmatites and felsic igneous rocks, such as granites, syenites, and rhyolites, and is also found in low-grade metamorphic rocks. In addition, it forms through chemical processes in certain sedimentary environments. Large crystals are found in Brazil, and in South Dakota, New Mexico, and California, USA.

**ONTARIO, CANADA**
*Bancroft, Ontario and Mont St.-Hilaire are two sources of albite.*

### PROPERTIES

| | |
|---|---|
| **GROUP** | Silicates – tectosilicates |
| **CRYSTAL SYSTEM** | Triclinic |
| **COMPOSITION** | NaAlSi$_3$O$_8$ |
| **COLOUR** | White, colourless |
| **FORM/HABIT** | Tabular |
| **HARDNESS** | 6–6½ |
| **CLEAVAGE** | Perfect, good |
| **FRACTURE** | Conchoidal to uneven, brittle |
| **LUSTRE** | Vitreous to pearly |
| **STREAK** | White |
| **SPECIFIC GRAVITY** | 2.6 |
| **TRANSPARENCY** | Translucent |
| **R.I.** | 1.4–1.5 |

## PLAGIOCLASES IN PORCELAIN

Chinese porcelain became highly prized in Europe after its introduction around 1300. In the 1570s attempts were made to produce porcelain as delicate-looking and tough as the Chinese by adding ground glass to the clay, but the results were unsatisfactory. It was not until 1707 in Saxony that Ehrenfried von Tschirnhaus discovered that if you mixed ground plagioclase, such as albite, with clay you could create porcelain of similar quality.

**PORCELAIN BOWL**
*Made in the late 14th century during the Ming dynasty, this Chinese porcelain bowl features a red design of a dragon.*

**THIN SECTION**
*This albite porphyroblast enclosing trails of graphite is in mica schist from As Sifah, Oman.*

*Twin lamellae*

*Trails of graphite*

*Mica*

*Vitreous to pearly lustre*

**TABULAR CRYSTALS**
*This specimen is of a large group of tabular white albite crystals, many of which are twinned.*

*Twinned tabular crystals*

**MIXED CUT**
*Some albite is transparent enough to facet into brilliant gemstones, such as this mixed cut.*

# OLIGOCLASE

THE MOST COMMON variety of the feldspar mineral plagioclase, oligoclase is a sodium potassium aluminosilicate. Its usual habit is massive or granular, although it does form tabular crystals, commonly twinned. It can be grey, white, greenish, yellowish, brown, or colourless. Oligoclase occurs in granite, granitic pegmatites, diorite, rhyolite, and other felsic igneous rocks. In metamorphic rocks, it occurs in high-grade, regionally metamorphosed schists and gneisses. Localities include Kimito, Finland; Lisbon, Portugal; and New Hampshire and New York, USA. Sunstone is a gemstone variety of oligoclase; when it is cut *en cabochon* the reflections from hematite inclusions give a reddish glow and a sparkling appearance. It occurs in Kragero, Norway, and in North Carolina, USA.

**LAKE BAIKAL**
*Sunstone oligoclase gems are found at Lake Baikal in Russia.*

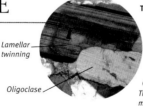

**THIN SECTION**
*The characteristic very fine lamellar twinning is clearly visible in this section of Sierra Nevada granite.*

*Lamellar twinning*

*Oligoclase*

**MASSIVE OLIGOCLASE**
*This typical specimen of massive oligoclase is from Penland, Mitchell County, North Carolina, USA.*

*Perfect cleavage*

*Vitreous lustre*

### PROPERTIES

| | |
|---|---|
| **GROUP** | Silicates – tectosilicates |
| **CRYSTAL SYSTEM** | Triclinic |
| **COMPOSITION** | (Na,Ca)Al$_2$Si$_2$O$_8$ |
| **COLOUR** | Grey, white |
| **FORM/HABIT** | Massive |
| **HARDNESS** | 6 |
| **CLEAVAGE** | Perfect |
| **FRACTURE** | Conchoidal to uneven, brittle |
| **LUSTRE** | Vitreous |
| **STREAK** | White |
| **SPECIFIC GRAVITY** | 2.6 |
| **TRANSPARENCY** | Translucent |
| **R.I.** | 1.54–1.55 |

**SUNSTONE PIN**
*Sunstone has tiny plate-like inclusions of hematite or goethite oriented parallel to one another.*

# ANDESINE

**THE ANDES**
*Andesine is named after the Andes Mountains.*

ANDESINE OCCURS WIDELY in igneous rocks of medium silica content, especially in the rock andesite (see p.46). It is an intermediate plagioclase, a sodium calcium aluminosilicate. Andesine often forms well-defined crystals, usually exhibiting multiple twinning. Other types of twinning also commonly occur.

Andesine can be massive, or occur as rock-bound grains. Good crystals are found in Miyagi and Toyama Prefectures, Japan; Marmato, Colombia; Trentino, Italy; Cannes, France; and Bodenmais, Germany.

## PROPERTIES

| | |
|---|---|
| **GROUP** | Silicates – tectosilicates |
| **CRYSTAL SYSTEM** | Triclinic |
| **COMPOSITION** | $NaAlSi_3O_8 – CaAl_2Si_2O_8$ |
| **COLOUR** | Grey, white |
| **FORM/HABIT** | Short prismatic |
| **HARDNESS** | $6–6\frac{1}{2}$ |
| **CLEAVAGE** | Perfect |
| **FRACTURE** | Conchoidal to uneven |
| **LUSTRE** | Subvitreous to pearly |
| **STREAK** | White |
| **SPECIFIC GRAVITY** | 2.7 |
| **TRANSPARENCY** | Transparent to translucent |

*Andesine phenocryst*

**THIN SECTION**
*An andesine crystal is visible here in lava of the rock andesite.*

*Andesite lava matrix*

*Andesine crystal*

*Triclinic crystals*

*Blue porphyry*

**ANDESINE CRYSTALS**
*This specimen has andesine crystals up to 2cm ($^3/_4$in) long in blue porphyry. It was found in Estérel, Var, France.*

# BYTOWNITE

**CHIHUAHUA**
*This locality in Mexico produces good-quality bytownite gems.*

THE RAREST PLAGIOCLASE, bytownite is named for the locality in which it was first recognized, Bytown (now Ottawa), Canada (although the material from Bytown was later reanalysed and found to be a mixture). Well-developed crystals are relatively uncommon, and are short prismatic to tabular, exhibiting the polysynthetic twinning characteristic of the plagioclases. Bytownite occurs in mafic and ultramafic igneous rocks, both intrusive and extrusive, and it is also found in stony meteorites. Gem-quality material is found at Nueva Casas Grandes, Chihuahua, Mexico, and Lakeview, Oregon, USA. Other localities include: Rhum Island, Scotland; Fiskenaesset, Greenland; Chester and Lebanon counties, Pennsylvania, USA; and Ottawa, Canada.

*Perfect cleavage*

*Vitreous lustre*

**MASSIVE BYTOWNITE**
*This specimen of bytownite is grey-brown. Bytownite is more commonly found as rock-bound grains.*

## PROPERTIES

| | |
|---|---|
| **GROUP** | Silicates – tectosilicates |
| **CRYSTAL SYSTEM** | Triclinic |
| **COMPOSITION** | $NaAlSi_3O_8 – CaAl_2Si_2O_8$ |
| **COLOUR** | White, grey |
| **FORM/HABIT** | Short prismatic to tabular |
| **HARDNESS** | $6–6\frac{1}{2}$ |
| **CLEAVAGE** | Perfect |
| **FRACTURE** | Uneven to conchoidal |
| **LUSTRE** | Vitreous to pearly |
| **STREAK** | White |
| **SPECIFIC GRAVITY** | 2.7 |
| **TRANSPARENCY** | Transparent to translucent |
| **R.I.** | 1.57–1.59 |

**RECTANGULAR**
*This step cut is of the yellow-orange bytownite sometimes called sunstone.*

**OVAL CUT**
*Transparent, colourless bytownite is sometimes cut in many-faceted shapes.*

# ANORTHITE

**HOKKAIDO**
*Hokkaido is one of several Japanese localities for anorthite.*

PURE ANORTHITE is uncommon. The calcium end member of the plagioclases, anorthite is calcium aluminosilicate and can contain up to 10 per cent albite. It takes its name from the Greek *anorthos*, meaning "not straight", in reference to its triclinic form. It develops well-formed crystals, which are short prismatic, and it can also be massive and granular. Its crystals are white, greyish, or reddish, brittle, and glassy. Anorthite is a major rock-forming mineral present in many mafic igneous rocks, both intrusive and extrusive, in contact metamorphic rocks, and in some meteorites. Specimen localities include: Tunaberg, Sweden; Labrador, Canada; the Ural Mountains, Russia; Trentino, Italy; Tamil Nadu, India; and Grass Valley, California, USA.

*Anorthite crystal*

*Vitreous lustre*

**PINK ANORTHITE**
*Here, pink anorthite crystals are associated with augite.*

## PROPERTIES

| | |
|---|---|
| **GROUP** | Silicates – tectosilicates |
| **CRYSTAL SYSTEM** | Triclinic |
| **COMPOSITION** | $CaAl_2Si_2O_8$ |
| **COLOUR** | White, grey, pink |
| **FORM/HABIT** | Short prismatic |
| **HARDNESS** | $6–6\frac{1}{2}$ |
| **CLEAVAGE** | Perfect |
| **FRACTURE** | Conchoidal to uneven, brittle |
| **LUSTRE** | Vitreous |
| **STREAK** | White |
| **SPECIFIC GRAVITY** | 2.7 |
| **TRANSPARENCY** | Transparent to translucent |

# LABRADORITE

THE CALCIC, MIDDLE-RANGE MEMBER of the plagioclase feldspars, labradorite seldom forms crystals, but when crystals do occur, they are tabular. It most often occurs in crystalline masses that can be microscopic or up to a metre or more across, and is commonly characterized by its "schiller" effect – a rich play of iridescent colours, principally blue, on cleavage surfaces. Crystals that display this effect are used as gemstones, cut *en cabochon*. The "schiller" effect is caused by the scattering of light from thin layers of a second feldspar that develops through internal chemical separation during the cooling of what was originally a single feldspar. The base colour of labradorite is generally blue or dark grey, but can be colourless or white.

**LABRADOR, CANADA**
*Labradorite was first identified at Paul Island, Labrador, by Moravian missionaries in 1770.*

**CAMEO**
*Labradorite is sometimes used as a carving material to produce unusual ornamental objects such as this cameo.*

Transparent labradorite can be yellow, orange, red, or green. Labradorite is a major or important constituent of certain medium-silica and silica-poor igneous and metamorphic rocks, including basalt, gabbro, diorite, andesite, and amphibolite. Gemstone labradorite comes from Madagascar, Finland, and Russia. Nearly transparent material with a beautiful schiller comes from southern India, and facetable material is found in Mexico and the USA. High-quality gemstone material from Finland is sometimes called spectrolite.

**PENDANT**
*A labradorite cabochon forms the centre of this modern silver pendant.*

Labradorite crystal

**THIN SECTION**
*In this specimen, a group of labradorite crystals are in olivine gabbro.*

**LAPLAND**
*This area produces a large quantity of highly coloured labradorite, known locally as spectrolite.*

## PROPERTIES

| | |
|---|---|
| **GROUP** | Silicates – tectosilicates |
| **CRYSTAL SYSTEM** | Triclinic |
| **COMPOSITION** | $NaAlSi_3O_8$ – $CaAl_2Si_2O_8$ |
| **COLOUR** | Blue, grey, white |
| **FORM/HABIT** | Usually massive |
| **HARDNESS** | 6–6½ |
| **CLEAVAGE** | Perfect |
| **FRACTURE** | Uneven to conchoidal |
| **LUSTRE** | Vitreous |
| **STREAK** | White |
| **SPECIFIC GRAVITY** | 2.7 |
| **TRANSPARENCY** | Transparent to translucent |
| **R.I.** | 1.56–1.57 |

**CHOKER**
*The polished oval of labradorite in this contemporary choker displays beautifully the rainbow iridescence of this gemstone.*

Vitreous lustre

"Schiller" effect

Polysynthetic twinning

Perfect cleavage

**CABOCHON**
*Labradorite is usually cut en cabochon to better display the iridescent colour play of the stone.*

**SUNSTONE**
*The orange feldspar found in Oregon and locally called sunstone is actually labradorite.*

**BLUE LABRADORITE**
*This beautifully coloured specimen shows the polysynthetic twinning that is characteristic of the plagioclase feldspars. The twinning is apparent as a series of parallel lines on broken faces.*

**"SCHILLER" EFFECT**
*The colour play is clearly visible in this specimen, in which orange, purple, and blue can be seen.*

241

# FELDSPATHOIDS

FELDSPATHOIDS ARE A GROUP of sodium- and potassium-bearing aluminosilicate minerals similar to the feldspars in chemical composition, but with physical and chemical properties that are intermediate between those of feldspars and zeolites. The feldspathoid structure consists of oxygen tetrahedra with silicon and aluminium atoms at their centres. Each oxygen atom is shared by the $SiO_4$ and $AlO_4$ tetrahedra, forming a three-dimensional framework. The open spaces in the framework can be occupied by carbonate, sulphate, or chlorine ions. Feldspathoids are mainly found in alkali-rich, silica-poor igneous and metamorphic rocks, where they occur in place of the feldspars that would have formed had there been more silica present.

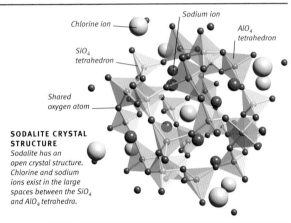

Chlorine ion

Sodium ion

$AlO_4$ tetrahedron

$SiO_4$ tetrahedron

Shared oxygen atom

**SODALITE CRYSTAL STRUCTURE**
Sodalite has an open crystal structure. Chlorine and sodium ions exist in the large spaces between the $SiO_4$ and $AlO_4$ tetrahedra.

## NEPHELINE

**EIFEL, GERMANY**
This is a locality for superb nepheline crystals.

THE MOST COMMON feldspathoid mineral, nepheline is an aluminosilicate of sodium and potassium. It takes its name from the Greek *nephele*, meaning "cloud", a reference to the fact that it becomes cloudy or milky when placed in strong acids. It is generally massive. When crystals are found, they are commonly hexagonal prisms, but may exhibit a variety of prism and pyramid shapes. Large crystals are often rough and pitted. Nepheline is also found as large, tabular phenocrysts. Usually white, often with a yellowish or greyish tint, it can also be grey or red-brown, depending on inclusions. It is a characteristic mineral of mafic plutonic rocks, where it may occur with spinel, perovskite, and olivine. In intermediate plutonics, such as nepheline syenites, it is often found with augite, aegirine, and low-silica amphiboles. It also occurs in some volcanics, and in some metamorphics such as nepheline gneisses. Good crystals are found at Davis Hill Mine, Bancroft, Ontario, and Mont St.-Hilaire, Canada; Eifel, Germany; Transvaal, South Africa; and in the USA at Magnet Cove, Arkansas, and in the Bearpaw and Judith Mountains of Montana. It occurs in particularly beautiful crystal form with mica, garnet, and sanidine feldspar on Monte Somma, Italy. Nepheline is sometimes mined for use in the manufacture of glass and ceramics.

Rock groundmass

**NEPHELINE CRYSTALS**
This specimen shows well-developed crystals of nepheline filling cavities in a rock groundmass.

Hexagonal nepheline prisms

Eudialyte

**THIN SECTION**
Nepheline crystals with regular outlines are enclosed here in eudialyte from a nepheline syenite gneiss from Brazil.

Nepheline crystal

Massive habit

Imperfect fracture

### PROPERTIES

| | |
|---|---|
| **GROUP** | Silicates – tectosilicates |
| **CRYSTAL SYSTEM** | Hexagonal |
| **COMPOSITION** | (Na,K)AlSiO$_4$ |
| **COLOUR** | White, grey, yellow, or red-brown |
| **FORM/HABIT** | Massive |
| **HARDNESS** | 5$^{1}/_2$–6 |
| **CLEAVAGE** | Poor |
| **FRACTURE** | Subconchoidal, brittle |
| **LUSTRE** | Vitreous to greasy |
| **STREAK** | White |
| **SPECIFIC GRAVITY** | 2.6 |
| **TRANSPARENCY** | Transparent to opaque |

**MASSIVE NEPHELINE**
This specimen of nepheline from Arkansas, USA, shows its most typical massive habit.

# LEUCITE

**MOUNT KILAMANJARO**
*This Tanzanian mountain is an important locality for leucite.*

TAKING ITS NAME from the Greek *leukos*, meaning "white", in allusion to its most common colour, leucite can also be grey or colourless. It is potassium aluminosilicate, and has until recently been considered a feldspathoid, although it is now classified as a zeolite. Its crystals can be up to 9cm (3¹/₂in) across. Leucite is cubic at high temperatures, forming trapezohedral crystals, but at lower temperatures it is tetragonal. It also forms massive and granular aggregates, and disseminated grains. Leucite occurs only in igneous rocks, especially those that are potassium-rich and silica-poor. It is found with natrolite, alkali-feldspar, nepheline, and analcime, and occurs worldwide. Important localities are Eifel, Germany; Monte Somma, Italy; and in the USA in the Leucite Hills, Wyoming, Hot Springs, Arkansas, the Bearpaw Mountains, Montana, and the Sierra Nevadas, California. Leucite is a source of alum, which is a widely used astringent.

**LEUCITE CRYSTALS**
*In this specimen, formed at a high temperature, fine trapezohedral crystals of leucite rest in cavities in a rock groundmass. This example is from Mount Vesuvius, Italy, one of the key localities for leucite.*

— *Trapezohedron*

**FERTILIZING CROPS**
*Because of its high potassium content, leucite is used as a fertilizer in Italy.*

*Rock groundmass*

### PROPERTIES

| | |
|---|---|
| **GROUP** Silicates – tectosilicates | |
| **CRYSTAL SYSTEM** Cubic/tetragonal | |
| **COMPOSITION** $KAlSi_2O_6$ | |
| **COLOUR** White, grey, or colourless | |
| **FORM/HABIT** Trapezohedral | |
| **HARDNESS** 5¹/₂–6 | |
| **CLEAVAGE** Poor | |
| **FRACTURE** Conchoidal, brittle | |
| **LUSTRE** Vitreous | |
| **STREAK** White | |
| **SPECIFIC GRAVITY** 2.5 | |
| **TRANSPARENCY** Transparent to translucent | |

# SODALITE

**KOLA PENINSULA**
*This part of Russia produces massive sodalite.*

NAMED IN 1811 for its high sodium content, sodalite is a feldspathoid mineral, a sodium aluminium silicate chloride. It usually forms massive aggregates or disseminated grains. Crystals are relatively rare, and are dodecahedral or octahedral. Sodalite principally occurs in silica-poor igneous rocks such as nepheline syenites and their associated pegmatites. It is sometimes found in volcanic ejecta and in contact metamorphosed limestones and dolomites. Massive sodalite is found in the Kola Peninsula, Russia; Eifel, Germany; Kishangarh, Rajasthan, India; Bancroft, Ontario, Canada; and in the USA at Litchfield, Maine, Red Hill, New Hampshire, and Magnet Cove, Arkansas. Its principal use is as a gemstone. Transparent material from Mont St.-Hilaire, Canada, has been faceted for collectors, and rare crystals are found on Mount Vesuvius, Italy.

### PROPERTIES

| | |
|---|---|
| **GROUP** Silicates – tectosilicates | |
| **CRYSTAL SYSTEM** Cubic | |
| **COMPOSITION** $Na_8Al_6Si_6O_{24}Cl$ | |
| **COLOUR** Grey, white, blue | |
| **FORM/HABIT** Massive | |
| **HARDNESS** 5¹/₂–6 | |
| **CLEAVAGE** Poor to distinct | |
| **FRACTURE** Uneven to conchoidal | |
| **LUSTRE** Vitreous to greasy | |
| **STREAK** White to light blue | |
| **SPECIFIC GRAVITY** 2.1–2.3 | |
| **TRANSPARENCY** Transparent to translucent | |
| **R.I.** 1.48 | |

**ART DECO CLOCK**
*Mother-of-pearl is edged by vivid blue sodalite in this Art Deco clock, which also stands on a sodalite base.*

**NECKLACE**
*This modern Egyptian necklace features beads made of sodalite (blue) and carnelian (red), as well as gold panel dividers with Egyptian motifs.*

*Vitreous lustre*

*Uneven fracture*

**MASSIVE SODALITE**
*This specimen shows the intense blue exhibited by much sodalite, sometimes causing it to be mistaken for lapis lazuli.*

*Massive habit*

**PEBBLE**
*Cheap and plentiful, sodalite is a popular mineral for tumble-polishing.*

**CABOCHON**
*When cut and polished en cabochon, sodalite makes an attractive and inexpensive gem.*

**OVAL BRILLIANT**
*Rare, semi-transparent sodalite is sometimes faceted for collectors.*

# LAZURITE

THE MINERAL LAZURITE is the main component of lapis lazuli and accounts for the stone's intense blue colour, although lapis lazuli also contains pyrite and calcite, and usually some sodalite and haüyne, too. Lazurite itself, a sodium calcium aluminosilicate sulphate, forms distinct crystals; it should not be confused with the phosphate lazulite. In lapis lazuli, lazurite is well dispersed. Distinct crystals were thought to be very rare until large numbers were brought out of the mines of Badakhshan, Afghanistan, in the 1990s. The best quality lapis lazuli is intense dark blue, with minor patches of white calcite and brassy yellow pyrite. Lapis lazuli was used as a gemstone for millennia by the Egyptians: the mask of Tutankhamun, for example, has lapis

lazuli inlay (see p.116). Other objects containing lapis lazuli, including scarabs, pendants, and beads date from much earlier, at least 3100BC. Powdered lapis lazuli was also used as a cosmetic (the first eye shadow), as a pigment, and as a medicine. In Sumeria, the tomb of Queen Pu-abi (2500BC) contained numerous gold and silver jewellery pieces richly adorned with lapis. In the 4th century BC, both the Chinese and the Greeks carved various lapis lazuli artefacts. Ancient references to "sapphire", the *sapphirus* of the Romans, usually, in fact, refer to lapis lazuli. Its modern name originates in the Persian word

**BADAKHSHAN**
*Lapis lazuli has been mined in Badakhshan, north-eastern Afghanistan, since antiquity. The mines here were the source of ancient Egyptian and Sumerian material.*

**SCARAB RING**
*This ancient Egyptian gold ring contains a scarab stone carved from lapis lazuli.*

**EARRING**
*This Egyptian lapis lazuli-inlaid earring comes from the tomb of Tutankhamun, and dates from the 14th century BC.*

**IMITATION CABOCHON**
*Synthetic lapis lazuli is made in France by Pierre Gilson. It is similar in composition to lazurite, but softer.*

**POLISHED CABOCHON**
*Cabochons of natural lapis lazuli frequently display visible white veining as well as irregular flecks of pyrite.*

**POLISHED SLAB**
*This slice of lazurite has been ground with grits to give a flat surface, and polished to show its intense colour to best effect.*

**LAPIS LAZULI BEAD NECKLACE**
*From ancient times lapis lazuli has been a popular choice for beads and other forms of ornamentation.*

Vitreous lustre

Dodecahedral crystals

## PROPERTIES

**GROUP** Silicates – tectosilicates
**CRYSTAL SYSTEM** Cubic
**COMPOSITION** $Na_3Ca(Al_3Si_3O_{12})S$
**COLOUR** Various intense shades of blue
**FORM/HABIT** Dodecahedral when crystallized
**HARDNESS** 5–5½
**CLEAVAGE** Indistinct
**FRACTURE** Uneven, brittle
**LUSTRE** Dull to vitreous
**STREAK** Bright blue
**SPECIFIC GRAVITY** 2.4
**TRANSPARENCY** Translucent to opaque
**R.I.** 1.5 average

**LAZURITE CRYSTALS**
*This lazurite specimen from Badakhshan, Afghanistan, shows superbly developed crystals in a matrix of calcite. The largest crystals are about 19mm (³/4in) long.*

Calcite matrix

**CHINESE SNUFF BOTTLE**
*Chinese snuff bottles were carved from many natural materials. Lapis lazuli was a very popular material, used for bottles, bowls, and intricate figurines.*

lazhuward, meaning "blue", and the Arabic word lazaward, meaning "heaven" or "sky", and came into use in Europe during the Middle Ages.

Lapis lazuli is relatively rare and commonly forms in crystalline limestones as a product of contact metamorphism. The mines in Afghanistan remain a major source. Lighter blue material comes from Chile, and some lapis is found in Italy, Argentina, the USA, and Tajikistan in Russia. Diopside, amphibole, feldspar, mica, apatite, titanite (sphene), and zircon may also occur in lapis lazuli, depending on its origin. It has been synthesized by Pierre Gilson in France.

## LORE OF LAPIS LAZULI

Much of the historical lore of lapis lazuli is difficult to untangle from the lore of other blue stones because of the problem of knowing exactly which blue stone was being referred to. In ancient Greece and Rome, there was a belief that *sapphirus*, probably lapis lazuli, cured eye diseases and set prisoners free. It was noted by the Greek physician Dioscorides, around AD55, that it was an antidote for snake venom; it was an even older Assyrian cure for melancholy.

### SPIRITUAL ASSOCIATIONS

Another widespread ancient belief was that it protected the wearer from the evil eye – possibly as its blue colour flecked with gold pyrite resembled the night sky, the dwelling place of God. Similarly, a medieval treatise suggests that "meditation upon stone carries the soul to heavenly contemplation".

To the Buddhists of antiquity, lapis lazuli brought peace of mind and equanimity, and was good for dispelling evil thoughts.

**CASKET**
*Containing the remains of Thomas á Becket, this English casket dates from around AD1190 and includes gold, lapis lazuli, and semiprecious stones.*

**CARVED BUDDHA**
*Due to the association of lapis lazuli with peace, it was a popular carving material for images of the Buddha. It is still a popular choice for modern religious objects.*

**WILTON DIPTYCH**
*The intense blue of lapis lazuli pigment survives in this medieval altarpiece, now in London's National Gallery. Blue pigments made from powdered lazurite turn green with age.*

*Powdered lapis lazuli*

**PIGMENT**
*Powdered lapis lazuli was used for hundreds of years to make the pigment ultramarine, which has now been replaced by a synthetic alternative.*

*Twinned crystals*

**CARVED TORTOISES**
*This modern lapiz lazuli carving of three tortoises and a frog comes from Afghanistan.*

"Strong bull, great of horns, perfect in form, with long, flowing beard, bright as lapis lazuli."

**ANCIENT ASSYRIAN HYMN TO THE MOON–GOD SIN**

*Cubic crystals*

**LAZURITE IN MATRIX**
*Lazurite dispersed in marble forms an attractive but inferior gem.*

**RENAISSANCE BROOCH**
*This brooch, created by Benvenuto Cellini (1500–71), is gold on a base of lapis lazuli surrounded by pearls. The scene depicted is from the Greek myth of Leda and the Swan.*

# THE STANDARD OF UR

THIS MYSTERIOUS OBJECT WAS FOUND IN THE ROYAL CEMETERY OF UR IN SOUTHERN IRAQ. TWO PANELS DEPICT SUMERIANS AT PEACE AND AT WAR IN A BEAUTIFUL MOSAIC OF LAPIS LAZULI, SHELL, AND RED LIMESTONE.

Dating from between 2600 and 2400BC, the Standard of Ur was crafted in a period when incredible treasures in gold, silver, bronze, and semiprecious stones were created. During this period Sumerian kings were buried along with their court officials, servants, and women, in order that they could continue their service in the next world.

When found, lying in the corner of a chamber above the right shoulder of a man, the original wooden frame of the Standard had decayed, and the bitumen acting as glue had disintegrated. Leonard Woolley, the excavator at Ur between 1922 and 1934, believed that it was carried on a pole as a standard, but it is now thought more likely that it formed the soundbox of a musical instrument. If so, the man it was buried with may well have been a court musician. The Standard was restored as a musical soundbox by the British Museum where it currently resides. It measures 21.6cm (8.4in) in height and 49.5cm (19.3in) in length, and the two main side panels are known as "War" and "Peace", reflecting the scenes depicted on each. The lapis lazuli was probably from Afghanistan, the principal ancient source of this precious stone, the chief constituent of which is lazurite (see pp.244–45).

**GRAVE NECKLACE**
*This necklace of gold, lapis lazuli, and carnelian was found in one of the richest tombs at Ur – the burial place of Pu-abi.*

**GOLD CUP**
*Found on the floor of the pit of the Queen's grave at Ur, this gold cup was one of four drinking vessels lying beside the Queen's sacrificial victims in the tomb. It was probably used by one of them to drink the poison that ended his life.*

**"STANDARD OF UR", PEACE SIDE**
*Animals, fish, and other goods are brought in procession to a banquet on the "Peace" panel (right). A musician plays a lyre to seated figures, who are drinking.*

**"STANDARD OF UR", WAR SIDE**
*The "War" panel (above) depicts chariots pulled by four donkeys and infantrymen carrying spears. Enemy soldiers are shown being killed with axes while others are paraded naked and presented to the King, who holds a spear.*

# HAÜYNE

**EIFEL**
*This part of Germany contains substantial haüyne deposits.*

NAMED FOR ONE of the pioneers of crystallography, French mineralogist René Just Haüy, haüyne is a sodium calcium aluminosilicate containing a sulphate radical. It is a feldspathoid mineral. Its crystals are dodecahedral or octahedral, and it is commonly found in rounded grains. Twinning is also common. It is blue, white, grey, yellow, green, or pink. Haüyne is one of the components of lapis lazuli, along with lazurite, calcite, pyrite, and sodalite. It is easily soluble in acids, leaving a silica gel. Haüyne is primarily found in silica-poor volcanic rocks, although it has also been found in a few metamorphic rocks. It is commonly associated with leucite, nepheline, and nosean (a sodium aluminosilicate, with which haüyne forms a partial solid-solution). The volcanic rocks of Germany, especially Niedermendig and Laacher See, Eifel, are some of its main sources. It was first identified in the volcanic rocks of Italy, and other localities include Siberia, Russia; Nanjing, China; Auvergne, France; and in the USA at Cripple Creek, Colorado, Edwards, New York, Winnett, Montana, and Lawrence County, South Dakota.

**BRILLIANT**
*Haüyne is difficult to cut so is rarely faceted. This small stone was faceted for a collector.*

## RENÉ JUST HAÜY

Haüy (1743–1822) was a priest who grew interested in crystallography in 1781 when he noticed that the fragments of a broken calcite crystal cleave along straight lines that meet at constant angles. In 1784 he showed that a crystal is built up of tiny units and that the shape of the crystal is determined by the shape of the units. In 1802 Haüy became a professor of mineralogy at the Museum of Natural History in Paris. He wrote a crystallography book at the request of Napoleon Bonaparte, as well as other key works.

**RENÉ JUST HAÜY**
*Haüy was a French mineralogist and the founder of the study of the structure and formation of crystals.*

*Haüyne crystals*

*Lava groundmass*

**HAÜYNE IN LAVA**
*This specimen of blue haüyne crystals is from Monte Somma, Italy.*

### PROPERTIES
**GROUP** Silicates – tectosilicates
**CRYSTAL SYSTEM** Cubic
**COMPOSITION** $Na_3Ca(Al_3Si_3O_{12})(SO_4)$
**COLOUR** Blue, white, grey, yellow, green, or pink
**FORM/HABIT** Dodecahedral, octahedral
**HARDNESS** 5½–6
**CLEAVAGE** Distinct
**FRACTURE** Uneven to conchoidal, brittle
**LUSTRE** Vitreous to greasy
**STREAK** Blue to white
**SPECIFIC GRAVITY** 2.5
**TRANSPARENCY** Transparent to translucent
**R.I.** 1.49–1.51

# HELVITE

**SAXONY**
*This part of Germany is a key locality for helvite crystals.*

TAKING ITS NAME from the Greek *helios*, meaning "sun", due to its colour, helvite forms crystals that are tetrahedral or pseudo-octahedral. It is also found in rounded aggregates. It is yellow to yellow-green, red-brown to brown, and darkens with weathering. Helvite is found in granitic and syenitic pegmatites, and in hydrothermal deposits. Crystals up to 12cm (5in) long come from Cosquín, Argentina. Other localities include Mont St.-Hilaire, Canada; Aquiles Serdán, Mexico; and Colorado, USA.

*Rock groundmass*

*Helvite*

### PROPERTIES
**GROUP** Silicates – tectosilicates
**CRYSTAL SYSTEM** Cubic
**COMPOSITION** $MnBe_3(SiO_4)_3S$
**COLOUR** Yellow, red-brown, or brown
**FORM/HABIT** Tetrahedral
**HARDNESS** 6–6½
**CLEAVAGE** Distinct
**FRACTURE** Conchoidal to uneven, brittle
**LUSTRE** Vitreous to resinous
**STREAK** White
**SPECIFIC GRAVITY** 3.2–3.4
**TRANSPARENCY** Translucent

**CRYSTALS**
*These crystals of yellow helvite on a rock groundmass are from Saxony, Germany.*

# CANCRINITE

**QUEBEC**
*Mont St.-Hilaire, Canada, is a cancrinite source.*

CANCRINITE IS A feldspathoid mineral. Its relatively rare crystals can be several centimetres across. It is usually fine-grained and massive, or in columnar masses. It is pale to dark yellow, pale orange, pale violet, or pink to purple. Cancrinite forms in syenites, in contact metamorphic zones, and in some pegmatites. Fine specimens come from the Lovozero Massif, Russia; Laacher See, Germany; and Litchfield, Maine, USA.

### PROPERTIES
**GROUP** Silicates – tectosilicates
**CRYSTAL SYSTEM** Hexagonal/trigonal
**COMPOSITION** $(NaCa)_8(Al_6Si_6)O_{24}(CO_3)_2·2H_2O$
**COLOUR** Pale to dark yellow, orange, violet, pink, or purple
**FORM/HABIT** Massive
**HARDNESS** 5–6
**CLEAVAGE** Perfect, poor
**FRACTURE** Uneven
**LUSTRE** Vitreous
**STREAK** White
**SPECIFIC GRAVITY** 2.5
**TRANSPARENCY** Transparent to translucent

*Cancrinite*   *Syenite*

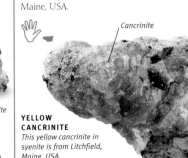

**YELLOW CANCRINITE**
*This yellow cancrinite in syenite is from Litchfield, Maine, USA.*

# ZEOLITES

ZEOLITES ARE A GROUP of more than 50 water–containing silicate minerals that have a particularly open type of crystal structure. They have a three–dimensional crystal framework of $SiO_4$ and $AlO_4$ tetrahedra that surrounds open cavities in which molecules of water or elements such as potassium, calcium, or sodium can fit. The cavities are interconnected by channels that run the entire length of the crystal, and are open at the crystal surface. These channels permit the relatively easy movement of ions (electrically charged atoms) between cavities, allowing zeolites to trap or exchange ions moving through them. Some zeolites are extensively used in water softeners because they have the capacity to exchange their sodium ions for the calcium ions present in hard water. Some zeolites can function as molecular sieves: they allow only smaller ions to pass through the channels to be absorbed. These zeolites are particularly useful in petroleum and gas refining (see p.251).

**LAUMONTITE**
The calcium ions in the laumontite channels can be exchanged for sodium ions, making it useful in water softeners.

$AlO_4$ tetrahedron

$SiO_4$ tetrahedron

Calcium ion

$SiO_4$ tetrahedron

Cavity

$AlO_4$ tetrahedron

**NATROLITE CHANNELS**
In natrolite, the tetrahedra form chains, creating linked cavities in which water and sodium molecules fit.

# NATROLITE

ALLUDING TO ITS SODIUM CONTENT, natrolite takes its name from the Greek *natrium*, which means "soda". It is a hydrated sodium aluminosilicate. Its crystals are generally long and slender, with vertical striations. They may appear pseudotetragonal, and grow up to a metre in length. Natrolite is also found in radiating masses of acicular crystals, and compact masses. It is both pyroelectric and piezoelectric. It can be pale pink, colourless, white, red, grey, yellow, or green. Some specimens fluoresce orange to yellow under ultraviolet light.

Natrolite is found in cavities or fissures in basaltic rocks, volcanic ash deposits, and veins in gneiss, granite, and other rock types. It is often associated with other zeolites, apophyllite, quartz, and heulandite. Unusually fine specimens come from Golden, British Columbia, Canada; Larne, Northern Ireland; Hegau, Germany; Mumbai, India; Canterbury, New Zealand; Cape Grim, Tasmania, Australia; and Bound Brook, New Jersey, and the Dallas Gem Mine, California, USA.

**TASMANIA**
Fine natrolite crystals come from Tasmania, Australia.

## WIDESPREAD ZEOLITE

One of the most widespread of the zeolites, natrolite was also one of the earliest studied of this mineral group. As a result, it was the first for which ion-exchange properties were discovered. It became widely used for softening water.

**NATROLITE IN CAVITY**
Typically natrolite is found in cavities or vesicles in basaltic rocks.

### PROPERTIES

| | |
|---|---|
| **GROUP** | Silicates – tectosilicates |
| **CRYSTAL SYSTEM** | Orthorhombic |
| **COMPOSITION** | $Na_2Al_2Si_3O_{10}\cdot2H_2O$ |
| **COLOUR** | Pale pink, colourless, white, grey, red, yellow, or green |
| **FORM/HABIT** | Acicular |
| **HARDNESS** | 5–5$^{1}/_{2}$ |
| **CLEAVAGE** | Perfect |
| **FRACTURE** | Uneven, brittle |
| **LUSTRE** | Vitreous to pearly |
| **STREAK** | White |
| **SPECIFIC GRAVITY** | 2.3 |
| **TRANSPARENCY** | Transparent to translucent |

**NATROLITE CRYSTALS**
This specimen comprises a radiating mass of slender, prismatic, transparent to translucent natrolite crystals.

Radiating crystals

Transparent to translucent crystals

**KLAPROTH**
German chemist and neurologist Martin Heinrich Klaproth named natrolite in 1803.

249

# GMELINITE

**COUNTY ANTRIM**
*Gmelinite is found in County Antrim, Northern Ireland.*

GMELINITE IS NAMED FOR German mineralogist and chemist C.G. Gmelin. Its crystals are hexagonal plates or hexagonal prisms. It can be colourless, white, pale yellow, greenish, orange, pink, and red. It occurs in silica-poor volcanic rocks, sodium-rich pegmatites, and marine basalts. It is often intergrown with chabazite (see p.252), and associated with other zeolites, quartz, aragonite, and calcite. It is widespread but found in only small amounts. Localities include Nova Scotia, Canada; Mornington Peninsula, Australia; and New Jersey, USA.

**CANADIAN GMELINITE**
*This specimen is from Two Islands, Nova Scotia, Canada.*

*Tabular crystals*

### PROPERTIES

| | |
|---|---|
| **GROUP** | Silicates – tectosilicates |
| **CRYSTAL SYSTEM** | Hexagonal |
| **COMPOSITION** | (Na, K, $Ca_{0.5})_4Al_8Si_{16}O_{48}\cdot22H_2O$ |
| **COLOUR** | Colourless to white, yellow, green, orange, or red |
| **FORM/HABIT** | Platy, tabular |
| **HARDNESS** | $4^1/_2$ |
| **CLEAVAGE** | Good |
| **FRACTURE** | Uneven to conchoidal, brittle |
| **LUSTRE** | Vitreous to dull |
| **STREAK** | White |
| **SPECIFIC GRAVITY** | 2.1 |
| **TRANSPARENCY** | Transparent to opaque |

# HARMOTOME

**SAXONY**
*Harmotome is found in the Harz Mountains.*

COMMONLY FORMING blocky crystals, this zeolite also occurs as multiple, pseudotetragonal, pseudo-orthorhombic, or cruciform twins. All forms appear glassy, and most vary from colourless to grey. Harmotome occurs in low-temperature hydrothermal veins, in volcanic rocks, and as an alteration product of barium-containing feldspars. It is widespread, with good specimens from Idar-Oberstein, Germany; Thunder Bay, Canada; and Taimyr, Russia.

*Blocky crystals*

### PROPERTIES

| | |
|---|---|
| **GROUP** | Silicates – tectosilicates |
| **CRYSTAL SYSTEM** | Monoclinic |
| **COMPOSITION** | $(Ba_{0.5},Ca_{0.5},K, Na)_5Al_5 Si_{11}O_{32}\cdot12H_2O$ |
| **COLOUR** | Colourless, white, grey |
| **FORM/HABIT** | Blocky, multiple twins |
| **HARDNESS** | $4^1/_2$–5 |
| **CLEAVAGE** | Good, fair |
| **FRACTURE** | Uneven to conchoidal, brittle |
| **LUSTRE** | Vitreous |
| **STREAK** | White |
| **SPECIFIC GRAVITY** | 2.5 |
| **TRANSPARENCY** | Transparent to opaque |

**WHITE CRYSTALS**
*These well-formed, blocky crystals of harmotome are set on a rocky groundmass.*

# LAUMONTITE

**INDIA**
*Mumbai, India, is one of the localities that has high-quality laumontite.*

LAUMONTITE is a common zeolite mineral. The French mineralogist François P.N.G. de Laumont collected the first specimen for study in 1785, and it was named for him. It occurs as simple prismatic crystals, or often as "swallowtail" twins, and also forms massive, columnar, fibrous, and radiating aggregates. Generally colourless, white, pink, or red, it can become powdery grey, pink, or yellow when partially dehydrated. Some specimens fluoresce white to yellow. Laumontite is typically found filling veins and vesicles in igneous rocks, and is also found in hydrothermal veins, pegmatites, and metamorphic rocks, and it is one of the more abundant zeolites present in sedimentary rocks. It is a widespread mineral, with excellent specimens coming from Mumbai, India; Alamos, Sonora, Mexico; and California, Oregon, and Washington, USA. The crystal structure of laumontite makes it especially useful in water softeners.

### PROPERTIES

| | |
|---|---|
| **GROUP** | Silicates – tectosilicates |
| **CRYSTAL SYSTEM** | Monoclinic |
| **COMPOSITION** | $CaAl_2Si_4O_{12}\cdot4H_2O$ |
| **COLOUR** | Colourless, white, red |
| **FORM/HABIT** | Prismatic |
| **HARDNESS** | 3–4 |
| **CLEAVAGE** | Perfect |
| **FRACTURE** | Uneven |
| **LUSTRE** | Vitreous, dull when dehydrated |
| **STREAK** | Colourless |
| **SPECIFIC GRAVITY** | 2.3 |
| **TRANSPARENCY** | Translucent |

*Prismatic crystal*

*Vitreous lustre*

## USE IN WATER SOFTENERS

All zeolites have a crystal structure composed of a lattice of interlinked channels of cavities in which ions can be exchanged or trapped. This open crystal structure is known as a molecular sieve (see p.252). In laumontite, sodium ions are exchanged for calcium ions, making it ideal for softening hard water.

**WATER SOFTENER**

**CRYSTALLINE LAUMONTITE**
*This specimen shows a mass of prismatic laumontite crystals.*

# HEULANDITE

**AURANGABAD, INDIA**
*This part of India is a major source of heulandite crystals.*

HEULANDITE IS A NAME given to a series of zeolite minerals. The group was named in 1822 for British collector and mineral dealer J.H. Heuland. It is a calcium sodium aluminosilicate hydrate. Crystals are elongated, tabular, and widest at the centre, creating a characteristic coffin shape. They can be up to 12cm (5in) long. Heulandite is usually colourless or white, but can also be red, grey, yellow, pink, green, or brown. Heulandite is a low-temperature zeolite found in a wide range of geological environments: with other zeolite minerals filling cavities in granites, pegmatites, and basalts; in metamorphic rocks; and in weathered andesites and diabases (dolerites). It is found worldwide, with notable localities being Aurangabad, India; Gunnedah, NSW, Australia; the Faroe Islands; Berufjördhur, Fassarfell, and Randafell, Iceland; Sakha (Yakutia), Russia; Paraná, Brazil; Kilpatrick Hills, Scotland; and Oregon and New Jersey, USA. The heulandite structure and the substitution of aluminium atoms for some of the silicon atoms give it excellent properties for use in water softeners and in petroleum refining.

*Pearly lustre*

*Coffin shape*

*Tabular crystals*

## PROPERTIES

| | |
|---|---|
| **GROUP** | Silicates – tectosilicates |
| **CRYSTAL SYSTEM** | Monoclinic |
| **COMPOSITION** | $CaAl_2Si_7O_{18}\cdot6H_2O$ |
| **COLOUR** | Colourless, white |
| **FORM/HABIT** | Tabular |
| **HARDNESS** | $3\frac{1}{2}-4$ |
| **CLEAVAGE** | Perfect |
| **FRACTURE** | Uneven, brittle |
| **LUSTRE** | Vitreous to pearly |
| **STREAK** | Colourless |
| **SPECIFIC GRAVITY** | 2.2 |
| **TRANSPARENCY** | Transparent to translucent |

## USE IN PETROL REFINING

Zeolites such as heulandite are widely used in petroleum refining as molecular sieves. The open channels of cavities in their crystal structure act as a microscopic sieve, giving them the ability to filter out particles of molecular size. This enables them to remove undesirable impurities, separate gas from liquid, and to filter various petroleum components. The chemical structure of heulandite makes it particularly useful in petrol refining. It is used to remove paraffins to improve the combustion properties of the petrol.

**PETROCHEMICAL PLANT**

**EXEMPLARY CRYSTALS**
*The classic coffin shape of heulandite crystals can be seen in this specimen from Berufjördhur, Austurland, Iceland.*

# PHILLIPSITE

**CANADA**
*Phillipsite occurs at Mont St.-Hilaire, Quebec.*

THE ZEOLITE PHILLIPSITE was named in 1825 for the Geological Society of London's founder, William Phillips. It is commonly found in multiple-twinned, pseudo-orthorhombic, prismatic crystals, blocky crystals, or radiating aggregates. It is usually colourless or white, but can also be pink, red, or light yellow.

It is a common, low-temperature zeolite filling cavities and fissures in basalt, and in ore veins. Localities include Oregon, USA; Giessen, Germany; Sicily, Italy; and Mont St.-Hilaire, Canada.

## PROPERTIES

| | |
|---|---|
| **GROUP** | Silicates – tectosilicates |
| **CRYSTAL SYSTEM** | Monoclinic |
| **COMPOSITION** | $K(Ca_{.5},Na)_2(Si_5Al_3)O_{16}\cdot6H_2O$ |
| **COLOUR** | Colourless, white, pink, red, or yellow |
| **FORM/HABIT** | Pseudo-orthorhombic, tetragonal, cubic |
| **HARDNESS** | $4-4\frac{1}{2}$ |
| **CLEAVAGE** | Distinct, indistinct |
| **FRACTURE** | Uneven, brittle |
| **LUSTRE** | Vitreous |
| **STREAK** | Colourless |
| **SPECIFIC GRAVITY** | 2.2 |
| **TRANSPARENCY** | Transparent to translucent |

*Rock groundmass*

*Blocky crystals*

**BLOCKY PHILLIPSITE**
*This mass of blocky, twinned phillipsite crystals covers a rock groundmass.*

# SCOLECITE

**HEBRIDES**
*Fine scolecite specimens occur in the Scottish Hebridean islands of Mull and Skye.*

SCOLECITE IS A ZEOLITE. Its crystals tend to be pseudotetragonal or pseudo-orthorhombic. It is both pyroelectric and piezoelectric. Although generally white or colourless, it can be pink, salmon, red, or green. It is a common hydrothermal zeolite, found in basalts, andesites, gabbros, gneisses, amphibolites, and contact metamorphic zones, often associated with quartz, other zeolites, calcite, and prehnite. Large crystals come from Pune, India; Rio Grande do Sul, Brazil; the Isles of Mull and Skye, Scotland; and Cowlitz County, Washington, USA.

## PROPERTIES

| | |
|---|---|
| **GROUP** | Silicates – tectosilicates |
| **CRYSTAL SYSTEM** | Monoclinic |
| **COMPOSITION** | $CaAl_2Si_3O_{10}\cdot3H_2O$ |
| **COLOUR** | Colourless, white, pink, red or green |
| **FORM/HABIT** | Thin prismatic |
| **HARDNESS** | $5-5\frac{1}{2}$ |
| **CLEAVAGE** | Perfect |
| **FRACTURE** | Uneven, brittle |
| **LUSTRE** | Vitreous |
| **STREAK** | Colourless |
| **SPECIFIC GRAVITY** | 2.3 |
| **TRANSPARENCY** | Transparent to opaque |

**SLENDER CRYSTALS**
*Scolecite crystals tend to be thin, elongated prisms, as here. They can reach 35cm (14in) in length.*

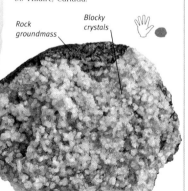

*Vertical striations*

*Prismatic crystal*

251

# MESOLITE

**PUNE, INDIA**
*Superb-quality mesolite comes from India.*

FORMING AS LONG, SLENDER NEEDLES, radiating masses, and less commonly, compact masses or fibrous stalactites, mesolite belongs to the zeolite group. Its name is from the Greek for "middle" and "stone", being chemically intermediate between natrolite and scolecite. Widespread, it is found in cavities in basalt and andesite, and in hydrothermal veins. Exceptional specimens occur in India; the Czech Republic, Antarctica, and the Faroe Islands; and in Washington, Oregon, and Colorado, USA.

**WHITE MESOLITE**
*Needles of white mesolite rest on a bed of green apophyllite crystals.*

**HAIR-LIKE TUFTS**
*When formed in hair-like tufts, mesolite is known as cotton stone.*

*Tufts of acicular crystals*

*Silky lustre*

### PROPERTIES
**GROUP** Silicates – tectosilicates
**CRYSTAL SYSTEM** Monoclinic
**COMPOSITION** $Na_2Ca_2(Al_6Si_9)O_{30}\cdot8H_2O$
**COLOUR** Colourless, white
**FORM/HABIT** Acicular
**HARDNESS** 5
**CLEAVAGE** Perfect
**FRACTURE** Uneven, brittle
**LUSTRE** Vitreous to silky
**STREAK** White
**SPECIFIC GRAVITY** 2.3
**TRANSPARENCY** Transparent to translucent

# THOMSONITE

**RUSSIA**
*Russia is one of many parts of the world where this mineral is found.*

THOMSONITE BELONGS TO the zeolite group. Its crystals can be bladed to blocky prismatic, or long, coarse acicular prisms. It can be colourless, white, pink, red, or yellow. Widespread, it occurs in cavities in basalts and syenites, and less often in contact-metamorphic zones and pegmatites. Localities include Italy, Scotland, Russia, Germany, the Czech Republic, Japan, and New Jersey, Oregon, and Colorado, USA.

**THOMAS THOMSON**
*The Scots chemist Thomson lent his name to the mineral in 1820.*

**ACICULAR CRYSTALS**
*Groups of radiating, acicular thomsonite crystals cluster on a basalt groundmass.*

*Radiating acicular crystals*

*Vitreous lustre*

### PROPERTIES
**GROUP** Silicates – tectosilicates
**CRYSTAL SYSTEM** Orthorhombic
**COMPOSITION** $NaCa_2(Al_5Si_5O_{20})\cdot6H_2O$
**COLOUR** Colourless, white
**FORM/HABIT** Lamellar or radiating aggregates
**HARDNESS** 5–5½
**CLEAVAGE** Perfect
**FRACTURE** Uneven to subconchoidal
**LUSTRE** Vitreous to pearly
**STREAK** White
**SPECIFIC GRAVITY** 2.3–2.4
**TRANSPARENCY** Transparent to translucent

# CHABAZITE

**IRELAND**
*Beautiful, large chabazite crystals can be found in Northern Ireland.*

A COMMON ZEOLITE MINERAL, chabazite is a sodium calcium aluminosilicate hydrate, its crystals forming distorted cubes or pseudorhombohedrons consisting of multiple twins. Crystals may also be prismatic, with twinning common in all forms. It can be colourless, white, cream, pink, red, orange, yellow, or brown. It takes its name from the Greek *chabazios* or *chalazios*, "hailstone". Chabazite is found in cavities in basalt or andesite, volcanic ash deposits, pegmatites, and granitic and metamorphic rocks. It is widespread, with splendid 2.5–5cm (1–2in) crystals occurring in Northern Ireland, Iceland, Germany, Scotland, the Czech Republic, Hungary, India, Canada, Australia, and in New Jersey and Oregon, USA.

### PROPERTIES
**GROUP** Silicates – tectosilicates
**CRYSTAL SYSTEM** Hexagonal/trigonal
**COMPOSITION** $(Na,Ca_{0.5},K)_4Al_4Si_8O_{24})\cdot12H_2O$
**COLOUR** Colourless, white, pink, orange, yellow or brown
**FORM/HABIT** Pseudocubic rhombohedral
**HARDNESS** 4–5
**CLEAVAGE** Indistinct
**FRACTURE** Uneven, brittle
**LUSTRE** Vitreous
**STREAK** White
**SPECIFIC GRAVITY** 2–2.2
**TRANSPARENCY** Transparent to translucent

**PSEUDOCUBIC CRYSTALS**
*This group of pseudocubic chabazite crystals comes from the Bay of Fundy, Nova Scotia, Canada.*

## SIEVE STRUCTURE

Zeolites, such as chabazite, have an open crystal structure that permits small molecules to pass through, as if through a sieve, while preventing the passage of larger molecules. This structure makes it particularly useful for filtering methane from the gases emitted by decaying, organic waste matter.

**MOLECULAR SIEVE**
*Small molecules can pass through the spaces between the radiating ions (red spheres).*

*Pseudocubic chabazite crystal*

# STILBITE

**HARZ**
*The German Harz Mountains are a source of many minerals, including stilbite.*

A HYDRATED SODIUM CALCIUM aluminosilicate, stilbite derives its name from the Greek "to shine", a reference to its vitreous to pearly lustre. Its crystals are tabular, commonly twinned, and also occur in sheaf-like "bow tie" aggregates. Usually colourless or white, it can be yellow, brown, salmon-pink, or red, and rarely green, blue, or black. Most stilbites are calcium-rich, but some are sodium-dominant.

A member of the zeolite group, stilbite occurs in mafic igneous rocks, granitic pegmatites, gneisses and schists, and hot spring deposits. It is widespread, with fine crystals, some exceeding 10cm (4in), occurring in Germany, Russia, Scotland, India, Australia, Mexico, Brazil, Iceland, Canada, and New Jersey, Pennsylvania, Oregon, and Washington, USA.

**REFINING PETROL**
*The sieve-like crystal structure of the zeolite stilbite enables it to separate hydrocarbons in the process of petroleum refining.*

## PROPERTIES

**GROUP** Silicates – tectosilicates
**CRYSTAL SYSTEM** Monoclinic
**COMPOSITION** $(Na,Ca_{0.5},K)_9$ $(Al_9Si_{27}O_{72})\cdot 28\ H_2O$
**COLOUR** Colourless, white, pink
**FORM/HABIT** Tabular
**HARDNESS** $3^1/2–4$
**CLEAVAGE** Perfect
**FRACTURE** Conchoidal, brittle
**LUSTRE** Vitreous to pearly
**STREAK** White
**SPECIFIC GRAVITY** 2.2
**TRANSPARENCY** Transparent to translucent

**BOW TIES**
*A group of stilbite crystals in "bow tie" aggregates.*

**TABULAR CRYSTALS**
*This group of white, tabular stilbite crystals on a rock groundmass comes from the Faroe Islands.*

*Pearly lustre*

*Tabular crystals*

---

# ANALCIME

**QUEBEC**
*Crystals from Quebec, Canada, can measure as much as 25cm (10in) across.*

FORMERLY GROUPED WITH the feldspathoids, analcime is now classified as a zeolite. Most crystals are trapezohedral. Its sodium can be replaced with significant amounts of potassium, and the sodium to aluminium ratio can vary considerably. Variations in the order-disorder of the sodium–aluminium portion of its structure can vary the structure enough that it can be classified in several other crystal systems.

Most samples are colourless or white, but some are yellow, brown, pink, red, or orange. Its name comes from the Greek *analkis*, "weak", a reference to its weak electrical charge when heated or rubbed. Some samples of analcime fluoresce under ultraviolet light.

Analcime occurs in seams and cavities in basalt, diabase, granite, and gneiss and in extensive beds formed by precipitation from alkaline lakes. Excellent crystals come from Italy, New Zealand, Scotland, England, the Faroe Islands, Iceland, Russia, and Australia, as well as Maine, Pennsylvania, New Jersey, Colorado, Oregon, and Washington, in the USA.

## PROPERTIES

**GROUP** Silicates – tectosilicates
**CRYSTAL SYSTEM** Cubic
**COMPOSITION** $Na(AlSi_2)O_6\cdot H_2O$
**COLOUR** White, colourless, yellow, brown red, or orange
**FORM/HABIT** Trapezohedral
**HARDNESS** $5–5^1/2$
**CLEAVAGE** None
**FRACTURE** Subconchoidal, brittle
**LUSTRE** Vitreous
**STREAK** White
**SPECIFIC GRAVITY** 2.3
**TRANSPARENCY** Transparent to translucent

## SILICA GEL

One of the main products created from analcime is silica gel, which takes the form of tiny spheres. The readiness with which these spheres absorb moisture from the air has led to their use as an industrial drying agent or desiccator.

Silica, silicon oxide, is a colourless vitreous solid. The rigid gel form is made by coagulating a solution of sodium silicate and heating it to drive off water. The gel itself is colourless but a blue cobalt salt is added to the spheres, which turns pink as they absorb water. When all the spheres are pink, indicating that they are saturated, the gel can be regenerated by heating it up.

Silica gel has been known since 1640, but came into prominence only in World War I, when it was used as an absorbent in gas masks.

**SILICA GEL SPHERES**
*The blue spheres turn pink as they become saturated with water.*

*Trapezohedral crystals*

*Vitreous lustre*

*Calcite crystals*

**TRAPEZOHEDRONS**
*This group of superbly crystallized analcime trapezohedrons from the Dean Quarry, Cornwall, England, rests on a bed of calcite crystals.*

253

# SCAPOLITE

**CABOCHON**
*Scapolite not quite transparent enough for faceting is cut en cabochon, as in this pale purple example.*

**STEP**
*The colour of this yellow scapolite is both enhanced and displayed by faceting in a step cut. Yellow scapolite can be confused with heliodor.*

**MIXED**
*This colourless scapolite is faceted in an oval mixed cut for maximum brilliance.*

SCAPOLITE WAS ORIGINALLY believed to be a single mineral, a calcium aluminosilicate. Now it is defined as a compositional series from calcium-bearing meionite to sodium-bearing marialite. Pure end members are not found in nature, although they have been synthesized. Natural scapolites seldom approach even 80 per cent of either end member. The name scapolite originated in the Greek *scapos*, meaning "rod" and *lithos*, "stone". Scapolite is still used in the gem trade to refer to any members of the scapolite group cut as gemstones.

Scapolite crystals are short to medium prismatic, often with prominent pyramidal and pinacoidal development, and can appear similar to those of vesuvianite (see p.295). They occur up to 25cm (10in) in length. Scapolite can also be granular and massive. It can be colourless, white, grey, yellow, orange, pink, or purple, and may exhibit chatoyancy, creating a cat's-eye effect when *en cabochon*. Scapolites are found principally in metamorphic rocks, particularly in regionally metamorphosed rocks such as marble, calcareous gneiss,

and greenschist. The largest crystals are usually found in marble. Scapolites form less commonly around igneous intrusions, especially those of feldspathic pegmatites and gabbros. Principal occurrences for large crystals are in Quebec and Ontario, Canada, and New York, USA. Other important sources are in Finland, Mexico, Italy, the Czech Republic, Germany, Russia, and Norway. Gem-quality material comes from China, Myanmar, Australia, Mozambique, Ghana, Chile, and Tanzania.

Several names have been given to scapolites in the past, but have now been discarded: wernerite was the former group name, and dipyre and mizzonite were names for intermediate members of the group. Marialite was named in 1866 by the German mineralogist vom Rath for his wife Maria. Meionite was named in 1801 by French mineralogist René Just Haüy. The word is derived from the Greek word *meion*, meaning "less", alluding to its pyramidal faces, which are less steep than those of vesuvianite, with which this mineral is sometimes confused.

*Prismatic scapolite crystal*

*Distinct cleavage*

*Vitreous lustre*

*Rock groundmass*

*Vitreous lustre*

*Massive habit*

*Massive habit*

**MARIALITE CRYSTALS**
In this specimen, pink prismatic marialite crystals have formed in a cluster on a rock groundmass.

**MASSIVE VARIETIES**
Both of these specimens are massive in habit. The meionite variety (right) is from the Limberg Quarry, Pargas, Finland.

**MEIONITE**

**MASSIVE SCAPOLITE**

## PROPERTIES

| | |
|---|---|
| **GROUP** | Silicates – tectosilicates |
| **CRYSTAL SYSTEM** | Tetragonal |
| **COMPOSITION** | Marialite $Na_4(Al_3Si_9O_{24})Cl$ Meionite $Ca_4(Al_6Si_6O_{24})(CO_3SO_4)$ |
| **COLOUR** | Colourless, white, grey, yellow, orange, or pink |
| **FORM/HABIT** | Prismatic |
| **HARDNESS** | 5–6 |
| **CLEAVAGE** | Good |
| **FRACTURE** | Uneven to conchoidal |
| **LUSTRE** | Vitreous |
| **STREAK** | White |
| **SPECIFIC GRAVITY** | 2.5– 2.7 |
| **TRANSPARENCY** | Transparent to opaque |
| **R.I.** | 1.53–1.60 |

# POLLUCITE

CONTAINING CALCIUM, SODIUM, rubidium, and lithium, pollucite is a complex hydrous aluminosilicate in the zeolite group. It is one of two minerals that were discovered in 1846 and named after the Gemini twins in Greek myth, Castor and Pollux. Pollucite was named for Pollux; castonite was named for the other twin, although it has since been renamed petalite (see p.268). Pollucite is commonly massive, or forms rounded, apparently corroded crystals. It is usually colourless or white, but can be pink, blue, or violet. Pollucite forms a solid-solution series with analcime (see p.253). It is found worldwide only in rare-earth-bearing granitic pegmatites, where it occurs with spodumene, petalite, quartz, lepidolite, eucryptite, and apatite. Crystals up to 60cm (24in) across have been found at Kamdeysh, Afghanistan, and it occurs at a number of other Afghan localities. The deposit at Bernic Lake, Manitoba, Canada, has an estimated 350,000 tonnes of mostly massive material, although crystals up to 2.5cm (1in) across are also found there. Gem material is found at Elba, Italy, and in the USA in Maine, Connecticut, South Dakota, and in California.

**GILGIT, PAKISTAN**
*Crystals up to 8cm (3in) across are found in Gilgit.*

**GEMINI TWINS**
*Pollucite was named for one of the two Gemini twins, Pollux, seen in this relief by Agostino di Duccio (1418–1481).*

*Massive habit*

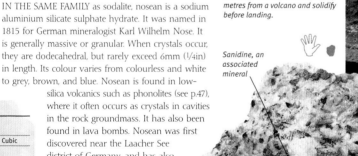

*Greasy lustre*

**CAESIUM ATOMIC CLOCK**
*Pollucite is a source of caesium, used in atomic clocks for ultra-precise time measurement.*

## PROPERTIES

| | |
|---|---|
| **GROUP** | Silicates – tectosilicates |
| **CRYSTAL SYSTEM** | Cubic |
| **COMPOSITION** | $(Cs,Na)(AlSi_2)O_6H_2O$ |
| **COLOUR** | Colourless, white, pink, blue or violet |
| **FORM/HABIT** | Massive |
| **HARDNESS** | $6\frac{1}{2}$–7 |
| **CLEAVAGE** | None |
| **FRACTURE** | Conchoidal to uneven |
| **LUSTRE** | Vitreous to greasy |
| **STREAK** | White |
| **SPECIFIC GRAVITY** | 2.7–3.0 |
| **TRANSPARENCY** | Transparent to translucent |
| **R.I.** | 1.51 |

**MASSIVE POLLUCITE**
*This massive pollucite specimen is from Buckfield, Maine, USA. Crystals with sharp faces are uncommon.*

**STEP**
*Colourless pollucite is cut for collectors in a variety of shapes. The step cut emphasizes its transparency.*

**CUSHION**
*Colourless pollucite can be cut in cushion or brilliant cuts for maximum sparkle. It is faceted mainly for collectors.*

# NOSEAN

IN THE SAME FAMILY as sodalite, nosean is a sodium aluminium silicate sulphate hydrate. It was named in 1815 for German mineralogist Karl Wilhelm Nose. It is generally massive or granular. When crystals occur, they are dodecahedral, but rarely exceed 6mm (1/4in) in length. Its colour varies from colourless and white to grey, brown, and blue. Nosean is found in low-silica volcanics such as phonolites (see p.47), where it often occurs as crystals in cavities in the rock groundmass. It has also been found in lava bombs. Nosean was first discovered near the Laacher See district of Germany, and has also been found in Cornwall, England; the Alban Hills, south of Rome, Italy; Cantal, France; the Cape Verde Islands; and in the USA at Cripple Creek, Colorado, and the La Sal Mountains of Utah.

**CRIPPLE CREEK, USA**
*Nosean occurs in this area of Colorado.*

## PROPERTIES

| | |
|---|---|
| **GROUP** | Silicates – tectosilicates |
| **CRYSTAL SYSTEM** | Cubic |
| **COMPOSITION** | $Na_8Al_6Si_6O_{24}(SO_4)\cdot H_2O$ |
| **COLOUR** | Colourless, white, grey, brown, or blue |
| **FORM/HABIT** | Massive, granular |
| **HARDNESS** | $5\frac{1}{2}$ |
| **CLEAVAGE** | Poor |
| **FRACTURE** | Uneven to conchoidal |
| **LUSTRE** | Vitreous |
| **STREAK** | Colourless |
| **SPECIFIC GRAVITY** | 2.3 |
| **TRANSPARENCY** | Transparent to translucent |

**LAVA BOMBS**
*Globules of molten rock containing nosean can be hurled hundreds of metres from a volcano and solidify before landing.*

*Sanidine, an associated mineral*

*Well-formed nosean crystals*

**NOSEAN CRYSTALS**
*Here, small, dodecahedral nosean crystals have formed within a rock groundmass of sanidine. The vitreous lustre appears on fresh crystal surfaces.*

255

# PHYLLOSILICATES

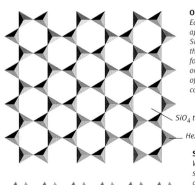

**OVERHEAD VIEW**
*Each phyllosilicate layer consists of approximately hexagonal rings formed of SiO₄ tetrahedra. Each tetrahedron shares three corners with other SiO₄ tetrahedra to form this arrangement. The fourth corner, or O atom, remains unshared. In the middle of the hexagon formed by the unshared corners, there is room for an OH group.*

*SiO₄ tetrahedron*

*Hexagonal ring*

**SIDE VIEW**
*When the sheet is viewed from the side, it is clear that all the unshared corners of the tetrahedra point in the same direction.*

*Unshared corner*

*SiO₄ tetrahedron*

PHYLLOSILICATES TAKE their name from the Greek *phyllum*, meaning "leaf". They are also referred to as sheet silicates or disilicates. They are composed of flat sheets of silicate tetrahedra, each tetrahedron consisting of a central silicon atom surrounded by four oxygen atoms that form the corners. The silicon atoms are also arranged at the corners of hexagons.

The chemical formulae for the various phyllosilicate minerals contains $Si_2O_5$ and its multiples, reflecting the fact that each silicate tetrahedron shares one oxygen atom with each of three other tetrahedra. The fourth oxygen atom is unshared.

# SERPENTINE

SERPENTINE IS A GROUP of at least 16 hydrous silicate minerals. Some are rich in magnesium, some in iron, and others in aluminium, manganese, lithium, nickel, zinc, or calcium. There are four major varieties, which can also be regarded as structural subgroups for the lesser varieties: chrysotile, a fibrous variety used as asbestos (see p.258); antigorite, a variety occurring in either corrugated plates or fibres; lizardite, a very fine-grained, platy variety; and amesite, which occurs in platy or pseudohexagonal, columnar crystals. Serpentine minerals generally occur as masses of tiny, intergrown crystals.

Although their chemistry is complex, the varieties are similar in appearance, and it is possible to describe the characteristics of the group as a whole. Serpentine is

**KYNANCE COVE, CORNWALL, ENGLAND**
*This area of serpentinite rocks, containing serpentine minerals, is in south-west England.*

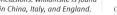

**WILLIAMSITE CABOCHON**
*This cabochon of williamsite has patchy coloration due to inclusions. Williamsite is found in China, Italy, and England.*

**VENUS OF GALGENBERG**
*Carved in serpentine about 30,000 years ago, this dancing figure is from Galgenberg, Austria.*

*Translucent*

named in allusion to its mottled appearance, which resembles a snake's skin. Its varieties may be yellow, green, greyish, white, or greenish-blue. Serpentine is always a secondary mineral – that is, derived from the chemical alteration of other minerals such as olivine (see p.299), amphiboles (see pp.278–85), or pyroxenes (see pp.270–77). Sometimes it entirely replaces the olivine in peridotites, or radically replaces other

*No cleavage*

*Greasy lustre*

## MINOAN CARVING

The Minoan civilization flourished on the island of Crete from about 3000BC to about 1100BC. In Minoan Crete, local serpentine was carved into vases, bowls, and other small objects, such as the elaborate seals for which the civilization is noted (see p.105). Unlike in other stone-carving cultures of the period, no large-scale sculptures were created. Also notable by their absence are carvings and inscriptions exaggerating the achievements of kings and rulers. Instead Minoan carvings tended to depict common rites and rituals. Among other civilizations of the ancient world, including the Egyptians and earlier Greek civilizations, soft local stone was used for carvings and vessels.

**SERPENTINE CUP**
*Known as the "chieftan cup", this serpentine cup – found on the Minoan site of Ayia Triada – depicts a prince receiving skins after an animal hunt.*

**BLOSSOM BOWL**
*Carved out of serpentine in a six-petalled form, this "blossom bowl" was a common style of Minoan vase.*

## PROPERTIES

**GROUP** Silicates – phyllosilicates

**CRYSTAL SYSTEM** Monoclinic

**COMPOSITION**
$(Mg,Fe,Ni)_3Si_2O_5(OH)_4$

**COLOUR** White, grey, yellow, green, or greenish-blue

**FORM/HABIT** Usually massive or pseudomorphous

**HARDNESS** 3½–5½

**CLEAVAGE** Perfect but not visible

**FRACTURE** Conchoidal to splintery

**LUSTRE** Subvitreous to greasy, resinous, earthy, dull

**STREAK** White

**SPECIFIC GRAVITY** 2.5–2.6

**TRANSPARENCY** Translucent to opaque

**R.I.** 1.55 – 1.56

**PRECIOUS SERPENTINE**
*This beautiful specimen is composed of a number of serpentine minerals. It is a high-quality piece of the kind often carved and sold as jade. It can be distinguished from jade by its softness.*

The silicate sheets are interlayered with sheets of other elements, principally metal atoms. The layers are stacked with unshared oxygen atoms towards the centre, and as a result these groups are weakly held together. Consequently, bonding is strong within the sheets, but the sheets are weakly bonded to each other – like a stack of leaves.

Phyllosilicates therefore tend to have a tabular, platy, or flaky habit, and a single pronounced cleavage along the layering. They also tend to have a low specific gravity, a relatively low hardness, and have considerable flexibility and elasticity of cleavage layers. The talc group of minerals, of which pyrophyllite is one, and the mica group are both good examples of phyllosilicates.

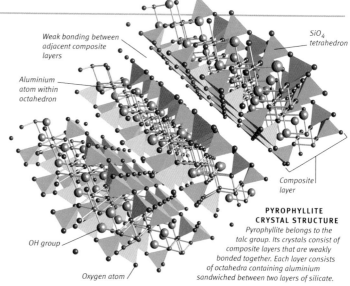

*Weak bonding between adjacent composite layers*

*Aluminium atom within octahedron*

*$SiO_4$ tetrahedron*

*Composite layer*

*OH group*

*Oxygen atom*

### PYROPHYLLITE CRYSTAL STRUCTURE
Pyrophyllite belongs to the talc group. Its crystals consist of composite layers that are weakly bonded together. Each layer consists of octahedra containing aluminium sandwiched between two layers of silicate.

minerals in other rock types; the new, altered rock is called serpentinite. It is characteristically found where highly altered, deep-seated rocks are exposed, such as along the crests and axes of great folds, in island arcs, and in Alpine mountain chains. Some serpentines take a high polish, and rich, translucent, apple-green varieties (sometimes called precious or noble serpentine) are carved and sold as "new jade". Other serpentines are quarried as ornamental stones, and are sometimes called serpentine marble. Serpentine is so widespread that it is impossible to list even the major occurrences, but huge quarrying operations take place in England, the USA, Germany, Canada, China, Afghanistan, South Africa, and Italy, to name but a few localities.

## LIZARDITE SERPENTINE

The lizardite variety of serpentine takes its name from its occurrence on the Lizard Peninsula of Cornwall, England. The peninsula is composed principally of Precambrian metamorphic rocks, including gneisses, schists, gabbros, and serpentines. The serpentine from this area has been quarried and carved for centuries.

**LIZARD BOX**
*This box made of lizardite is topped, appropriately, with a silver lizard.*

**SERPENTINE BOULDERS**
*This outcrop of serpentine boulders is in the Siskiyou National Forest, Oregon.*

**SERPENTINE MINERALS**
*These examples are three of the 16 in the serpentine mineral group, showing a variety of habits and colours.*

*Chrysotile*

*Fine-grained texture*

*Fibrous texture*

**SERPENTINITE GROUNDMASS**

**LIZARDITE**

**ANTIGORITE**

URAL MOUNTAINS
*Russia's Ural Mountains are
a treasure-house of mineral
deposits, including large
amounts of chrysotile.*

# CHRYSOTILE

THE MOST IMPORTANT asbestos mineral, chrysotile is the fibrous variety of the magnesium silicate mineral serpentine (see p.257) and accounts for about 95 per cent of all asbestos in commercial use. Chrysotile is in fact three distinct minerals: clinochrysotile, orthochrysotile, and parachrysotile. They are chemically identical, but orthochrysotile and parachrysotile have orthorhombic rather than monoclinic crystals. They are indistinguishable in hand specimens, and clinochrysotile and orthochrysotile may even occur within the same fibre. Individual chrysotile fibres are white and silky, but aggregate fibres in veins are usually green or yellowish. The mineral can even take on a golden appearance – its name is derived from the Greek for "hair of gold". Chrysotile fibres are actually tubes in which the structural layers of the mineral are rolled in the form of a spiral.

Chrysotile principally occurs as veins in altered peridotite, along with other serpentine minerals. Fibres are generally oriented across the vein and are less than 1.3cm ($\frac{1}{2}$in) long. Large deposits occur in Quebec, Canada, and in the Ural Mountains of Russia. It is also found in England, Scotland, Switzerland, Serbia, Australia, Mexico, South Africa, Zimbabwe, Swaziland, and in various locations in the USA.

## PROPERTIES

**GROUP** Silicates – phyllosilicates
**CRYSTAL SYSTEM** Monoclinic
**COMPOSITION**
$Mg_3Si_2O_5(OH)_4$
**COLOUR** White, green, yellowish, golden
**FORM/HABIT** Fibrous
**HARDNESS** 2–3
**CLEAVAGE** Perfect
**FRACTURE** None
**LUSTRE** Subresinous to greasy
**STREAK** White
**SPECIFIC GRAVITY** 2.6
**TRANSPARENCY** Translucent to opaque

**CHRYSOTILE FIBRES**
*The fibrous nature of chrysotile is apparent in this specimen, in which a mass of flexible chrysotile crystals rests on a rock groundmass.*

Mass of thin fibres

Bent and broken fibres

Greasy lustre

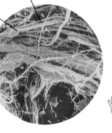

Broken fibre

Chrysotile fibre

**THIN SECTION**
*This false-colour scanning electron micrograph of chrysotile shows masses of intertwining tubes of fibres lying roughly parallel to one another.*

## ASBESTOS FIBRES

Asbestos is not a mineral name, but a general term applied to any of several minerals that readily separate into long, flexible fibres. Aside from chrysotile, the other asbestos minerals are amphiboles, and include the fibrous forms of actinolite, tremolite, anthophyllite, and riebeckite. Asbestos fibre is freed by crushing its surrounding rock, and the fibres are then further separated, usually by a blowing process. At one time, the longest fibres were spun into yarn, and the shorter fibres used in paper, millboard, and asbestos-cement building material. Asbestos was also widely used in brake linings, gaskets, and insulation; and in roofing shingles and floor and ceiling tiles. Spun asbestos fabrics were used for safety apparel and for theatre curtains. It was discovered in the 1970s that inhalation of some forms of asbestos fibres could result in a fatal form of lung cancer. Most countries now ban asbestos products.

**ASBESTOS TAILINGS**
*Asbestos tailings (waste products) are an environmental hazard in some locations.*

**ASBESTOS WORKERS**
*Protective clothing is worn for the safe removal of asbestos from buildings, ships, and other structures.*

# KAOLINITE

**KAOLINITE MINE**
*This kaolinite mine at Gunheath, St. Austell, is at the heart of the china clay industry in Cornwall, England. Kaolinite is also mined in China, Chile, Germany, Russia, the Czech Republic, France, and Brazil.*

CLAY MINERALS ARE FAR removed, at least in their outward appearance, from more attractive and glamorous minerals, such as gold and diamonds. Yet, by providing the raw material for clay, pottery, and tiles, they have played a vital part in the progress of human civilization. Important among these vital minerals is kaolinite. Its name originates from the hill in China (Kao-ling) where it was mined for centuries.

Kaolinite is an aluminium silicate hydrate that forms microscopic pseudohexagonal plates in compact or granular masses and in mica-like piles. Crystals range in size from about 0.1 micrometre to 10 micrometres long. Three other minerals, dickite, nacrite, and halloysite, are chemically identical to kaolinite but crystallize in the monoclinic system. All four have been found together, and are often visually indistinguishable.

Kaolinite is a natural product of the alteration of mica and plagioclase and alkali feldspars under the influence of water, dissolved carbon dioxide, and organic acids. It often occurs with other clay minerals, quartz, mica, sillimanite, tourmaline, and rutile. The ore of kaolinite is kaolin, millions of tonnes of which are used annually. Kaolin is used in both raw and refined form in agriculture; as a filler in some food (such as chocolate); mixed with pectin as an antidiarrhoeal; as a binder in talcum and other powders; as a filler and strengthener in rubber; as a paint extender; and as a dusting agent in foundry operations.

**NIGERIAN MASK**
*This late 19th- or early 20th-century helmet mask of a maiden spirit is mainly wood and kaolin.*

## CHINA CLAY

Kaolinite is the principal component of china clay, the soft white clay that is an essential ingredient in the manufacture of china and porcelain. When mixed with 20 to 35 per cent water, it can be moulded under pressure, and the shape is retained after the pressure is removed. Thus intricate ceramic products such as tableware and statuary can be formed or sculpted from it. It was the essential ingredient of Chinese porcelain. It began to be manufactured in Europe in the 17th century, creating a huge china-clay mining industry. Large deposits are thought to have formed when feldspar was eroded from rocks such as granite and deposited in lake beds, where it weathered into kaolin.

**CHINESE PORCELAIN**
*These two porcelain bowls, painted with delicate blue designs, were made during the Ming Dynasty (1368–1644).*

**CHINA CLAY MINE, CORNWALL**
*The discovery of china clay in Cornwall in 1746 made possible the English porcelain industry.*

**ROLLS OF PAPER**
*Kaolin is used in the making of paper. Glossy paper has a high kaolin content.*

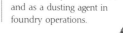

**IMPURE KAOLINITE**
*Kaolinite can be found mixed with other oxides, such as the iron oxide that imparts an orange colour to this specimen.*

*Earthy lustre*

*Quartz*

*Crystal shape of feldspar*

### PROPERTIES

| | |
|---|---|
| **GROUP** | Silicates – phyllosilicates |
| **CRYSTAL SYSTEM** | Triclinic |
| **COMPOSITION** | $Al_2Si_2O_5(OH)_4$ |
| **COLOUR** | White when pure |
| **FORM/HABIT** | Massive |
| **HARDNESS** | 2–2½ |
| **CLEAVAGE** | Perfect |
| **FRACTURE** | Unobservable |
| **LUSTRE** | Earthy |
| **STREAK** | White |
| **SPECIFIC GRAVITY** | 2.6 |
| **TRANSPARENCY** | Opaque |

**CORNISH KAOLINITE**
*The kaolinite in this specimen from Goonbarrow Pit, Cornwall, England, is derived from weathered potassium feldspar.*

# TALC

TALC IS A COMMON MINERAL, easily
distinguishable by its extreme softness.
Crystals are rare; talc is most commonly
found in foliated, fibrous, or massive
aggregates. It can be white, colourless,
pale to dark green, and yellowish to
brown, and is often found mixed with
other minerals, such as serpentine
and calcite. Dense, high-purity talc is
called steatite. Talc is a metamorphic
mineral found in veins and
in magnesium-rich rocks. It is
often associated with serpentine,
tremolite, and forsterite and occurs as
an alteration product of ultramafic
igneous rocks. It is widespread, and found
in most areas of the world where low-
grade metamorphism occurs. Talc is used
in lubricants, leather dressings, as well as
in ceramics, such as toilets,
paint, paper, roofing
materials, plastic, and
rubber.

*Greasy lustre*

*Translucent*

**METAMORPHIC
MINERAL**
*The lines in this
specimen were
produced by stress
during metamorphism.*

## PROPERTIES

**GROUP** Silicates –
phyllosilicates
**CRYSTAL SYSTEM** Triclinic
or monoclinic
**COMPOSITION**
$Mg_3Si_4O_{10}(OH)_2$
**COLOUR** White, colourless,
green, yellow to brown
**FORM/HABIT** Foliated and
fibrous masses
**HARDNESS** 1
**CLEAVAGE** Perfect
**FRACTURE** Uneven to
subconchoidal
**LUSTRE** Pearly to greasy
**STREAK** White
**SPECIFIC GRAVITY** 2.8
**TRANSPARENCY** Translucent

**TOILETRIES**
*Finely ground talc is the
principal component of talcum
powder and other toiletries.*

**RELIQUARY**
*This Buddhist
reliquary, with its
narrow base and
neck, and feather-like
design on the sides,
is made of steatite.*

## SOAPSTONE

The name soapstone is given to compact
masses of talc and other minerals due to
their soapy or greasy feel. Soapstone has
been used since ancient times for carvings,
ornaments, and utensils. In more recent
times, the Inuit people of Canada have
become known for their soapstone
carvings of birds and animals.

**LION–DOG**
*This Chinese lion–dog
forms part of a soapstone
seal, or "chop".*

# PYROPHYLLITE

**BISBEE, USA**
*One of several
locations in
the USA for
pyrophyllite
is Bisbee in
Arizona.*

PYROPHYLLITE, an aluminium silicate hydroxide,
rarely forms distinct crystals. It is usually found in
granular masses of flattened lamellae, but is often so
fine grained as to appear textureless. It is sometimes
found in coarse laths and radiating aggregates. It can
be colourless, white, cream, brownish-green, pale blue,
or grey. Its name is from the Greek for "fire" and
"leaf", referring to its tendency to exfoliate when
heated. Pyrophyllite forms by the low-grade
metamorphism of aluminium-rich sedimentary rock,
such as bauxite. It was carved by the
ancient Chinese into small images and
ornaments. It is widespread, with
commercial deposits in
California and North
Carolina, USA;
Canada; and South
Africa. Coarse,
radiating, lath-like
clusters are found
in Brazil and
Georgia, USA.

## COMMERCIAL USES

Pyrophyllite is mined commercially for use as a filler in paints and
rubber, in cosmetics, and in dusting powders. Some "talcum" powder,
may, in fact, be pyrophyllite. It has good insulating properties because
of its high melting point and low electrical conductivity,
and is more useful than talc in heat resistors. The main
world suppliers for industrial pyrophyllite are
China, Korea, and the USA.

**COSMETIC USES**
*Finely powdered
pyrophyllite gives a
sheen to lipsticks.*

**FIRE BRICKS**
*The very good
insulating properties
of pyrophyllite mean
that it is widely used
in heat resistors.*

**PYROPHYLLITE STARS**
*This aggregate of pyrophyllite
displays radiating groups of
laths, with associated quartz.*

*Radiating mass
of pyrophyllite
crystals*

*Quartz
crystals*

## PROPERTIES

**GROUP** Silicates –
phyllosilicates
**CRYSTAL SYSTEM** Triclinic
or monoclinic
**COMPOSITION**
$Al_2Si_4O_{10}(OH)_2$
**COLOUR** White, colourless,
brown-green, pale blue, grey
**FORM/HABIT** Compact
masses
**HARDNESS** 1–2
**CLEAVAGE** Perfect
**FRACTURE** Uneven
**LUSTRE** Pearly to dull
**STREAK** White
**SPECIFIC GRAVITY**
2.7–2.9
**TRANSPARENCY** Transparent
to translucent

# MUSCOVITE

MUSCOVITE, ALSO CALLED common mica, potash mica, or isinglass (the latter name from its use in Russia for window panes), is the most common member of the mica group. Typically tabular with a hexagonal or pseudohexagonal outline, crystals also commonly occur as lamellar masses and fine-grained aggregates. It is usually colourless or silvery-white, but it can also be brown, light grey, pale green, or rose-red. It is an extremely common rock-forming mineral, and occurs in metamorphic rocks, in granites, and in veins and pegmatites. A single crystal mined in India was 3m (10ft) in diameter and 5m (16ft) long.

**MICA WINDOW**
*Thin, transparent sheets of mica make good windows that are hard to break.*

**HOPEWELL MICA**
*This bird's foot was cut from mica in 500BC– AD500 near Iowa, USA.*

## THE MICA GROUP

The micas are among the principal rock-forming minerals. They are classic phyllosilicates: sheets of tightly bonded tetrahedra that are poorly bonded to each other. Minerals of the mica group have perfect cleavage, allowing them to be split into thin, flexible leaves. Micas are potassium aluminosilicate hydrates with sodium, lithium, magnesium, or iron also present in certain varieties.

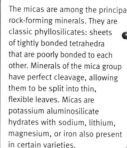

*Strong bond between tetrahedra and octahedra*

*Weak bond between K⁺ ions and tetrahedra*

**CRYSTAL STRUCTURE OF MUSCOVITE**

*Tabular, pseudohexagonal crystal*

*Leaves of mica visible*

*Apatite*

**FLAT CRYSTALS**
*This specimen includes a group of tabular muscovite crystals and green crystals of apatite.*

### PROPERTIES

| | |
|---|---|
| **GROUP** | Silicates – phyllosilicates |
| **CRYSTAL SYSTEM** | Monoclinic |
| **COMPOSITION** | $KAl_2(Si_3Al)O_{10}(OH,F)_2$ |
| **COLOUR** | Colourless, silvery-white, pale green, rose, brown |
| **FORM/HABIT** | Tabular |
| **HARDNESS** | 2½ |
| **CLEAVAGE** | Perfect basal |
| **FRACTURE** | Uneven |
| **LUSTRE** | Vitreous |
| **STREAK** | Colourless |
| **SPECIFIC GRAVITY** | 2.8 |
| **TRANSPARENCY** | Transparent to translucent |

# GLAUCONITE

**ANTRIM**
*Glauconite is in this county of Northern Ireland.*

GLAUCONITE IS A MEMBER of the mica group, named in 1828 from the Greek *glaukos*, meaning "blue-green", an allusion to its colour. Widespread, it usually occurs as rounded aggregates and pellets of fine-grained, scaly particles. It forms in shallow marine environments. The sedimentary rock greensand is so-called because of the green colour imparted by glauconite.

### PROPERTIES

| | |
|---|---|
| **GROUP** | Silicates – phyllosilicates |
| **CRYSTAL SYSTEM** | Monoclinic |
| **COMPOSITION** | $(K,Na)(Mg,Al,Fe)_2(Si,Al)_4O_{10}(OH)_2$ |
| **COLOUR** | Yellow-green, green, blue-green |
| **FORM/HABIT** | Rounded aggregates |
| **HARDNESS** | 2 |
| **CLEAVAGE** | Perfect basal |
| **FRACTURE** | Uneven |
| **LUSTRE** | Dull to earthy |
| **STREAK** | N/D |
| **SPECIFIC GRAVITY** | 2.4–2.9 |
| **TRANSPARENCY** | Translucent to opaque |

*Aggregate of small grains*

**FINE GRAINS**
*This typical massive aggregate includes glauconite grains.*

*Dull lustre*

# PHLOGOPITE

**STERLING HILL**
*Phlogopite occurs at this mine in New Jersey, USA.*

PHLOGOPITE BELONGS to the mica group. Its name derives from the Greek *phlogos*, "fire-like", alluding to the red-brown colour of some varieties. Its crystals are tabular to pseudohexagonal, and it usually occurs as platy aggregates. It can be colourless, pale yellow, and reddish to dark brown, often with a coppery look. Some specimens exhibit asterism (see p.152). Widespread, phlogopite is found in metamorphosed limestones and in igneous rocks.

### PROPERTIES

| | |
|---|---|
| **GROUP** | Silicates – phyllosilicates |
| **CRYSTAL SYSTEM** | Monoclinic |
| **COMPOSITION** | $KMg_3AlSi_3O_{10}(OH)_2$ |
| **COLOUR** | Colourless, pale yellow to brown |
| **FORM/HABIT** | Tabular or pseudohexagonal |
| **HARDNESS** | 2½–3 |
| **CLEAVAGE** | Perfect basal |
| **FRACTURE** | Uneven |
| **LUSTRE** | Pearly to submetallic |
| **STREAK** | Brownish-white |
| **SPECIFIC GRAVITY** | 2.9 |
| **TRANSPARENCY** | Transparent to translucent |

*Prismatic, twinned phlogopite crystals*

**PHLOGOPITE CRYSTALS**
*These pseudohexagonal phlogopite crystals are in a rock groundmass.*

261

**PIKE'S PEAK, USA**
*This Colorado locality yields good biotite specimens.*

# BIOTITE

BIOTITE, ALSO CALLED black mica, is very widespread and is common in both igneous and metamorphic rocks. It forms large crystals in granites and granite pegmatites, and forms tabular to short prismatic crystals, often pseudohexagonal in cross section. It is usually black or brown in colour, but can be pale yellow to tan, or bronze. It readily cleaves into thin, flexible sheets. Notable localities include Mull, Scotland; Bancroft, Ontario, Canada; and, in the USA, Pike's Peak, Colorado, Adirondak Mountains, New York, and King's Mountain, North Carolina.

**JEAN BIOT (1774–1862)**
*Biotite was named in 1847 for the French physicist Jean Biot, who studied the optical properties of the micas.*

*Vitreous lustre*

*Tabular crystal*

### PROPERTIES

| | |
|---|---|
| **GROUP** | Silicates – phyllosilicates |
| **CRYSTAL SYSTEM** | Monoclinic |
| **COMPOSITION** | $K(Mg,Fe)_3(AlSi_3)O_{10}(OH,F)_2$ |
| **COLOUR** | Black, brown, pale yellow, tan, or bronze |
| **FORM/HABIT** | Tabular |
| **HARDNESS** | $2\frac{1}{2}$–3 |
| **CLEAVAGE** | Perfect basal |
| **FRACTURE** | Uneven |
| **LUSTRE** | Vitreous to submetallic |
| **STREAK** | White to pale |
| **SPECIFIC GRAVITY** | 2.7–3.4 |
| **TRANSPARENCY** | Transparent to translucent |

**DARK MICA**
*The pseudohexagonal outline of many crystals can be seen clearly in this specimen of biotite in calcite from Frontenac County, Canada.*

# LEPIDOLITE

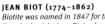

**DIXON, NEW MEXICO, USA**
*Deposits of lepidolite can be found in New Mexico.*

THIS LIGHT MICA CONTAINS a substantial portion of lithium and is the Earth's most common lithium-bearing mineral. Its name is from the Greek *lepidos*, meaning "scale", and *lithos*, "stone". Crystals may appear pseudohexagonal. Lepidolite is also found as botryoidal or reniform masses, and as fine- to coarse-grained interlocking plates. Its perfect cleavage yields thin, flexible sheets. Although typically pale lilac, it can be colourless, pale yellow, or grey. It is one of the few minerals containing significant quantities of rubidium, which is used to determine the age of rocks more than 10 million years old. It occurs almost exclusively in granitic pegmatites, where it is associated with other lithium minerals, beryl, topaz, quartz, tourmaline, and spodumene. Major occurences are in the Ural Mountains, Russia; Skuleboda, Sweden; Coolgardie, Western Australia; Penig, Saxony, Germany; the Czech Republic; Bernic Lake, Canada; Itinga, Brazil; Honshu, Japan; and at various US localities. It is economically important as a major source of lithium, and is used to make glass and enamels.

## HIGH-TECH USES OF LITHIUM

Lithium, a soft, white, lustrous metal, is the lightest of the solid elements. One of the latest uses of lithium is in electro-optical ceramics – materials that are optical transparent and with optical properties that vary with applied voltage. Lithium niobate ($LiNbO_3$) and lithium tantalate ($LiTaO_3$) are used in switches, modulators, and demodulators for high-speed optical communications.

**WEATHER BALLOON**
*Lithium hydride is used to generate hydrogen to fill weather balloons.*

*Vitreous lustre*

### PROPERTIES

| | |
|---|---|
| **GROUP** | Silicates – phyllosilicates |
| **CRYSTAL SYSTEM** | Monoclinic |
| **COMPOSITION** | $K(Li,Al)_3(AlSi_3)O_{10}(OH,F)_2$ |
| **COLOUR** | Pale lilac, colourless, yellow, or grey |
| **FORM/HABIT** | Tabular to short prismatic |
| **HARDNESS** | $2\frac{1}{2}$–$3\frac{1}{2}$ |
| **CLEAVAGE** | Perfect basal |
| **FRACTURE** | Uneven |
| **LUSTRE** | Vitreous to pearly |
| **STREAK** | Colourless |
| **SPECIFIC GRAVITY** | 3.0 |
| **TRANSPARENCY** | Transparent to translucent |

*Tabular lepidolite crystal*

**DISTINCTIVE COLOUR**
*Numerous violet lepidolite crystals protrude from this pegmatite specimen. The pseudohexagonal outline of many crystals is clearly visible.*

# VERMICULITE

**CZECH REPUBLIC**
*Significant amounts of vermiculite are found here.*

THE NAME VERMICULITE is applied to a group of silicate minerals in which various chemical substitutions occur in the molecular structure. The term is also used to refer to a type of clay. Vermiculite may be completely interlayered with other micas and clay-like minerals. It usually forms tabular, pseudohexagonal crystals, or platy aggregates. When heated to nearly 300°C (572°F), vermiculite can expand quickly and strongly to 20 times its original thickness. It occurs as large pseudomorphs replacing biotite; as small particles in soils and ancient sediments; at the interface between felsic, intrusive rocks and mafic rocks; and by hydrothermal alteration of iron-bearing micas. Large deposits occur in Mount Strangways, Australia; Transvaal, South Africa; Fazzan, Libya; Coahuila, Mexico; Prayssac, France; Nova Ves, the Czech Republic; and in the USA at Bare Hills, Maryland; Macon County, North Carolina; and Pinal County, Arizona.

*Pseudo-hexagonal outline*

*Foliated habit*

**VERMICULITE LAYERS**
*This specimen of vermiculite was mined in Pennsylvania, USA.*

### PROPERTIES

| | |
|---|---|
| **GROUP** | Silicates – phyllosilicates |
| **CRYSTAL SYSTEM** | Monoclinic |
| **COMPOSITION** | $(Mg,Fe,Al)_3(Al,Si)_4O_{10}(OH)_2 \cdot 4H_2O$ |
| **COLOUR** | Grey-white, golden-brown |
| **FORM/HABIT** | Foliated, expanded crystals |
| **HARDNESS** | 1–2 |
| **CLEAVAGE** | Perfect |
| **FRACTURE** | Uneven |
| **LUSTRE** | Oily to earthy |
| **SPECIFIC GRAVITY** | 2.6 |
| **TRANSPARENCY** | Translucent |

## USES OF VERMICULITE

Expanded vermiculite is very light. It is widely used in concrete and plaster, for thermal and acoustic insulation, as a packing medium, a filler/extender in paper, paint, or plastics, for growing seeds, and as a soil conditioner.

**GROWING MEDIUM**
*Vermiculite is used in horticulture as it retains water and offers good aeration.*

**INSULATION**
*Vermiculite is widely used in the lofts of houses for thermal insulation.*

---

# FUCHSITE

**URAL MOUNTAINS**
*Russia's Urals contain fuchsite deposits.*

THE CHROMIUM-BEARING VARIETY of muscovite mica, fuchsite is generally found in concentric, curved aggregates and as flakes. Its perfect cleavage yields thin, flexible plates. The intensity of its colour depends on chromium concentration, which may be up to 25 per cent. Fuchsite is found where hydrothermal solutions have replaced carbonates in gold deposits, and in metamorphic rocks in corundum and staurolite schists. It is found in New Hampshire, USA; and Manitoba, Canada.

### PROPERTIES

| | |
|---|---|
| **GROUP** | Silicates – phyllosilicates |
| **CRYSTAL SYSTEM** | Monoclinic |
| **COMPOSITION** | $K(Cr,Al)_2(AlSi_3)O_{10}(OH,F)_2$ |
| **COLOUR** | Light to medium green |
| **FORM/HABIT** | Curved aggregates |
| **HARDNESS** | $2^{1}/_{2}$ |
| **CLEAVAGE** | Perfect basal |
| **FRACTURE** | Uneven |
| **LUSTRE** | Vitreous to pearly |
| **STREAK** | White |
| **SPECIFIC GRAVITY** | 2.8–2.9 |
| **TRANSPARENCY** | Transparent to opaque |

**MEXICAN FUCHSITE**
*This specimen shows characteristic curved aggregates.*

*Curved aggregate*

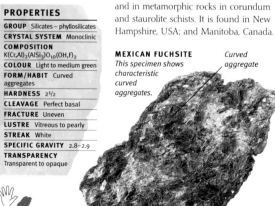

---

# ZINNWALDITE

**PIKE'S PEAK, USA**
*Zinnwaldite is found in this part of Colorado.*

ZINNWALDITE BELONGS TO the mica group. Its crystals are tabular to short prismatic, often pseudohexagonal in shape. It is found in hydrothermal veins and granitic pegmatites. Specimens come from Altenburg, Germany; Zinnwald, the Czech Republic; Baveno, Italy; many localities in Japan; Virgem de Lapa, Brazil; and, in the USA at Pike's Peak, Colorado, and Amelia Court House, Virginia.

### PROPERTIES

| | |
|---|---|
| **GROUP** | Silicates – phyllosilicates |
| **CRYSTAL SYSTEM** | Monoclinic |
| **COMPOSITION** | $K(LiFeAl)_3(AlSi_3)O_{10}(F,OH)_2$ |
| **COLOUR** | Grey-brown |
| **FORM/HABIT** | Tabular to short prismatic |
| **HARDNESS** | $2^{1}/_{2}$–4 |
| **CLEAVAGE** | Perfect |
| **FRACTURE** | Uneven |
| **LUSTRE** | Vitreous to pearly |
| **STREAK** | White |
| **SPECIFIC GRAVITY** | 2.9–3.0 |
| **TRANSPARENCY** | Transparent to translucent |

*Vitreous lustre*

**CZECH ZINNWALDITE**
*This specimen is from a classic locality at Cinovec, Severocesk, the Czech Republic.*

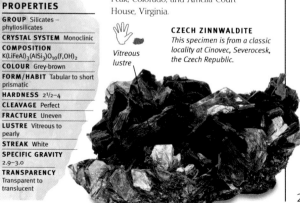

263

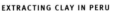

# CLAY MINERALS

IT IS VITAL TO MAKE A DISTINCTION between the common use of the word "clay", to describe a very fine-grained, rather plastic rock, and clay minerals. The potter working with "clay" may be working with a material made up of several clay minerals, as well as with particles of other minerals that are "clay-sized" – less than 0.005mm long. Clay rocks and clay minerals share some properties, such as their ability to swell.

**GREEN RIVER CANYON, USA**
*Cracked, drying, sedimentary clay lines the banks of the Green River in Utah.*

### STRUCTURE OF CLAY MINERALS
The clay minerals are a group of sheet silicate minerals that crystallize primarily in the monoclinic system. They are generally white, buff, or yellowish when pure, and only rarely form visible crystals. The various clay minerals are hard to tell apart in hand specimens. Seen under a microscope, most crystals of clay minerals are pseudohexagonal plates. Clay minerals generally consist of two structural units: a layer of silicon-oxygen tetrahedra arranged in a hexagonal network in two dimensions; and two layers of close-packed oxygen or hydroxyl ions. This second component creates sites in which metals, usually sodium, calcium, aluminium, magnesium, or iron, are located.

Most clays are the result of the weathering of feldspars or other aluminium-rich minerals. Different clay minerals can result from the same parent rock depending

**BENTONITE**

**HALLOYSITE**

**ILLITE**

**BROWN NONTRONITE**

**GREEN NONTRONITE**

**CLAY VARIETIES**
*Above is a selection of clay minerals. Although these specimens appear different from each other, in practice clay minerals are often hard to tell apart as hand specimens.*

**EXTRACTING CLAY IN PERU**
*A man extracts clay to be used in the manufacture of tiles in Chinipampa, Peru. Tile and mud-brick are major building materials in much of the world.*

on the climate and the amount, chemistry, and flow of water involved in the weathering process. Clay minerals are widespread in sedimentary rocks, such as mudstones and shales, in marine sediments, and in soils.

**MAYAN TERRACOTTA SCULPTURE**
*The ancient Mesopotamian world used clay extensively for terracotta sculpture as well as for pottery.*

### TYPES AND USES
There are more than a dozen clay minerals, and different sources include different minerals. Those usually defined as clay minerals are montmorillonite, nontronite, kaolinite (see p.259), dickite, nacrite, illite, glauconite, and celadonite. Sometimes chlorites, vermiculite (see p.263), polygorskite, and pyrophyllite are also included. Clay rocks usually contain a mixture of clay minerals. Sedimentary clay, a clastic sediment formed at the bottom of rivers, lakes, and seas, is composed mainly of illite and kaolinite, but can contain many other minerals, including other clay minerals. It is used for

**THIN SECTION**
*The platy, pseudohexagonal nature of clay minerals is clearly visible here. The platy characteristic gives clay its rather plastic qualities.*

Thin plates

Quartz

brick-making. The clay rock bentonite, derived mainly from volcanic ash, has a large component of the clay mineral montmorillonite, and is used in the petroleum industry as drilling mud – a heavy suspension consisting of clays, chemical additives, and weighting materials, employed in rotary drilling. China clay, formed by the deep weathering of granites, is composed mainly of kaolinite, but may also contain significant portions of other clay minerals. It is used in the manufacture of porcelain, paint fillers, and pharmaceuticals among other things. Ball clay is an impure china clay that also contains a substantial amount of illite and other minerals. It is used in high-quality ceramics.

**TERRACOTTA DISH**
*As terracotta (literally "baked earth"), clay has been used for millennia in a variety of ceramics.*

**RED ROOFS**
*Clay is widely used for tiles, here on the steep roofs of houses in Lübeck, Germany. It has been a popular choice for centuries wherever it is locally abundant.*

Clusters of pseudohexagonal montmorillonite crystals

## PROPERTIES OF MONTMORILLONITE

**GROUP** Silicates – phyllosilicates

**CRYSTAL SYSTEM** Monoclinic

**COMPOSITION** $(Na,Ca)(Al,Mg)_2Si_4O_{10}(OH)_2 \cdot nH_2O$

**COLOUR** White to buff, pale yellow, pink

**FORM/HABIT** Platy

**HARDNESS** 1–2

**CLEAVAGE** Perfect

**FRACTURE** Uneven

**LUSTRE** Earthy

**STREAK** White

**SPECIFIC GRAVITY** 2.1

**TRANSPARENCY** Translucent to opaque

**MONTMORILLONITE**
*Montmorillonite is a clay mineral that is most commonly found in deposits of bentonite. This specimen includes visible crystals – these are rare among clay minerals.*

**MINGUN PAYA**
*The uncompleted pagoda of Mingun Paya in Myanmar was begun in the 18th century by King Bodawpaya, and would have been 150m (500ft) tall. It is still the world's largest all-brick structure.*

# THE USES OF BRICK

After flint, clay was the next most important mineral in the early history of humanity. The use of clay in pottery-making long pre-dates recorded human history. Gigantic buildings, such as the ziggurats of Mesopotamia (including the Tower of Babel), were built from mud-brick, and the evidence suggests that the builders of the first stone pyramid took inspiration from the earlier mud-brick examples. But the use of mud-brick was far from limited to dramatic buildings. Entire cities such as Ur, Babylon, Nineveh, and virtually all the houses scattered across the ancient Mesopotamian countryside were built from mud-brick.

## MODERN MUD–BRICK

Unfired mud-brick is far from an antiquated building material. In the same areas of the world where ancient mud-brick buildings were erected, mud-brick is still the cheapest and most readily available building material. In dry climates in North and South America mud-brick, called adobe, is still a material of choice for even the most modern dwellings.

The properties that make clay ideal for brick-making – plasticity when wet and coherence when dry – make it ideal for other purposes such as the lining and sealing of dams, canals, and reservoirs, and it is also an important component of Portland cement.

## FIRED BRICK

The everyday mud-bricks in the ancient world were sun-dried, but those used for more important buildings were fired. By the 1st century BC, the Romans were making massive use of fired bricks to create structures that stand to this day, such as the massive Baths of Caracalla, in Rome. The Byzantines inherited Roman techniques of brickwork, and from them knowledge was passed on to the Ottoman Turks. The Italians were the next to adopt fired-brick architecture which became common in Italy in the 11th century. Over the next two centuries the knowledge spread to the rest of Europe and finally to the Americas.

**DJENNE MOSQUE**
*The great mosque in this ancient trading city in southern Mali is made from mud-brick.*

**BRICKMAKERS**
*Workers in Dhaka, Bangladesh, stack raw bricks made from river-bottom mud ready for firing before use.*

# CHAMOSITE

CHAMOSITE IS USUALLY MASSIVE, often occurring in earthy aggregates. Its crystals can be tabular, pseudohexagonal, or pseudorhombohedral. It is found mainly in sedimentary ironstones such as chamosite oolites, in marine clays, and in some low-grade metamorphosed ironstones. Localities include France, Poland, the Czech Republic, Germany, Western Australia, Japan, England, South Africa, and, in the USA, Pennsylvania, Arkansas, Colorado, Michigan, and Maine.

## PROPERTIES

| | |
|---|---|
| **GROUP** | Silicates – phyllosilicates |
| **CRYSTAL SYSTEM** | Monoclinic |
| **COMPOSITION** | $Fe_3(Fe_2Al)(Si_3AlO_{10})(OH)_8$ |
| **COLOUR** | Green to greenish-black |
| **FORM/HABIT** | Massive |
| **HARDNESS** | $2^1/_2$–3 |
| **CLEAVAGE** | Perfect |
| **FRACTURE** | Uneven |
| **LUSTRE** | Oily to vitreous |
| **STREAK** | White green to grey |
| **SPECIFIC GRAVITY** | 3.0–3.3 |
| **TRANSPARENCY** | Translucent |

*Massive habit*

*Earthy lustre*

**MASSIVE CHAMOSITE**
*This is a specimen of massive chamosite. Crystals are rarely seen.*

# CAVANSITE

**PUNE, INDIA**
*Pune has the best specimens of cavansite.*

CAVANSITE AND ITS POLYMORPH pentagonite are both hydrous calcium vanadium silicates. This mineral was named after some of the elements it contains: calcium (Ca), vanadium (V), and silicon (Si). Cavansite forms aggregates of prismatic crystals, and also occurs as platelets and in rosettes. It is blue to greenish-blue in colour, and is found in basalts in association with zeolites, and in tuffs. The finest specimens of cavansite come from Pune, India; cavansite also occurs at Lake Owyhee State Park, Oregon, USA, where it was discovered, and near Globe, also in Oregon.

## PROPERTIES

| | |
|---|---|
| **GROUP** | Silicates – phyllosilicates |
| **CRYSTAL SYSTEM** | Orthorhombic |
| **COMPOSITION** | $Ca(VO)Si_4O_{10}\cdot4H_2O$ |
| **COLOUR** | Blue to greenish-blue |
| **FORM/HABIT** | Prismatic |
| **HARDNESS** | 3–4 |
| **CLEAVAGE** | Good |
| **FRACTURE** | Conchoidal |
| **LUSTRE** | Vitreous |
| **STREAK** | White |
| **SPECIFIC GRAVITY** | 2.2–2.3 |
| **TRANSPARENCY** | Transparent to translucent |

*Heulandite*

*Cavansite rosette*

**CAVANSITE ROSETTE**
*This cavansite rosette on heulandite is from Wagholi Quarry, Pune, India.*

*Rock groundmass*

# CHRYSOCOLLA

**CORNWALL**
*Chrysocolla is found with copper in this part of England.*

THE TERM CHRYSOCOLLA originates in ancient times, and was applied by the Greek philospher Theophrastus in 315BC to various materials used in soldering gold. The name is derived from the Greek *chrysos*, meaning "gold", and *kolla*, meaning "glue". Chrysocolla is commonly very fine grained and massive. Crystals are very rarely seen, but can form botryoidal, radiating aggregates. It is frequently intergrown with other minerals such as quartz, chalcedony, or opal to yield a harder, more resilient gemstone variety. Chrysocolla intergrown with turquoise and malachite from Israel is called Eilat stone. Chrysocolla forms as a decomposition product of copper minerals, especially in arid regions. It is found worldwide. Localities include Cornwall, England, Israel, Mexico, the Czech Republic, Australia, the Congo, and New Mexico, Arizona, and Utah, USA. Gemstone pieces sometimes exceed 2.3kg (5lb) in weight.

**GREEN CHRYSOCOLLA**
*Chrysocolla is intergrown here with turquoise and malachite.*

## PROPERTIES

| | |
|---|---|
| **GROUP** | Silicates – phyllosilicates |
| **CRYSTAL SYSTEM** | Orthorhombic |
| **COMPOSITION** | $Cu_2H_2(Si_2O_5)(OH)_4\cdot nH_2O$ |
| **COLOUR** | Blue, blue-green |
| **FORM/HABIT** | Massive |
| **HARDNESS** | 2–4 |
| **CLEAVAGE** | None |
| **FRACTURE** | Uneven to conchoidal |
| **LUSTRE** | Vitreous to earthy |
| **STREAK** | Pale blue, tan, grey |
| **SPECIFIC GRAVITY** | 2.0–2.4 |
| **TRANSPARENCY** | Translucent to nearly opaque |

**CHRYSOCOLLA WITH AZURITE**
*Chrysocolla is copper silicate hydroxide. It is commonly associated with malachite and azurite.*

*Chrysocolla*

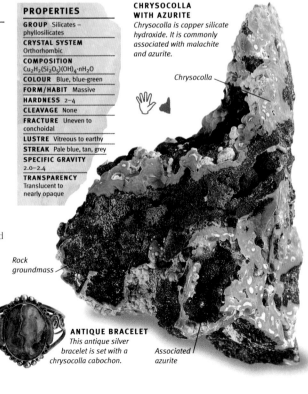

*Rock groundmass*

*Associated azurite*

**PEBBLE**
*This tumble-polished pebble is bright blue chrysocolla.*

**CABOCHON**
*Green chrysocolla is seen here within a reddish iron oxide.*

**CABOCHON**
*Blue and green chrysocolla are intermixed here.*

**ANTIQUE BRACELET**
*This antique silver bracelet is set with a chrysocolla cabochon.*

# CLINOCHLORE

**BAHIA, BRAZIL**
*Fine clinochlore crystals come from Bahia.*

**KÄMMERERITE**
*This variety is also known as chromian clinochlore.*

REFERRING TO ITS OPTICAL SYMMETRY and colour, clinochlore takes its name from the Greek *klinen*, meaning "to incline", and *chloros*, "green". Crystals are tabular to blocky, with a tabular pseudohexagonal or pseudo-orthorhombic habit. Clinochlore forms a solid-solution series with chamosite. It forms as a product of low-grade regional and contact metamorphism, as a hydrothermal alteration product, in marine clays and soils, in amygdales in volcanic rocks, and in fissure veins. Localities include Pajsberg, Sweden; Quebec, Canada; Banff, Scotland; Valais, Switzerland; Bahia, Brazil; Marienberg, Germany; Russia's Ural Mountains; and, in the USA, in Brewster, New York, Lancaster County, Pennsylvania, and Lowell, Vermont.

*Tabular crystals*

*Rock groundmass*

**CLINOCHLORE CRYSTALS**
*These tabular clinochlore crystals rest on a rock groundmass.*

## PROPERTIES

| | |
|---|---|
| **GROUP** | Silicates – phyllosilicates |
| **CRYSTAL SYSTEM** | Monoclinic |
| **COMPOSITION** | $Mg_3(Mg_2Al)(Si_3AlO_{10})(OH)_8$ |
| **COLOUR** | Green |
| **FORM/HABIT** | Tabular |
| **HARDNESS** | $2^{1}/_2$ |
| **CLEAVAGE** | Perfect |
| **FRACTURE** | Uneven |
| **LUSTRE** | Vitreous to dull |
| **STREAK** | Pale yellow |
| **SPECIFIC GRAVITY** | 2.6–2.9 |
| **TRANSPARENCY** | Translucent |

# APOPHYLLITE

*Prismatic habit*

**ISLE OF SKYE**
*This Scottish island has apophyllite crystals.*

ONCE CONSIDERED TO BE a single mineral, apophyllite is now divided into two distinct species – fluorapophyllite and hydroxyapophyllite. They form a solid-solution series in which flourine can predominate over oxygen and hydrogen, and *vice versa*. Specimens whose precise chemical composition has not been established are still called apophyllite. It is commonly found as glassy prismatic, blocky, or tabular white to greyish crystals. Colourless and green material from India is sometimes faceted as a collector's gem. Its name comes from the Greek *apo*, meaning "to get", and *phyllazein*, meaning "leaf", a reference to the way in which the mineral separates into flakes or layers when it is heated.

Apophyllite frequently occurs with zeolite minerals in basalt, and less commonly in cavities in granites. It is also found in metamorphic rocks, and in hydrothermal deposits. Apophyllite is a widespread mineral. Crystals up to 20cm (8in) in length are found at Bento Gonsalves, Brazil. Other localities include Mont St.-Hilaire, Canada; St. Andreasberg, Germany; the Isle of Skye, Scotland; Guanajuato, Mexico; India, and, in the USA, Paterson, New Jersey, Ashe County, North Carolina, Lebanon, Pennsylvania, and Centreville, Virginia.

**PINK CRYSTAL**
*This superb example of a pink apophyllite crystal is 6cm ($2^{1}/_2$in) long.*

**APOPHYLLITE CRYSTALS**
*In this specimen from Rahuri, India, blocky green crystals of apophyllite rest on a groundmass of stilbite.*

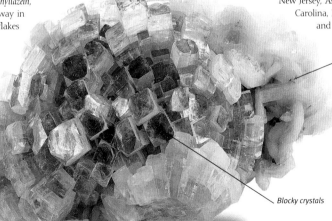

*Apophyllite*
*Rock groundmass*
*Stilbite*
*Blocky crystals*

## PROPERTIES

| | |
|---|---|
| **GROUP** | Silicates – phyllosilicates |
| **CRYSTAL SYSTEM** | Tetragonal |
| **COMPOSITION** | $KCa_4Si_8O_{20}$ $(F,OH)\cdot8H_2O$ (fluorapophyllite) |
| **COLOUR** | Colourless, pink, green, or yellow |
| **FORM/HABIT** | Tabular, prismatic |
| **HARDNESS** | $4^{1}/_2$–5 |
| **CLEAVAGE** | Perfect |
| **FRACTURE** | Uneven, brittle |
| **LUSTRE** | Vitreous |
| **STREAK** | Colourless |
| **SPECIFIC GRAVITY** | 2.3–2.4 |
| **TRANSPARENCY** | Transparent to translucent |

**MINES IN INDIA**
*Pune and Mumbai in India are prolific producers of apophyllite. Vendors sell trays of specimens on street corners.*

*Apophyllite*
*Apophyllite*
*Basalt*
*Stilbite*

*Basalt groundmass*
*Apophyllite*

**PINK APOPHYLLITE**

**GREEN APOPHYLLITE**

**REDDISH-PINK APOPHYLLITE**

**COLOURS OF APOPHYLLITE**
*As well as being pink and green, apophyllite can be cream, pale yellow, or reddish. Most commonly it is colourless.*

**MINAS GERAIS**
*Petalite is among the many minerals found in this area of Brazil.*

# PETALITE

A LITHIUM ALUMINIUM SILICATE, petalite was named in 1800 for the Greek word for "leaf", a reference to its perfect cleavage, which allows it to peel off in thin, leaf-like layers. Lithium was first recognized as a new chemical element in 1817 by Johan August Arfvedson from its occurrence in this mineral. Petalite is rarely found as individual crystals and most commonly occurs as aggregates. It is colourless to greyish–white, and occasionally light pink or green. It forms in granitic pegmatites along with albite, quartz, and lepidolite. It is widespread, with notable occurrences in Canada, Sweden, Italy, Russia, Australia, Zimbabwe, and California and Maine, USA. Facet-grade petalite is found in Brazil, yielding collectors' stones of up to 50 carats. Petalite is generally cut only for collectors as it is too fragile to wear as jewellery.

*Perfect cleavage*

*Vitreous lustre*

**PROPERTIES**

| | |
|---|---|
| **GROUP** | Silicates – phyllosilicates |
| **CRYSTAL SYSTEM** | Monoclinic |
| **COMPOSITION** | LiAlSi$_4$O$_{10}$ |
| **COLOUR** | Colourless to greyish-white, pink, or green |
| **FORM/HABIT** | Usually as aggregates |
| **HARDNESS** | 6$^1$/$_2$ |
| **CLEAVAGE** | Perfect basal |
| **FRACTURE** | Subconchoidal |
| **LUSTRE** | Vitreous |
| **STREAK** | White |
| **SPECIFIC GRAVITY** | 2.4 |
| **TRANSPARENCY** | Transparent to translucent |
| **R.I.** | 1.50–1.51 |

**LEAF-LIKE LAYERS**
*This translucent, pinkish specimen shows the perfect cleavage and vitreous lustre typical of petalite.*

**EMERALD**
*Particularly brilliant pieces of gem petalite can be faceted into striking emerald cuts.*

**MIXED**
*This petalite gem has a triangular mixed cut, which is ideal for bringing out its brilliance.*

**CUSHION**
*This fine specimen of gem petalite has a cushion mixed cut. Petalite is generally cut only for collectors.*

## LITHIUM BATTERIES

Along with spodumene and lepidolite, petalite is an important source of lithium. Its high chemical activity makes possible the tiny batteries used in personal pagers, heart pacemakers, and cameras. Lithium batteries are also used in military equipment and for aerospace applications. Experimentation with lithium compounds such as lithium-titanium disulphide promises a new generation of high-efficiency batteries.

**LITHIUM BATTERIES**

---

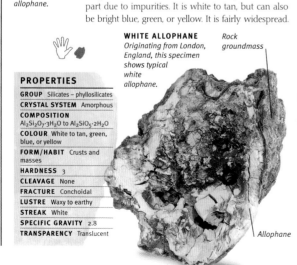

**HOKKAIDO**
*This locality in Japan is one of many for allophane.*

# ALLOPHANE

A WEATHERING PRODUCT of volcanic ash or a hydrothermal alteration product of feldspars, allophane is almost invariably mixed with other minerals such as opal, limonite, and gibbsite. An amorphous aluminosilicate hydrate mineral, allophane has a variable chemical composition, in part due to impurities. It is white to tan, but can also be bright blue, green, or yellow. It is fairly widespread.

**WHITE ALLOPHANE**
*Originating from London, England, this specimen shows typical white allophane.*

*Rock groundmass*

*Allophane*

**PROPERTIES**

| | |
|---|---|
| **GROUP** | Silicates – phyllosilicates |
| **CRYSTAL SYSTEM** | Amorphous |
| **COMPOSITION** | Al$_2$Si$_2$O$_7$·3H$_2$O to Al$_2$SiO$_5$·2H$_2$O |
| **COLOUR** | White to tan, green, blue, or yellow |
| **FORM/HABIT** | Crusts and masses |
| **HARDNESS** | 3 |
| **CLEAVAGE** | None |
| **FRACTURE** | Conchoidal |
| **LUSTRE** | Waxy to earthy |
| **STREAK** | White |
| **SPECIFIC GRAVITY** | 2.8 |
| **TRANSPARENCY** | Translucent |

---

**BOHEMIA**
*Okenite occurs in Bohemia in the Czech Republic, among other localities.*

# OKENITE

NAMED FOR THE GERMAN NATURALIST Lorenz Oken in 1828, Okenite is a calcium silicate hydrate. Its crystals are fibrous or bladed, and lath-like. It is sometimes found in spherical masses of radiating fibres. Okenite is often found associated with zeolites, apophyllite, quartz, prehnite, and calcite in basalts. It is found in many localities, including Mumbai, India, where there are fine, white, fur-like aggregates.

**ACICULAR OKENITE**
*This spherical mass of acicular okenite with laumontite is from Mumbai, Maharashtra, India.*

*Laumontite*

**PROPERTIES**

| | |
|---|---|
| **GROUP** | Silicates – inosilicates |
| **CRYSTAL SYSTEM** | Triclinic |
| **COMPOSITION** | Ca$_5$Si$_9$O$_{23}$·9H$_2$O |
| **COLOUR** | Colourless, white, pale yellow, or blue |
| **FORM/HABIT** | Fibrous, bladed |
| **HARDNESS** | 4$^1$/$_2$–5 |
| **CLEAVAGE** | Perfect |
| **FRACTURE** | Splintery |
| **LUSTRE** | Pearly to vitreous |
| **STREAK** | White |
| **SPECIFIC GRAVITY** | 2.3 |
| **TRANSPARENCY** | Transparent to translucent |

*Spherical mass of okenite crystals*

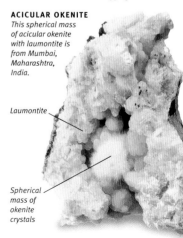

# PREHNITE

**HARZ MOUNTAINS**
*Prehnite occurs in these mineral-packed German mountains.*

NAMED IN 1789 for its discoverer, Hendrik von Prehn, a Dutch military officer, prehnite commonly occurs as globular, spherical, or stalactitic aggregates of fine to coarse crystals. Rare individual crystals are short prismatic to tabular, many with curved faces. It can be pale to mid-green, tan, pale yellow, grey, or white. Prehnite is often found lining cavities in volcanic rocks, associated with zeolites and calcite, and in mineral veins in granite. Crystals measuring several centimetres come from Canada. It is also found in many other localities worldwide. Transparent material from Australia and Scotland is faceted for collectors. Some fibrous material is cut *en cabochon* to show a cat's-eye effect. Other cut prehnite has been marketed under the name Cape emerald.

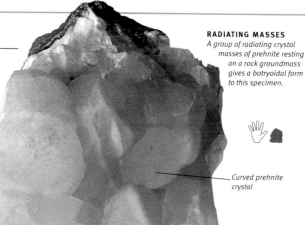

**RADIATING MASSES**
*A group of radiating crystal masses of prehnite resting on a rock groundmass gives a botryoidal form to this specimen.*

*Curved prehnite crystal*

*Vitreous lustre*

### PROPERTIES

| | |
|---|---|
| **GROUP** | Silicates – phyllosilicates |
| **CRYSTAL SYSTEM** | Orthorhombic |
| **COMPOSITION** | $Ca_2Al_2Si_3O_{10}(OH)_2$ |
| **COLOUR** | Green, yellow, tan, or white |
| **FORM/HABIT** | Tabular, botryoidal masses |
| **HARDNESS** | 6–6½ |
| **CLEAVAGE** | Distinct basal |
| **FRACTURE** | Uneven, brittle |
| **LUSTRE** | Vitreous |
| **STREAK** | White |
| **SPECIFIC GRAVITY** | 2.9 |
| **TRANSPARENCY** | Transparent to translucent |
| **R.I.** | 1.61–1.64 |

**CABOCHON**
*Translucent prehnite is often polished en cabochon. Some show a cat's eye.*

**STEP**
*This semitransparent prehnite has been faceted in a step cut to good effect.*

**MIXED**
*This yellowish faceted prehnite is not fully transparent, but it is still a beautiful gem.*

**BRILLIANT**
*This faceted oval prehnite shows good colour and transparency.*

**GREEN PREHNITE**
*Prehnite occurs in several colours. Here green, spherical masses can be seen on a rock groundmass.*

# SEPIOLITE

**ONTARIO**
*Deposits of massive sepiolite are found in localities in this part of Canada.*

A COMPACT, EARTHY, clay-like, often porous, mineral, sepiolite is a magnesium silicate hydrate. It takes its name from its resemblance to the light and porous bone of the cuttlefish *Sepia*. It is perhaps better known by its popular name of meerschaum, from the German for "sea-foam". Porous aggregates appear to be softer than its crystals, and many are light enough to float on water. It is usually found in nodular masses of interlocking fibres, which give it a toughness that belies its mineralogical softness. It is usually white or grey, and may be tinted yellow, brown, or green. Sepiolite is an alteration product of rocks such as serpentinite and magnesite. It is found as irregular nodules in alluvial deposits near Eskisehir, Turkey, the most important commercial deposit.

**SEPIOLITE BEADS**
*These intricately fashioned beads have been carved out of sepiolite.*

### PROPERTIES

| | |
|---|---|
| **GROUP** | Silicates – phyllosilicates |
| **CRYSTAL SYSTEM** | Orthorhombic |
| **COMPOSITION** | $Mg_4Si_6O_{15}(OH)_2{\cdot}6H_2O$ |
| **COLOUR** | White, grey, pinkish |
| **FORM/HABIT** | Massive |
| **HARDNESS** | 2–2½ |
| **CLEAVAGE** | Good but rarely seen |
| **FRACTURE** | Uneven |
| **LUSTRE** | Dull to earthy |
| **STREAK** | White |
| **SPECIFIC GRAVITY** | 2.1–2.3 |
| **TRANSPARENCY** | Opaque |

**TOBACCO PIPES**
*Meerschaum sepiolite is soft when first extracted, but it hardens on drying. Its chief use is for tobacco pipes, which are often intricately carved. As they are smoked, meerschaum pipes develop a lovely golden-brown patina.*

*Dull lustre*

**MASSIVE SEPIOLITE**
*This specimen of massive sepiolite shows a characteristic dull, earthy lustre.*

269

# SINGLE-CHAIN INOSILICATES

INOSILICATES TAKE THEIR NAME from the Greek word for "thread". They are sometimes referred to as chain silicates. Single–chain inosilicates are composed of chains of silica tetrahedra, and contain the $Si_2O_6$ silica group in their chemical formulae. Many of these silicates are fibrous in nature. The pyroxenes are a major mineral group within the single–chain inosilicates, and all share structural and chemical similarities. There are over 20 pyroxenes, and they are important rock–forming minerals. They are distinguished from the amphiboles by their two perfect cleavages at about 87 and 93 degrees to each other, those of the amphiboles being at about 124 and 56 degrees.

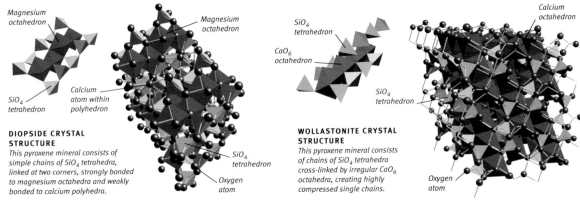

Magnesium octahedron

Magnesium octahedron

$SiO_4$ tetrahedron

Calcium atom within polyhedron

$SiO_4$ tetrahedron

Oxygen atom

**DIOPSIDE CRYSTAL STRUCTURE**
*This pyroxene mineral consists of simple chains of $SiO_4$ tetrahedra, linked at two corners, strongly bonded to magnesium octahedra and weakly bonded to calcium polyhedra.*

$SiO_4$ tetrahedron

$CaO_6$ octahedron

$SiO_4$ tetrahedron

Calcium octahedron

**WOLLASTONITE CRYSTAL STRUCTURE**
*This pyroxene mineral consists of chains of $SiO_4$ tetrahedra cross-linked by irregular $CaO_6$ octahedra, creating highly compressed single chains.*

Oxygen atom

## PIGEONITE

**MULL, SCOTLAND**
*This island off the west of Scotland has deposits of pigeonite.*

A MEMBER OF THE pyroxene group of minerals, pigeonite is generally found as rock-bound grains, with well-formed crystals being relatively rare. It is brown, purplish-brown, or greenish-brown to black in colour. An iron-rich variety is sometimes called ferropigeonite.

Pigeonite is found in many lavas and smaller intrusive rock bodies, where it is the dominant pyroxene, and it is an important component of andesites and dolerites (diabases). It is a widespread mineral, with notable localities at Skaergaard, Greenland; Mull, Scotland; Labrador, Canada; Mount Wellington, Tasmania; and in the USA at Pigeon Point, Minnesota, Goose Creek, Virginia, and Lambertville, New Jersey. It is also found in meteorites, and in the mare basalts on the Moon. These are areas of basalt which were once thought to have been seas. *Mare* is the Latin for "sea".

**THIN SECTION**
*As the pigeonite in this gabbro cooled it exsolved its calcium content in two generations of calcium-rich augite to create the branches and needles of a pine tree.*

**PIGEON POINT**
*Pigeonite is named after the place where it was discovered: Pigeon Point, Minnesota, USA.*

Perfect cleavage

**MARE IMBRIUM**
*Maria, the large, dark, relatively flat areas of the Moon once believed to be seas, are basalts containing pigeonite.*

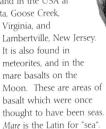

**PURPLE-BROWN PIGEONITE**
*This specimen of pigeonite comes from the Kovdor Pit, Kola Peninsula, Russia.*

Cleavage mass

### PROPERTIES

| | |
|---|---|
| **GROUP** | Silicates – inosilicates |
| **CRYSTAL SYSTEM** | Monoclinic |
| **COMPOSITION** | $(Mg,Fe,Ca)_2(Si_2O_6)$ |
| **COLOUR** | Brown to black |
| **FORM/HABIT** | Granular |
| **HARDNESS** | 6 |
| **CLEAVAGE** | Good |
| **FRACTURE** | Uneven to conchoidal, brittle |
| **LUSTRE** | Vitreous |
| **STREAK** | White to pale brown |
| **SPECIFIC GRAVITY** | 3.2–3.5 |
| **TRANSPARENCY** | Semi-transparent |

# ENSTATITE

A COMMON MINERAL in the pyroxene family, enstatite is magnesium silicate. Its name comes from the Greek *enstates*, meaning "opponent" – it is used as a refractory to line ovens and kilns, thus it is an "opponent" of heat. Enstatite forms a solid-solution series with ferrosilite – $(Fe,Mg)_2(Si_2O_6)$ – in which iron substitutes for magnesium. Bronzite is an intermediate variety of enstatite, and is brown with a metallic lustre. Enstatite generally occurs as rock-bound grains, or in massive aggregates. When it develops well-formed crystals, they tend to be short prismatic, often with complex terminations. It is found less often as fibrous masses of parallel, acicular crystals. It is colourless, pale yellow, or pale green, and becomes darker with increasing iron content, turning greenish-brown to black. Enstatite is a very widespread mineral, commonly occurring in mafic and ultramafic igneous rocks, and in meteorites. While it is an important rock-forming mineral, its only commercial use is as a gemstone. Localities producing gem-quality enstatite are Mysore, India (for star-enstatite); Canada (iridescent enstatite); the gem gravels of Myanmar and Sri Lanka (facet-grade material); and Arizona, USA.

## PROPERTIES

| | |
|---|---|
| **GROUP** | Silicates – inosilicates |
| **CRYSTAL SYSTEM** | Orthorhombic |
| **COMPOSITION** | $Mg_2Si_2O_6$ |
| **COLOUR** | Colourless, yellow, green, brown, or black |
| **FORM/HABIT** | Massive |
| **HARDNESS** | 5–6 |
| **CLEAVAGE** | Good to perfect |
| **FRACTURE** | Uneven |
| **LUSTRE** | Vitreous |
| **STREAK** | Grey to white |
| **SPECIFIC GRAVITY** | 3.1–3.9 |
| **TRANSPARENCY** | Translucent to opaque |
| **R.I.** | 1.65–1.68 |

**MASSIVE ENSTATITE**
*This massive, grey-brown specimen of enstatite comes from Bamle, Telemark, Norway. Its dark colour indicates it is from the ferrosilite end of the solid-solution series.*

*Prism face*

*Uneven fracture*

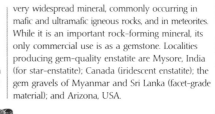

**FIBROUS ENSTATITE**
*Enstatite is sometimes found as fibrous masses of parallel, acicular crystals, as are many of the single-chain inosilicates.*

**CABOCHON**
*Enstatite is one of the gems that produces a cat's-eye effect when cut en cabochon.*

**MIXED**
*Facet-grade enstatite, here in an oval mixed cut, is recovered from the gravels of Myanmar and Sri Lanka.*

# AEGIRINE

AEGIRINE WAS NAMED in 1835 for Aegir, the Scandinavian god of the sea, a consequence of its being first discovered in Norway. A member of the pyroxene mineral group, aegirine is a sodium iron silicate. Its crystal habit tends to be prismatic and acicular. The prism faces are often lustrous and striated, while the faces of terminations are often etched and dull. Its colour is dark green, reddish-brown, or black. It forms a solid-solution series with hedenbergite and diopside (see p.272).

Aegirine is found worldwide in mafic igneous rocks, especially in syenites and syenitic pegmatites. It is also found in metamorphosed iron-rich sediments, and in schists and fluid-enhanced metamorphic rocks. Notable localities include Kongsberg, Buskerud, Norway; Mont St-Hilaire, Canada; Westland, New Zealand; and in the USA at Magnet Cove, Arkansas, and in the Cuyuna Range of Minnesota.

## PROPERTIES

| | |
|---|---|
| **GROUP** | Silicates – inosilicates |
| **CRYSTAL SYSTEM** | Monoclinic |
| **COMPOSITION** | $NaFe(Si_2O_6)$ |
| **COLOUR** | Dark green, red-brown, or black |
| **FORM/HABIT** | Prismatic |
| **HARDNESS** | 6 |
| **CLEAVAGE** | Good to perfect |
| **FRACTURE** | Uneven |
| **LUSTRE** | Vitreous |
| **STREAK** | Yellow-green to pale green |
| **SPECIFIC GRAVITY** | 3.5–3.6 |
| **TRANSPARENCY** | Translucent to opaque |

*Vertical striations*

*Groundmass*

*Vitreous lustre*

**AEGIRINE CRYSTAL**
*This specimen is a long prismatic crystal in a rocky groundmass. Its characteristic vertical striations can easily be seen.*

# DIOPSIDE

**TYROL**
*Diopside occurs in the rocks of these mountains in Austria.*

A MEMBER OF THE pyroxene family, diopside can be colourless, but is more often bottle-green, brownish-green, or light green. Crystals are equant to prismatic, or less commonly tabular, fibrous masses, or masses of large, tabular crystals. Bright green diopside, coloured by chromium, is known as chrome diopside. Violet-blue crystals, coloured by manganese, are found in Italy and the USA. Diopside occurs in kimberlites and olivine basalts, metamorphosed siliceous limestones and dolomites, and in iron-rich contact-metamorphic rocks.

## PROPERTIES

**GROUP** Silicates – inosilicates
**CRYSTAL SYSTEM** Monoclinic
**COMPOSITION** CaMg(Si$_2$O$_6$)
**COLOUR** White, pale to dark green, violet-blue
**FORM/HABIT** Equant to prismatic
**HARDNESS** 6
**CLEAVAGE** Distinct in two directions at almost right angles
**FRACTURE** Uneven
**LUSTRE** Vitreous
**STREAK** White to pale green
**SPECIFIC GRAVITY** 3.3
**TRANSPARENCY** Transparent to translucent
**R.I.** 1.66–1.72

Quartz

*Prismatic diopside crystal*

Rock groundmass

## DIOPSIDE IN METEORITES

**METEORITE**
*A meteorite burns as it falls through the Earth's atmosphere.*

Diopside is found in small amounts in many chondrite meteorites (see pp.74–75). Chondrites are principally composed of small (around 1mm) spheres of varying composition known as chondrules with a fine-grained matrix in between. They are thought to be among the earliest-formed materials in existence in the entire Solar System.

**CHONDRITE**

**PRISMATIC DIOPSIDE**
*This specimen of diopside comes from St. Marcel, Valle d'Aosta, Italy.*

**VIOLANE**
*Blue crystalline diopside is sometimes called violane.*

**EMERALD**
*This example of an emerald-cut diopside has a good, dark green colour.*

**STEP**
*This medium-green diopside has been faceted in a rectangular step cut.*

# HEDENBERGITE

**MOUNT ETNA**
*Hedenbergite occurs in many places, including Mount Etna on the island of Sicily.*

**BERZELIUS**
*Chemist Berzelius named the silicate for his colleague Ludvig Hedenberg.*

THE PYROXENE HEDENBERGITE is a calcium iron silicate, which forms a complete solid-solution series with diopside (see above), in which magnesium completely replaces iron. Its crystals are equant to prismatic but more common habits are massive, bladed or lamellar. Hedenbergite is common in metamorphosed siliceous limestones and dolomites, in thermally metamorphosed iron-rich sediments, and in some igneous rocks. It is found around the world.

## PROPERTIES

**GROUP** Silicates – inosilicates
**CRYSTAL SYSTEM** Monoclinic
**COMPOSITION** CaFe(Si$_2$O$_6$)
**COLOUR** Pale to dark green or brownish-green to greenish-black
**FORM/HABIT** Equant to prismatic
**HARDNESS** 6
**CLEAVAGE** Distinct in two directions at almost right angles
**FRACTURE** Uneven to conchoidal
**LUSTRE** Vitreous to resinous
**STREAK** Pale green to tan
**SPECIFIC GRAVITY** 3-4
**TRANSPARENCY** Translucent to nearly opaque

Vitreous lustre

**BLADED CRYSTALS**
*Masses of bladed hedenbergite crystals form radiating bundles.*

# AUGITE

**HARZ**
*This mountain range in Germany has good examples of augite.*

A SILICATE OF calcium, magnesium, iron, titanium, and aluminium, augite is the most common pyroxene mineral. It occurs chiefly as short, prismatic crystals, and as large, cleavable masses. Augite is common in basalts, gabbros, andesites, and various other dark-coloured igneous rocks. It is also found in some metamorphic rocks, and is a common constituent of lunar basalts and some meteorites. Notable crystal localities are in Germany, the Czech Republic, Italy, Russia, Japan, Mexico, Canada, and in New York and Utah, USA.

## PROPERTIES

**GROUP** Silicates – inosilicates
**CRYSTAL SYSTEM** Monoclinic
**COMPOSITION** (Ca,Na)(Mg,Fe,Ti,Al)(Al,Si)$_2$O$_6$
**COLOUR** Greenish-black to black, dark green, brown
**FORM/HABIT** Short prismatic
**HARDNESS** 5½–6
**CLEAVAGE** Distinct in two directions at almost right angles
**FRACTURE** Uneven to subconchoidal
**LUSTRE** Vitreous to dull
**STREAK** Pale brown to greenish-grey
**SPECIFIC GRAVITY** 3.3
**TRANSPARENCY** Translucent to nearly opaque

**CRYSTALS**
*These two short prismatic augite crystals were found in the Czech Republic.*

Short prismatic crystal

# SPODUMENE

ALTHOUGH SPODUMENE was discovered in 1877 in Brazil, it was not until 1879 that the gemstone varieties kunzite and hiddenite were recognized as being the same mineral. A member of the pyroxene group, spodumene is a lithium aluminosilicate. It was named from the Greek *spodumenos*, "reduced to ashes", an allusion to its ash-grey colour. Kunzite (coloured by manganese), is pink or lilac, and hiddenite (coloured by chromium) is emerald-green. Gem material of any in the group usually shows strong pleochroism – two different shades of the body colour when viewed from different directions. Gemstones are cut with great care in order to show the best colour through the top surface of the stone.

Spodumene is typically found in lithium-bearing granite pegmatite dykes, often with other lithium-bearing minerals such as eucryptite and lepidolite. Common spodumene is an important ore of lithium and is employed in making ceramics as well as having many other uses. One of the largest single crystals of any mineral ever found was a spodumene. It was 14.3m

**HIDDENITE, USA**
*The gem hiddenite is found only in this part of North Carolina.*

**KUNZITE**
*Gem-quality pink to lavender spodumene is called kunzite. This 10cm (4in) crystal from Brazil is typical gem material.*

**HIDDENITE**
*The rare green variety of the gem spodumene is called hiddenite. This crystal is from Hiddenite, North Carolina, USA.*

(47ft) long, weighed 90 tonnes, and came from South Dakota, USA. Gem-quality spodumene crystals are typically considerably smaller, seldom being more than a kilogram in mass. Spodumene is widespread in occurrence. California, USA, Brazil, Pakistan, and Afghanistan are major sources of kunzite, and hiddenite occurs with emerald in North Carolina, USA.

Kunzite and hiddenite are cut as gems more for collectors than for the public, because their perfect cleavage makes them fragile, and their colour often fades when they are exposed to sunlight.

Kunzite is named after the American gemmologist G.F. Kunz, who first described it in 1902, and hiddenite is named after W.E. Hidden, who discovered it in North Carolina in 1879. Spodumene is alternatively called *triphane*, Greek for "three appearances", a reference to its cleavage and parting surfaces.

**PICASSO KUNZITE**
*This stunning necklace of baroque pearls features a 396.3-carat faceted kunzite gem from Brazil. It was designed by Paloma Picasso in 1986 for Tiffany's centenary.*

**HEART-SHAPED KUNZITE**
*The best colour in kunzite is across the end of the crystal. Unusual cuts like the heart are often used to preserve as much expensive material as possible.*

**COMMON SPODUMENE**
*Most spodumene that is mined as an industrial source of lithium occurs as common opaque crystals, such as the crystal shown here.*

Prismatic habit

Vertical striations

**CUSHION-CUT KUNZITE**
*Like the heart, the cushion shape is another good choice for faceting the end of a crystal.*

**STEP-CUT HIDDENITE**
*Hiddenite crystals tend to be elongated and so are commonly faceted in step cuts.*

**OCTAGONAL STEP CUT**
*Step cuts are often used in elongated crystals to save material.*

## INDUSTRIAL USES OF LITHIUM

Spodumene is a major source of lithium. As well as its use in batteries (see p.268), lithium is widely used in industry. Metallic lithium is used for hardening aluminium, lead, and other soft metals. Lithium hydride is used in the conversion of aldehydes, ketones, and carboxylic esters to alcohols; lithium hydroxide is used to convert stearic and other fatty acids for use in lubricating greases; and lithium fluoride is used as a fluxing agent in enamels and glasses.

**VENTILATION SYSTEM**
*Lithium compounds are used to remove carbon dioxide in breathing apparatus.*

**GLASS BOTTLES**
*Lithium fluoride is used during the manufacture of glass to strengthen it.*

### PROPERTIES

| | |
|---|---|
| **GROUP** | Silicates – inosilicates |
| **CRYSTAL SYSTEM** | Monoclinic |
| **COMPOSITION** | $LiAl(Si_2O_6)$ |
| **COLOUR** | Grey, white to green, pink, colourless |
| **FORM/HABIT** | Prismatic |
| **HARDNESS** | $6^{1}/_2$–7 |
| **CLEAVAGE** | Perfect |
| **FRACTURE** | Subconchoidal to splintery |
| **LUSTRE** | Vitreous |
| **STREAK** | White |
| **SPECIFIC GRAVITY** | 3.0–3.2 |
| **TRANSPARENCY** | Transparent to translucent |
| **R.I.** | 1.66–1.67 |

# JADEITE

THERE ARE TWO DIFFERENT minerals that are commonly called "jade": jadeite and nephrite. Jadeite is a mineral in its own right, a pyroxene; nephrite, an amphibole (see pp.280-81), is a variety of tremolite or actinolite. Jadeite is made of interlocking, blocky, granular crystals, whereas nephrite crystals are fibrous. These two differing textures can sometimes help distinguish between them: nephrite often appears fibrous or silky; jadeite commonly has a more sugary or granular texture. Crystals of jadeite do occur, but they are rare. They are usually found in hollows within massive material, and are short prismatic in habit. Jadeite appears in a number of colours whereas nephrite has a much more limited colour range. Pure jadeite is white. Its other colours include green, coloured by iron; lilac, coloured by manganese and iron; and pink, brown, red, blue, black, orange, and yellow, coloured by inclusions of other minerals. Emerald-green jadeite, coloured by chromium, is called imperial jade.

Jadeite generally occurs in metamorphic rocks with a higher-pressure origin than nephrite, although some has been found in lower pressure metamorphic rocks.

Although generally recovered as alluvial pebbles and boulders, it is also found in the rocks in which it originally formed. It is widespread in metamorphic rocks formed at subduction zones. Weathered jadeite typically develops a brown skin, which is often incorporated into carvings. It frequently has a dimpled "orange-peel" surface when polished.

Myanmar is a major source of jadeite, and in particular imperial jade. Other sources are in Japan and California, USA. Ancient tools and weapons of jadeite are found in Myanmar, in the same way that nephrite objects have been discovered in China. A very small amount of jadeite found its way into China over the centuries, and from the 18th century it was imported in large quantities from Myanmar and soon displaced nephrite as a favoured carving material. Because of its value, other green stones are often misnamed "jade".

**MYANMAR**
*Myanmar has been a source and exporter of jadeite for hundreds of years; ancient jadeite implements are found there.*

**IMPERIAL JADE**
*This cabochon shows the colour and translucency characteristic of the finest imperial jade.*

**JADEITE SPHERE**
*Jadeite is worked into many shapes and forms, including this sphere of mottled green.*

**JADEITE BEADS**
*Darkly iridescent grey-green jadeite stones have been cut into beads of various sizes, and strung together to make this appealing necklace.*

## MESOAMERICAN JADE

For the Indians of Mexico, Central, and South America, jadeite had a similar cultural value to nephrite in China. It was a symbol of water and the burgeoning of plant life. Known as *chalchihuitl*, it was more precious than gold. The Olmecs were the first Mesoamericans to discover and carve jade – perhaps 3,000 years ago. Across Mexico and Central America it was used in the most precious objects: masks, depictions of the gods, and ritual items.

### POWERFUL AND SACRED STONE

Jadeite was cast into sacred wells as a offering to the gods. It was sometimes inlaid in the heart in sacred sculptures, and a piece of jadeite placed in the mouth of a deceased nobleman was believed to serve as his or her heart in the afterlife. Jade grave goods were essential for members of the nobility in most Mesoamerican cultures. They were often "killed" – ritually broken – so they could accompany their owner into the afterlife. When powdered and mixed with herbs, jadeite was used to treat a fractured skull, fevers, and to resurrect the dying. Mesoamerican jadeite principally came from sources in Guatemala and Costa Rica.

**JAGUAR MASK**
*The jaguar god represented power, night, war, and sacrifice to the Mayans, who created this terrifying jadeite mask.*

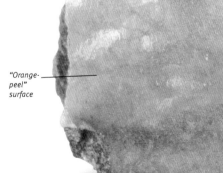

*Greasy lustre*

*"Orange-peel" surface*

**MAYAN MASK**
*This Pre-Columbian feline mask of a man from Oaxaca is made from red jadeite.*

**MEXICAN MASK**
*This Olmec mask of blue jadeite dates from 800–400BC.*

**MEXICAN MASK**
*This 18th-century jadeite mask has the smooth surface achieved by modern abrasives.*

### DRAGON VASE
*This exquisite carving (right) makes use of a pale green to lavender colour variation to highlight the dragon motif.*

### SNUFF BOTTLE
*Snuff bottles, such as this one (above), were produced in China in the 18th and 19th centuries; particularly prized were those carved from jadeite.*

Polished surface

## JADEITE REACHES EUROPE
Until the late 16th century, virtually all European jade was nephrite. When the Spanish reached Mexico, they discovered that the Aztecs prized a green stone that was similar in appearance to, and believed to be the same stone as, European jade. They were told by the Aztecs that this stone cured internal ailments, especially those of the liver, spleen, and kidneys. This stone was brought back to Europe, along with the belief in its healing powers. The Spanish called the stone *piedra de ijada* ("loin stone"), which was mistranslated into French as *pierre de jade*, from which the word "jade" is derived. The jade from South America was believed to be the same as the Old World jade until 1863, when a Chinese carving was analysed and discovered to be a different stone. The new stone was given the name jadeite, meaning derived from "jade".

### MOCTEZUMA AND CORTÉS
*In this 16th-century painting, Moctezuma kneels before the Spanish Conquistador Hernando Cortés. Moctezuma offered him gifts made of jadeite and turquoise but Cortés refused them, considering them worthless.*

Brown rind

"Thank God they're only after the
# gold and silver – they
don't know about jade."

### THE AZTEC EMPEROR MOCTEZUMA, AFTER ENCOUNTERING CORTÉS

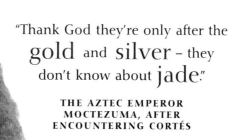

### WEATHERED JADEITE
*The lavender colouring of this massive jadeite specimen is caused by manganese and iron impurities. The mineral is partly polished to show its colour and is covered with the brown "rind" characteristic of weathered jadeite.*

JADEITE COBBLE

LILAC JADEITE

IMPERIAL JADE

### PROPERTIES
**GROUP** Silicates – inosilicates
**CRYSTAL SYSTEM** Monoclinic
**COMPOSITION** $Na(Al,Fe)Si_2O_6$
**COLOUR** White, green, lilac, pink, brown, orange, yellow, red, blue, or black
**FORM/HABIT** Massive, crystals rare
**HARDNESS** 6–7
**CLEAVAGE** Good
**FRACTURE** Splintery
**LUSTRE** Vitreous to greasy
**STREAK** White
**SPECIFIC GRAVITY** 3.2–3.4
**TRANSPARENCY** Transparent to translucent
**R.I.** 1.66–1.68

### JADEITE COLOURS
*The cobble (top) has a "window" cut in it to show the green of the jadeite beneath the "rind". The imperial jade (above) has been polished to bring out its colour.*

# ASTROPHYLLITE

USUALLY OCCURRING as bladed crystals radiating from a common centre (a form known as stellated), astrophyllite is a hydrous sodium potassium iron manganese titanium silicate. It derives its name from the Greek *astron*, meaning "star", and *phyllon*, "leaf". When it cleaves, it forms very brittle, thin leaves.

Astrophyllite is golden-yellow to dark brown. It forms in cavities in igneous rocks, such as alkali granites, and especially in syenites and syenite pegmatites. It is also found in gneisses. Astrophyllite is usually associated with albite, aegirine, arfvedsonite, zircon, nepheline, titanite, and other minerals. Specimens come from its place of discovery in Norway, as well as localities in Spain, Russia, Tajikistan, Egypt, Guinea, Canada, South Africa, and Arkansas, Colorado, and Washington, USA.

**PIKE'S PEAK**
*This mineral-rich area of Colorado, USA, contains deposits of astrophyllite.*

*Submetallic lustre*

*Radiating groups of crystals*

**BLADED CRYSTALS**
*This mass of radiating, bladed astrophyllite crystals shows a typical, leaf-like cleavage.*

## PROPERTIES

| | |
|---|---|
| **GROUP** | Silicates – inosilicates |
| **CRYSTAL SYSTEM** | Triclinic |
| **COMPOSITION** | $(K,Na)_3(Fe,Mn)_7Ti_2(Si_8O_{24})(O,OH)_7$ |
| **COLOUR** | Golden-yellow to dark brown |
| **FORM/HABIT** | Bladed, stellated |
| **HARDNESS** | 3 |
| **CLEAVAGE** | Perfect |
| **FRACTURE** | Uneven, brittle |
| **LUSTRE** | Submetallic to pearly |
| **STREAK** | Yellow |
| **SPECIFIC GRAVITY** | 3.4 |
| **TRANSPARENCY** | Translucent in thin laminae |

# WOLLASTONITE

A CALCIUM SILICATE, wollastonite occurs as rare tabular crystals or massive, coarse-bladed, foliated, or fibrous masses. Its colours are white, grey, or pale green. It has six known structural variations, one of which is monoclinic. The variations are indistinguishable in hand specimens. Wollastonite occurs as a result of the contact metamorphism of limestones, and is formed in igneous rocks from their contamination by carbonaceous inclusions. It also appears in regionally metamorphosed rocks in slates, phyllites, and schists. It is often accompanied by other calcium-containing silicates such as diopside, tremolite, grossular garnet, and epidote. Deposits are found in Italy, Romania, Mexico, Canada, Northern Ireland, Japan, Australia, and in Utah, Michigan, California, New York, and Arizona, USA.

**VESUVIUS**
*Mount Vesuvius, Italy, is one of several localities in Italy for wollastonite.*

## INDUSTRIAL USES

Wollastonite is used in ceramic products, including floor and wall tiles, ceramic electrical insulators, and porcelain fixtures. It is also used in welding-rod coatings and in paints, and as a replacement for asbestos.

Wollastonite fuses at fairly low temperatures, making it an excellent flux. It is electrically nonconductive, and it has virtually no volatile materials within it to interfere with its various applications. As a result, most automotive and aviation spark plugs have insulators made from wollastonite.

**ELECTRICAL INSULATOR**
*Ceramic electrical insulators use wollastonite due to its low conductivity.*

**CERAMIC TILES**
*Wollastonite is a raw material for many tiles.*

## PROPERTIES

| | |
|---|---|
| **GROUP** | Silicates – inosilicates |
| **CRYSTAL SYSTEM** | Triclinic |
| **COMPOSITION** | $CaSiO_3$ |
| **COLOUR** | White, grey, green |
| **FORM/HABIT** | Tabular, massive |
| **HARDNESS** | $4\frac{1}{2}$–5 |
| **CLEAVAGE** | Perfect |
| **FRACTURE** | Uneven to splintery |
| **LUSTRE** | Vitreous to silky |
| **STREAK** | White |
| **SPECIFIC GRAVITY** | 2.9 |
| **TRANSPARENCY** | Translucent in thin laminae |

**W.H. WOLLASTON**
*Wollastonite was named for this English mineralogist and noted scientist in 1818. Wollaston discovered palladium and rhodium.*

*Splintery fracture*

*Fibrous mass*

**SILKY LUSTRE**
*This mass of coarse-bladed, parallel crystals of wollastonite reveals a splintery fracture and silky lustre.*

# RHODONITE

**URALS**
*Rhodonite is found in the Ural Mountains in Russia.*

TAKING ITS NAME from the Greek *rhodon*, meaning "rose", which is this mineral's typical colour, rhodonite is a manganese silicate. It occurs as rounded crystals, masses, or grains, and is often coated or veined black with manganese oxides. Rhodonite is found in various manganese ores, often with rhodochrosite, or as a metamorphic product of rhodochrosite. It has been used as a manganese ore in India, but it is more often mined as a gem and ornamental stone. Sources are Brazil, Canada, Sweden, Russia, England, and the USA. Crystals are found in Peru, New Jersey, USA, and Australia.

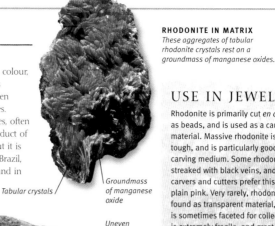

**RHODONITE IN MATRIX**
*These aggregates of tabular rhodonite crystals rest on a groundmass of manganese oxides.*

Tabular crystals / Groundmass of manganese oxide

## USE IN JEWELLERY

Rhodonite is primarily cut *en cabochon*, as beads, and is used as a carving material. Massive rhodonite is relatively tough, and is particularly good as a carving medium. Some rhodonite is streaked with black veins, and many carvers and cutters prefer this to the plain pink. Very rarely, rhodonite is found as transparent material, which is sometimes faceted for collectors. It is extremely fragile, and great care must be taken during faceting.

**RHODONITE EARRINGS AND BOX**
*The rhodonite box makes a feature of the black streaking in the mineral. The earrings use cabochon beads of rhodonite.*

## PROPERTIES

| | |
|---|---|
| **GROUP** | Silicates – inosilicates |
| **CRYSTAL SYSTEM** | Triclinic |
| **COMPOSITION** | $(Mn,Ca)_5(Si_5O_{15})$ |
| **COLOUR** | Pink to rose-red |
| **FORM/HABIT** | Tabular |
| **HARDNESS** | 6 |
| **CLEAVAGE** | Perfect |
| **FRACTURE** | Conchoidal to uneven |
| **LUSTRE** | Vitreous |
| **STREAK** | White |
| **SPECIFIC GRAVITY** | 3.5–3.7 |
| **TRANSPARENCY** | Translucent |
| **R.I.** | 1.71–1.73 |

Vitreous lustre / Uneven fracture

**MASSIVE RHODONITE**
*This specimen of rough rhodonite shows the intense coloration and fine texture of the best gem-quality material.*

**POLISHED PEBBLE**
*Medium-grade rhodonite is a popular material for tumble-polishing, and is widely sold.*

# PECTOLITE

**MONT ST.-HILAIRE**
*This Canadian mountain is one of the localities for pectolite.*

PECTOLITE'S COMPACT STRUCTURE is responsible for its name, which is derived from the Greek *pektos*, meaning "congealed" or "well put together". Its crystals are elongated and flattened, but it more often occurs as acicular sprays or radially fibrous masses. It may be white, pale tan, or pale blue. Some specimens are triboluminescent – that is, they give off light with friction. Fibrous varieties yield dangerously sharp splinters. Pectolite is a hydrothermal mineral found in cavities in basaltic and andesitic igneous rocks, in basalt cavities associated with zeolites, and in mica peridotites. Pectolite occurs widely in Canada, England, and the USA. Other specimen localities include Italy, the Dominican Republic, Greenland, Russia, and Japan.

Radiating aggregates of crystals

Acicular habit /

## PROPERTIES

| | |
|---|---|
| **GROUP** | Silicates – inosilicates |
| **CRYSTAL SYSTEM** | Triclinic |
| **COMPOSITION** | $NaCa_2(Si_3O_8)(OH)$ |
| **COLOUR** | White, tan, blue |
| **FORM/HABIT** | Acicular |
| **HARDNESS** | $4^1/_2$–5 |
| **CLEAVAGE** | Perfect |
| **FRACTURE** | Uneven, brittle |
| **LUSTRE** | Vitreous to silky |
| **STREAK** | White |
| **SPECIFIC GRAVITY** | 2.8–2.9 |
| **TRANSPARENCY** | Translucent |

**ACICULAR CRYSTALS**
*This mass of radiating groups of acicular pectolite crystals shows clearly the mineral's tendency to break into sharp splinters.*

**LARIMAR**
*Some pectolite from the Dominican Republic and Bahamas is sky-blue; it has the trade name Larimar.*

# DOUBLE-CHAIN INOSILICATES

THIS GROUP includes two of the most common types of silicate minerals: the amphiboles and the pyroxenes. The basic structural unit of double-chain inosilicates consists of two interconnected chains of silica tetrahedra. The amphiboles are distinguished from the similar-looking pyroxenes by their two perfect cleavages at approximately 124 and 56 degrees to each other; those of the pyroxenes are at about 87 and 93 degrees. In general, amphibole minerals tend to be long and slender, while pyroxenes are short and stout.

$SiO_4$ tetrahedron

Hydrogen atom

Magnesium octahedron

**TREMOLITE DOUBLE CHAIN**

Magnesium octahedron

$SiO_4$ tetrahedron

Calcium polyhedron

Hydrogen ion

**TREMOLITE STRUCTURE**
In tremolite crystals, double chains of $SiO_4$ tetrahedra are linked by magnesium octahedra and irregular calcium polyhedra.

## HORNBLENDE

**NORWEGIAN FJORD**
Well-formed hornblende crystals are found in Norway.

THE NAME HORNBLENDE is applied to what are now recognized as a group of minerals, but only detailed chemical analysis makes it possible to tell them apart. The two end-member hornblendes are both calcium-rich and monoclinic in crystal structure. They are ferrohornblende and magnesiohornblende, in which magnesium substitutes for iron in the chemical composition. To the collector or the geologist in the field, they are both simply known as hornblende. Other elements can also appear in the crystal structure of the group as a whole, especially chromium, titanium, and nickel, the concentrations of which are one indicator of metamorphic grade. Crystals of hornblende are most commonly bladed and unterminated, and often show a pseudohexagonal cross section. Well-formed crystals are short to long prismatic. The mineral also occurs as cleavable masses, and as radiating groups. Its colours are green, dark green, and brownish-green to black. Hornblende occurs widely in metamorphic rocks, especially in amphibolites, hornblende schists, gneisses, and mafic igneous rocks. A few localities for well-formed crystals are Oslo, Norway; Renfrew, Ontario, Canada; and throughout the Piedmont metamorphic province of Pennsylvania and Delaware, USA. It also occurs with ruby in the Harts Range, Northern Territory, Australia.

**SINGLE CRYSTAL**
This single, short prismatic, double-terminated crystal is from Lukov, Severocesky Kraj, the Czech Republic.

Prism face

Termination

Prismatic hornblende crystal

Vitreous lustre

Plagioclase

Twinned hornblende crystal

Quartz

**THIN SECTION**
Here a lozenge-shaped hornblende crystal, with twinning, is visible in granodiorite.

Vertical striations

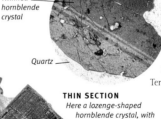

**HORNBLENDE CRYSTALS**
This specimen comprises a particularly fine group of prismatic hornblende crystals embedded in a rock groundmass.

### PROPERTIES

| | |
|---|---|
| **GROUP** | Silicates – inosilicates |
| **CRYSTAL SYSTEM** | Monoclinic |
| **COMPOSITION** | eg: $Ca_2(Fe^2,Mg)_4(Al,Fe3)(Si_7Al)O_{22}(OH,F)_2$ |
| **COLOUR** | Green, black |
| **FORM/HABIT** | Prismatic |
| **HARDNESS** | 5–6 |
| **CLEAVAGE** | Perfect |
| **FRACTURE** | Uneven, brittle |
| **LUSTRE** | Vitreous |
| **STREAK** | White to grey |
| **SPECIFIC GRAVITY** | 3.1–3.3 |
| **TRANSPARENCY** | Translucent to opaque |

# TREMOLITE

**ALPS**
*Tremolite occurs in some rocks found in the European Alps.*

TREMOLITE IS AN ABUNDANT and widespread amphibole that forms a solid-solution series (see p.91) with ferro-actinolite. Iron can substitute in increasing amounts for magnesium. The name actinolite is used for intermediate compositions. Both minerals have the same structure, and their colours vary with increasing iron content from colourless to white, grey, grey-green (tremolite), green, dark green and nearly black. Their crystals are short to long prismatic when well-formed, but are more commonly found in unterminated bladed crystals, in parallel aggregates of bladed crystals, or in radiating groups. Tremolite and actinolite form thin, parallel, flexible fibres up to 25cm (10in) in length, which have found commercial use as asbestos. These amphiboles are the product of both thermal and regional metamorphism. Good crystals come from Kantiwa, Afghanistan, and Edwards, New York, USA. Actinolite occurs at Monte Redondo, Portugal, and Washington, USA.

**THIN SECTION**
*This tremolite, with an acicular habit, is in quartz from a siliceous dolomitic limestone.*

— *Quartz*
— *Tremolite*
— *Tremolite*

*Thin, prismatic crystals*

**ACTINOLITE**
*This specimen is formed of a group of thin, prismatic crystals of actinolite. Its dark colour is due to its high iron content.*

## PROPERTIES

**GROUP** Silicates – inosilicates
**CRYSTAL SYSTEM** Monoclinic
**COMPOSITION**
$Ca_2(Mg,Fe^2)_5Si_8O_{22}(OH)_2$
**COLOUR** Colourless, white, grey green to green-black
**FORM/HABIT** Bladed
**HARDNESS** 5–6
**CLEAVAGE** Perfect
**FRACTURE** Splintery, brittle
**LUSTRE** Vitreous to silky
**STREAK** White
**SPECIFIC GRAVITY** 2.9–3.4
**TRANSPARENCY** Transparent to translucent

*Feather-like aggregate of tremolite crystals*

*Vitreous to silky lustre*

**TREMOLITE CRYSTALS**
*This specimen has plumose aggregates of white, bladed tremolite crystals.*

# ANTHOPHYLLITE

**QUEBEC**
*Long crystals of anthophyllite occur in this part of Canada.*

THE AMPHIBOLE anthophyllite was named from the Latin *anthophyllum*, meaning "clove", in reference to its colour. The iron and magnesium content is variable in its structure. Iron-rich anthophyllite is called ferroanthophyllite; when sodium is present, it becomes sodium-anthophyllite. Anthophyllites are clove-brown to dark brown, pale green, grey, or white. They are usually found in columnar to fibrous masses. Crystals are uncommon, and are prismatic and usually unterminated. Anthophyllite is commonly produced by regional metamorphism of ultramafic rocks. It is an important component of some crystalline schists and gneisses, and is found worldwide. Some important localities are Glen Urquhart, Scotland; Miass, Russia; and Macon County, North Carolina, and Kamiah, Idaho, USA. Anthophyllite is a minor asbestos mineral.

**RADIATING CRYSTALS**
*This specimen is a mass of fibrous, radiating crystals of anthophyllite.*

**THIN SECTION**
*Here, large, bladed prisms of anthophyllite can be seen in a gneiss.*

*Cordierite*

*Anthophyllite*

## PROPERTIES

**GROUP** Silicates – inosilicates
**CRYSTAL SYSTEM**
Orthorhombic
**COMPOSITION**
$(Mg,Fe)_7Si_8O_{22}(OH)_2$
**COLOUR** Clove- to dark brown, pale green, grey, or white
**FORM/HABIT** Massive, fibrous
**HARDNESS** 5½–6
**CLEAVAGE** Perfect, imperfect
**FRACTURE** Uneven
**LUSTRE** Vitreous
**STREAK** Colourless to grey
**SPECIFIC GRAVITY** 2.8–3.6
**TRANSPARENCY**
Transparent to nearly opaque

*Mass of fibrous radiating crystals*

*Vitreous lustre*

279

# NEPHRITE

**KOBUK RIVER, ALASKA**
*This US locality is a major source of nephrite, with large deposits of good-quality material.*

**GREEN DROPLET**
*This mid-green nephrite gem is homogeneous in colour. It has been cut en cabochon in the form of a droplet, and polished.*

**BLACK DROPLET**
*This polished cabochon droplet is made of iron-rich black nephrite, from Wyoming, USA.*

**DAGGER HANDLE**
*This subtly carved antique dagger handle is made from fine, light-coloured nephrite.*

NEPHRITE IS ONE OF TWO different minerals commonly known as jade. The other is jadeite (see pp.274–75), a pyroxene. Nephrite is more common and widespread than jadeite. Technically, "nephrite" is not a mineral name, but the name applied to the tough, compact form of either the amphibole tremolite (see p.279) or actinolite. Both are calcium magnesium silicate hydroxides, and structurally identical, except that in actinolite some of the magnesium is replaced by iron. Nephrite is composed of a mat of tightly interlocking fibres, creating a stone tougher than steel. Where nephrite is found, especially in China and New Zealand, this toughness brought it into use very early on for tools and weapons. Its colour varies with its composition: dark green when iron rich; cream-coloured when magnesium rich. White nephrite is essentially pure tremolite, and is sometimes called "mutton-fat jade".

The geologist A.G. Werner gave it the mineral name nephrite in 1780, from the Latin *nephrus*, meaning "kidney", alluding to the use of this form of "jade" in Europe to treat kidney disease. The term "jade" evolved from the Spanish *piedra de ijada*, meaning "loin stone", a stone of similar appearance to nephrite brought from the New World by the Spanish, which was, in fact, jadeite. Nephrite is formed in metamorphic environments, especially in metamorphosed ultramafic rocks where it is associated with talc and serpentine, and in regionally metamorphosed areas where dolomites have been intruded by mafic rocks.

Large deposits of nephrite are found on the Kobuk River in Alaska, USA; in British Columbia, Canada; in Xinjing Autonomous Region, probably the original Chinese source; and near Lake Wakatipu, Otago, New Zealand, the Maori source. It is also found near Lander, Wyoming, and in Mariposa and Siskiyou counties, and in Monterey and Morro

*Waxy lustre*

**JADE TIKI**
*Hei tikis are worn by the Maori of New Zealand and are passed down through generations.*

"Better to be **shattered jade** than unbroken **pottery**."

**CHINESE PROVERB**

## ANCIENT CHINESE JADE CARVING

**RITUAL PRISM**
*This Chinese ritual prism was created during the 3rd millennium BC.*

In ancient China, jade was called *yu*, and was believed to embody *yang*, or cosmic energy. As a result, jade was attributed life-giving qualities. In the social order, jade became the very embodiment of sovereignty and power. Taoists believed that it guaranteed immortality, and Confucius regarded jade as an important ritual material. From Confucian belief evolved a unique system of ritual jade to validate the hierarchical structure of Chinese feudal society: the king was entitled to the *gui* jades named *zhen*, and the five classes of the nobility were assigned the stones called *huan*, *xin*, *gong*, *gu*, and *pu*. The wearing of jade came to virtually dominate court ritual by the first centuries AD. Burial customs, especially those of the nobility, included the use of jade. Gold and jade were placed in all nine body openings, and other objects of jade frequently accompanied the deceased. The most extravagant use of jade was the construction of a burial suit of gold and jade (see pp.282–83) in the hope of preserving the body from decay. For the living, jade was taken internally in the belief that it regenerated the body.

**PI DISC**
*This decorative disc from AD25–220 includes a grain pattern and dragon-shaped knob. Pi discs usually accompanied burials.*

**DRAGON**
*Intricately carved, this jade dragon pendant was made in about 500BC.*

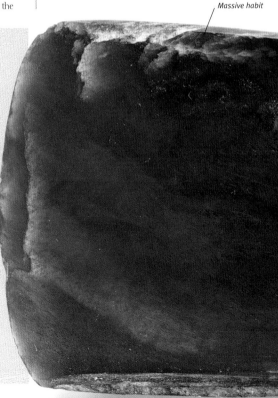

*Massive habit*

Bays, California, USA; in the Lake Baikal area of Siberia; and in Ch'unch'on, South Korea. In addition, it is found in Australia, Mexico, Brazil, Taiwan, Zimbabwe, Italy, Poland, Germany, and Switzerland.

**NEPHRITE HIPPO**
*This hippopotamus carved from nephrite has hooves cast in gold. It was made by the Faberge workshop, St. Petersburg, in 1908–17.*

## IDEAL CARVING STONE

Nephrite's tough structure makes it ideal for carving, and virtually no distinction was made until relatively recently between nephrite and jadeite when referring to "jade". Nephrite has been carved by the Chinese for more than 3,000 years, and it was only about 200 years ago that a preference developed there for Burmese jadeite. Other stones are similar to jade in appearance, though they are usually of inferior hardness, and some incorporate "jade" in their name: "soft jade", "Korean jade", and "Transvaal jade". None of these are jade – neither nephrite nor jadeite. Serpentine is widely carved in China today, and is the stone most often mistaken for (and sold as) jade.

**CARVING JADE**
*This woman is one of many highly skilled people in China who carve designs in nephrite and similar stones.*

**WEAVING TOOLS**
*These three weaving implements from New Zealand make use of the durability of nephrite. The two larger tools were probably used to make fishing nets.*

### PROPERTIES

| | |
|---|---|
| **GROUP** | Silicates – inosilicates |
| **CRYSTAL SYSTEM** | Monoclinic |
| **COMPOSITION** | $Ca_2(Mg,Fe)_5(Si_8O_{22})(OH)_2$ |
| **COLOUR** | Cream, light to dark green |
| **FORM/HABIT** | Massive |
| **HARDNESS** | 6½ |
| **CLEAVAGE** | Perfect |
| **FRACTURE** | Splintery, brittle |
| **LUSTRE** | Dull to waxy |
| **STREAK** | White |
| **SPECIFIC GRAVITY** | 2.9–3.4 |
| **TRANSPARENCY** | Translucent to nearly opaque |
| **R.I.** | 1.61–1.63 |

**NEPHRITE AXEHEAD**
*This axehead of New Zealand nephrite makes full use of the stone's toughness. Tools such as this were among the first uses of nephrite. It was even called "axe stone" in some cultures.*

*Splintery fracture*

**NEPHRITE BOULDER**
*This small boulder of nephrite has been sliced through and polished to reveal its quality.*

*Translucent surface*

*Waxy lustre*

*Splintery fracture*

# CHINESE BURIAL SUITS

## IN ANCIENT CHINA JADE WAS BELIEVED TO PRESERVE THE DECEASED BODY FROM DECAY. THIS LED TO THE CREATION OF SOME OF THE MOST UNUSUAL STONE ARTEFACTS CREATED BY MAN: JADE BURIAL SUITS.

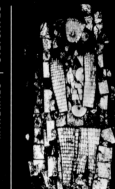

**MISSING PRINCESS**
*Princess Tou Wan's suit was found in a disarticulated state. Nothing but a pinch of dust remained of her body.*

Jade burial suits were made for the most part during the Han Dynasty (206BC–AD220), and consisted of several thousand small jade plaques sewn together with metal wires. The nature and value of the metal used was determined by the deceased's social rank. Only royalty could wear suits sewn with gold thread; nobles were allowed suits sewn with silver thread; and lesser nobles were permitted bronze thread. The most elaborate suit ever made was that of Prince Nanyue (died c.122BC); found in Guangzhou, it consisted of 4,000 jade plaques sewn together with gold wire.

In 1955, two tombs were discovered in separate caves dug into cliffs at Mancheng in the province of Hebei, 145km (90 miles) south-west of Beijing. Surrounded by over 10,000 objects of art, they contained the coffins of Prince Liu Shen and his wife, Princess Tou Wan. Their bodies, laid there almost 2,000 years ago, were encased

### LABOUR OF DEVOTION
*The restored suit of Princess Tou Wan consists of 2,156 jade plaques sewn together with about 700g (24oz) of gold thread. It is estimated that, using the technology of the Han period, such a suit would have taken 10 man-years to construct.*

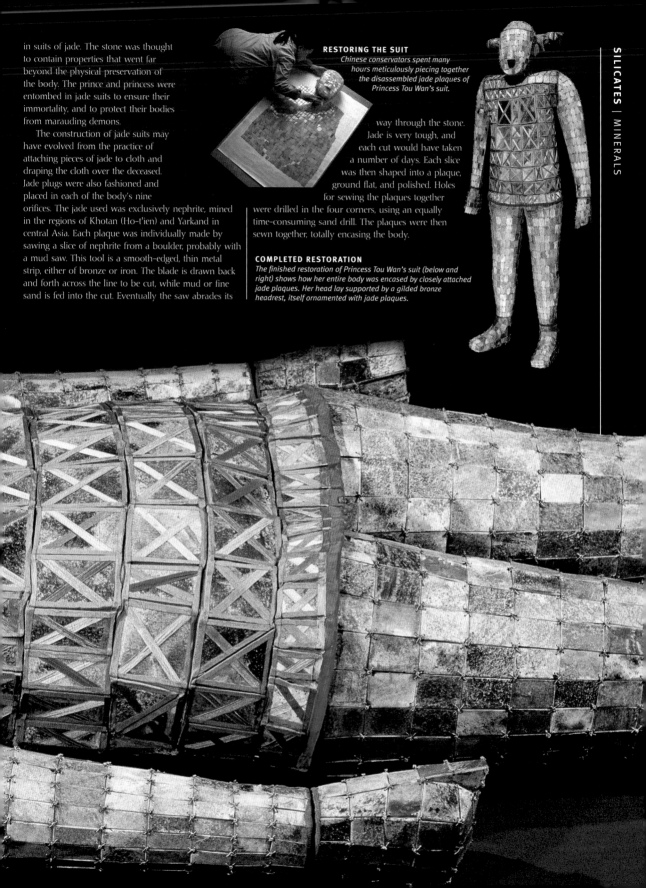

in suits of jade. The stone was thought to contain properties that went far beyond the physical preservation of the body. The prince and princess were entombed in jade suits to ensure their immortality, and to protect their bodies from marauding demons.

The construction of jade suits may have evolved from the practice of attaching pieces of jade to cloth and draping the cloth over the deceased. Jade plugs were also fashioned and placed in each of the body's nine orifices. The jade used was exclusively nephrite, mined in the regions of Khotan (Ho-t'ien) and Yarkand in central Asia. Each plaque was individually made by sawing a slice of nephrite from a boulder, probably with a mud saw. This tool is a smooth-edged, thin metal strip, either of bronze or iron. The blade is drawn back and forth across the line to be cut, while mud or fine sand is fed into the cut. Eventually the saw abrades its

### RESTORING THE SUIT
*Chinese conservators spent many hours meticulously piecing together the disassembled jade plaques of Princess Tou Wan's suit.*

way through the stone. Jade is very tough, and each cut would have taken a number of days. Each slice was then shaped into a plaque, ground flat, and polished. Holes for sewing the plaques together were drilled in the four corners, using an equally time-consuming sand drill. The plaques were then sewn together, totally encasing the body.

### COMPLETED RESTORATION
*The finished restoration of Princess Tou Wan's suit (below and right) shows how her entire body was encased by closely attached jade plaques. Her head lay supported by a gilded bronze headrest, itself ornamented with jade plaques.*

# RICHTERITE

**MONT ST.-HILAIRE**
*Transparent green crystals occur in this region of Quebec, Canada.*

NAMED IN 1865 FOR THE German mineralogist Theodore Richter, the amphibole mineral richterite is a sodium calcium magnesium silicate. Where iron replaces the magnesium in the structure, the mineral is called ferrorichterite; if fluorine replaces the hydroxyl, it is fluororichterite. Richterite crystals are long and prismatic, or prismatic to fibrous aggregates, or rock-bound crystals. Richterite is brown, greyish-brown, yellow, brownish- to rose-red, or pale to dark green. Richterite occurs in thermally metamorphosed limestones, and in contact metamorphic zones. It also occurs as a hydrothermal product in mafic igneous rocks, and in manganese-rich ore deposits. Localities include Mont St.-Hilaire, Quebec, and Wilberforce and Tory Hill, Ontario, Canada; Långban and Pajsberg, Sweden; West Kimberley, Western Australia; Sanka, Myanmar; and, in the USA, at Iron Hill, Colorado; Leucite Hills, Wyoming; and Libby, Montana.

## PROPERTIES

| | |
|---|---|
| **GROUP** | Silicates – inosilicates |
| **CRYSTAL SYSTEM** | Monoclinic |
| **COMPOSITION** | $Na(Ca,Na)Mg_5Si_8O_{22}(OH)_2$ |
| **COLOUR** | Brown, yellow, red, or green |
| **FORM/HABIT** | Prismatic |
| **HARDNESS** | 5–6 |
| **CLEAVAGE** | Perfect |
| **FRACTURE** | Uneven, brittle |
| **LUSTRE** | Vitreous |
| **STREAK** | Pale yellow |
| **SPECIFIC GRAVITY** | 3.0–3.5 |
| **TRANSPARENCY** | Transparent to translucent |

*Quartz groundmass*

*Richterite crystals*

**RICHTERITE PRISMS**
*Prismatic richterite crystals are scattered throughout a groundmass of quartz.*

# GLAUCOPHANE

**ZERMATT**
*This area of Switzerland is a source of glaucophane.*

GLAUCOPHANE AND ITS associated minerals are known as the glaucophane facies, and their presence indicates that metamorphism has occurred under particular temperature and pressure conditions. Glaucophane is a common amphibole mineral, a sodium magnesium aluminium silicate, which falls under the general term asbestos. Where iron replaces the magnesium it is known as ferroglaucophane. Its crystals are slender and prismatic, and it can also be massive, fibrous, and granular. It is grey, lavender-blue, or bluish-black, and fuses readily to form a green glass. Its name is derived from the Greek *glaukos*, meaning "bluish-green", and *phainesthai*, "to appear". Glaucophane occurs in schists formed by the low-temperature, high-pressure metamorphism of sodium-rich sediments, or by the introduction of sodium into the metamorphic process. It is often accompanied by epidote, almandine, chlorite, and jadeite. It is widespread, with Switzerland, Wales, Scotland, Greece, Italy, Japan, and Colorado, USA, among its localities.

## PROPERTIES

| | |
|---|---|
| **GROUP** | Silicates – inosilicates |
| **CRYSTAL SYSTEM** | Monoclinic |
| **COMPOSITION** | $Na_2(Mg_3Al_2)Si_8O_{22}(OH)_2$ |
| **COLOUR** | Bluish-grey to black |
| **FORM/HABIT** | Slender prismatic |
| **HARDNESS** | 6 |
| **CLEAVAGE** | Distinct |
| **FRACTURE** | Uneven to conchoidal |
| **LUSTRE** | Vitreous to pearly |
| **STREAK** | Grey-blue |
| **SPECIFIC GRAVITY** | 3.2 |
| **TRANSPARENCY** | Transparent to translucent |

**ANGLESEY**
*Just off the north-west coast of Wales, UK, Anglesey has fine examples of glaucophane.*

**ITALIAN GLAUCOPHANE**
*This specimen from Polloni, Piedmont, Italy, shows glaucophane with fuchsite, rutile, and pyrite.*

*Dark blue-green glaucophane*

*Pyrite*

# RIEBECKITE

RIEBECKITE IS A SODIUM IRON SILICATE in the amphibole family. It forms prismatic, striated crystals, their colour depending on the concentration of iron in the structure. An asbestiform variety of riebeckite is named crocidolite, and is commonly called blue asbestos. Riebeckite is found in felsic igneous rocks, such as granites and syenites, and in felsic volcanics, especially sodium-rich rhyolites.

Crocidolite is of metamorphic origin and is formed under moderate temperature and pressure from ironstones. Crocidolite occurs at Robertstown, South Australia; Henan Province in China; Cochabamba, Bolivia; and South Africa. Silica-replaced crocidolite occurs in several colours, and is the golden-brown gem tiger-eye and the blue hawk's eye. Its principal source is the South African asbestos deposit..

## PROPERTIES

| | |
|---|---|
| **GROUP** | Silicates – inosilicates |
| **CRYSTAL SYSTEM** | Monoclinic |
| **COMPOSITION** | $Na_2(Fe^{2+}{}_3 Fe^{3+}{}_2)Si_8O_{22}(OH)_2$ |
| **COLOUR** | Dark blue, black |
| **FORM/HABIT** | Prismatic |
| **HARDNESS** | 6 |
| **CLEAVAGE** | Perfect |
| **FRACTURE** | Uneven |
| **LUSTRE** | Vitreous, silky |
| **STREAK** | Blue-grey |
| **SPECIFIC GRAVITY** | 3.3–3.4 |
| **TRANSPARENCY** | Transparent to translucent |
| **R.I.** | 1.54–1.55 |

## ASBESTOS

Asbestos is the name given to the thin, flexible fibres of some minerals, such as riebeckite and tremolite (see p.279). Its silky fibres have been spun into cloth and mixed with other minerals to form wallboard, roofing, and weatherproofing for houses. Originally valued for its fireproof qualities and its ability to withstand electricity and strong acid, it was later discovered that its minute fibres can cause cancer when inhaled. Mining has now ceased globally for blue asbestos. (See also p.258.)

**ASBESTOS FIREPROOF GEAR**
*Asbestos cloth was once made into fireproof clothing for firefighters.*

Group of prismatic crystals

Vertical parallel striations

*Fibrous crystals*

**BLUE ASBESTOS**

*Rock groundmass*

**TIGER-EYE**
*Silica-saturated crocidolite forms the gemstone tiger-eye.*

**HAWK'S EYE**
*Silica-saturated blue crocidolite becomes the gemstone hawk's eye.*

**RIEBECKITE CRYSTALS**
*The long, striated crystals characteristic of riebeckite are clearly visible in this specimen.*

# ARFVEDSONITE

ARFVEDSONITE IS A MEMBER of the amphibole group. When lithium and magnesium replace iron in the structure, it forms the mineral eckermannite. Arfvedsonite forms well-developed crystals less often than the other amphiboles; it is generally found as massive aggregates. When crystals do occur, they are short to long prismatic or tabular, and often twinned. Arfvedsonite fuses readily, leaving a magnetic glass globule. It occurs in felsic igneous rocks and their associated pegmatites, often with other amphiboles and pyroxenes, such as aegirine and augite. It is also found in regional metamorphics. Arfvedsonite is relatively widespread; crystals occur in Colorado and New Hampshire, USA; the Kola Peninsula, Russia; around Oslo, Norway; and at Mont St.-Hilaire, Canada.

## PROPERTIES

| | |
|---|---|
| **GROUP** | Silicates – inosilicates |
| **CRYSTAL SYSTEM** | Monoclinic |
| **COMPOSITION** | $Na_3(Fe^{2+}{}_4 Fe^{3+})Si_8O_{22}(OH)_2$ |
| **COLOUR** | Blue-black to black |
| **FORM/HABIT** | Massive aggregates |
| **HARDNESS** | 5½–6 |
| **CLEAVAGE** | Perfect |
| **FRACTURE** | Uneven |
| **LUSTRE** | Dull to vitreous |
| **STREAK** | Pale to dark blue-grey |
| **SPECIFIC GRAVITY** | 3-4 |
| **TRANSPARENCY** | Translucent |

Rock groundmass

Vitreous lustre

## DISCOVERY

Arfvedsonite was discovered near Ilimaussaq, Greenland, and recognized as a new mineral in 1923. It was named for the Swedish chemist Johan Arfvedson (1792–1841), who discovered the element lithium in 1817. It is still found in Greenland, at Ilimaussaq and Narssarssuk.

**GREENLAND**

**ARFVEDSONITE FRAGMENTS**
*This specimen shows a mass of arfvedsonite crystal fragments set in a rock groundmass.*

# CYCLOSILICATES

CYCLOSILICATES TAKE THEIR name from cyclo, meaning "circle", and are so named because their crystals consist of closed, ring-like circles of tetrahedra that share corners. They are sometimes also referred to as ring silicates. Each tetrahedron shares two of its oxygen atoms with other tetrahedra, and the rings thus formed may have three members (such as benitoite), four members (such as axinite), or six members (such as beryl). Their chemical formulae will each have some multiple of $SinO3n$, where "n" reflects the number of rings.

**BERYL**

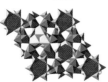

**TOURMALINE**

**RING CRYSTAL STRUCTURES**
*Both beryl and tourmaline contain six-fold rings of $SiO_4$ tetrahedra.*

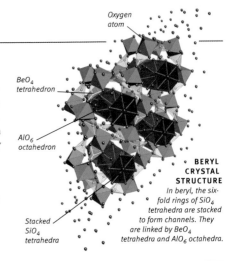

Oxygen atom

$BeO_4$ tetrahedron

$AlO_6$ octahedron

Stacked $SiO_4$ tetrahedra

**BERYL CRYSTAL STRUCTURE**
*In beryl, the six-fold rings of $SiO_4$ tetrahedra are stacked to form channels. They are linked by $BeO_4$ tetrahedra and $AlO_6$ octahedra.*

## DIOPTASE

**KAZAKHSTAN**
*Superb dioptase specimens come from Kazakhstan.*

DIOPTASE CAN SUPERFICIALLY resemble emerald – sufficiently so that crystals mined from the rich deposit in Kazakhstan were wrongly identified as such when sent to Tsar Paul in 1797. History does not record his reaction. Were it not for its softness and good cleavage, dioptase would make a superb gemstone to rival emerald in colour. Its prismatic crystals, often with rhombohedral terminations, can be highly transparent, and it is from this that its name is derived: *dia*, from the Greek for "through", and *optazein*, meaning "visible", or "to see". Transparent examples can be weakly pleochroic, and intensely coloured specimens can be translucent. Dioptase also occurs in crystalline and massive aggregates. It forms where copper veins have been altered by oxidation. Superb specimens still come from Kazakhstan; and also from Iran, Namibia, the Congo, Argentina, Chile, and the USA.

### PROPERTIES

| | |
|---|---|
| **GROUP** | Silicates – cyclosilicates |
| **CRYSTAL SYSTEM** | Hexagonal/trigonal |
| **COMPOSITION** | $CuSiO_2(OH_2)$ |
| **COLOUR** | Emerald- to blue-green |
| **FORM/HABIT** | Prismatic |
| **HARDNESS** | 5 |
| **CLEAVAGE** | Perfect |
| **FRACTURE** | Uneven to conchoidal |
| **LUSTRE** | Vitreous to greasy |
| **STREAK** | Pale greenish-blue |
| **SPECIFIC GRAVITY** | 3.3 |
| **TRANSPARENCY** | Transparent to translucent |

**DIOPTASE CRYSTALS**
*These fine, prismatic dioptase crystals are on quartz, on chrysocolla. The specimen comes from Kaokoveld, Namibia.*

Dioptase crystals

Vitreous lustre

*Chrysocolla groundmass*

## SUGILITE

**INDIA**
*Sugilite deposits are found in the Madhya Pradesh region of India.*

DISCOVERED IN 1944, sugilite was recognized as a mineral only in 1976. It is named for Japanese petrologist Ken-ici Sugi, its co-discoverer. A sodium potassium lithium silicate hydrate, it contains variable amounts of iron, manganese, and aluminium. It is usually massive or granular, but when crystals do occur, they are small (less than 2cm/$^3$/$_4$in) across and prismatic. It is pale to deep pink, brownish-yellow, or purple. The pink to purple coloration is caused by manganese; the pink material is aluminium-rich, and the purple is iron-rich. Sugilite forms in metamorphosed manganese deposits, and as xenoliths in marble. It is found at Mont St.-Hilaire, Canada; Iwagi Island, Japan; Kuruman, South Africa; and Faggiona, Italy.

Purple coloration

### PROPERTIES

| | |
|---|---|
| **GROUP** | Silicates – cyclosilicates |
| **CRYSTAL SYSTEM** | Hexagonal |
| **COMPOSITION** | $KNa_2(Fe,Mn,Al)_2Li_3Si_{12}O_{30}\cdot H_2O$ |
| **COLOUR** | Pink, brown-yellow, or purple |
| **FORM/HABIT** | Massive |
| **HARDNESS** | $5^1/_2$–$6^1/_2$ |
| **CLEAVAGE** | Poor |
| **FRACTURE** | Subconchoidal |
| **LUSTRE** | Vitreous |
| **STREAK** | White |
| **SPECIFIC GRAVITY** | 2.7–2.8 |
| **TRANSPARENCY** | Translucent to opaque |
| **R.I.** | 1.60–1.61 |

**MASSIVE SUGILITE**
*Purple sugilite, such as this specimen from Kuruman, Northern Cape Province, South Africa, is used for gemstones.*

Massive habit

**CABOCHON**
*Sugilite is always cut en cabochon when used as a gemstone. It is relatively new to the gemstone market.*

**PEBBLE**
*Valued for their vivid purple colour, sugilite pebbles are sometimes polished in rock tumblers.*

# CORDIERITE

CORDIERITE WAS NAMED for French geologist Pierre L.A. Cordier, who first described it in 1813. It can be blue, violet-blue, grey, or blue-green. Gem-quality blue cordierite is known as iolite, and because of its colour is also called "water sapphire". Cordierite is pleochroic, exhibiting three colours. Crystals are prismatic, and the best blue colour is seen down their length; their pleochroism is so pronounced that they may appear colourless when viewed across the crystal. It occurs in high-grade, thermally metamorphosed rocks. It is also found in schists and gneisses, and more rarely in pegmatites and quartz veins. In addition to Sri Lanka and Myanmar, there is a major source of iolite near Madras, India. Flawless crystals up to 5cm (2in) are found on Garnet Island, Northwest Territories, Canada. Gem material also comes from Madagascar, Tanzania, and South Africa.

**MYANMAR**
*Iolite is found in the gem gravels of Myanmar and Sri Lanka.*

## PROPERTIES

| | |
|---|---|
| **GROUP** | Silicates – cyclosilicates |
| **CRYSTAL SYSTEM** | Orthorhombic |
| **COMPOSITION** | $(Mg,Fe)_2Al_4Si_5O_{18}$ |
| **COLOUR** | Blue, blue-green, grey-violet |
| **FORM/HABIT** | Short prismatic, granular |
| **HARDNESS** | $7-7\frac{1}{2}$ |
| **CLEAVAGE** | Moderate to poor |
| **FRACTURE** | Conchoidal to uneven |
| **LUSTRE** | Vitreous to greasy |
| **STREAK** | White |
| **SPECIFIC GRAVITY** | 2.6 |
| **TRANSPARENCY** | Transparent to translucent |
| **R.I.** | 1.53–1.55 |

**PANNING FOR GEMSTONES**
*Among the gem gravels of northern Myanmar, gemstones such as iolite are found by traditional panning.*

**CORDIERITE CRYSTALS**
*This group of short prismatic cordierite crystals occurs in a rock groundmass. Common cordierite is used as a heat-resistant material.*

Cordierite crystal

**CRYSTAL IN MATRIX**

Cordierite crystal

Rock groundmass

**IOLITE**
*Iolite is sometimes faceted into cubes to demonstrate its strong pleochroism.*

**STEP**
*Some dark gemstones can be faceted in various ways to best display their colour.*

**MIXED**
*Iolite shows good colour and brilliance with small facets.*

---

# BENITOITE

DISCOVERED IN 1907 by a prospector who mistook it for sapphire, benitoite is barium titanium silicate. It is generally blue to dark blue, although it can be colourless, white, or, rarely, pink. It is strongly dichroic, with blue, cut stones appearing blue or colourless when seen from different angles. It has strong colour dispersion, making it nearly as brilliant as diamond. It is found in hydrothermally altered serpentine and in veins in schist. Gem-quality crystals are generally small, with cut stones seldom exceeding 5 carats. Its principal gem source is San Benito Mine, California, where it was first discovered; pink crystals have been found at a nearby mine. It is also found at Esneux, Belgium; Niigata Prefecture, Japan; and Magnet Cove, Arkansas, USA.

**SAN BENITO COUNTY, USA**
*Benitoite was named after its place of discovery.*

## PROPERTIES

| | |
|---|---|
| **GROUP** | Silicates – cyclosilicates |
| **CRYSTAL SYSTEM** | Hexagonal |
| **COMPOSITION** | $BaTiSi_3O_9$ |
| **COLOUR** | Blue, colourless, pink |
| **FORM/HABIT** | Platy |
| **HARDNESS** | $6\frac{1}{2}$ |
| **CLEAVAGE** | Imperfect |
| **FRACTURE** | Conchoidal to uneven |
| **LUSTRE** | Vitreous |
| **STREAK** | White |
| **SPECIFIC GRAVITY** | 3.7 |
| **TRANSPARENCY** | Transparent to translucent |
| **R.I.** | 1.76–1.80 |

Benitoite

Natrolite

**CRYSTALS IN MATRIX**

**BENITOITE CRYSTALS**
*These specimens from the San Benito Mine show good crystal form and colour intensity.*

Pyramidal crystal

**CRYSTAL**

**SAN BENITO MINE**
*In San Benito, California, this mine is both the discovery locality and the main source for gem benitoite.*

**BENITOITE IN A MATRIX**
*This group of intense blue dipyramidal benitoite crystals rests on a rock matrix, along with natrolite.*

Natrolite

Pyramidal benitoite crystals

**BRILLIANT**
*This brilliant cut benitoite shows good blue colour looking straight into the stone.*

**MIXED**
*Seen side-on, the blue colour has nearly disappeared, demonstrating the dichroism of benitoite.*

**PALA, CALIFORNIA**
*Some of the world's most beautiful tourmaline specimens come from the Pala district, San Diego, California, USA.*

**TOURMALINE STAMP**

**PLASTER IMPRESSION**

**ALEXANDER THE GREAT, STAMP AND PLASTER IMPRESSION**
*This is a superb intaglio in zoned purple-yellow tourmaline depicting the head of Alexander the Great, dating to the 3rd or 2nd century BC.*

# TOURMALINE

TOURMALINE IS THE NAME GIVEN to a family of borosilicate minerals of complex and variable composition, but all members have the same basic crystal structure. There are 11 species in the tourmaline group, including elbaite, dravite, schorl, and liddicoatite. Numerous varieties are also recognized, including indicolite (blue), achroite (colourless), rubellite (pink or red), and verdelite (green), and these variety names can apply to more than one tourmaline species. Only the major ones are discussed here. Elbaite forms a solid-solution series with dravite, and dravite forms a solid-solution series with elbaite and with schorl.

## ABUNDANT GEMSTONE

Crystals of tourmaline are generally prismatic. Coloured crystals are very strongly dichroic, and frequently display colour zoning. Tourmaline is abundant, and its best-formed crystals are usually found in pegmatites and in metamorphosed limestones in contact with granitic magmas. Tourmaline minerals are resistant to weathering, so they accumulate in gravel deposits – the origin of its name is the Singhalese word turamali, meaning "gem pebbles". For the same reason, tourmaline is an accessory mineral in some sedimentary rocks. Gem-quality tourmaline occurs in numerous localities. Tourmaline's piezoelectric properties mean that it is also an important industrial mineral. It is employed in pressure devices such as depth-sounding equipment and other apparatus that detect and measure variations in pressure. It is also used in optical devices for polarizing light.

Although many of its transparent varieties are valued as gems, most tourmaline is dark, opaque, and not particularly attractive except as well-formed mineral specimens. Probably the most common tourmaline is schorl, a black, opaque, iron-rich mineral. Its prismatic crystals may reach several metres long. Dravite is a very dark-coloured (usually brown) tourmaline, rich in magnesium. It is sometimes cut as a gem, and its colour can be lightened by heat-treatment. It is found in the gem gravels of Sri Lanka, and in the USA.

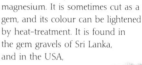

**BUDDHA PENDANT**
*Tourmaline is a favourite material of gem carvers, who make pieces such as this rubellite pendant.*

*Quartz crystal*

*Albite*

## MOUNT MICA MINES

Although tourmaline is a relatively common mineral, gem-quality tourmaline is almost exclusively limited to pegmatites (see p.36). In 1820, elbaite of exceptional clarity was discovered and mined at Mount Mica in eastern Maine, USA. Although the mine never produced the same quantities as the Pala, California, mines (the Pala mines produced 120 tonnes between 1902 and 1911), Maine tourmaline was more highly prized by cutters and carvers. The Mount Mica mines were revitalized in 1972 when another rich pocket of tourmaline was found. Although the deposits are depleted, mining continues, and tourmaline is still the state gem of Maine.

**TOURMALINE MINER**
*In this photograph from the end of the 19th century, Loren Merrill, a mining entrepreneur, poses in one of the largest pockets of tourmaline ever found at Mount Mica.*

**BLACK, GREEN, AND PINK**

**WATERMELON**

**INDICOLITE**

**WATERMELON IN CROSS SECTION**

## TOURMALINE COLOURS

*The specimens shown here are a sample of the great variety that exists in the tourmaline family. Coloration can vary lengthways in a crystal as in the black and green crystal, which also contains pink at one end, or in cross section, as in the watermelon crystals. It can also be constant, as in the blue indicolite.*

*Elbaite crystal*

*zoned green at base*

Canada, Mexico, Brazil, and Australia. Liddicoatite provides a small amount of gem material, principally from the Czech Republic, Russia, and Madagascar. A pink tourmaline cabochon has been discovered set in a gold ring of Nordic origin dating from AD1000.

### VARIETY OF COLOURS

Elbaite provides the most gemstone material. Usually green, it can have many different colours, including pink or red, blue, and green; it can also be colourless. Yellow-green is the most common of all gem tourmaline colour varieties. Emerald-green is much rarer and more valuable, and, until the 18th century, it was often confused with emerald. Most emerald-green stones come from Brazil, Tanzania, and Namibia.

The most dramatic of the colour-zoned gems is the "watermelon" tourmaline, which, when sliced across the crystal, shows a red or pink centre surrounded by a rim of green. Some crystals are pink at one end and green at the other.

The superb red and green crystals from the Pala district, San Diego, USA, the colour-zoned watermelon crystals from Brazil, and the magnificent red prismatic crystals from Madagascar and Mozambique are among the most stunningly beautiful gem materials. Other significant gem deposits are found in Maine, USA; Russia; Afghanistan; Pakistan; Nigeria; and Namibia.

### PROPERTIES

| | |
|---|---|
| **GROUP** | Silicates – cyclosilicates |
| **CRYSTAL SYSTEM** | Hexagonal/trigonal |
| **COMPOSITION** | $Na(Mg,Fe,Li,Mn,Al)_3Al_6(BO_3)_3Si_6 \cdot O_{18}(OH,F)_4$ |
| **COLOUR** | Black, green, brown, red, blue, yellow, or pink |
| **FORM/HABIT** | Prismatic, acicular |
| **HARDNESS** | $7–7^{1}/_{2}$ |
| **CLEAVAGE** | Indistinct |
| **FRACTURE** | Uneven to conchoidal |
| **LUSTRE** | Vitreous |
| **STREAK** | Colourless |
| **SPECIFIC GRAVITY** | 3.0–3.2 |
| **TRANSPARENCY** | Transparent to translucent |
| **R.I.** | 1.61–1.64 |

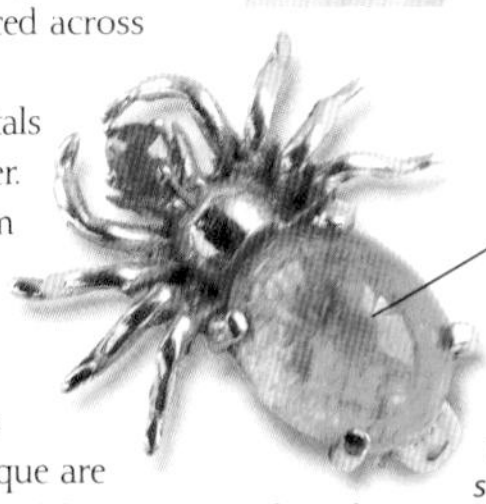

### INDICOLITE STEP

*This emerald-cut, greenish-blue tourmaline gemstone is an example of indicolite. It comes from a deposit in Paraiba, Brazil.*

### RUBELLITE STEP

*The rich red colour of some rubellite is well illustrated by this superb translucent stone faceted in an emerald cut.*

### WATERMELON STEP

*This stone is faceted from watermelon tourmaline to show both colours of this type of gem.*

### MIXED CUSHION

*Orange coloration in tourmaline is uncommon. This orange-brown gem, faceted in a mixed cushion cut, is dravite.*

### ROUND BRILLIANT

*This brilliant-cut colourless tourmaline stone is a rare variety of elbaite, known as achroite.*

*Pink tourmaline*

### SPIDER BROOCH

*This attractive spider has two tourmaline cabochons set into gold.*

*Broken crystal of elbaite*

*Schorl*

### SCHORL

*In this specimen, short prismatic crystals of schorl are on a rock groundmass with scalenohedra of calcite.*

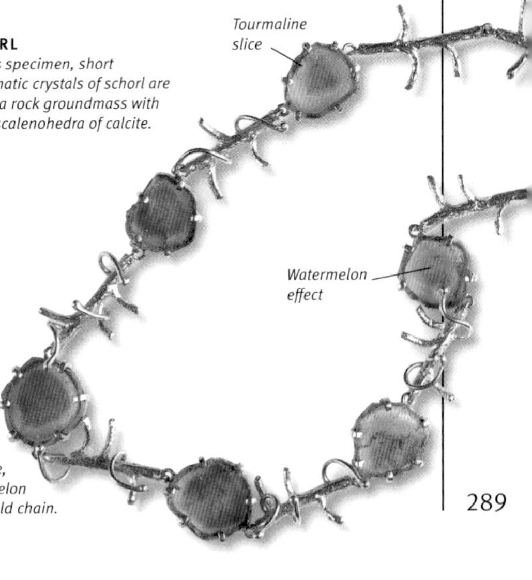

*Tourmaline slice*

*Calcite*

*Rock groundmass*

*Watermelon effect*

### ELBAITE CRYSTALS

*In this specimen from the type locality of the island of Elba, Italy, elbaite rests on a groundmass of quartz and albite.*

### SLICE NECKLACE

*This modern design uses slices of tourmaline, cut to reveal the watermelon effect, on an intricate gold chain.*

# BERYL: AQUAMARINE

FEW PEOPLE HAVE EVER heard of the mineral beryl, but almost everyone has heard of its principal gemstone varieties – emerald (see pp.292–93) and aquamarine, the most common variety. Before 1925 its solitary use was as a gemstone, but from then on many important uses were found for beryllium. Since then, common beryl – a beryllium aluminium silicate – has been widely sought as the ore of this rare element. Other gemstone varieties (see opposite) are heliodor (golden yellow), morganite (pink), and goshenite (colourless). Common beryl ranges in colour from tan to pale green, pale or sky blue, or yellowish. It is a minor constituent of many granitic rocks and their associated pegmatite dykes, of mica schists, and of gneisses. The name "beryl" is derived from the Greek *beryllos*, applied to many green stones.

Much beryl production is a by-product in the mining of feldspar and mica, and no large deposits have been found. However, beryl crystals, which normally occur as columnar, hexagonal prisms, are found from time to time: a 200-tonne crystal was found in Brazil, and a crystal 5.8m (19ft) long and 1.5m (5ft) in diameter was discovered in South Dakota, USA.

Aquamarine, meaning "sea water", is almost universally found in cavities in pegmatites or in alluvial deposits, and forms larger and clearer crystals than emerald. One transparent crystal from Brazil weighed 110kg (245lb). In the 19th century, sea-green aquamarine was highly valued; today sky-blue crystals imparted by traces of iron are preferred. In ancient times, aquamarine amulets engraved with Poseidon were thought to protect sailors.

**URAL MOUNTAINS**
*The Ural Mountains of Russia have rich deposits of gemstones, including aquamarine.*

## PROPERTIES

| | |
|---|---|
| **GROUP** | Silicates – cyclosilicates |
| **CRYSTAL SYSTEM** | Hexagonal |
| **COMPOSITION** | $Be_3Al_2Si_6O_{18}$ |
| **COLOUR** | Green, blue, yellowish |
| **FORM/HABIT** | Prismatic |
| **HARDNESS** | $7\frac{1}{2}$–8 |
| **CLEAVAGE** | Indistinct |
| **FRACTURE** | Uneven to conchoidal |
| **LUSTRE** | Vitreous |
| **STREAK** | White |
| **SPECIFIC GRAVITY** | 2.6–2.8 |
| **TRANSPARENCY** | Transparent to translucent |
| **R.I.** | 1.57–1.58 |

**AQUAMARINE CRYSTAL**
*This shows the often complex termination seen on beryl crystals.*

**CABOCHON**
*When cut en cabochon, some aquamarines – like this gemstone – show a cat's-eye effect.*

**RECTANGULAR STEP**
*Step cuts are frequently used for aquamarine to show off the deep blue colour.*

**SQUARE CUSHION**
*Prismatic crystals like those of aquamarine are usually cut in squares or rectangles to preserve the gemstone material.*

**OVAL MIXED**
*This aquamarine has been facetted in a mixed cut. The effect – as with all gemstones – is to deepen the colour of the stone.*

**GREENISH AQUAMARINE**

**BERYL CRYSTALS**
*These crystals are prismatic. The greenish aquamarine can be heat-treated to turn it blue.*

Prism face

Complex termination

Prism face

**BERYL CRYSTAL**

**NUCLEAR FUEL ELEMENT**
*Beryllium is used to regulate the flow of neutrons in beryllium moderated reactors.*

## MOUNT ANTERO, USA

At just over 4,250m (14,000ft), Mount Antero, in the Rocky Mountains of Colorado, USA, is the highest gemstone locality in North America. Although the site is famous for its aquamarine, lesser amounts of phenakite and other pegmatite minerals are also found here. Because of its high altitude, it is accessible to collectors for only two to three months of the year.

**ANTERO CRYSTAL**
*This crystal was discovered at Mount Antero.*

Prismatic crystal

Vitreous lustre

Rock groundmass

**AQUAMARINE CRYSTALS**
*This mass of prismatic aquamarine crystals is from Dusso, Karakoram Range, northern Pakistan.*

# MORGANITE

**MINAS GERAIS, BRAZIL**
*This is a prime locality for morganite.*

MORGANITE IS GEM-QUALITY beryl that is coloured pink, rose-lilac, peach, orange, or pinkish yellow, by the presence of cesium or manganese. Morganite crystals often show colour banding, with a sequence from blue near the base to nearly colourless in the centre, to peach or pink at the termination. Morganite is also dichroic, and tends to form squat, tabular crystals. It is commonly found in lithium-rich pegmatites with lepidolite and tourmaline. It comes from a number of localities in Minas Gerais, Brazil, with some crystals reaching 25kg (55lb) in weight. Other important localities include Pala, California, USA; Muiane, Mozambique; Elba, Italy; and several localities in Madagascar. Morganite is almost always faceted, and stones with a yellow or orange tinge are sometimes heat-treated to improve the pink colour.

**J. PIERPONT MORGAN**
*Morganite is named for the American banker and gem enthusiast J. Pierpont Morgan.*

## PROPERTIES

| | |
|---|---|
| **GROUP** | Silicates – cyclosilicates |
| **CRYSTAL SYSTEM** | Hexagonal |
| **COMPOSITION** | $Be_3Al_2Si_6O_{18}$ |
| **COLOUR** | Pink, rose-lilac, peach, orange, or pinkish-yellow |
| **FORM/HABIT** | Tabular |
| **HARDNESS** | $7\frac{1}{2}$–8 |
| **CLEAVAGE** | Indistinct |
| **FRACTURE** | Uneven to conchoidal |
| **LUSTRE** | Vitreous |
| **STREAK** | White |
| **SPECIFIC GRAVITY** | 2.6–2.8 |
| **TRANSPARENCY** | Transparent to translucent |
| **R.I.** | 1.57–1.58 |

**MORGANITE ROUGH**
*This piece of gem morganite has been broken from a larger crystal. Its pink colour comes from manganese.*

Morganite band

Heliodor band

**COLOUR-BANDED CRYSTAL**

Tourmaline inclusions

**TABULAR CRYSTAL**

**MORGANITE CRYSTALS**
*Morganite tends to form squat, tabular crystals that are often colour-banded.*

**ROUND BRILLIANT**
*This is a richly coloured morganite brilliant cut. Brazil produces pink morganite crystals such as this.*

**OVAL MIXED**
*An oval mixed cut highlights the lighter shade of morganite. Some pale morganite is from California.*

**MIXED DROPLET**
*This droplet-shaped morganite, has been facetted in what is known as a mixed pendaloque.*

---

# GOSHENITE

**PAKISTAN**
*Northern Pakistan has deposits of goshenite.*

THE COLOURLESS VARIETY of beryl, goshenite is named after Goshen, Massachusetts, USA, where it was first found. It is the least common of the gem beryls. As with most of the other beryls, goshenite is principally found in pegmatites. Current sources are the Ural Mountains of Russia; Minas Gerais, Brazil; Dassu, Pakistan; and Ampangabe, Madagascar. The German word for spectacles, *brille*, may derive from the word "beryl", given that colourless beryl was once used for spectacle lenses.

## PROPERTIES

| | |
|---|---|
| **GROUP** | Silicates – cyclosilicates |
| **CRYSTAL SYSTEM** | Hexagonal |
| **COMPOSITION** | $Be_3Al_2Si_6O_{18}$ |
| **COLOUR** | Colourless |
| **FORM/HABIT** | Prismatic |
| **HARDNESS** | $7\frac{1}{2}$–8 |
| **CLEAVAGE** | Indistinct |
| **FRACTURE** | Uneven to conchoidal |
| **LUSTRE** | Vitreous |
| **STREAK** | White |
| **SPECIFIC GRAVITY** | 2.6–2.8 |
| **TRANSPARENCY** | Transparent to opaque |
| **R.I.** | 1.57–1.58 |

**FANCY CUT**
*Gem cuts with many facets bring out the sparkle in colourless material such as goshenite.*

Pinacoid face

Pyramid face

**CRYSTAL**
*This goshenite crystal shows a hexagonal outline and a flat termination.*

---

# HELIODOR

**NORTH CAROLINA**
*This US state is rich in heliodor.*

HELIODOR HAS CRYSTALS that are generally columnar hexagonal prisms. It is found in granitic pegmatites, and its name derives from the Greek *helios*, meaning "sun". The Ural Mountains of Russia produce the best-quality stones. Heliodor is also found in Nigeria, Namibia, Brazil, and the Ukraine. Localities in the USA are Connecticut, Maine, North Carolina, and Pennsylvania.

## PROPERTIES

| | |
|---|---|
| **GROUP** | Silicates – cyclosilicates |
| **CRYSTAL SYSTEM** | Hexagonal |
| **COMPOSITION** | $Be_3Al_2Si_6O_{18}$ |
| **COLOUR** | Yellow or golden yellow |
| **FORM/HABIT** | Prismatic |
| **HARDNESS** | $7\frac{1}{2}$–8 |
| **CLEAVAGE** | Indistinct |
| **FRACTURE** | Uneven to conchoidal |
| **LUSTRE** | Vitreous |
| **STREAK** | White |
| **SPECIFIC GRAVITY** | 2.6–2.8 |
| **TRANSPARENCY** | Transparent to translucent |
| **R.I.** | 1.57–1.58 |

**RECTANGULAR STEP**
*This rectangular step-cut heliodor shows a fine, golden-yellow colour. Iron is the colouring agent.*

Prismatic crystal with complex termination

**PRISMATIC CRYSTAL**
*This prismatic crystal of heliodor is on a rock groundmass.*

Rock groundmass

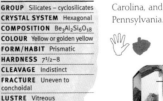

291

# BERYL: EMERALD

**URAL MOUNTAINS, RUSSIA**
*These mineral-rich mountains have been mined for emeralds for nearly two centuries.*

EMERALD IS THE GRASS-GREEN variety of the mineral beryl; other colours of beryl are aquamarine, morganite, goshenite, and heliodor (see pp.290–91). Its name originates in the Greek word *smaragdos*, which seems to have been given to a number of green stones besides emerald. The "emerald" of the Scriptures probably meant carbuncle, a garnet, and *smaragdus*, referred to by Pliny, almost certainly included several species. Thus historical references to "emerald" cannot be assumed to indicate the modern mineral.

## SOURCES OF EMERALDS

Several rich deposits of emeralds in Colombia were discovered and exploited after the Spanish conquest of South America. The deposits at Muzo and Coscuez are still being worked. The emeralds occur within thin veins of white calcite or quartz, in dark shale, and in a black bituminous limestone. After the discovery of the Colombian deposits, emeralds were shipped from South America in huge quantities. A glut of emeralds in Europe ensued, triggering a brisk trade of the gemstones to the Middle East and India, where they were especially favoured by the Mogul rulers. In about 1830, emeralds were discovered in the Ural Mountains of Russia, where they occur in mica or chlorite schist. They have also been found in mica schist in Austria, in granite in Norway, and in a pegmatite vein in New South Wales,

**CHIMU FUNERARY MASK**
*This South American gold mask has emerald beads for eyes.*

**MUZO EMERALD MINE**
*The emerald mine near Muzo, Colombia, is in a region of rugged mountains.*

Australia. Other sources include Brazil, South Africa, Zimbabwe, Pakistan, Zambia, and North Carolina, USA.

Emerald's green colour results from trace chromium. Trace vanadium may also be present, and beryls coloured by vanadium alone

**CRYSTAL IN QUARTZ**

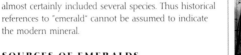

**CRYSTAL IN MICA SCHIST**

**CRYSTALS IN PEGMATITE**

**EMERALD CRYSTAL**
*This unusually large, walnut-sized crystal is in the Oxford University mineral collection, England. It is from Santa Fé de Bogotá, Colombia.*

Prism face

**PRISMATIC CRYSTAL**

**EMERALD CRYSTALS**
*These specimens illustrate several modes of occurrence for emerald, with associated minerals, and an unusually long, prismatic emerald crystal.*

## PROPERTIES

| | |
|---|---|
| **GROUP** | Silicates – cyclosilicates |
| **CRYSTAL SYSTEM** | Hexagonal |
| **COMPOSITION** | $Be_3Al_2Si_6O_{18}$ |
| **COLOUR** | Green |
| **FORM/HABIT** | Prismatic |
| **HARDNESS** | $7^{1}/_{2}$–8 |
| **CLEAVAGE** | Imperfect |
| **FRACTURE** | Uneven to conchoidal |
| **LUSTRE** | Vitreous |
| **STREAK** | White |
| **SPECIFIC GRAVITY** | 2.6–3.0 |
| **TRANSPARENCY** | Transparent to translucent |
| **R.I.** | 1.57–1.60 |

**TRAPICHE EMERALD**
*The star-like pattern in this emerald cabochon is carbonaceous material trapped during its crystallization.*

can be emerald-green, but there are differences in opinion as to whether or not they should be called emerald. Flawless emeralds are rare, so various treatments have been devised to hide or disguise flaws. Cameos, intaglios, and beads can make the best of a flawed stone, and inferior stones are sometimes oiled to fill and disguise cracks and enhance colour. In faceted stones, the "emerald-cut" was devised to minimize loss of valuable material. Basically rectangular, to fit the shape of the normally prismatic emerald crystals, the stone is cut in parallel steps with truncated corners to best display and enhance the colour.

After considerable effort, emerald was finally synthesized in 1937. Synthetic emeralds are currently manufactured in the United States, and appear very similar to natural crystals, rivalling them in colour and beauty. The world's greatest collection of natural emeralds is held in the Republic of Bogotá Bank in Columbia. The largest crystal in the collection weighs 1,795 carats (360g/12³/₄oz).

## THE LORE OF EMERALD

To the Egyptians, emeralds were a symbol of fertility and life. The Aztecs called emerald *quetzalitzli* and associated it with the *quetzal*, a bird with long green plumage – a symbol of seasonal renewal. In Europe, alchemists regarded emerald as the stone of Mercury (Hermes) – messenger of the gods and conductor of the souls of the dead.

When held in the mouth, emerald was believed to be a cure for dysentery and was worn as a preventive for epilepsy. It was also said to assist women at childbirth, drive away evil spirits, and protect the chastity of the wearer. It was held to have great medicinal value if administered internally, and in particular it was said to be good for the eyesight. In the 17th century, Anselmus de Boot, physician to the Holy Roman Emperor, recommended an amulet of emerald to prevent panic, cure fever, and stop bleeding. Unfortunately, emerald was also considered to be an enemy of sexual passion. Albertus Magnus, writing in the 13th century, noted that when King Bela of Hungary embraced his wife, his magnificent emerald broke into three pieces!

**SPANISH INQUISITION NECKLACE**
*The rich, velvety colour and exceptional clarity make the central, barrel-shaped emerald in this necklace (which weighs about 45 carats) among the world's finest.*

## ANCIENT EMERALD MINES

As early as 1300BC, emeralds were being mined in Upper Egypt at Jabal Sukayt and Jabal Zabgrah near the Red Sea coast, east of Aswan. Most emeralds used in historical jewellery would have been from these mines, which, after the conquest of Egypt by Alexander the Great, became known as "Cleopatra's Mines". These workings were rediscovered around 1817, but now yield only poor-quality emeralds. These emeralds occur in mica schist and talc schist (see p.83). Only one other source was known in antiquity: Habachtal, near Salzburg, Austria.

**CLEOPATRA'S MINES**
*This 19th-century photograph was taken when Cleopatra's Mines were still being worked.*

**THE HYLLE JEWEL**
*This Lombardic, late-14th-century initial M with annunciation figures is set with cabochon emeralds, rubies, pearls, and diamonds.*

"A livelier emerald twinkles in the grass, A purer sapphire melts into the sea."

**TENNYSON,** MAUD

**EMERALD RING**
*This classic emerald-cut emerald mounted in platinum is flanked by two brilliant-cut diamonds in this elegant 1940s-style ring.*

**MOGUL PENDANT**
*In this 17th-century Indian pendant, the hexagonal emerald has been carved and set in a gold mount.*

**SYNTHETIC**
*This synthetic Gilson emerald pendeloque has a good green colour with characteristic veil-like inclusions.*

**EMERALD CUT**
*The emerald cut was specifically designed for emeralds to minimize loss of material in cutting.*

**MIXED CUT**
*This mixed-cut emerald shows the cloudier material that is sometimes faceted.*

**CABOCHON**
*This emerald has an unusual octagonal cut – a cabochon top half and a faceted bottom half.*

**ART DECO BRACELET**
*Carved 25-carat emeralds are set in platinum and interspersed with 20-carat mixed-cut diamonds in this French Art Deco bracelet.*

**POLISHED PEBBLE**
*Emerald too cloudy to facet or too light in colour for cabochons is sometimes tumble polished.*

# SOROSILICATES

THERE ARE MORE THAN 70 sorosilicate minerals, although only the crystalline epidote–group minerals and vesuvianite are common. Sorosilicates consist of double-tetrahedral groups in which one oxygen atom is shared by two tetrahedra. This sharing gives the group its name: *soro* which is derived from the Latin for "sister".

*AlO₆ octahedron*

*SiO₄ tetrahedron*

*Iron octahedron*

*Calcium ion*

**EPIDOTE CRYSTAL STRUCTURE**
In epidote, paired and isolated SiO₄ tetrahedra are linked by AlO₆ tetrahedra to form chains. Calcium and iron fill the cavities in the framework.

## HEMIMORPHITE

**DERBYSHIRE**
*This English county is a classic locality for hemimorphite.*

HEMIMORPHITE IS ONE of two minerals formerly called calamine in the USA. Its name is derived from the Greek *hemi*, meaning "half", and *morphe*, "form", a reference to its crystal form. Hemimorphite crystals are double-terminated prisms with a differently shaped termination at each end, a property shown by very few other minerals. Hemimorphite can also be botryoidal, massive, granular, fibrous, or form encrustations. Usually colourless or white, it can also be pale yellow, pale green, or sky-blue. It is a secondary mineral formed in the alteration zone of zinc deposits. Well-crystallized specimens come from Algeria, Namibia, Germany, Mexico, Spain, and many localities in the USA.

**WHITE HEMIMORPHITE**
*Clusters of radiating, tabular hemimorphite crystals encrust a rock groundmass in this specimen.*

*Crystal clusters*

*Rounded masses*

*Rock groundmass*

*Rock groundmass*

### PROPERTIES

**GROUP** Silicates – sorosilicates
**CRYSTAL SYSTEM** Orthorhombic
**COMPOSITION** Zn₄Si₂O₇(OH)₂·H₂O
**COLOUR** Colourless, white, yellow, blue, or green
**FORM/HABIT** Asymmetric, prismatic, tabular, botryoidal
**HARDNESS** 4½–5
**CLEAVAGE** Perfect, good, poor
**FRACTURE** Uneven, brittle
**LUSTRE** Vitreous
**STREAK** White
**SPECIFIC GRAVITY** 3.4–3.5
**TRANSPARENCY** Transparent to translucent

**GREEN HEMIMORPHITE**
*This group of green hemimorphite crystals forms rounded aggregates on a rock groundmass.*

## DANBURITE

**DANBURY, USA**
*Danburite was discovered in 1839 in Danbury, Connecticut.*

DANBURITE IS CALCIUM BOROSILICATE. Its crystals are glassy and prismatic, resembling topaz, but it is easily distinguished from topaz by its poor cleavage. It can also be granular. It is colourless, amber, yellow, grey, pink, or yellow-brown. Some specimens fluoresce blue to blue-green. Generally a moderate to low-temperature contact-metamorphic mineral, danburite is also found in ore deposits formed at relatively high temperatures. It is sometimes found in pegmatites and in evaporites. It occurs associated with cassiterite, chalcopyrite, smoky quartz, anhydrite, axinite, fluorite, and corundum. Localities include Uri, Switzerland; Presov, Slovakia; and Russell and DeKalb, New York, and Ramona, California, USA. Danburite is cut as a gemstone, both faceted and *en cabochon*. Gem danburite comes from Dalnegorsk,

Russia, in crystals 30cm (12in) long and 10cm (4in) in diameter; Mogok, Myanmar; Charcas, Mexico; and Imalo, Madagascar. A 108-carat faceted stone is in the collection of the Geological Museum in London.

### PROPERTIES

**GROUP** Silicates – sorosilicates
**CRYSTAL SYSTEM** Orthorhombic
**COMPOSITION** CaB₂Si₂O₈
**COLOUR** Colourless, yellow
**FORM/HABIT** Prismatic
**HARDNESS** 7–7½
**CLEAVAGE** Indistinct
**FRACTURE** Subconchoidal to uneven
**LUSTRE** Vitreous to greasy
**STREAK** White
**SPECIFIC GRAVITY** 3.0
**TRANSPARENCY** Transparent to translucent
**R.I.** 1.63–1.64

*Transparent crystal*

**BRILLIANT**
*Transparent danburite is faceted into gems for collectors. This gemstone is from Myanmar.*

*Vitreous lustre*

**DANBURITE CRYSTAL**
*This specimen is a transparent single crystal. Although danburite has crystals that are like those of topaz, they lack the brilliance of topaz.*

# AXINITE

**CORNWALL**
*Fine axinite is found in this part of south-west England.*

THIS GROUP OF MINERALS takes its name from the axehead shape of its crystals. There are four minerals in the group: ferro-axinite, the most common (described by the formula in the Properties box); magnesio-axinite where magnesium replaces the iron in ferro-axinite; manganaxinite, where manganese replaces the iron in ferro-axinite; and tinzenite, intermediate in composition between ferro-axinite and manganaxinite. Axinite occurs not only as axehead-shaped crystals but also as rosettes, and in massive and granular forms. The most familiar colour of axinite is clove-brown; varieties can also be grey to bluish-grey, honey-, grey-, or golden-brown; pink, violet-blue, yellow, orange, or red. Colour is not reliable for identification, and crystals of the four are essentially indistinguishable in hand specimens. Axinite is most commonly found in low-temperature and contact metamorphic rocks, and also in mafic igneous rocks. It is found worldwide; gem axinite comes from the USA, Russia, and Australia.

## PROPERTIES

| | |
|---|---|
| **GROUP** | Silicates – sorosilicates |
| **CRYSTAL SYSTEM** | Triclinic |
| **COMPOSITION** | $Ca_2FeAl_2(BSi_4O_{15})(OH)$ |
| **COLOUR** | Clove-brown, grey to bluish-grey, pink, violet, yellow, orange, or red |
| **FORM/HABIT** | Axe-shaped crystals |
| **HARDNESS** | $6^{1}/_{2}$–7 |
| **CLEAVAGE** | Good, poor |
| **FRACTURE** | Uneven to conchoidal, brittle |
| **LUSTRE** | Vitreous |
| **STREAK** | Colourless to light brown |
| **SPECIFIC GRAVITY** | 3.2–3.3 |
| **TRANSPARENCY** | Transparent to translucent |
| **R.I.** | 1.67–1.70 |

**CRYSTALS**
*In this unusual specimen, a smaller, wedge-shaped axinite crystal has grown on to a larger axinite crystal.*

**STEP**
*Axinite is faceted as a gem, although fine-cut stones are small, seldom exceeding 5 carats.*

**BRILLIANT**
*This fine, brilliant-cut axinite is in an unusual and delicate shade of violet.*

*Tabular, wedge-shaped crystals*

*Vitreous lustre*

**AXINITE CRYSTALS**
*This mass of well-formed, transparent, tabular, wedge-shaped axinite crystals has formed on a rock groundmass.*

---

# ILVAITE

**TYROL**
*The Italian Tyrol is a source of ilvaite.*

NAMED FOR THE LOCATION of its discovery, the Italian island of Elba – *Ilva* in Latin – ilvaite frequently occurs in short to long prismatic crystals, commonly striated vertically. It is also coarsely crystalline, massive, or granular in habit. It fuses to a magnetic bead, and is soluble in hydrochloric acid. Ilvaite is generally found in contact metamorphic zones with zinc, copper, and iron ores. It is occasionally found in syenites. Large crystals come from Elba, and from Dalnegorsk, Russia. It is also found at St. Andreasberg, Lower Saxony, Germany; Saitama and Niigata prefectures, Japan; and many localities in the USA.

## PROPERTIES

| | |
|---|---|
| **GROUP** | Silicates – sorosilicates |
| **CRYSTAL SYSTEM** | Monoclinic or orthorhombic |
| **COMPOSITION** | $CaFe_3OSi_2O_7(OH)$ |
| **COLOUR** | Black to greyish-black |
| **FORM/HABIT** | Short to long prismatic |
| **HARDNESS** | $5^{1}/_{2}$–6 |
| **CLEAVAGE** | Distinct, indistinct |
| **FRACTURE** | Uneven, brittle |
| **LUSTRE** | Submetallic to dull |
| **STREAK** | Greenish to brownish-black |
| **SPECIFIC GRAVITY** | 4.0 |
| **TRANSPARENCY** | Opaque |

*Striations along crystal*

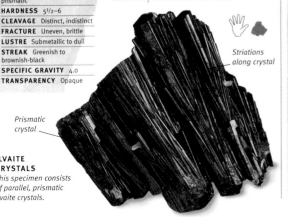

*Prismatic crystal*

**ILVAITE CRYSTALS**
*This specimen consists of parallel, prismatic ilvaite crystals.*

---

# VESUVIANITE

**MOUNT VESUVIUS**
*Vesuvianite is named for its place of discovery in Italy.*

FORMERLY CALLED IDOCRASE, vesuvianite has crystals that are prismatic and glassy, and are usually coloured green or chartreuse, but can be yellow to brown, yellow-green, red, black, blue, or purple. Numerous elements may substitute in the structure, including tin, lead, manganese, chromium, zinc, and sulphur. Vesuvianite is formed by the contact metamorphism of impure limestones, and it is found in marbles and granulites, frequently accompanied by grossular, wollastonite, diopside, and calcite. Gemstone localities include Sakha, Siberia; and, in the USA, Sanford and Auburn, Maine, Olmsteadville, New York, Franklin, New Jersey, and San Benito County, California.

## PROPERTIES

| | |
|---|---|
| **GROUP** | Silicates – sorosilicates |
| **CRYSTAL SYSTEM** | Tetragonal or monoclinic |
| **COMPOSITION** | $Ca_{10}(Mg,Fe)_2Al_4(SiO_4)_5(Si_2O_7)_2(OH,F)_4$ |
| **COLOUR** | Green, yellow |
| **FORM/HABIT** | Prismatic |
| **HARDNESS** | $6^{1}/_{2}$ |
| **CLEAVAGE** | Poor |
| **FRACTURE** | Subconchoidal to uneven, brittle |
| **LUSTRE** | Vitreous to resinous |
| **STREAK** | White to pale greenish-brown |
| **SPECIFIC GRAVITY** | 3–4 |
| **TRANSPARENCY** | Transparent to translucent |
| **R.I.** | 1.70–1.75 |

*Vertical striations*

**VESUVIANITE CRYSTALS**
*This superb specimen consists of prismatic vesuvianite crystals.*

**CABOCHON**
*Vesuvianite is both faceted and cut en cabochon for collectors.*

**CUSHION**
*Transparent vesuvianite is faceted for collectors but is too soft to wear.*

# EPIDOTE

**SALIDA, USA**
*Fine epidote crystals can be found in this area of the USA.*

EPIDOTE FORMS COLUMNAR prisms or thick tabular crystals with faces finely striated parallel to the crystal's length. It derives its name from the Greek *epidosis*, "increase", because one side of the prism is always longer than the others. A calcium aluminium iron silicate hydroxide, epidote can also be acicular, massive, and granular. This mineral is most easily recognized by its characteristic colour – light to dark pistachio-green. Pleochroism (the appearance of different colours when viewed from different directions) is strong. Tawmawlite is a bright green chromium-rich variety. Epidote occurs widely in low-grade regionally metamorphosed rocks, and as a product of hydrothermal alteration of plagioclase feldspar. Good crystals are found in Canada, France, Myanmar, Norway, Peru, and Colorado and California, USA. Gem-quality crystals are found in Austria and Pakistan.

*Vitreous lustre*

## PROPERTIES

| | |
|---|---|
| **GROUP** | Silicates – sorosilicates |
| **CRYSTAL SYSTEM** | Monoclinic |
| **COMPOSITION** | $Ca_2Al_2(Fe,Al)(SiO_4)(Si_2O_7)O(OH)$ |
| **COLOUR** | Pistachio-green |
| **FORM/HABIT** | Short to long prismatic |
| **HARDNESS** | 6–7 |
| **CLEAVAGE** | Good |
| **FRACTURE** | Uneven to splintery |
| **LUSTRE** | Vitreous |
| **STREAK** | Colourless or greyish |
| **SPECIFIC GRAVITY** | 3–4 |
| **TRANSPARENCY** | Translucent |
| **R.I.** | 1.74–1.78 |

**EPIDOTE CRYSTALS**
*This superb group of striated epidote crystals, some reaching 2.5cm (1in) in length, shows typical prismatic development.*

*Striated prismatic crystals*

**MIXED**
*Epidote is sometimes found in transparent crystals, which are cut for collectors.*

**TABLE**
*This dark brown epidote is faceted in a rectangular, table cut.*

**UNAKITE NECKLACE**
*Rock that is made up primarily of epidote may be polished or tumbled and sold as unakite.*

# ZOISITE

**SWISS ALPS**
*One locality for zoisite is the Swiss Alps.*

CRYSTALS OF ZOISITE ARE prismatic and vertically striated and can be yellowish-green, green, white, green-brown, or grey. The most valuable is tanzanite, coloured sapphire-blue by vanadium. A massive, pinkish-red variety is called thulite. Zoisite is a calcium aluminium silicate hydroxide, and belongs to the epidote group. It is characteristic of regional metamorphism and of hydrothermal alteration of igneous rocks. It is also found in quartz veins and pegmatites.

Localities include Spain, Germany, Scotland, and Japan. Thulite occurs in Norway, Italy, and the USA.

## PROPERTIES

| | |
|---|---|
| **GROUP** | Silicates – sorosilicates |
| **CRYSTAL SYSTEM** | Orthorhombic |
| **COMPOSITION** | $Ca_2Al_3(SiO_4)_3(OH)$ |
| **COLOUR** | Yellow-green , white |
| **FORM/HABIT** | Prismatic |
| **HARDNESS** | 6–7 |
| **CLEAVAGE** | Perfect |
| **FRACTURE** | Conchoidal to uneven, brittle |
| **LUSTRE** | Vitreous |
| **STREAK** | White |
| **SPECIFIC GRAVITY** | 3.2–3.4 |
| **TRANSPARENCY** | Transparent to translucent |
| **R.I.** | 1.69–1.70 |

*Deep vertical striations*

*Prismatic crystals*

**ZOISITE CRYSTALS**
*This specimen of ordinary zoisite shows a typical prismatic shape and vertical striations.*

## TANZANITE

The dark blue variety of zoisite, tanzanite, is sometimes mistaken for sapphire. It is frequently heat treated to remove any brown patches and enhance its colour. Tanzanite crystals have distinct pleochroism, and show grey, purple, or blue depending on the angle from which they are viewed. The faceter must orient the stone carefully for the best colour. Cut tanzanites may appear more violet in incandescent light. Tanzanite comes from Tanzania and Pakistan.

**CRYSTAL**
*Tanzanite sometimes forms superb, violet-blue crystals.*

**STEP CUT**
*Step cuts are popular for dark-coloured stones such as tanzanite.*

**MIXED CUT**
*This fine blue faceted tanzanite is about 1 carat.*

**CABOCHON**
*This pinkish-red cabochon coloured by manganese is known as thulite.*

**POLISHED SLAB**
*This polished thulite slab is dotted with pink. It is used as a decorative stone.*

# CLINOZOISITE

**ONTARIO**
*This part of Canada produces good specimens of clinozoisite.*

A CALCIUM ALUMINIUM silicate hydroxide, clinozoisite is a member of the epidote group and is the monoclinic polymorph of zoisite. It forms prismatic, elongated crystals, and can also be granular, massive, and fibrous. It is colourless, yellowish-green, yellowish-grey, or rarely, rose to red. It is common in regionally metamorphosed rocks, in felsic igneous rocks, and as an alteration product of plagioclase. Localities include Canada, Mexico, Austria, Switzerland, Italy, and the USA.

### PROPERTIES

| | |
|---|---|
| **GROUP** | Silicates – sorosilicates |
| **CRYSTAL SYSTEM** | Monoclinic |
| **COMPOSITION** | $Ca_2Al_3(SiO_4)_3(OH)$ |
| **COLOUR** | Grey, yellow, rose |
| **FORM/HABIT** | Prismatic |
| **HARDNESS** | $6\frac{1}{2}$ |
| **CLEAVAGE** | Perfect |
| **FRACTURE** | Uneven, brittle |
| **LUSTRE** | Vitreous |
| **STREAK** | White |
| **SPECIFIC GRAVITY** | 3.2–3.4 |
| **TRANSPARENCY** | Transparent to translucent |

**ROSE CLINOZOISITE**
*Clinozoisite with thulite, from Prägraten, Austria.*

Vitreous lustre

# PIEMONTITE

**PIEDMONT**
*Piemontite was discovered in this upland part of northern Italy.*

PIEMONTITE (also spelled piedmontite) was named for the locality of its discovery in Piedmont, Italy. It is a calcium manganese aluminosilicate hydrate, and forms elongated prismatic crystals. Manganese is the cause of its reddish-brown coloration. It is found in low-grade, regionally metamorphosed rocks, oxidized volcanic rocks, and as a hydrothermal alteration product in manganese deposits. Specimens come from Scotland, Pakistan, Italy, Japan, Egypt, New Zealand, and New Mexico and California, USA.

### PROPERTIES

| | |
|---|---|
| **GROUP** | Silicates – sorosilicates |
| **CRYSTAL SYSTEM** | Monoclinic |
| **COMPOSITION** | $Ca_2(Al,MnFe)_3(SiO_4)_3OH)$ |
| **COLOUR** | Reddish-brown, black, red |
| **FORM/HABIT** | Prismatic |
| **HARDNESS** | 6 |
| **CLEAVAGE** | Perfect |
| **FRACTURE** | N/A |
| **LUSTRE** | Vitreous |
| **STREAK** | Cherry-red |
| **SPECIFIC GRAVITY** | 3.5 |
| **TRANSPARENCY** | Translucent to opaque |

**PRISMATIC CRYSTALS**
*This specimen is from the type locality, St. Marcel, Valle d'Aosta, Italy.*

Vitreous lustre

# ALLANITE

**URALS**
*Russia's Ural Mountains are a good source of allanite.*

A MEMBER OF THE EPIDOTE group, allanite-(Ce) contains cerium and other rare-earth elements and is much more common than yttrium-rich allanite-(Y). Crystals are generally tabular to long prismatic. It can also be granular or occur as embedded grains. Allanite is a common accessory mineral in many igneous and metamorphosed igneous rocks and is weakly radioactive. It is commonly associated with other rare-earth-bearing minerals. Noteworthy occurrences are in Greenland, France, Norway, Madagascar, Russia, Finland, Canada, and the USA. Crystals up to 45cm (18in) in length are found near Los Angeles, California.

### PROPERTIES

| | |
|---|---|
| **GROUP** | Silicates – sorosilicates |
| **CRYSTAL SYSTEM** | Monoclinic |
| **COMPOSITION** | (Ce,Ca,Y)₂ $(Al,Fe,Fe)_3(SiO_4)_4(OH)$ |
| **COLOUR** | Light brown to black, yellow, green |
| **FORM/HABIT** | Tabular to long prismatic |
| **HARDNESS** | $5\frac{1}{2}$–6 |
| **CLEAVAGE** | Imperfect |
| **FRACTURE** | Conchoidal to uneven |
| **LUSTRE** | Resinous, greasy, or submetallic |
| **STREAK** | Light brown |
| **SPECIFIC GRAVITY** | 3.4–4.2 |
| **TRANSPARENCY** | Transparent to translucent |

**PRISMATIC CRYSTALS**
*These long prismatic, parallel crystals of allanite-(Y) are from Ytterby, Stockholm, Sweden.*

Parallel crystals

Resinous lustre

# KORNERUPINE

**SRI LANKA**
*Sri Lanka's gem gravels contain kornerupine as well as many other minerals.*

NAMED FOR DANISH GEOLOGIST A.D. Kornerup, kornerupine is a magnesium iron aluminium borosilicate. It forms striated prismatic crystals of 5cm (2in) and more, which are dark to sea-green, but can also be white, cream, colourless, blue, pink, greenish-yellow, or black. It is sometimes mistaken for tourmaline. Kornerupine occurs in silica-poor, aluminium-rich metamorphic rocks, often in association with cordierite, sillimanite, and corundum. It is a relatively uncommon mineral, with less than 60 known localities worldwide. Crystals up to 23cm (9in) long come from the Fiskenaesset region of Greenland, some of which are gem quality. Other gem-quality material comes from Itrongay and Betroka, Madagascar, and from the gem gravels of Sri Lanka. Gem material is not plentiful, and it is cut mainly for collectors.

### PROPERTIES

| | |
|---|---|
| **GROUP** | Silicates – neosilicates |
| **CRYSTAL SYSTEM** | Orthorhombic |
| **COMPOSITION** | $(Mg,Fe,Al)_{10}(Si,Al,B)_5(O,OH,F)_{22}$ |
| **COLOUR** | Green, white, blue |
| **FORM/HABIT** | Long prismatic |
| **HARDNESS** | $6\frac{1}{2}$–7 |
| **CLEAVAGE** | Distinct prismatic |
| **FRACTURE** | N/A |
| **LUSTRE** | Vitreous |
| **STREAK** | White |
| **SPECIFIC GRAVITY** | 3.3–3.5 |
| **TRANSPARENCY** | Transparent to translucent and opaque |
| **R. I.** | 1.66–1.68 |

**RECTANGULAR STEP**
*In some localities kornerupine occurs in gem quality, and is cut for collectors.*

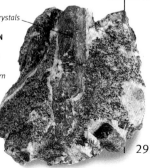

Prismatic crystals

**CRYSTALS IN MATRIX**
*This specimen is from Harts Range, Northern Territory, Australia.*

# NESOSILICATES

NESOSILICATES HAVE ISOLATED groups of silica tetrahedra. *Nesos* is Greek for "island", a reference to the fact that the silica tetrahedra are not attached to each other. Another name for this group is orthosilicates. The sides of the crystals are typically equal in length and they have relatively poor cleavage. Molecules within the structure tend to be densely packed, resulting in a relatively high specific gravity and hardness. Nesosilicates have an $SiO_4$ chemical unit in their formulae. Olivine and the garnets all belong to this group of silicates.

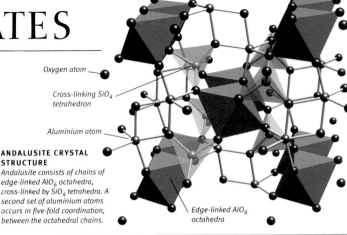

Oxygen atom

Cross-linking $SiO_4$ tetrahedron

Aluminium atom

### ANDALUSITE CRYSTAL STRUCTURE
Andalusite consists of chains of edge-linked $AlO_6$ octahedra, cross-linked by $SiO_4$ tetrahedra. A second set of aluminium atoms occurs in five-fold coordination, between the octahedral chains.

Edge-linked $AlO_6$ octahedra

# OLIVINE

**ZEBIRGET, RED SEA**
*Olivine was introduced to Europe from this island by the Crusaders in the Middle Ages.*

WHILE THE NAME OLIVINE may be unfamiliar, most people will have heard of its gemstone variety: peridot, which has been mined for over 3,500 years on the former John's Island in the Red Sea, now called Zebirget. "Olivine" is applied to any mineral belonging to the forsterite-fayalite solid-solution series, in which iron and magnesium substitute freely in the structure. Fayalite
($Fe_2SiO_4$) is the iron end member, and forsterite ($Mg_2SiO_4$) is the magnesium end member. The most common intermediate olivines have compositions near the forsterite end of the series, and most peridot is about 90 per cent forsterite. The olivine group also includes the rare mineral liebenbergite (Ni, Mg)$_2SiO_4$) and the less rare tephroite ($Mn^{2+}_2SiO_4$).

## OLIVINE PROPERTIES
Crystals are tabular, often with wedge-shaped terminations – although well-formed crystals are rare. Olivine can also be massive or granular. Forsterite-fayalite olivines are usually yellowish-green but can be yellow, brown, or grey. In addition to the usual green colour of common olivine – a result of the small amounts of iron that are almost invariably present – its colour can vary from white for pure forsterite to black for fayalite, but both are very unusual.

## OLIVINE IN THE CRUST
Olivine is inferred to be a major component of the Earth's upper mantle, and as such is probably one of the most abundant mineral constituents of the planet. Olivine has also been found in some lunar rocks and in stony and stony-iron meteorites. It is generally the first mineral to crystallize from rocks with relatively low silica content, thus its composition reflects to some extent that of its parent magma. Thick accumulations can occur as a result of olivine

### VOLCANIC BOMB
*Volcanoes emitting mafic lavas sometimes explosively hurl peridot-rich "bombs".*

### PERIDOT CRYSTAL
*This beautiful peridot crystal originates from Sapat, near Naran, Pakistan. Pakistan is the principal source for gem-quality olivine. Peridot is generally found where there has been little opportunity for alteration, such as in dry climates or in recently formed rocks such as ultramafic lavas.*

Secondary clay minerals

Natural fracture

# MONTICELLITE

MONTICELLITE WAS NAMED in 1831 for the Italian mineralogist Teodoro Monticelli. Crystals are rare, but when found they are short prismatic and often rounded. It is more often found as grains. Monticellite forms in contact metamorphic zones in limestone, and is sometimes found in mafic igneous rocks. It is associated with forsterite, magnetite, apatite, biotite, vesuvianite, and wollastonite. Localities include Bancroft, Ontario, Canada; Nassau, Germany; Siberia, Russia; Kushiro, Hokkaido Prefecture, Japan; Hobart, Tasmania, Australia; and several sources in the USA.

**MOUNT VESUVIUS**
*Monticellite is found on this Italian volcano.*

Monticellite  Calcite

**BROWN MONTICELLITE**
*Brown monticellite occurs with green apatite and calcite in this specimen from Arkansas, USA.*

Apatite

## PROPERTIES

**GROUP** Silicates – nesosilicates
**CRYSTAL SYSTEM** Orthorhombic
**COMPOSITION** $CaMgSiO_4$
**FORM/HABIT** Granular
**HARDNESS** 5½
**CLEAVAGE** Poor
**FRACTURE** Conchoidal
**COLOUR** Colourless, white, brown
**STREAK** White
**LUSTRE** Vitreous to oily
**SPECIFIC GRAVITY** 3.0
**TRANSPARENCY** Transparent to translucent

---

**NECKLACE AND BROOCH**
*Faceted orange citrine and pale yellow peridot gems are set in this necklace and brooch.*

**TIFFANY BROOCH**
*A central oval-cut green peridot is surrounded by a band of topaz and an outer border of green enamel in this elegant brooch.*

crystals settling through a body of magma that is still partly magnetic. Olivine occurs most commonly in mafic and ultramafic igneous rocks such as basalt and peridotite. In general, the forsterite (magnesium-rich) end of the olivine series forms in low- and very low-silica environments, including metamorphosed limestones and dolomite, while the fayalite (iron-rich) olivines form in more silica-rich conditions. Common olivine is very widespread; peridot comes from China, Myanmar, Brazil, Australia, South Africa, Norway, and from Hawaii and Arizona, USA.

In the presence of water at low temperatures, olivine readily undergoes hydrothermal alteration, forming serpentine, talc, chlorite, or magnetite. Common olivine has a high melting point, and is used in the manufacture of heat-resistant bricks.

*Transparent olivine crystal*

## THE LORE OF PERIDOT

Peridot beads were made by the Egyptians as early as 1580–1350BC. Peridot was considered a symbol of the Sun from ancient times to the Middle Ages, and an early Greek manuscript informs us that it confers royal dignity on its bearer. To be protected from evil spirits, the owner of a peridot should have it pierced, strung on the hair of an ass, and tied around the left arm – according to the 11th-century French bishop Marbodius, at least. Strangely for such an ancient gem, there is relatively little other esoteric lore associated with it. In all likelihood, this is because it was grouped with other green stones as Pliny's *smaragdus* stone (see p.292).

**TOPKAPI THRONE**
*This gold throne in Istanbul's Topkapi Palace is decorated with 955 peridot cabochons.*

**SOLID-SOLUTION SERIES**
*The olivine group includes a solid-solution series with iron-rich fayalite and magnesium-rich forsterite as its end members.*

*Crystal face*

*Light colour from magnesium*

**FORSTERITE**

*Dark colour from iron*

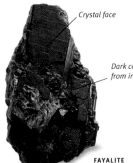

**FAYALITE**

## PROPERTIES

**GROUP** Silicates – nesosilicates
**CRYSTAL SYSTEM** Orthorhombic
**COMPOSITION** $(Mg,Fe)_2SiO_4$
**COLOUR** Green, yellow, brown, white, or black
**FORM/HABIT** Tabular, massive, granular
**HARDNESS** 6½–7
**CLEAVAGE** Imperfect
**FRACTURE** Conchoidal
**LUSTRE** Vitreous
**STREAK** White
**SPECIFIC GRAVITY** 3.3–4.3
**TRANSPARENCY** Transparent to translucent
**R.I.** 1.64–1.69

**POLISHED PEBBLE**
*Small pieces of peridot are tumble polished for use in Baroque jewellery.*

**STEP**
*Dark green peridot is faceted in step cuts to reveal and display its colour.*

**MIXED**
*Peridot that is a lighter green is cut with a multitude of facets to deepen its colour.*

# THE GARNET GROUP

GARNETS ARE WIDESPREAD MINERALS, particularly abundant in metamorphic rocks. Although they form in many different colours, garnets are easy to recognize because they are generally found as well-developed crystals in a cubic crystal form. There are 15 garnet species, all conforming to the general formula $A_3B_2(SiO_4)_3$ where A can be calcium, ferrous iron, magnesium, or manganese, and B can be aluminium, ferric iron, chromium, manganese, silicon, titanium, zirconium, or vanadium. A complete solid–solution series exists between almandine and both pyrope and spessartine, and between grossular and both andradite and uvarovite. The name an individual garnet specimen is given is the name of the end member that makes up the largest percentage of its composition.

**MULTI-GEM SAUTOIR**
*This neck chain is made of red spinel and pyrope garnet beads. A diamond and rock-crystal rondelle suspends a tassel of similar design.*

**DODECAHEDRON**

**TRAPEZOHEDRON**

**DODECAHEDRON AND TRAPEZOHEDRON**

**TRAPEZOHEDRON AND DODECAHEDRON**

**FORMS OF GARNET CRYSTALS**
*Garnet crystals are all cubic, usually occurring as dodecahedrons, trapezohedrons, or combinations of the two.*

## PYROPE

NAMED IN 1803 from the Greek for "fire" and "to appear", pyrope garnet has also been called Bohemian garnet, or cape ruby. Its colour can be dark red, violet-red, rose-red, or reddish-orange, depending on its composition. Iron, chromium, titanium, and manganese all substitute in the structure, and all are colouring agents to some degree. High-chromium pyrope can exhibit an alexandrite (colour-change) effect (see p.159). Its crystals are dodecahedral and trapezohedral, although it is most often found in rounded grains or pebbles. Pyrope is a high-pressure mineral found in metamorphic rocks and in very high-pressure igneous rocks such as peridotites and kimberlites. Although it is relatively widespread, large cut gems (of more than 10 carats) are uncommon.

**BOHEMIA**
*This area of the Czech Republic is famous for its pyrope crystals.*

**BOHEMIAN GARNET EARRING**
*Beautiful garnet jewellery comes from Bohemia, where pyropes as large as hens' eggs have been found.*

*Rock groundmass*

**PROPERTIES**

| | |
|---|---|
| **GROUP** Silicates – nesosilicates | |
| **CRYSTAL SYSTEM** Cubic | |
| **COMPOSITION** $Mg_3Al_2(SiO_4)_3$ | |
| **COLOUR** Red | |
| **FORM/HABIT** Dodecahedral, trapezohedral | |
| **HARDNESS** 7–7$\frac{1}{2}$ | |
| **CLEAVAGE** None | |
| **FRACTURE** Conchoidal, brittle | |
| **LUSTRE** Vitreous | |
| **STREAK** White | |
| **SPECIFIC GRAVITY** 3.6 | |
| **TRANSPARENCY** Transparent to translucent | |
| **R.I.** 1.73–1.76 | |

**PYROPE IN MATRIX**
*This specimen from Mexico includes several pyrope garnets in a matrix. Most pyrope is found as pebbles in placer deposits with other gems.*

**PYROPE CRYSTALS**
*Gem-quality pyrope is usually recovered as stream-rounded pebbles.*

*Pyrope crystal*

**TAPLOW BUCKLE**
*Anglo-Saxon craftsmen combined gold and garnets into objects of stunning beauty, such as this 6th-century buckle set with garnet.*

*Garnet inlay*  *Garnet cabochon*

**OVAL STEP**
Step cuts are favoured for pyrope gems because of their deep colour

**STEP**
Emerald cuts are the favoured step cut for pyrope because of the stone's brittleness.

**MIXED**
Mixed cuts can bring sparkle to the deep colour of pyrope gemstones.

**BRILLIANT**
This brilliant cut shows some of the darkening that can occur on pyrope.

# SPESSARTINE

**ELDORADO BAR, USA**
*This part of Montana has gem-quality spessartine.*

NAMED FOR SPESSART, BAVARIA, Germany, the type locality, spessartine garnet is manganese aluminium silicate. Its crystals are dodecahedral or trapezohedral, and its colours can be pale yellow when nearly pure, to orange or deep red. Pure spessartine is relatively rare – it is almost always mixed with some amount of almandine, giving the orange to red coloration. Pure spessartine is ordinarily found in manganese-rich metamorphic rocks, and in granites and pegmatite veins. Gem-quality spessartine is rare, and stones are cut more for collectors than for use in jewellery. The Rutherford Mine in Virginia, USA, produces some of the largest spessartines, one of the heaviest weighing 6,720 carats (1.34kg or nearly 3lb). A 709-carat faceted spessartine is also reported from Brazil. An intermediate spessartine-pyrope composition is sometimes called Malaya garnet.

## PROPERTIES

| | |
|---|---|
| **GROUP** | Silicates – nesosilicates |
| **CRYSTAL SYSTEM** | Cubic |
| **COMPOSITION** | $Mn_3Al_2(SiO_4)_3$ |
| **COLOUR** | Yellow, orange, red |
| **FORM/HABIT** | Dodecahedral, trapezohedral |
| **HARDNESS** | $7-7^1/_2$ |
| **CLEAVAGE** | None |
| **FRACTURE** | Conchoidal, brittle |
| **LUSTRE** | Vitreous |
| **STREAK** | White |
| **SPECIFIC GRAVITY** | 4.2 |
| **TRANSPARENCY** | Transparent to translucent |
| **R.I.** | 1.79–1.81 |

**NECKLACE**
*This unusual necklace makes dramatic use of rare, pale yellow spessartine gems.*

**EARRINGS**
*Orange spessartine garnet cabochons are the focus of these delicate drop earrings.*

**STEP**
*This octagonal step-cut spessartine has liquid inclusions under the edge facets.*

**BRILLIANT**
*Facet-grade spessartine comes from Madagascar, Sri Lanka, Brazil, Australia, and the USA.*

**CABOCHON**
*Spessartines not transparent enough to facet still make attractive cabochon gems.*

Dodecahedral crystal

Rock groundmass

**SPESSARTINE**
*In this specimen from Norway, well-formed dodecahedral crystals encrust a rock groundmass.*

# ALMANDINE

**BROKEN HILL**
*Huge almandine crystals are found in this part of Australia.*

THE MOST COMMON garnet, almandine is an iron aluminium silicate. The name comes from Alabanda (now Araphisar), Turkey, where it has been cut since antiquity. Its crystals are dodecahedral and trapezohedral, and its colour tends to be a pinkier red than other garnets. Rutile needles are sometimes present, and when cut *en cabochon*, it often shows a four-rayed star. Almandine is found worldwide, in many places in well-formed crystals weighing 4kg ($8^1/_2$lb) or more. When it occurs in metamorphic rocks, its presence indicates the grade of metamorphism. It also occurs widely in igneous rocks, and is occasionally found as inclusions in diamond. It is mined for abrasives, and cut extensively for gems. It is somewhat brittle, and cut stones tend to chip on the edges.

**FOLIATE BROOCH**
*This antique gold brooch (left) uses three faceted almandine gems and a seed pearl to create a central flower motif.*

**ALMANDINE-TOPPED DOUBLET**
*Dark red almandine gemstones have been set on glass to create this classic doublet (right).*

**CABOCHON**
*This almandine cabochon has a hollow back to lighten its deep red colour and let more light through. It contains black inclusions.*

**BRILLIANT**
*The red of almandines is often too dark for faceted stones. This lighter specimen has been cut in a round brilliant to enhance the fiery red colour.*

## PROPERTIES

| | |
|---|---|
| **GROUP** | Silicates – nesosilicates |
| **CRYSTAL SYSTEM** | Cubic |
| **COMPOSITION** | $Fe_3Al_2(SiO_4)_3$ |
| **COLOUR** | Red, pink |
| **FORM/HABIT** | Dodecahedral, trapezohedral |
| **HARDNESS** | $7-7^1/_2$ |
| **CLEAVAGE** | None |
| **FRACTURE** | Subconchoidal, brittle |
| **LUSTRE** | Vitreous |
| **STREAK** | White |
| **SPECIFIC GRAVITY** | 4.3 |
| **TRANSPARENCY** | Transparent to translucent |
| **R.I.** | 1.76–1.83 |

Modified dodecahedron

**ALMANDINE CRYSTAL**
*This almandine in schist from near Wrangell, Alaska, USA, shows a modified dodecahedral form.*

Schist

# ANDRADITE

ANDRADITE, A TYPE OF GARNET, is a calcium iron silicate, which has several variety names, each related to its colour. The yellowish variety is called topazolite, because of its resemblance to topaz; yellowish-green or emerald-green forms are called demantoid or, commercially, Uralian emerald; black andradite is called melanite. Other colours are brownish-red, brownish-yellow, and greyish-green. The green of demantoid is caused by the presence of chromium; yellow to black is due to titanium. Andradites make spectacular gems, with greater colour dispersion (separation of light into colours) than diamonds. Andradite is commonly found with grossular in contact-metamorphosed limestone, and in mafic igneous rocks. Gem demantoid comes from the Ural and Chutosky Mountains of Russia, and Valmalenco, Italy; topazolite from Saxony, Germany, and California, USA.

**BRILLIANT MELANITE**
This melanite gem, faceted in a brilliant cut, is a typical black, opaque stone. Melanite rarely shows any transparency.

**MIXED**
The high colour dispersion of andradite results in brilliant faceted stones, as seen in this mixed cut.

**MELANITE**
This andradite specimen, from Arendal, Aust-Agder, Norway, is the variety called melanite.

**DEMANTOID GARNET RING**
Cut with many triangular facets, andradite has eye-catching sparkle and brilliance.

*Vitreous lustre*

*Black colouring*

## PROPERTIES

| | |
|---|---|
| **GROUP** | Silicates – nesosilicates |
| **CRYSTAL SYSTEM** | Cubic |
| **COMPOSITION** | $Ca_3Fe_2(SiO_4)_3$ |
| **COLOUR** | See text |
| **FORM/HABIT** | Dodecahedral, trapezohedral |
| **HARDNESS** | $6\frac{1}{2}$–7 |
| **CLEAVAGE** | None |
| **FRACTURE** | Conchoidal, brittle |
| **LUSTRE** | Vitreous |
| **STREAK** | White |
| **SPECIFIC GRAVITY** | 3.8 |
| **TRANSPARENCY** | Transparent to translucent |
| **R.I.** | 1.85–1.89 |

---

# GROSSULAR

**TANZANIA**
Gem-quality green grossular comes from Tanzania.

IN REFERENCE TO its gooseberry-green colour, the calcium aluminum silicate grossular is named for the Latin *grossularia*, "gooseberry". Its predominantly dodecahedral crystals can also be white, colourless, pink, cream, orange, red, honey, brown, or black. Crystals up to 13cm (5in) are known. When reddish-brown, it is called hessonite (see opposite). Grossular is ordinarily found in both regional and thermal calcium-rich metamorphic rocks, and occasionally in meteorites. Massive greenish grossular is sometimes marketed under the name South African, or Transvaal, jade. Most grossular is opaque to translucent, but transparent, pale to emerald-green faceting material (marketed as tsavorite in the USA and tsavolite in Europe) comes from Kenya and Tanzania.

**FROG BROOCH**
This sparkling, delicate brooch combines diamonds with a line of green grossular gems along the frog's back.

*Grossular crystal*

**GROSSULAR ON DIOPSIDE**
These grossular crystals, from Piedmont, Italy, are on a groundmass of diopside.

## PROPERTIES

| | |
|---|---|
| **GROUP** | Silicates – nesosilicates |
| **CRYSTAL SYSTEM** | Cubic |
| **COMPOSITION** | $Ca_3Al_2(SiO_4)_3$ |
| **COLOUR** | See text |
| **FORM/HABIT** | Dodecahedral |
| **HARDNESS** | $6\frac{1}{2}$–7 |
| **CLEAVAGE** | None |
| **FRACTURE** | Conchoidal |
| **LUSTRE** | Vitreous |
| **STREAK** | White |
| **SPECIFIC GRAVITY** | 3.6 |
| **TRANSPARENCY** | Transparent to translucent |
| **R.I.** | 1.69–1.73 |

**PINK POLISHED SLAB**
This slab cut through a large grossular crystal shows colour-zoning within the crystal.

**HORSE STATUE**
This stylized mythical horse with an elaborate saddle and head is carved out of mottled grossular.

**COLOUR RANGE**
Grossular comes in a wide range of colours, as these gems demonstrate.

*Diopside*

# HESSONITE

THE REDDISH-BROWN variety of grossular (see opposite), hessonite is a calcium aluminium silicate. Mostly found in dodecahedral crystals, its colour is due to manganese and iron inclusions. Its popular name is "cinnamon stone". Hessonite is found principally in calcareous metamorphosed rocks. The most important source of gem-quality hessonite is in the gem gravels or metamorphic rocks of Sri Lanka, but other localities also produce superb material: Lowell, Vermont, USA; the Jeffrey Mine, Quebec, Canada; Coahuila and Zacatecas, Mexico; the Isle of Mull, Scotland; and the Ala Valley, Piedmont, Italy. The Romans and ancient Greeks carved cameos and intaglios, and cut cabochons from hessonite.

**HESSONITE CLUSTER**
*The dodecahedral crystals of cinnamon stone, or hessonite, cluster in this superb specimen.*

Striated crystal

Twinned crystals

**ART-DECO BROOCH**
*This brooch is designed as a microphone with a cushion-shaped hessonite garnet surrounded with diamonds, on a diamond "stand".*

Vitreous lustre

## PROPERTIES

| | |
|---|---|
| **GROUP** | Silicates – nesosilicates |
| **CRYSTAL SYSTEM** | Cubic |
| **COMPOSITION** | $Ca_3Al_2(SiO_4)_3$ |
| **COLOUR** | Reddish-brown |
| **FORM/HABIT** | Dodecahedral |
| **HARDNESS** | 7 |
| **CLEAVAGE** | None |
| **FRACTURE** | Conchoidal |
| **LUSTRE** | Vitreous |
| **STREAK** | White |
| **SPECIFIC GRAVITY** | 3.7 |
| **TRANSPARENCY** | Transparent to translucent |
| **R.I.** | 1.73–1.75 |

**LALIQUE PLAQUE**
*In this 1905 design, circular-cut hessonite, peridot, and garnet gems accent the main motif of a maiden's head in profile with cascading, plum-coloured, enamelled hair.*

**ROUND MIX**
*The colour of lighter hessonite is enhanced by a multitude of small facets.*

**OVAL MIX**
*Deeper-coloured hessonites show their colour most effectively through broader facets.*

# UVAROVITE

THE RAREST OF ALL the gem garnets is uvarovite, the calcium chromium silicate. It was named in 1832 for the Russian nobleman Count Uvarov. Its predominantly dodecahedral crystals are almost always too small to be cut. Uvarovite's brilliant green colour comes from chromium – the same colouring agent that is in ruby and emerald, which is known to have an inhibiting effect on crystal growth. It is probable that uvarovite's small crystals are a result of this. Uvarovite is found in chromium-bearing igneous and metamorphic rocks. Some of the largest crystals come from Outokumpu, Finland. It is also found in Russia's Ural Mountains, the Bushveld Complex, South Africa, California, USA, and Silesia, Poland.

**UVAROVITE CRYSTAL**
*This dodecahedral uvarovite crystal is in a skarn matrix.*

Dodecahedral crystal

Skarn matrix

Crust of uvarovite crystals

Rock groundmass

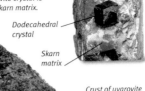

## PROPERTIES

| | |
|---|---|
| **GROUP** | Silicates – nesosilicates |
| **CRYSTAL SYSTEM** | Cubic |
| **COMPOSITION** | $Ca_3Cr_2(SiO_4)_3$ |
| **COLOUR** | Bright green |
| **FORM/HABIT** | Dodecahedral, trapezohedral |
| **HARDNESS** | $6\frac{1}{2}$–7 |
| **CLEAVAGE** | None |
| **FRACTURE** | Uneven to conchoidal, brittle |
| **LUSTRE** | Vitreous |
| **STREAK** | White |
| **SPECIFIC GRAVITY** | 3.8 |
| **TRANSPARENCY** | Transparent to translucent |
| **R.I.** | 1.86–1.87 |

**TINY CRYSTALS**
*A crust of tiny uvarovite dodecahedrons covers a rock groundmass. This specimen is from Zalostoc, Mexico.*

# ZIRCON

ONE OF THE FEW STONES to approach diamond in fire and brilliancy, zircon is such a superb gem due to its high refractive index and colour dispersion. Its colourless stones can closely resemble diamonds, and they have been intentionally and mistakenly substituted for them. Zircon exhibits double refraction, and bottom facets seen through the top of a cut stone will appear double. Diamond does not share this property, and so this is a useful test for distinguishing between the two gemstones; zircon is also significantly softer than diamond, and cut stones will show wear on the edges of the facets.

Zircon has been known since antiquity, and takes its name from the Arabic *zargun*, derived in turn from the Persian *zar*, meaning "gold", and *gun*, "colour". It has been mined for over 2,000 years from the gem gravels of Sri Lanka, and was used as a gemstone in Greece and Italy as far back as the 6th century AD. It forms prismatic to dipyramidal crystals, which can be colourless, yellow, grey,

**MYANMAR**
*Much of the world's supply of gem zircon comes from the gem gravels of Myanmar, formerly Burma.*

**RECTANGULAR STEP**
*This colourless zircon cut in a rectangular step cut was produced by heat treating reddish-brown material.*

**OVAL BRILLIANT**
*Green zircons, such as this oval brilliant, are often stones in which the trace elements have undergone partial decay.*

**CUSHION BRILLIANT**
*Golden-yellow zircons such as this cushion brilliant are produced when blue zircon is reheated in the presence of oxygen.*

**ROUND BRILLIANT**
*Blue zircon is produced by heat-treating brown gem zircon. Virtually all commercial blue zircon is heat-treated.*

**ZIRCON BRACELET**
*Colourless faceted zircon gems alternate with orange gems in this modern gold-plated bracelet.*

**ZIRCON RINGS**
*In the ring above, the yellow zircon in the centre is flanked by pale blue zircons. The ring to the left sparkles with a deep blue zircon.*

## PROPERTIES

| | |
|---|---|
| **GROUP** | Silicates – nesosilicates |
| **CRYSTAL SYSTEM** | Tetragonal |
| **COMPOSITION** | $ZrSiO_4$ |
| **COLOUR** | Colourless, brown, red, yellow, orange, blue, green |
| **FORM/HABIT** | Prismatic to dipyramidal |
| **HARDNESS** | 7½ |
| **CLEAVAGE** | Imperfect |
| **FRACTURE** | Uneven to conchoidal |
| **LUSTRE** | Adamantine to oily |
| **STREAK** | White |
| **SPECIFIC GRAVITY** | 4.6–4.7 |
| **TRANSPARENCY** | Transparent to opaque |
| **R.I.** | 1.93–1.98 |

**AFGHAN ZIRCON**
*This specimen of zircon crystals, up to 3cm (1¼in) long, in a feldspar and biotite matrix is from Afghanistan.*

*Twinned zircon crystal*

*Feldspar and biotite groundmass*

## USING ZIRCON TO DATE ROCKS

Zircon crystals are extremely hard, allowing them to survive in many types of rock. They are also resistant to chemical and physical change. These characteristics make them ideal for dating very old rocks radiometrically. Radioactive isotopes, such as uranium, decay at a known rate over time, forming what are known as daughter isotopes. This means that by measuring the number of daughter isotopes in a zircon crystal within a rock, and comparing that number with the original number of atoms, it is possible to determine how much decay has taken place and estimate the age of the crystal and therefore the rock.

**DATING CRYSTALS**
*Here, a scientist uses an ion probe generator to date zircon crystals.*

*Small zircons*

*Biotite*

green, brown, blue, and red. Single crystals can reach a considerable size: examples weighing up to 2kg (5.5lb) and 4kg (10lb) have been found in Australia and Russia, respectively. Its coloured varieties have been given other names in the past, although these are now obsolete: the transparent red variety has been called hyacinth (jacinth); clear and colourless zircon from Sri Lanka has been called Matura diamond; and the name "jargon" or "jargoon", derived from the Arabic *zargun*, has been applied to all other gem colours.

Green zircon represents the green foliage of the kalpa tree, a gemstone tree that is a symbolic Hindu offering to the gods. Elsewhere in the East, wearing zircon was believed to endow the wearer with wisdom, honour, and riches. It was also an amulet for travellers.

## INDUSTRIAL USES OF ZIRCON

Zircon is a silicate mineral, zirconium silicate, and is the principal source of the rare metal zirconium, a highly corrosion-resistant metal used to coat the interiors of nuclear reactors. Zircon crystals almost always contain traces of the radioactive elements hafnium, uranium, and thorium. These eventually destroy the crystal structure, which can often be restored by heating (see panel, above right). Zircon is also useful because it has a high melting point and is therefore used to

make foundry sand, heat-resistant materials, and ceramics. Because of its hardness, zircon is also used to make industrial abrasives.

Zircon is widespread as a minor constituent in silica-rich igneous rocks; it also occurs in metamorphic rocks. It is resistant to weathering, and because of its relatively high specific gravity it concentrates in stream and river gravels, and in beach deposits. Gem varieties are mostly recovered from stream gravels in Sri Lanka, Thailand, Myanmar, Australia, and New Zealand. Beach sands are a major source of commercial production, particularly in India, Australia, Brazil, and Florida, USA. Other key localities are Canada, Mexico, Norway, France, and Pakistan.

# HEAT-TREATING ZIRCONS

Heat-treating minerals to change their colour or improve their clarity has been a widespread practice for centuries. It may well have begun with zircon. Brown stones from Thailand and Sri Lanka turn blue or colourless when heated; in blue stones that have lost their colour reheating will re-establish the blue colour. The process itself does not require expensive apparatus. The zircons to be treated are placed in a clay pot intermixed with fine sand; the pot is then heated in the coals of a campfire for a period determined by long experience. After slow cooling, the newly coloured stones are separated from the sand, and cut. Blue zircon reheated in the presence of oxygen changes to golden-yellow, the source of the name *zargun*. Because the word is of ancient origin, so, too, must be the practice of heat-treating.

**HEAT-TREATED CUT GEM**

**UNTREATED CRYSTALS**

**HEAT-TREATED CRYSTALS**

**FLORAL BROOCH**
*This floral brooch, in an Indian-style silver setting, contains colourless, blue, yellow and brown faceted zirons.*

Syenite groundmass

Zircon crystal

**ZIRCON IN SYENITE**

Pyramidal terminations

Prism face

**SINGLE CRYSTAL**

Feldspar

Prismatic crystals

**PRISMATIC CRYSTALS**

### ZIRCON CRYSTALS
*Zircon crystals are most commonly found in igneous rocks, especially syenite, and in association with feldspar, garnet or monazite. They conform to the tetragonal system of symmetry and are usually prismatic.*

Biotite

### PANNING
*This woman is panning for zircon gems in river gravels in the ex-Khmer Rouge heartland of Pailin in western Cambodia.*

# TOPAZ

**URAL MOUNTAINS**
*The finest display specimens are from Russia's Urals, where sky-blue crystals may reach 30cm (12in) or more in length.*

THE LEGENDARY ISLAND of Topazios, just off the coast of Egypt in the Red Sea and now called Zebirget, is one source of the name topaz. It is also thought to derive from *tapaz*, the Sanskrit word for "fire". Many authorities believe that the stone called topaz today was unknown to the ancients and that the name *topazos* referred to olivine (peridot, see p.298–99) as Topazios was a well-known locality for that gem. It follows that the "topaz" in the Old Testament may also have been peridot. The gem was used in ancient Egypt, Greece, and Rome, the Roman source being faraway Sri Lanka.

**NECKLACE**
*This 1950s' gold-plated necklace combines rectangular-cut, prong-set sherry topaz gems with circular cuts of amber.*

## PROPERTIES

**GROUP** Silicates – nesosilicates
**CRYSTAL SYSTEM** Orthorhombic
**COMPOSITION** $Al_2SiO_4(F,OH)_2$
**COLOUR** Colourless, blue, yellow, pink, brown, green
**FORM/HABIT** Prismatic
**HARDNESS** 8
**CLEAVAGE** Perfect basal
**FRACTURE** Subconchoidal to uneven
**LUSTRE** Vitreous
**STREAK** Colourless
**SPECIFIC GRAVITY** 3.4–3.6
**TRANSPARENCY** Transparent to translucent
**R.I.** 1.62–1.63

**ZEBIRGET**
*Zebirget (also spelled Zabargad) in the Red Sea is on the site of the ancient Topazios, said to be the earliest source of topaz.*

### WIDE VARIETY OF COLOURS

Topaz is aluminium silicate fluoride hydroxide. Its well-formed crystals have a characteristic lozenge-shaped cross section, and striations parallel to their length. They are found in a wide range of colours, with the sherry-yellow stones from Brazil being particularly valuable. Pink topaz is even more valuable, but natural pink stones are rare. In 1750 a Parisian jeweller discovered that yellow Brazilian topaz becomes pink on exposure to a moderate heat, and most pink topaz today is heat-treated yellow material. Some blue topaz is almost indistinguishable from aquamarine with the naked eye. However, much of the deep blue topaz on the market is derived from irradiated and heat-treated colourless topaz. The natural colour in many cases is unstable: for example, brown topaz from Siberia and Utah, USA, is prone to bleaching by sunlight.

Colourless topaz is sufficiently refractive that when brilliant-cut, it has been mistaken for diamond. The Braganza diamond (1,640 carats) in the Portuguese crown during the 17th century was thought then to be the largest diamond ever found but is now thought to have been a colourless topaz.

**COLOURS OF TOPAZ**
*In addition to the well-known sherry colour, topaz is also found in yellow, orange, pink, brown, shades of blue, and occasionally green.*

**YELLOW TOPAZ**

**BROWN TOPAZ CRYSTAL**

**NATURAL BLUE CRYSTAL**

**BLUE TOPAZ**

*Albite*

**OCTAGONAL STEP**
*The colour of this octagonal step-cut gem is a natural topaz blue.*

**OVAL STEP**
*The blue colour of this oval step cut has probably been enhanced.*

**OVAL STEP**
*Topaz occasionally comes in pink, as in this superb oval step cut.*

**TEARDROP MIXED**
*The golden-yellow of this teardrop mixed cut is that most often associated with topaz.*

**BRILLIANT**
*The brilliance of this nearly colourless topaz has been enhanced by many extra facets.*

Topaz is formed by fluorine-bearing vapours given off during the last stages of the crystallization of various igneous rocks, typically occurring in cavities in rhyolites, granites, pegmatite dykes, and hydrothermal veins. It is also found in alluvial deposits as water-worn pebbles. Perfect cleavage causes crystals to break easily, with the result that part of a crystal often remains in the matrix.

Aside from Russia, Brazil and Nigeria are major producers of topaz. Beautiful crystals come from Japan, and the first modern use of the term "topaz" occurred in 1737 when applied to the abundant yellow crystals from Saxony, Germany. Gem topaz is also found in Colorado, Texas, and California, USA; the Cairngorm Mountains of Scotland; Sri Lanka; Myanmar; Australia; Tasmania; Pakistan; and Mexico. Common topaz is found in a number of localities worldwide.

Yellow andradite garnet is known as "topazolite" due to its resemblance to topaz. The name "topaz" is also applied to other stones – often to increase the market appeal of much cheaper material. Citrine – yellow quartz – has been called "false topaz", "quartz topaz", and "Madeira topaz"; yellow sapphire was long referred to as "Oriental topaz"; and smoky quartz has been marketed as "smoky topaz" or "smoky topaz quartz".

# BRAZILIAN TOPAZ

Brazil is a famous locality for topaz, as well as other gems. The Brazilian state of Minas Gerais is the world's largest producer of topaz, and the world's largest preserved crystal, weighing 271kg (596lb), is from there. The well-known sherry-yellow crystals come from Ouro Preto, Minas Gerais, where they occur in a kaolinitic matrix. The Ouro Preto deposit was discovered in 1735, and literally tonnes of topaz have been recovered from it. A 350-tonne topaz crystal was reported in 1944 from Mugui, Espirito Santo, also in Brazil.

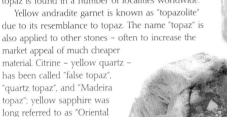

**SHERRY TOPAZ**
*The Ouro Preto deposit is the world's largest commercial source of sherry topaz.*

**BRAZILIAN PRINCESS**
*The faceted Brazilian Princess is 14.5cm (nearly 6in) across and weighs 21,005 carats.*

**ORIGINAL ROUGH GEM**
*The Brazilian Princess was faceted from this 34kg (75lb) crystal.*

**BROOCH**
*Two sherry-coloured topaz gems have been used to form the thistles in this brooch based on the famous Scottish emblem.*

*Prismatic crystal*

*Small albite crystals*

*Single crystal of microcline*

*Stream-rounded surface*

*Termination face*

**TOPAZ CRYSTAL**
*This superb prismatic, pinkish-brown-coloured topaz from Afghanistan is over 8cm (3in) tall, and weighs nearly 0.5kg (1lb). It rests, along with albite, on a microcline crystal.*

**TOPAZ PEBBLES**
*Like other dense gemstones, stream-rounded topaz accumulates in placer deposits. Placer topaz is mined in Nigeria and Brazil.*

# TITANITE

FORMERLY CALLED SPHENE, titanite is one of the few stones with a colour dispersion higher than that of diamond, making faceted stones cut from its transparent crystals fiery and brilliant. Unfortunately, these are almost exclusively cut as gemstones for collectors, as the mineral is generally too brittle and soft to be used in jewellery.

Titanite is a calcium titanium silicate mineral that may also contain significant amounts of iron, and minor amounts of thorium or uranium. The name sphene originates from the Greek *sphen*, meaning "wedge", a reference to the typical wedge shape of its crystals. It is also found in prismatic crystals, and can be massive, lamellar, or compact. Gem-quality crystals occur in yellow, green, or brown, and it can also be black, pink, red, blue, or colourless. This wide colour range is a result of the number of other elements that can substitute in its structure in addition to iron, thorium, and uranium. Tin and aluminium are but two that can make up as much as 10 per cent of the mineral's composition. Titanite is strongly pleochroic and is also double refractive; this is seen as doubling of the back facets, adding to the brilliance of faceted stones. Titanite is widely distributed as a minor component of silica-rich igneous rocks and associated pegmatites, gneisses, and schists. Notable occurrences of the mineral are in Austria, Italy, Norway, Switzerland, Madagascar, Canada, Mexico, Brazil, Sweden, Germany, Russia, Pakistan, and New Jersey and New York, USA.

**AUSTRIA**
*Excellent titanite specimens come from the Tyrol area of Austria.*

**RECTANGULAR STEP**
*This is a faceted, yellow-green, rectangular step-cut titanite. Although it has high fire, titanite is cut mainly for collectors.*

**OVAL STEP**
*This brownish-yellow titanite has been faceted in an oval step cut. Although beautiful, titanite is too soft to wear.*

**CUSHION MIXED**
*The top of this orange-yellow stone is cut with triangular facets, and the bottom in rectangular steps.*

**TRIANGULAR**
*This triangular yellow titanite shows the brilliance for which the stone is noted.*

**TITANITE RING**
*Faceted titanites, such as the brilliant cut set in this gold ring, have superb fire and intense colours.*

**CROSS**
*This cross features heart-shaped titanite gems, set within diamonds.*

## PROPERTIES

| | |
|---|---|
| **GROUP** | Silicates – nesosilicates |
| **CRYSTAL SYSTEM** | Monoclinic |
| **COMPOSITION** | $CaTiSiO_5$ |
| **COLOUR** | Yellow, green, brown, black, pink, red, or blue |
| **FORM/HABIT** | Wedge-shaped or prismatic |
| **HARDNESS** | 5–5½ |
| **CLEAVAGE** | Imperfect |
| **FRACTURE** | Conchoidal |
| **LUSTRE** | Vitreous to greasy |
| **STREAK** | White |
| **SPECIFIC GRAVITY** | 3.5–3.6 |
| **TRANSPARENCY** | Transparent to translucent |
| **R.I.** | 1.84–2.03 |

**TITANITE CRYSTALS**
*These interpenetrating, wedge-shaped titanite crystals, 3.5cm (1½in) long, are from Russia.*

Wedge-shaped crystal

Vitreous lustre

Imperfect cleavage

Wedge-shaped crystal

Twinned crystals

**CRYSTALS IN MATRIX**

Wedge-shaped crystal

**TRANSPARENT CRYSTALS**

**CRYSTAL GROUPS**
*On the far left is a mass of wedge-shaped titanite crystals in a matrix. The specimen on the left is a group of large transparent crystals, which are also wedge-shaped.*

Rock groundmass

# ANDALUSITE

Prismatic crystal

**ANDALUSIA**
*Andalusite is named for the Spanish locality of its discovery.*

ANDALUSITE CRYSTALS are commonly prismatic with a square cross section. The mineral can also be massive, or have elongated or tapered crystals. It is pink to reddish-brown, white, grey, violet, yellow, green, or blue. Gem-quality andalusite is strongly pleochroic with yellow, green, and red within the same stone depending on the colour of the stone and the direction of viewing. A yellowish-grey variety called chiastolite occurs as long prisms enclosing symmetrical wedges of carbonaceous material. These, in cross section, make a cross. Chiastolite takes its name from the Greek *chiastos*, meaning "cross". Andalusite is found locally in low-grade metamorphic rocks, and in regional metamorphic rocks,

where it is associated with corundum, kyanite, cordierite, and sillimanite. It is also found rarely in granites and granitic pegmatites. Crystal localities include Belgium, South Australia, Russia, Germany, and the USA. Gem-quality andalusite is found in Minas Gerais, Brazil, and the gem gravels of Sri Lanka.

**BROWN CRYSTALS**
*This fine group of prismatic andalusite crystals on a rock groundmass has a vitreous lustre.*

Prismatic crystal

Quartz groundmass

**ANDALUSITE CRYSTALS**
*This group of prismatic andalusite crystals from the Austrian Tyrol is in a groundmass of quartz.*

**CHIASTOLITE CROSS SECTION**
*Twinned crystals of andalusite, called chiastolite, are sometimes worn as charms.*

**RECTANGULAR STEP**
*Transparent andalusite is relatively uncommon. It is faceted into gems for collectors.*

## PROPERTIES

| | |
|---|---|
| **GROUP** | Silicates – nesosilicates |
| **CRYSTAL SYSTEM** | Orthorhombic |
| **COMPOSITION** | $Al_2OSiO_5$ |
| **COLOUR** | Pink, brown, white, grey, violet, yellow, green, or blue |
| **FORM/HABIT** | Prismatic |
| **HARDNESS** | $6\frac{1}{2}$–$7\frac{1}{2}$ |
| **CLEAVAGE** | Good to perfect, poor |
| **FRACTURE** | Conchoidal |
| **LUSTRE** | Vitreous |
| **STREAK** | White |
| **SPECIFIC GRAVITY** | 3.2 |
| **TRANSPARENCY** | Transparent to nearly opaque |
| **R.I.** | 1.63–1.64 |

# SILLIMANITE

**MULL**
*Crystals of sillimanite come from the island of Mull, off the west coast of Scotland.*

OCCURRING IN LONG, SLENDER, glassy crystals, or in blocky, poorly terminated prisms, sillimanite is an aluminium silicate. Commonly colourless to white, it can also be pale yellow to brown, pale blue, green, or violet. It is distinctly pleochroic with yellowish-green, dark green, and blue within the same stone when seen from different angles. Sillimanite is characteristic of high-temperature, regionally and thermally metamorphosed, clay-rich rocks, and is one of three polymorphs of $Al_2SiO_5$ (see kyanite, p.310). A common metamorphic mineral, it is often found with corundum, kyanite, and cordierite. Crystals come from Mull, Scotland; Vltava, the Czech Republic; Assam, India; La Romaine, Quebec, Canada; and, in the USA, Macon County, North Carolina, and Chester, Connecticut. Common sillimanite is used in heat-resistant ceramics and car spark plugs.

## PROPERTIES

| | |
|---|---|
| **GROUP** | Silicates – nesosilicates |
| **CRYSTAL SYSTEM** | Orthorhombic |
| **COMPOSITION** | $Al_2OSiO_5$ |
| **COLOUR** | Colourless, white, pale yellow, blue, green, or violet |
| **FORM/HABIT** | Prismatic to acicular |
| **HARDNESS** | 7 |
| **CLEAVAGE** | Perfect |
| **FRACTURE** | Uneven |
| **LUSTRE** | Silky |
| **STREAK** | White |
| **SPECIFIC GRAVITY** | 3.2–3.3 |
| **TRANSPARENCY** | Transparent to translucent |
| **R.I.** | 1.66–1.68 |

**SILLIMAN**
*Sillimanite was named for Professor Benjamin Silliman, a geologist and chemist who founded the American Journal of Science.*

Fibrous structure

**FIBROLITE**
*The columnar, fibrous form of sillimanite is called fibrolite. It is sometimes cut en cabochon, as here.*

**CUSHION MIXED**
*Facet-grade sillimanite occurs in the gem gravels of Sri Lanka and Myanmar, and in Brazil.*

Perfect cleavage

**FIBROUS SILLIMANITE**
*This mass of parallel, fibrous crystals of sillimanite is from West Chester, Pennsylvania, USA.*

Silky lustre

**BAHIA**
*Gem-quality kyanite crystals are found in this part of Brazil.*

# KYANITE

NAMED FROM THE GREEK *kyanos*, "dark blue", for its colour, kyanite occurs principally as elongated, flattened blades, and less commonly as radiating, columnar aggregates. It is usually blue and blue-grey, but can also be green or colourless. Kyanite's former name, disthene (meaning "two strengths"), refers to its characteristic variable hardness. It is about $4\frac{1}{2}$ when scratched parallel to the C (long) axis, but 6 when scratched perpendicular to this. Kyanite is formed during the regional metamorphism of clay-rich sediments. It is one of three polymorphs of $Al_2SiO_5$. The other two are sillimanite and andalusite. Each polymorph indicates metamorphism within particular temperature and pressure ranges. Gem occurrences include Bahia, Brazil; the St. Gotthard region of Switzerland; and Yancy County, North Carolina, USA.

**SPARK PLUGS**
*Kyanite is a major raw material for the aluminium silicate mullite, which is used in spark plugs and heat-resistant porcelains.*

**KYANITE BLADES**
*This specimen of kyanite with quartz, from northern Brazil, shows the characteristic elongated, bladed habit of kyanite crystals.*

**KYANITE CRYSTALS**
*These single crystals of kyanite show prismatic development.*

Vitreous lustre

Long, bladed crystal

Rock groundmass

### PROPERTIES

| | |
|---|---|
| **GROUP** | Silicates – nesosilicates |
| **CRYSTAL SYSTEM** | Triclinic |
| **COMPOSITION** | $Al_2SiO_5$ |
| **COLOUR** | Blue, green |
| **FORM/HABIT** | Bladed |
| **HARDNESS** | $4\frac{1}{2}$; 6 (see text) |
| **CLEAVAGE** | Perfect |
| **FRACTURE** | Splintery |
| **LUSTRE** | Vitreous |
| **STREAK** | Colourless |
| **SPECIFIC GRAVITY** | 3.6 |
| **TRANSPARENCY** | Transparent to translucent |
| **R.I.** | 1.71–1.73 |

**THIN SECTION**
*Kyanite crystals can be seen in this section in different orientations, revealing different cleavage patterns.*

Quartz

Kyanite

Kyanite showing twinning and cleavage

**STERLING HILL**
*Willemite occurs in this mine in New Jersey, USA.*

# WILLEMITE

WILLEMITE, NAMED FOR KING WILLEM I of the Netherlands, is a zinc silicate. Usually massive, it sometimes forms short, prismatic crystals or fibrous aggregates. Its is light to yellow-green, yellow-brown, red-brown, or colourless, and fluoresces bright green under ultraviolet light. Willemite is found in the oxidized zones of zinc deposits, and in metamorphosed limestones. It is often found with hemimorphite, smithsonite, franklinite, and zincite. Localities include Mont St.-Hilaire, Canada; Långban, Sweden; Attica, Greece; Mumbwa, Zambia; Tsumeb, Namibia; the Flinders Ranges, South Australia; and Pima, Arizona, USA. The most important site is the Franklin Mine, New Jersey, USA (see p.158). Willemite is locally an important zinc ore.

### PROPERTIES

| | |
|---|---|
| **GROUP** | Silicates – nesosilicates |
| **CRYSTAL SYSTEM** | Hexagonal/trigonal |
| **COMPOSITION** | $Zn_2SiO_4$ |
| **COLOUR** | Green, red-brown |
| **FORM/HABIT** | Massive |
| **HARDNESS** | $5–5\frac{1}{2}$ |
| **CLEAVAGE** | Good |
| **FRACTURE** | Conchoidal to uneven |
| **LUSTRE** | Vitreous to resinous |
| **STREAK** | Colourless |
| **SPECIFIC GRAVITY** | 4.0 |
| **TRANSPARENCY** | Transparent to translucent |

**TROOSTITE**
*This variety of willemite, known as troostite, is from the Franklin Mine in New Jersey, USA. Specimens from Franklin show particularly strong green fluorescence.*

Prismatic willemite crystal

Vitreous to resinous lustre

**EARLY TV TUBE**
*Because of its green fluorescence, willemite was used as a phosphor in early colour television tubes.*

# STAUROLITE

**TAOS, USA**
*Cruciform twins
of staurolite are
found in this part
of New Mexico.*

NAMED FROM THE GREEK *stauros*, "cross", for its
characteristic cross-like twinned form, staurolite is a
widespread mineral. It occurs with garnet,
tourmaline, and kyanite or sillimanite in
mica schists and gneisses and other
regionally metamorphosed
aluminium-rich rocks. It
forms only under a specific range of
temperatures and pressures and is thus very
useful in determining the conditions under
which the metamorphic rock formed.
Staurolite's cruciform twins, known as "fairy
crosses", are made into ornaments and
amulets. They are found in
Fannin, Georgia, and Taos,
New Mexico, USA;
Brittany, France; and
Rubelita,
Brazil.

*Prismatic staurolite
crystal*

*Twinned
staurolite
crystals*

**STAUROLITE IN
MICA SCHIST**

*Kyanite*

*Schist*

*Staurolite*

**STAUROLITE
CRYSTALS**
*These staurolite crystals from
Taos, New Mexico show several
forms of twinning.*

**KYANITE-
STAUROLITE SCHIST**
*In this specimen staurolite
occurs with kyanite in regionally
metamorphosed schist.*

## PROPERTIES

| | |
|---|---|
| **GROUP** | Silicates – nesosilicates |
| **CRYSTAL SYSTEM** | Monoclinic |
| **COMPOSITION** | $(Fe,Mg)_4Al_{17}(Si,Al)_8O_{45}(OH)_3$ |
| **COLOUR** | Brown |
| **FORM/HABIT** | Pseudo-orthorhombic prismatic |
| **HARDNESS** | $7–7\frac{1}{2}$ |
| **CLEAVAGE** | Distinct |
| **FRACTURE** | Conchoidal |
| **LUSTRE** | Vitreous to resinous |
| **STREAK** | Colourless to grey |
| **SPECIFIC GRAVITY** | 3.7 |
| **TRANSPARENCY** | Transparent to opaque |
| **R.I.** | 1.74–1.75 |

**TRAPEZE**
*Staurolite can be
transparent enough
to facet, as in this
trapeze-cut stone.*

---

# PHENAKITE

**PIKE'S PEAK**
*This area of
Colorado, USA, is
a principal locality
for crystals of
phenakite.*

A RARE BERYLLIUM MINERAL, phenakite was
named in 1833 from the Greek for "deceiver", alluding
to it being mistaken for quartz. Its crystals are
predominantly rhombohedral, and less commonly
short and prismatic. Phenakite occurs in pegmatites,
granites, and mica schists. Large crystals are found
near Yekaterinburg in the Ural Mountains of Russia,
and in Colorado, USA. Transparent
crystals are faceted for collectors. Its
indices of refraction are higher than
topaz and its brilliance approaches that
of diamond. A stream-pebble of phenakite
weighing 1,470 carats was found in Sri
Lanka, and faceted to a 569-carat oval.

## PROPERTIES

| | |
|---|---|
| **GROUP** | Silicates – nesosilicates |
| **CRYSTAL SYSTEM** | Hexagonal/trigonal |
| **COMPOSITION** | $Be_2SiO_4$ |
| **COLOUR** | Colourless, white |
| **FORM/HABIT** | Rhombohedral |
| **HARDNESS** | $7\frac{1}{2}–8$ |
| **CLEAVAGE** | Indistinct |
| **FRACTURE** | Conchoidal |
| **LUSTRE** | Vitreous |
| **STREAK** | Colourless |
| **SPECIFIC GRAVITY** | 3 |
| **TRANSPARENCY** | Transparent or translucent |
| **R.I.** | 1.65–1.67 |

*Vitreous
lustre*

*Twinned crystal*

**RHOMBOHEDRAL
PHENAKITE**
*This phenakite crystal from
the Ural Mountains of
Russia shows multiple
parallel twinning.*

---

# EUCLASE

**URAL
MOUNTAINS**
*Russia's southern
Ural mountains
contain gem-
quality euclase.*

EUCLASE TAKES ITS NAME from the Greek *eu*,
"good", and *klasis*, "fracture", in reference to its perfect
cleavage. It forms striated prisms, often with complex
terminations. Generally white or colourless, it can
also be pale green or pale to deep blue. It occurs in
low-temperature hydrothermal veins, granitic
pegmatites, and in some
metamorphic schists
and phyllites. Transparent euclase is faceted
for collectors. Gem euclase comes from
Minas Gerais and several other localities
in Brazil; and Park County, Colorado, USA.

**PRISMATIC CRYSTAL**
*This striated prismatic
euclase crystal has a
conchoidal fracture at
its base.*

*Prismatic
crystals*

*Striated
crystal*

*Rock
groundmass*

## PROPERTIES

| | |
|---|---|
| **GROUP** | Silicates – nesosilicates |
| **CRYSTAL SYSTEM** | Monoclinic |
| **COMPOSITION** | $BeAlSiO_4(OH)$ |
| **COLOUR** | Colourless, white, blue, or green |
| **FORM/HABIT** | Prismatic |
| **HARDNESS** | $7\frac{1}{2}$ |
| **CLEAVAGE** | Perfect |
| **FRACTURE** | Conchoidal, brittle |
| **LUSTRE** | Vitreous |
| **STREAK** | White |
| **SPECIFIC GRAVITY** | 3.0 |
| **TRANSPARENCY** | Transparent to translucent |
| **R.I.** | 1.65–1.67 |

**BLUE MASS**
*This mass of
well-developed,
blue prismatic
crystals is
on a rocky
groundmass.*

# HUMITE

NAMED IN 1813 for the English mineral and art collector Sir Abraham Hume (1749–1838), humite is a silicate of magnesium and iron. It is generally found in granular masses, with well-formed crystals being rare. Humite is yellow to dark orange or reddish-orange in colour. Manganese substitutes for iron in the structure to form a complete solid-solution with manganhumite.

Humite occurs in contact- and regionally metamorphosed limestones and dolomites, where it is often found with cassiterite, hematite, mica, tourmaline, quartz, and pyrite. Humite occurs worldwide, but a few noteworthy locations are Persberg and elsewhere, Sweden; Isle of Skye, Scotland; Mount Vesuvius, Italy; Valais, Switzerland; and in the USA at Brewster, New York, Franklin, New Jersey, and Sterling Hill, New Jersey (manganhumite).

*Yellowish-brown humite crystals*

*Rock groundmass*

| PROPERTIES | |
|---|---|
| **GROUP** | Silicates – nesosilicates |
| **CRYSTAL SYSTEM** Orthorhombic | |
| **COMPOSITION** $(Mg,Fe)_7(SiO_4)_3(F,OH)_2$ | |
| **COLOUR** | Yellow to dark orange |
| **FORM/HABIT** | Granular |
| **HARDNESS** | 6–6½ |
| **CLEAVAGE** | Poor |
| **FRACTURE** | Subconchoidal to uneven |
| **LUSTRE** | Vitreous |
| **STREAK** | Yellow to orange |
| **SPECIFIC GRAVITY** | 3.2–3.3 |
| **TRANSPARENCY** Transparent to translucent | |

**HUMITE CRUST**
*In this specimen, a crust of yellowish-brown humite crystals with accessory mica covers a rock groundmass.*

# NORBERGITE

DISCOVERED IN and named after Norberg, Sweden, norbergite is hydrous magnesium silicate. It is a member of the humite group of minerals. It rarely forms crystals, and is usually granular in habit. It is light yellowish-brown, white, or rose in colour, fluorescing yellow under shortwave ultraviolet light. Norbergite occurs in contact-metamorphic and regionally metamorphosed rocks where there is contact between magnesium-rich sedimentary rocks and granites. It is often found with forsterite, diopside, phlogopite, and brucite. Localities include Franklin, New Jersey, and Edenville, New York, USA; Bihar, India; Pitkyaranta, Russia; Mount Vesuvius, Italy; and Lyangar, Tajikistan.

*Norbergite encrustation*

**NORBERGITE CRUST**
*Yellow norbergite encrusts this specimen from Canada.*

| PROPERTIES | |
|---|---|
| **GROUP** | Silicates – nesosilicates |
| **CRYSTAL SYSTEM** Orthorhombic | |
| **COMPOSITION** | $Mg_3SiO_4(F,OH)_2$ |
| **COLOUR** Light yellowish-brown, white, or rose | |
| **FORM/HABIT** | Granular |
| **HARDNESS** | 6–6½ |
| **CLEAVAGE** | Distinct |
| **FRACTURE** | Uneven to subconchoidal, brittle |
| **LUSTRE** | Vitreous |
| **STREAK** | Yellowish |
| **SPECIFIC GRAVITY** | 3.1–3.2 |
| **TRANSPARENCY** | Transparent to translucent |

# CHONDRODITE

A MEMBER of the humite group, chondrodite takes its name from the Greek for "granule", a reference to its usual occurrence as isolated grains. Crystals are blocky and 2.5cm (1in) or more long. Chondrodite is yellow, orange, brownish-red, or greenish-brown in colour. It forms in contact-metamorphosed limestones and dolomites, kimberlites, ultramafic rocks, and marbles. It occurs worldwide, but a few noteworthy localities are Skye, Scotland; Cardiff, Canada; Passau, Germany; Kamioka, Japan; Nordmark, Sweden; Slyudanka, Russia; and various locations in the USA. Chondrodite should not be confused with chondrite, which is a type of meteorite.

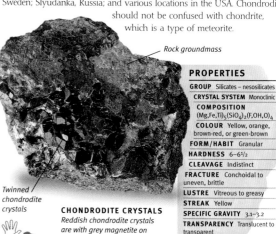

*Rock groundmass*

*Twinned chondrodite crystals*

**CHONDRODITE CRYSTALS**
*Reddish chondrodite crystals are with grey magnetite on a groundmass.*

| PROPERTIES | |
|---|---|
| **GROUP** | Silicates – nesosilicates |
| **CRYSTAL SYSTEM** | Monoclinic |
| **COMPOSITION** $(Mg,Fe,Ti)_5(SiO_4)_2(F,OH,O)_4$ | |
| **COLOUR** Yellow, orange, brown-red, or green-brown | |
| **FORM/HABIT** | Granular |
| **HARDNESS** | 6–6½ |
| **CLEAVAGE** | Indistinct |
| **FRACTURE** Conchoidal to uneven, brittle | |
| **LUSTRE** | Vitreous to greasy |
| **STREAK** | Yellow |
| **SPECIFIC GRAVITY** | 3.1–3.2 |
| **TRANSPARENCY** | Translucent to transparent |

# DATOLITE

**TASMANIA**
*Good datolite crystals are found here.*

A RELATIVELY UNCOMMON mineral, datolite is occasionally cut as a collectors' gemstone. Its name comes from the Greek word for "to divide", a reference to the granular occurrence of some of its varieties. Its crystals are generally platy to short prismatic, but may also be blocky or in spherical aggregates. Other habits are massive, compact, and cryptocrystalline. It can be colourless, white, grey, yellowish, pale pink, greenish-white, or green. Datolite occurs as veins and cavity linings in mafic igneous rocks and in metallic-ore veins. It is also found in gneisses and dolerites (diabases). Some notable deposits exist in Charcas, Mexico; Miyazaki Prefecture, Japan; Dalnegorsk, Russia; St. Andreasberg, Germany; Moravia, the Czech Republic; and in the USA in New Jersey, Virginia, Massachusetts, and Connecticut.

## PROPERTIES

| | |
|---|---|
| **GROUP** | Silicates – nesosilicates |
| **CRYSTAL SYSTEM** | Monoclinic |
| **COMPOSITION** | $CaBSiO_4(OH)$ |
| **COLOUR** | Colourless, white, yellowish, pale pink, or green |
| **FORM/HABIT** | Platy to short prismatic |
| **HARDNESS** | $5–5\frac{1}{2}$ |
| **CLEAVAGE** | Imperfect |
| **FRACTURE** | Uneven to subconchoidal, brittle |
| **LUSTRE** | Vitreous to greasy |
| **STREAK** | White |
| **SPECIFIC GRAVITY** | 2.9–3.0 |
| **TRANSPARENCY** | Transparent to translucent |

*Pyramid face*

*Prism face*

**DATOLITE CRYSTAL**
*This semi-transparent, colourless single crystal of datolite shows prism and pyramid faces.*

# DUMORTIERITE

**BAHIA, BRAZIL**
*Gem-quality dumortierite occurs here and in Mexico, South Africa, and Japan.*

DUMORTIERITE IS generally found as fibrous aggregates of radiating crystals, or it can be massive. Individual crystals are acicular prismatic. Its usual colours are pinkish-red or violet to blue, but it can be brown or greenish. Dumortierite occurs in pegmatites, in aluminium-rich metamorphic rocks, and in rocks that are metamorphosed by boron-bearing vapour derived from hot, intruding bodies of granite. Localities include Kutná Hora, the Czech Republic; Tvedestrand, Norway; the Kounrad Mine, Kazakhstan; and Fremont County, Colorado, Petaca, New Mexico, and Dillon, Montana, USA.

## PROPERTIES

| | |
|---|---|
| **GROUP** | Silicates – nesosilicates |
| **CRYSTAL SYSTEM** | Orthorhombic |
| **COMPOSITION** | $(Al,Fe)_7(BO_3)(SiO_4)_3O_3$ |
| **COLOUR** | Pinkish-red, bluish-violet, brown, or greenish |
| **FORM/HABIT** | Fibrous aggregates |
| **HARDNESS** | 7–8 |
| **CLEAVAGE** | Distinct, imperfect |
| **FRACTURE:** | Uneven, brittle |
| **LUSTRE** | Vitreous |
| **STREAK** | White or bluish-white |
| **SPECIFIC GRAVITY** | 3.2–3.4 |
| **TRANSPARENCY** | Transparent to translucent |
| **R.I.** | 1.66–1.72 |

**BLUE DUMORTIERITE**
*Bluer shades result from titanium substitution for iron.*

**CABOCHON**
*Dumortierite is frequently cut en cabochon, as here, or polished to make decorative stones.*

*Massive form*

# CHLORITOID

**SHETLAND ISLES**
*Chloritoid is found in this Scottish island group.*

NAMED FOR ITS RESEMBLANCE to chlorite, chloritoid is usually massive or foliated, often with curved plates. Its crystals are pseudohexagonal or tabular. Chloritoid occurs in low- to medium-grade, metamorphosed, fine-grained sediments, and is a useful mineral in the interpretation of the metamorphic environment. It is also found in lavas, tuffs, and rhyolites. It is a widespread mineral: localities include Ottre, Belgium; Hualien, Taiwan; and Pennsylvania, and North Carolina, USA.

## PROPERTIES

| | |
|---|---|
| **GROUP** | Silicates – nesosilicates |
| **CRYSTAL SYSTEM** | Monoclinic or triclinic |
| **COMPOSITION** | $(Fe,Mg,Mn)_2Al_4Si_2O_{10}(OH)_4$ |
| **COLOUR** | Dark to greyish-green |
| **FORM/HABIT** | Foliated, massive |
| **HARDNESS** | $6\frac{1}{2}$ |
| **CLEAVAGE** | Perfect |
| **FRACTURE** | Uneven |
| **LUSTRE** | Vitreous to subadamantine |
| **STREAK** | White |
| **SPECIFIC GRAVITY** | 3.5–3.8 |
| **TRANSPARENCY** | Translucent |

**CHLORITOID CRYSTALS**
*This specimen is from Ile de Groix, Morbihan, France.*

*Chloritoid*

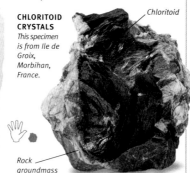

*Rock groundmass*

# GADOLINITE

**RUSSIA**
*Gadolinite deposits are found the Kola Peninsula, Russia.*

GADOLINITE EXISTS IN TWO FORMS: yttrian gadolinite, the most common; and cerian gadolinite. Lanthanum, scandium, dysprosium, and neodymium can be present, and gadolinite is an important source of all of these. It is generally found in compact masses, but when crystals occur, they are prismatic. It is green, blue-green, or rarely pale green when fresh, brown or black when altered. Gadolinite is relatively widespread in granite and granitic pegmatites. Large crystals are found at Ytterby, Sweden and Iveland, Norway. Other localities include: Japan, Germany, and the USA.

## PROPERTIES

| | |
|---|---|
| **GROUP** | Silicates – nesosilicates |
| **CRYSTAL SYSTEM** | Monoclinic |
| **COMPOSITION** | $Y_2FeBe_2Si_2O_{10}$ |
| **COLOUR** | Green, blue-green |
| **FORM/HABIT** | Compact masses |
| **HARDNESS** | $6\frac{1}{2}–7$ |
| **CLEAVAGE** | None |
| **FRACTURE** | Conchoidal, brittle |
| **LUSTRE** | Vitreous to greasy |
| **STREAK** | Greenish-grey |
| **SPECIFIC GRAVITY** | 4.4–4.8 |
| **TRANSPARENCY** | Opaque to nearly transparent |

**GADOLINITE CRYSTAL**
*This single, prismatic crystal is from Texas, USA.*

*Prismatic crystal*

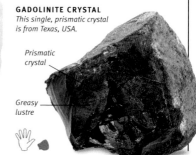

*Greasy lustre*

313

# ORGANIC GEMS

GENERATED by organic (biological) processes, organic gems may or may not be crystalline, and in some cases contain the same mineral matter, such as calcite or aragonite, as that generated through inorganic processes. Organic gems are judged by the same criteria as gemstones of mineral origin: beauty and durability. They were popular in ancient times because they are softer than minerals and easily worked by primitive methods.

The first discovered use of bone and ivory in carvings was during the Upper Palaeolithic Period, from 40,000 to 10,000 years ago. These were probably not for decoration, but give a good indication of the first awareness of the working properties of these organic materials. The first items for adornment appear during the Mesolithic Period (8000 to 2700BC) and were produced from shell, bone, and ivory. Organic gem materials were popular trade items: Baltic amber artefacts and salt-water shells are found over a thousand kilometres from their places of origin; and freshwater pearls were widely valued and traded.

**PENDANT**
*Seashells have long been valued objects for decoration.*

**SHELL BEACH**
*This profusion of conch shells covers a pebble beach on the island of Grand Bahama in the Caribbean.*

**BALTIC COAST**
*The Baltic coast has been a source of amber for at least three millennia. It still produces 90 per cent of the world's amber.*

**CAUGHT IN RESIN**
*As resin from trees dried 40–60 million years ago, insects and spiders sometimes became trapped in the sticky substance and were fossilized with it.*

# AMBER

AMBER IS FOSSILIZED RESIN, principally from extinct coniferous trees, although amber-like substances from earlier trees are known. It is generally found in association with lignite coal, itself the fossilized remains of trees and other plant material. Amber and partially fossilized resins are sometimes given mineral-like names depending on where they are found, their degree of fossilization or the presence of other chemical components. For example, resin from the London clay and resembling copal resin is called copalite. At least 12 other names are applied to minor variants. The word "electricity" is derived from the Greek name for amber, *electrum*. This is because amber can acquire an electric change when rubbed, a property that was described in about 600BC by Thales, and which is one of the most useful for identifying amber. For several thousand years, the largest source of amber has been the extensive deposits along the Baltic coast, extending intermittently

## PROPERTIES

**GROUP** Organics – hydrocarbons
**CRYSTAL SYSTEM** None
**COMPOSITION** Hydrocarbon $(C,H,O)$
**FORM/HABIT** Amorphous
**HARDNESS** $2–2^{1}/_{2}$
**CLEAVAGE** None
**FRACTURE** Conchoidal
**COLOUR** Yellow, sometimes brownish or reddish
**STREAK** White
**LUSTRE** Resinous
**SPECIFIC GRAVITY** 1.1
**TRANSPARENCY** Transparent to translucent
**R.I.** 1.54–1.55

**FLY IN FRESH RESIN**
*Resin oozing from trees, such as in this modern example, often traps insects, other small arthropods, and plant remains, which over time may became fossilized.*

*Polished amber*

**GEORGIAN NECKLACE**
*Amber has long been valued for jewellery, partly due to its light weight and warmth to the touch.*

from Gdánsk right around to the coastlines of Denmark and Sweden. It is both mined and recovered from Baltic shores after heavy storms. It has been widely traded since ancient times, and a cup carved from amber was discovered in a British Bronze Age burial. Perhaps the largest single use of amber was the creation of the "Amber Room" in Catherine the Great's palace in Russia, a huge room totally lined and decorated with cut amber. The amber from this room was pillaged by the Nazis in World War II, and was probably destroyed in a fire; however, the room has now been completely restored using newly cut amber (see pp.316–17).

## THE LORE OF AMBER

Amber is said to represent the dividing line between individual and cosmic energy, the individual's soul and the universal soul. It has been used to symbolize divinity, and a face the colour of amber is often seen on representations of saints and heroes. The Greek god Apollo wept tears of amber when he was banished from Olympus. The electrical properties of amber led to claims that amber rosaries and amulets act as condensers of current, discharging excessive energy in those who wear the amulet or handle the beads. It is also said that if a man keeps a piece of amber on him, he will never be betrayed by sexual impotence!

**CHINESE EAR ORNAMENT**
*This Chinese ear ornament is carved in the shape of a panda bear. Dehydration of the amber over the years has caused the surface to crack.*

**POLISHED AMBER**
*A characteristic golden orange colour, this amber bead is transparent with a resinous lustre.*

**"SUN-SPANGLED" BEAD**
*This polished bead contains cracks that may have been caused by trapped and flattened drops of water or by heat treatment.*

*Translucent crystal*

*Resinous lustre*

**BROOCH**
*This embossed silver brooch of naturalistic form is set with a yellow Baltic amber cabochon and three suspended amber drops.*

*Conchoidal fracture*

**AMBER NODULE**
*Transparent to translucent, most amber is golden yellow to golden orange, and usually occurs as nodules or small, irregularly shaped masses. When first found, it can have a dull and pitted surface.*

**NECKLACE AND EARRINGS**
*Small pieces of amber can be tumble-polished to create attractive baroque stones for necklaces and earrings.*

## TEARS OF THE HELIADES

According to Greek mythology, the god Phaethon lost control of his father Zeus's Sun chariot and drove it too close to the Earth, setting it ablaze. In fury, Zeus struck Phaethon dead with a lightning bolt. His body fell into the River Eridanus and was buried on its shores by the nymphs of the stream. Later, Phaethon's three sisters, the Heliades, found his grave and wept day and night. Their wasting bodies took root, becoming transformed into trees, and their tears hardened into drops of amber.

**THE GRIEVING HELIADES**
*Entitled* The Sisters of Phaethon, *this dramatic oil on slate was painted by Santi di Tito in 1572.*

315

# THE AMBER ROOM

DESCRIBED AS THE EIGHTH WONDER OF THE WORLD
AFTER ITS INSTALLATION IN THE CATHERINE PALACE IN
1765, THE AMBER ROOM SURVIVED THE 1917 RUSSIAN
REVOLUTION BUT ITS INTRICATELY CARVED PANELS
MYSTERIOUSLY DISAPPEARED DURING WORLD WAR II.

**KING FREDERICK I**
*The original Amber Room was the inspiration of King Frederick I of Prussia, who died two years after it was installed at his Berlin palace in 1711.*

The Amber Room began its extraordinary history in Prussia. A Danish craftsman, Gottfried Wolffrani, was commissioned by King Frederick I of Prussia to carve an amber chamber for the royal palace at Charlottenburg. Wolffrani was replaced by the Danzig craftsmen Ernst Schacht and Gottfried Turow, who finished the amber panels in 1711. They were installed, not at Charlottenburg Palace, but in the smoking room of the royal palace in Berlin.

In 1716 Tsar Peter I went to Berlin to forge an alliance with Frederick's son and successor, Frederick William I, against the Swedish king, Charles XIII. To mark their successful agreement, the Prussian king gave the tsar the Amber Room. Packed into 18 crates, the panels were carried by horse and wagon to St. Petersburg in April, 1717. There they remained until, in 1755, Empress Elisabeth I had them moved to the beautiful Catherine Palace at Tsarskoye Selo.

The room intended for the amber panels was larger than their original chamber in Berlin, so the carving of ten more panels was commissioned in the same style. By 1763 the enlarged Amber Room was complete.

The ornate panels alternated with 24 Venetian-mirrored rectangular columns which rested on amber bases. A mirror mounted in amber, a gift to Empress Elisabeth from King Frederick II, was also incorporated in the design. To unite the baroque and rococo features, the floor was inlaid with mother-of-pearl; gilded rocaille adorned the white doors; and four Florentine mosaics embellished the rich amber walls.

The Amber Room survived the 1917 Russian Revolution untouched, but in 1941 it was stripped by the invading German army and the amber panels shipped to Königsberg Castle in eastern Prussia. In the spring of 1945 the crates containing the Amber Room were almost certainly consumed by a fire that destroyed much of the castle, although many believed that the missing treasure was moved from the castle and hidden elsewhere by the Nazis. However, after many failed searches, finally, in 1979, the Russian government commissioned the rebuilding of the Amber Room in the Catherine Palace. It was completed in 2003.

**PRUSSIAN EAGLE**
*Seen on a panel of the Amber Room, this eagle is an emblem of the Prussian king, a reminder that the room was originally commissioned to grace a Prussian palace.*

**FAITHFUL REPRODUCTION**
*The recreation of the historic Amber Room began in 1982 and required the services of a team of specialized amber carvers. In the mid-1990s, the work was threatened by a lack of funds, but the project was saved by a large donation from a German sponsor.*

# JET

GENERALLY CLASSIFIED AS A LIGNITE COAL, jet has a high carbon content and a layered structure. It is black to dark brown, and sometimes contains tiny inclusions of pyrite, which have a metallic lustre. It tends to occur in rocks of marine origin, perhaps derived from waterlogged driftwood or other plant material. In this respect it differs from ordinary lignite in that lignite usually forms through coalification (see opposite) of peaty deposits on land. Jet can occur in distinct beds, such as those at Whitby, England. Other jet localities are in Spain, France, Germany, Poland, India, Turkey, Russia, and the USA.

Jet has been carved for ornamental purposes since prehistoric times – examples have been found in prehistoric caves. The Romans carved jet, which is light to wear, into bangles

**NEW MEXICO**
*The area of Raton, New Mexico, USA, is a source of jet, much of it incorporated into Native American jewellery.*

## PROPERTIES

| | |
|---|---|
| **GROUP** | Organics – hydrocarbons |
| **CRYSTAL SYSTEM** | None |
| **COMPOSITION** | Various |
| **FORM/HABIT** | Amorphous |
| **HARDNESS** | 2½ |
| **CLEAVAGE** | None |
| **FRACTURE** | Conchoidal |
| **COLOUR** | Black, brown |
| **STREAK** | Black to dark brown |
| **LUSTRE** | Velvety to waxy |
| **SPECIFIC GRAVITY** | About 1.3 |
| **TRANSPARENCY** | Opaque |
| **R.I.** | 1.64–1.68 |

**BAUBLE**
*This bauble faceted from jet shows the high polish and detailed shaping that has made jet a popular material for at least two millennia.*

and beads; some of this jet came from Whitby as early as the 1st century AD. In medieval times powdered jet drunk in water or wine was believed to have medicinal properties. In India and the Mediterranean, carved jet amulets were believed to protect against the evil eye, and Irish women traditionally burned jet to ensure the safety of their husbands when away from home. Native Americans continue to use jet in jewellery.

**NATIVE AMERICAN EAGLE**
*Jet has been shaped and carved in the form of an eagle and then inset with turquoise in this Native American pendant.*

**TURKISH BEADS**
*This necklace of drilled oblong jet beads was crafted in eastern Turkey. Highly polished, the beads have an attractive lustre.*

**JET BROOCH**
*Dating from about 1860, this jet brooch shows the intricacy that can be achieved when carving jet.*

*Conchoidal fracture*

*Amorphous habit*

*Black to brown colour*

**FINE-GRAINED JET**
*This specimen shows the velvety lustre and conchoidal fracture characteristic of high-quality jet. Pieces such as this are suitable for carving.*

*Velvety lustre*

*Fossil ammonite*

**FOSSILS IN JET**
*The ammonite and bivalve fossils in this jet specimen testify to its marine origin.*

## MOURNING JEWELLERY

Because of its sombre colour, jet has been used in mourning for hundreds of years, and has a long history of religious association. During the Middle Ages, jet carvings were sold to pilgrims at holy sites in Spain, and it has traditionally been fashioned into rosaries for monks. After the death of her husband Prince Albert in 1861, Queen Victoria went into a 40-year period of mourning. She set the trend for wearing dark, sombre clothes and jewellery, and for spending long periods in mourning. Jet from Whitby in the north of England was used extensively, although other black stones such as hematite were also incorporated into some pieces.

**DOVE PENDANT CARVED FROM JET**

**WHITBY BEACH**
*The beach at Whitby, Yorkshire, England, is famous for its jet.*

**POLISHED**
*Gems made from jet can be polished to give an opaque, velvety lustre to the surface.*

**ROSE CUT**
*The uppermost convex surface of this gem has been faceted, enlivening its dull surface.*

**QUEEN VICTORIA IN MOURNING**
*Queen Victoria wore jet jewellery during her long period of mourning.*

# COPAL

COPAL IS THE TERM used for resins obtained from various tropical trees. It can be collected from living trees or from accumulations in the soil beneath the trees, or mined if it is buried. Copals from different sources can have similar physical properties but different chemical properties. They are of the same approximate hardness as amber, but differ from amber in that they are still wholly or partially soluble in organic solvents. Buried copal is the nearest to amber in durability, and is in many cases virtually indistinguishable from it. Fossil copal from the London blue clay has been called copalite or Highgate resin. The name copal itself probably comes from *nahuatl copalli*, meaning "resin". The island of Zanzibar (a part of Tanzania) is a major source of buried copal. Copal also comes from China, Brazil, and other South American countries. It is used in making varnishes, lacquers, inks, and linoleum.

## PROPERTIES

| | |
|---|---|
| **GROUP** | Organics – hydrocarbons |
| **CRYSTAL SYSTEM** | None |
| **COMPOSITION** | Various |
| **FORM/HABIT** | Amorphous |
| **HARDNESS** | 2–2½ |
| **CLEAVAGE** | None |
| **FRACTURE** | Conchoidal |
| **COLOUR** | Colourless, yellow |
| **STREAK** | White |
| **LUSTRE** | Resinous |
| **SPECIFIC GRAVITY** | About 1.1 |
| **TRANSPARENCY** | Transparent to translucent |

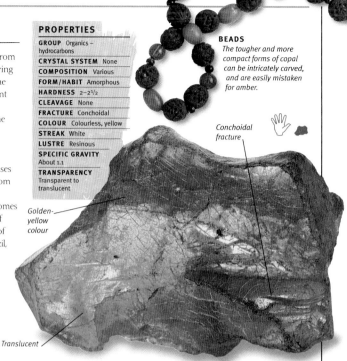

**BEADS**
*The tougher and more compact forms of copal can be intricately carved, and are easily mistaken for amber.*

*Conchoidal fracture*

*Golden-yellow colour*

*Translucent*

**KAURI GUM**
*Kauri gum is derived from the Kauri conifer of the family Araucariaceae, and is dug in New Zealand from sites that were previously forested.*

**COPAL NUGGET**
*This specimen of copal closely resembles amber; some copal is used in jewellery as an amber substitute.*

# ANTHRACITE

**COAL SEAMS**
*Horizontal seams of coal are clearly visible in this cliff.*

THE MOST HIGHLY metamorphosed variety of coal, anthracite, also called hard coal, contains the highest percentage of fixed carbon (about 90–98 per cent) and the lowest percentage of volatile matter of all coals. Anthracites are black, hard, and brittle, have a brilliant, almost metallic lustre, and form sharp, conchoidal fractures. Coal is made up of an irregular mixture of different chemical compounds called macerals, which are analogous to minerals in inorganic rocks. Unlike minerals, they have no fixed chemical composition and no definite crystalline structure. When burned, anthracite gives the greatest heat value and the lowest emissions, and easily sustains combustion once ignited. Unfortunately it is also the least common and most expensive form of coal.

## PROPERTIES

| | |
|---|---|
| **ORIGIN** | Metamorphic rock |
| **GROUP** | Organics – hydrocarbons |
| **CRYSTAL SYSTEM** | None |
| **COMPOSITION** | Various |
| **FORM/HABIT** | Amorphous |
| **HARDNESS** | 2–2½ |
| **CLEAVAGE** | None |
| **FRACTURE** | Conchoidal |
| **COLOUR** | Black |
| **STREAK** | Black |
| **LUSTRE** | Nearly metallic |
| **SPECIFIC GRAVITY** | About 1.1 |
| **TRANSPARENCY** | Opaque |

## COALIFICATION

When vegetable matter decays in the absence of oxygen, its carbon content increases, forming peat. Over time, the peat is transformed to lignite coal as a result of pressure exerted by sedimentary materials that accumulate over the peat deposits. Increasing pressure and temperature transform lignite to bituminous coal. Finally, at the highest degree of coal metamorphism, this becomes anthracite coal.

**PEAT**

**LIGNITE COAL**

**BITUMINOUS COAL**

**PEAT BEDS**
*Extensive peat beds such as these were the origin of today's coal deposits.*

*Conchoidal fracture*

*Near-metallic lustre*

**SHINY COAL**
*Hard and clean to the touch, anthracite is naturally shiny. It takes a brilliant polish, and is used for decorative as well as practical purposes. It is often sold as "jet".*

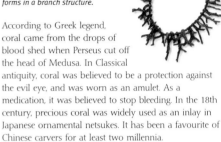

**CORAL REEF**
*Coral reefs comprise colonies of marine animals called coral polyps, which form branching structures as they grow.*

# CORAL

CORAL IS THE SKELETAL MATERIAL generated by sea-dwelling coral polyps. For most corals, this material is calcium carbonate, but in the case of black and golden corals, it is a horn-like substance called conchiolin. Coral has a dull lustre when recovered, but can take a bright polish. It is sensitive to even mild acids, and can become dull with extensive wear. Red and pink precious corals are found in the warm seas around Japan and Malaysia, in the Mediterranean, and in African coastal waters. Black coral comes from the West Indies, Australia, and around the Pacific Islands.

Coral is used in carvings and beads, and cut as cabochons for use in jewellery. Red coral appears as an ornament in Western European shields and helmets in the Iron Age, later to be replaced by red enamel.

**CORAL NECKLACE**
*This Native American necklace is made from small, polished branches and tiny beads of red coral. This coral forms in a branch structure.*

According to Greek legend, coral came from the drops of blood shed when Perseus cut off the head of Medusa. In Classical antiquity, coral was believed to be a protection against the evil eye, and was worn as an amulet. As a medication, it was believed to stop bleeding. In the 18th century, precious coral was widely used as an inlay in Japanese ornamental netsukes. It has been a favourite of Chinese carvers for at least two millennia.

## PROPERTIES

| | |
|---|---|
| **GROUP** | Organics – carbonate |
| **CRYSTAL SYSTEM** | Trigonal, orthorhombic, amorphous |
| **COMPOSITION** | $CaCO_3$ / conchiolin |
| **FORM/HABIT** | Coral-shaped |
| **HARDNESS** | $3\frac{1}{2}$ |
| **CLEAVAGE** | None |
| **FRACTURE** | Hackly |
| **COLOUR** | Red, pink, black, blue, golden |
| **STREAK** | White |
| **LUSTRE** | Dull to vitreous |
| **SPECIFIC GRAVITY** | 2.6–2.7 |
| **TRANSPARENCY** | Opaque |
| **R.I.** | 1.49–1.66 |

**FOSSILIZED CORAL**
*Fossil corals provide unusual and interesting material for cabochons.*

*Monkey climbing a tree*

**CORAL CARVING**
*Red coral is tough and compact, allowing complex carvings to be made from it.*

**RED CABOCHON**
*Red coral takes a brilliant polish and, when cut en cabochon, finds many jewellery uses for its vivid red colour.*

**BLACK CABOCHON**
*Some black corals are as tough and compact as red forms, taking an excellent polish, and providing attractive gemstones.*

**RED CORAL**
*A wood-grain pattern can clearly be seen on the branches of this red coral from the Mediterranean.*

*Coral branch*

*Living chambers*

## CORAL REEFS

A coral reef forms a ridge or hummock in shallow ocean areas. Corals are the most important part of the reef, and generally form its main structural framework. The coral polyps divide again and again, growing into colonies that can be up to several metres in diameter, becoming so large and heavy that only storms disturb them. Molluscs, sea urchins, calcareous algae, and microscopic protozoa also contribute to the reef, and provide fragments that wash or fall into the gaps between corals. Sheet-like growths of the algae and protozoa also bind and cement the reef together. A reef becomes limestone by the slow dissolution, redeposition, recrystallization, and chemical transformation of reef material.

**SOFT CORAL**
*Live coral on reefs is endangered by sea-level rises and warmer waters. As it dies, the reef disintegrates.*

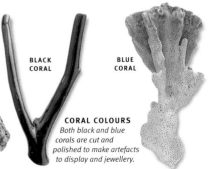

**BLACK CORAL**

**BLUE CORAL**

**CORAL COLOURS**
*Both black and blue corals are cut and polished to make artefacts to display and jewellery.*

# SHELL

LIKE CORAL, SHELL IS MINERAL matter generated by biological processes. It comes in a huge variety of sizes, shapes, and colours. The mineral component of shells is either calcite or aragonite, both forms of calcium carbonate. Shell forms a hard outer covering of many molluscs, and it is secreted in layers by cells in the mantle, a skinlike tissue in the mollusc's body wall. Different groups of molluscs are characterized by the number of calcareous layers, the composition of the layers (aragonite or aragonite and calcite), and their arrangement (in sheets, for example). The results are distinct microstructures that have differing mechanical properties

and, in some shells, differing colours. Shells for carving or other ornamentation can be marine or freshwater. In the late 18th and through the 19th century, the use of pearly shells in button making increased with the mechanization of production. Marine shells were used exclusively at first, but in the 1890s freshwater mussel shells found along the Mississippi River and its tributaries came into use. The demand for seashell became so great that mother-of-pearl (see p.322) was more sought after than the pearls themselves. Shells with different coloured layers have been carved into cameos since antiquity. Shell is also used in inlays, beads, and other decorative items.

**MONTERREY BAY**
*This bay in California, USA, has been a source of abalone shell for centuries. Harvesting in the bay is now restricted.*

**RAW MATERIALS**
*Some shells, like the spider conch, are used widely. Others, like the tortoiseshell, are banned in most countries.*

**SPIDER CONCH**

**ABALONE PENDANT**
*Polished pieces of abalone shell are widely sold as inexpensive but attractive jewellery.*

**HAWKSBILL TORTOISESHELL**

## PROPERTIES

| | |
|---|---|
| **GROUP** | Organics – carbonate |
| **CRYSTAL SYSTEM** | Trigonal, orthorhombic, amorphous |
| **COMPOSITION** | CaCO$_3$ |
| **FORM/HABIT** | Shell-shaped |
| **HARDNESS** | 2½ |
| **CLEAVAGE** | None |
| **FRACTURE** | Conchoidal |
| **COLOUR** | Red, pink, brown, blue, golden |
| **STREAK** | White |
| **LUSTRE** | Dull to vitreous |
| **SPECIFIC GRAVITY** | About 1.3 |
| **TRANSPARENCY** | Translucent to opaque |

**NATIVE AMERICAN PENDANT**
*This spondyllus shell has been partly encrusted with turquoise, mother-of-pearl, and jet to make a pendant.*

*Iridescent mother-of-pearl*

**ABALONE SHELL**
*Found in warm seas worldwide, abalone shells such as this one from New Zealand, are noted for their multi-coloured, iridescent mother-of-pearl lining. Abalone and other iridescent shells are cut and polished to display their brilliant rainbow colours.*

## TORTOISESHELL

Formed from keratin and containing no calcium carbonate, tortoiseshell was a product of the hawksbill turtle of Indonesia and the West Indies. It has been used in a wide range of products, from combs to tea caddies. When heated, the shell becomes soft and can be formed like plastic. Today it has largely been replaced by synthetic tortoiseshell, and many countries now prohibit the import of tortoiseshell goods, owing to depletion of the species from which they are made.

**OCEANIC MASK**
*Carved from tortoiseshell, this mask is from the Torres Straits area.*

**TEA CADDY**
*The simple lines of this tea caddy allow the patterns within the tortoiseshell to be the main feature of the box.*

**TIGER COWRIE CAMEO**
*The different coloured layers of this cowrie shell make an unusual and beautiful cameo.*

**THE BIRTH OF VENUS**
*Botticelli famously depicted the emergence of Venus from a scallop shell in his painting of about 1485.*

# PEARL

**MUSSELS WITH PEARLS**
*Pearls occur in a variety of mollusc species, and in a wide range of localities in both fresh and sea water.*

*Mother-of-pearl-coated metal castings*

**PEARL BUDDHAS**
*Metal castings of Buddha were placed into this living shell, which then formed a mother-of-pearl layer over them.*

PEARL IS A CONCRETION formed by a mollusc and consisting of the same material as the mollusc's shell, which is principally the mineral aragonite (see p.180). In addition to aragonite, the shell contains small amounts of conchiolin, a horn-like organic substance; together these are called nacre, or mother-of-pearl. The finest pearls are those produced by molluscs whose shells are lined with mother-of-pearl; such molluscs are limited to certain species of salt-water oysters and freshwater clams. The shell-secreting cells are located in a layer of the mollusc's body tissue called the mantle. When a foreign particle enters the mantle, the cells build up more or less concentric layers of pearl around it to protect the mantle. Baroque pearls are irregularly shaped pearls that have grown in muscular tissue; blister pearls are those that grow adjacent to the shell and are flat on one side.

Pearls are valued by their translucence, lustre, play of surface colour, and shape. The most valuable are spherical or drop-like, with a deep lustre and good colour play. In the jewellery industry, salt-water pearls are commonly referred to as Oriental pearls; those produced by freshwater molluscs are called freshwater pearls. Cultured pearls are those that have been grown on a pearl farm. tiny sphere of mother-of-pearl is implanted in the mantle of the nacre-producing mollusc,

**CANNING MERMAN**
*This late-16th-century jewel uses four baroque pearls, including one to form the body.*

which is then placed in the waters of the farm. The oyster forms a pearl around the tiny sphere, which is then harvested up to two years later (see pp.324–25).
The colour of pearls varies with the mollusc and its environment. A pearl can be any delicate shade from black to white, cream, grey, blue, yellow, green, lavender, and mauve. Rose-tinted Indian pearls are particularly prized. Pearls vary from the size of tiny seeds – called seed pearls – to one large baroque pearl that weighed 90g (3oz). Some of the finest pearls are found in the

**BOMBAY BUNCH**
*Mumbai (formerly Bombay), India, has been an important centre of the pearl trade for centuries. Pearls are sorted by size and strung together to sell as a lot.*

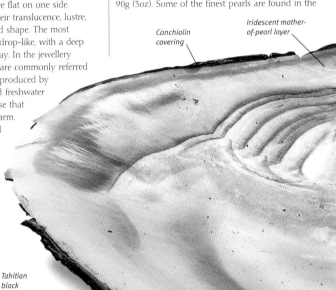

*Conchiolin covering*

*Iridescent mother-of-pearl layer*

**MARINE-CULTURED PEARLS**
*Marine-cultured pearls can be grown to uniform sizes and colours, depending on the chemistry of the local sea water.*

*Pearly lustre*

*Uniform colour*

*Tahitian black pearl*

*Black mother-of-pearl*

*Black blister pearl*

**MARINE-CULTURED PEARL NECKLACE**
*The pearls in this exquisite contemporary necklace of marine-cultured pearls show uniformity of colour and size.*

**MARINE BLACK-LIPPED OYSTER AND BLACK PEARL**
*The black-lip shell, a mollusc that produces black mother-of-pearl, also produces black cultured pearls.*

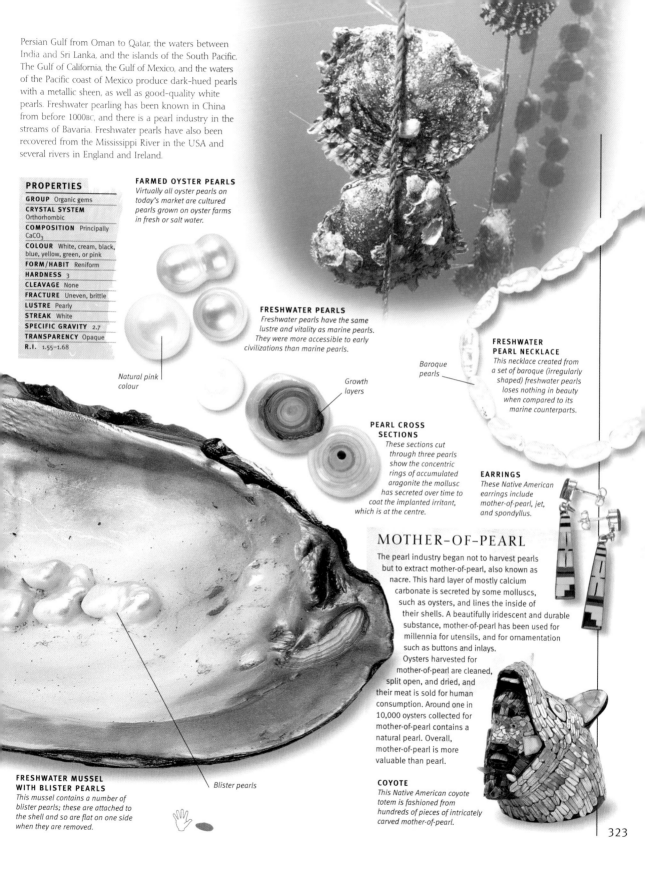

Persian Gulf from Oman to Qatar, the waters between India and Sri Lanka, and the islands of the South Pacific. The Gulf of California, the Gulf of Mexico, and the waters of the Pacific coast of Mexico produce dark-hued pearls with a metallic sheen, as well as good-quality white pearls. Freshwater pearling has been known in China from before 1000BC, and there is a pearl industry in the streams of Bavaria. Freshwater pearls have also been recovered from the Mississippi River in the USA and several rivers in England and Ireland.

## PROPERTIES

| | |
|---|---|
| **GROUP** | Organic gems |
| **CRYSTAL SYSTEM** | Orthorhombic |
| **COMPOSITION** | Principally $CaCO_3$ |
| **COLOUR** | White, cream, black, blue, yellow, green, or pink |
| **FORM/HABIT** | Reniform |
| **HARDNESS** | 3 |
| **CLEAVAGE** | None |
| **FRACTURE** | Uneven, brittle |
| **LUSTRE** | Pearly |
| **STREAK** | White |
| **SPECIFIC GRAVITY** | 2.7 |
| **TRANSPARENCY** | Opaque |
| **R.I.** | 1.55–1.68 |

**FARMED OYSTER PEARLS**
*Virtually all oyster pearls on today's market are cultured pearls grown on oyster farms in fresh or salt water.*

*Natural pink colour*

**FRESHWATER PEARLS**
*Freshwater pearls have the same lustre and vitality as marine pearls. They were more accessible to early civilizations than marine pearls.*

*Growth layers*

**PEARL CROSS SECTIONS**
*These sections cut through three pearls show the concentric rings of accumulated aragonite the mollusc has secreted over time to coat the implanted irritant, which is at the centre.*

*Baroque pearls*

**FRESHWATER PEARL NECKLACE**
*This necklace created from a set of baroque (irregularly shaped) freshwater pearls loses nothing in beauty when compared to its marine counterparts.*

**EARRINGS**
*These Native American earrings include mother-of-pearl, jet, and spondyllus.*

## MOTHER-OF-PEARL

The pearl industry began not to harvest pearls but to extract mother-of-pearl, also known as nacre. This hard layer of mostly calcium carbonate is secreted by some molluscs, such as oysters, and lines the inside of their shells. A beautifully iridescent and durable substance, mother-of-pearl has been used for millennia for utensils, and for ornamentation such as buttons and inlays.

Oysters harvested for mother-of-pearl are cleaned, split open, and dried, and their meat is sold for human consumption. Around one in 10,000 oysters collected for mother-of-pearl contains a natural pearl. Overall, mother-of-pearl is more valuable than pearl.

**FRESHWATER MUSSEL WITH BLISTER PEARLS**
*This mussel contains a number of blister pearls; these are attached to the shell and so are flat on one side when they are removed.*

*Blister pearls*

**COYOTE**
*This Native American coyote totem is fashioned from hundreds of pieces of intricately carved mother-of-pearl.*

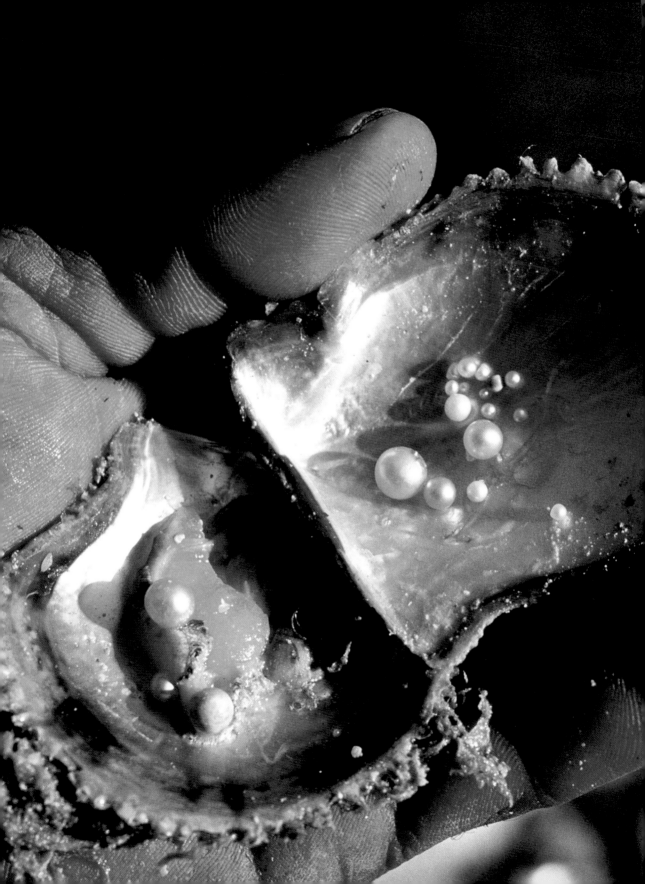

# CULTIVATING PEARLS

NATURALLY HIGHLY VARIABLE AND SCARCE, PEARLS
WERE TRADITIONALLY RECOVERED BY DIVING. TODAY
THE WIDESPREAD CULTIVATION OF OYSTER PEARLS HAS
BROUGHT MORE PERFECTLY ROUND, OFTEN LARGER
GEMS TO THE JEWELLERY MARKET.

The cultivation of pearls in freshwater mussels is thought to have begun in 13th-century China. These pearls were "blister" pearls – hemispherical pearls formed between the mussel and its shell. The production of whole cultured pearls, and its subsequent industry, was started in the 1890s by the Japanese Mikimoto Kokichi. After long experimentation, he discovered that a very small mother-of-pearl bead introduced into the tissue of a mollusc that produced mother-of-pearl – a process known as "seeding" – stimulated it to produce a perfect, round pearl.

Today, pearl farms can be found in sea water and freshwater. When immature pearl oysters raised in containers are two to three years old, they are implanted with a tiny sphere of mother-of-pearl. Taken to coastal waters or to deeper freshwater, the oysters are then suspended in wire nets or contained in some other way so that growth takes place in natural conditions. Divers tend the growing oysters, ensuring they have enough plankton to feed on and are not overcrowded, until they are ready for harvesting from 13 months to two years later, and their pearls extracted. Japan, northern Australia, and to a lesser extent Fiji are major producers of marine cultured pearls. The Mississippi River in the USA is a major site of freshwater pearl production.

A pearl's colour – and hence its value – depends upon the waters from which it comes. Pearls from Japan are cream or white with greenish tones; Persian Gulf pearls are usually cream; black or reddish-brown pearls come from Mexico; pink pearls are from Sri Lanka; Australian pearls are white with greenish or bluish shades; and pearls from the Gulf of Panama are golden-brown. Cultured pearls from Japanese and Australian waters are the most popular.

**EARLY PEARL FISHING**
*This miniature by Flemish artist Jehan de Grise shows pearl fishing in the gulf around Malabar, India. It is taken from the manuscript* Travels of Marco Polo *(dated c.AD1400).*

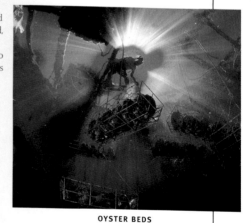

**OYSTER BEDS**
*In modern cultured pearl farming, seeded oysters or other pearl-bearing molluscs are suspended in baskets, and tended by divers using the latest diving technology.*

**A HANDFUL OF PEARLS**
*Perfectly spherical pearls can be produced by a pearl oyster if the seed implanted is also a perfect sphere. The size depends on the size of the seed and the length of time the oyster is allowed to grow on the farm.*

# FOSSILS

PLANTS | INVERTEBRATES | VERTEBRATES

# HOW FOSSILS ARE FORMED

**BURGESS SHALE**
*Soft-bodied, Middle Cambrian fauna are preserved in this shale in the Canadian Rockies.*

A FOSSIL IS A REMNANT, IMPRESSION, OR TRACE of an organism that has lived in a past geologic age. Some have been preserved in fine–grained sedimentary rock, such as limestone or shale. The most common fossils found are of plants and animals that once lived in a sea or lake. Typically, after an organism dies, the soft parts decompose, leaving only hard parts – the shell, teeth, bones, or wood. Buried in layers of sediment, they gradually turn to stone.

## STAGES OF FOSSILIZATION

Fossilization is a hit-or-miss process since normally the body of a dead creature decays, leaving no trace. The main factor is whether or not the plant or animal dies in a location where fossilization can take place. On a sea bed, where sediments are accumulating rapidly, the chances are good; in an environment where there is little or no sediment accumulation before the plant or animal completely decomposes, the chances are poor. Consequently, the fossil record is highly biased towards marine organisms with hard parts, such as shells. Where fossilization does occur, it goes through the same stages each time, but how long it takes can vary greatly.

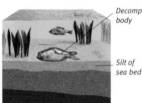

Decomposing body / Silt of sea bed

Remains are trapped in sediment

Additional layers / Skeleton replaced by minerals

Excavation frees fossil / Erosion partly reveals fossil

**1 SILT ACCUMULATION**
*A dead animal's body must be rapidly covered by silt. If it remains exposed it will simply disintegrate on the surface.*

**2 MINERALIZATION**
*Enveloped by sediment, the soft tissue dissolves but minerals are absorbed by the pores of the skeleton.*

**3 CONSOLIDATION**
*The pressure of successive layers of silt hardens both the fossilized skeleton and the surrounding silt into rock.*

**4 EXPOSURE**
*Massive land upheaval lifts the rock layer to the surface. Wind and rain reveal the fossil.*

## UNALTERED PRESERVATION

Under exceptional conditions organisms can be fossilized in an unaltered state. Minute creatures called diatoms belonging to the kingdom Protista have skeletons of silicon dioxide, which often remain unaltered. Marine- and lake-dwelling invertebrate animals, such as corals, molluscs, brachiopods, and bryozoans, have a calcareous skeleton or shell, which may be found fundamentally unaltered, even in rocks of great age. Under even rarer circumstances, organic matter that was rapidly sealed from the air or the attack of other organisms can remain virtually intact for millions of years. Insects, small animals, and plant remains sealed in amber are the classic example. It has even been suggested that DNA might be recovered from some of them.

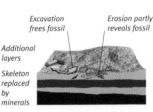

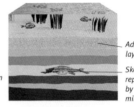

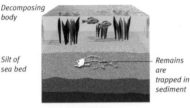

Individual plant (corallite)

**FOSSILIZED DIATOMS**
*Diatoms are tiny single-celled animals. Their skeletons of silicon dioxide undergo fossilization with little or no structural alteration.*

**FOSSIL CORAL CABOCHON**
*Some coral skeletons remain essentially unchanged during fossilization.*

**FOSSILIZED AND MODERN SHELLS**
*A number of modern molluscs evolved sufficiently long ago to have fossilized equivalents. The fossils vary in character, depending on the materials in which they were originally buried. Their coloration is very rarely preserved (see mussel fossil, right).*

Chalk preservation

**TRITON FOSSIL AND SHELL**

Limestone preservation

**BRACHIOPOD FOSSIL AND SHELL**

Retained coloration

**MUSSEL FOSSIL AND SHELL**

# MINERAL PRESERVATION

Most organisms are subjected to changes during fossilization. In a process known as permineralization, water seeping through the rock deposits calcium carbonate or other mineral salts in the pores of the shell or bone, fossilizing the remains. In some cases, the organic matter is completely replaced by minerals as it decays. This process is known as mineralization, or replacement, and can be seen in petrified wood. The bones of vertebrates, and some shells, are made of calcium phosphate, and may undergo the same processes. In other cases, circulating acid solutions dissolve the original shell or bone, leaving a cavity of identical shape. Circulating calcareous or siliceous solutions may then deposit new material in the cavity, creating a cast of the original shell. The original impression, the "negative" of the cast, is called the mould.

*Pyrite replacement*

*Fossilized bark*

### FOSSILIZED WOOD
*Also known as petrified wood, fossilized wood can be subjected to such a high degree of mineral replacement, literally cell by cell, that the detail of its original cell structure is preserved exactly.*

### PYRITIZED AMMONITE
*Sediments containing iron and sulphur occasionally form pyrite during lithification. If animal remains are present, pyrite may replace them.*

# TRACE FOSSILS

Organisms leave behind them traces in the form of tracks, trails, or even borings. Fossilized footprints of dinosaurs preserved at Glen Rose, Texas, USA, and at Lark Quarry, Winton, Queensland, Australia, reveal much about the weight and speed of the dinosaurs. Tracks of feeding and burrowing worms reveal how those animals lived.

### FOSSILIZED WORM TRACKS
*Fossils of soft-bodied animals such as worms are not found, but their fossilized tracks in rocks provide evidence of their existence.*

### DINOSAUR TRACKS
*Tracks have yielded information about dinosaur species – what they weighed, their mobility and length of stride, and how many travelled in a herd.*

# FOSSIL IMPRESSIONS

Many fossils are simply the impression of a plant or animal left in the rock. These are created when an object, such as a leaf, is buried in sediment. During or after the hardening of the sediment, the leaf decays and disappears, leaving its imprint. Occasionally the imprint of soft tissue – such as dinosaur skin – is discovered. In very rare instances, even imprints of internal organs have been preserved in rock.

### FOSSILIZED LEAVES
*The intricately fossilized impressions left in rock by leaves provide paleobotanists with a wealth of information about the structures of extinct plant species.*

# PRESERVING SOFT TISSUE

The soft parts of animals or plants are very rarely fossilized. Insects trapped in amber before they could decompose, and mammoths frozen intact in ice, are rare instances where soft tissue is preserved. In 1997, at Jarkov in Siberia, Russia, a 23,000-year-old mammoth was found frozen in permafrost – the most perfectly preserved mammoth found to date. The excavated permafrost block was airlifted to an ice cave for study.

### INSECT IN AMBER
*Amber perfectly preserves insects from decomposition. Mosquitoes can still have the blood of their victims inside them.*

### MAMMOTH TUSKS
*These enormous tusks from the Jarkov mammoth are being prepared for transportation on a sledge.*

# THE FOSSIL RECORD

**ROCK-DATING TOOL**
*This fossil ammonite provides an accurate age-date for the rock in which it is found.*

THE COMPLEX HISTORY OF LIFE ON EARTH preserved in fossils worldwide, is known as the fossil record. By studying fossils, geologists can identify patterns of progressive changes over time that represent the evolution of a particular group or species. Geologists have been able to assign relative ages to the strata of rocks in which specific fossils occur, and consequently divide the history of the Earth into a sequence of periods, collectively known as Geologic Time. Fossils provided some of the earliest clues to the existence of plate tectonics, confirming that the continents once comprised a continuous landmass.

## GEOLOGIC TIME

The vast interval of time occupied by the Earth's geologic history is known as Geologic Time. It extends from the age of the oldest known rocks – about 3.9 billion years ago – to the present day. Geologic Time is the Earth's history represented by and recorded in rock strata. The units of geologic time were originally established largely through the study of the fossil record and stratigraphy – the correlation and classification of rock strata. Due to evolutionary changes over geologic time, with the appearance of new species and the disappearance of old, certain types of organisms are characteristic of particular parts of the geologic record. Their fossils provide the means by which divisions in the record are made. The

geologic history of various geographic areas, and finally of the Earth as a whole, has been reconstructed through correlating the strata in which specific types of fossils occur. Recent advances in radiometric dating methods (see below) have permitted the relative geologic timescale derived from the fossil record to be assigned absolute dates.

The fossil record also provides evidence that the Earth's surface consists of immense, moving plates, and that the continents once fitted together in a single, huge landmass. The correlation of rock strata and their fossils on different continents suggested that the strata may once have been continuous, and also dated their times of separation.

**DIVISIONS OF GEOLOGIC TIME**
*The divisions of geologic time provide a calendar against which to set events in the Earth's history. The divisions also provide a universal frame of reference for geologists, climatologists, biologists, and paleontologists. In descending order of duration, these divisions are the eon, era, period, and epoch. These units of time, and the point of appearance of some of the organisms from the fossil record that characterizes them, are shown below.*

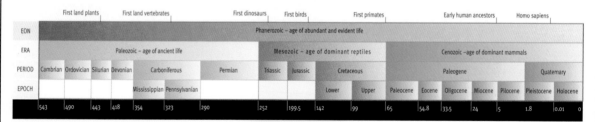

| | First land plants | | First land vertebrates | | | First dinosaurs | | First birds | | First primates | | Early human ancestors | | Homo sapiens | |
|---|---|---|---|---|---|---|---|---|---|---|---|---|---|---|---|
| EON | Phanerozoic – age of abundant and evident life | | | | | | | | | | | | | | |
| ERA | Paleozoic – age of ancient life | | | | | | Mesozoic – age of dominant reptiles | | | | Cenozoic – age of dominant mammals | | | | |
| PERIOD | Cambrian | Ordovician | Silurian | Devonian | Carboniferous | | Permian | Triassic | Jurassic | Cretaceous | | Paleogene | | | Quaternary | |
| EPOCH | | | | | Mississippian | Pennsylvanian | | | | Lower | Upper | Paleocene | Eocene | Oligocene | Miocene | Pliocene | Pleistocene | Holocene |

| 543 | 490 | 443 | 418 | 354 | 323 | 290 | 252 | 199.5 | 142 | 99 | 65 | 54.8 | 33.5 | 24 | 5 | 1.8 | 0.01 | 0 |

## INDEX FOSSILS

Some fossils occur within limited time spans and yet are widespread enough to identify geological relationships over large areas. Even if the type of rock differs, or the strata are at different levels, if certain specific combinations of fossils are present in two distinct locations (see below), then geologists know that both strata are of similar age. Such fossils are known as index fossils. The order of strata and the fossil record reveal the relative age of different rock strata, but not in absolute terms how old they are.

**INDEX TRILOBITE**
*Paradoxides is a useful trilobite for dating Middle Cambrian Rocks.*

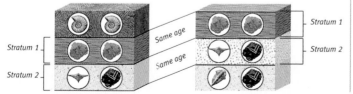

Stratum 1

Same age

Stratum 1

Stratum 2

Same age

Stratum 2

## RADIOMETRIC DATING

Radiometric dating is a method of providing an absolute date for some minerals, rocks, and fossils. Radioisotopes of some elements, such as lead, carbon, and zircon, decay at a known rate. So, by measuring the amount of decayed isotopes of one of these elements within a particular mineral, rock, or fossil, and comparing it with the amount yet to decay, an absolute date of formation of the material can be established.

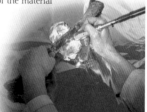

**RADIOCARBON DATING**
*This dating technique is used for most fossils. However, it is useful only for very recent life forms in which bone or tissue remains unfossilized.*

# PALEOECOLOGY

As well as recording the biological history of life on Earth, the fossil record preserves the climatic history of the planet. The climatic conditions experienced by an area are reflected both in the rocks themselves and in the fossils preserved in them. For example, certain species of coral require warm, shallow water, and certain types of flowering plants will grow only in colder, drier environments. However, climatic conditions cannot be inferred by the presence of only one type of fossil. All of the fossils present (the assemblage) are taken into account in the assessment.

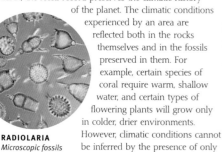

**RADIOLARIA**
*Microscopic fossils of radiolarian skeletons are good indicators of ocean environments.*

## FOSSIL FUELS

All fossil fuels contain carbon, and were formed from the remains of organic matter as a result of geologic processes. They include coal, petroleum, natural gas, shale oil, and bitumen. Around 90 per cent of all the energy used in the industrially developed nations is produced from fossil fuels. Our reliance on them creates two major problems: depletion and pollution. Both of these have potentially serious consequences for our future on Earth and for all other life on our planet.

**WILLIAM SMITH**
*This Englishman's geologic map of England and Wales (1815) was the father of all geologic maps.*

### DEPLETION

The term depletion does not simply mean the exhaustion of existing deposits. There are strict limits, derived from the current state of technology, that determine how much of each deposit is economically recoverable. For example, a seam of coal 3m (10ft) thick is uneconomic to be recovered if it is buried under 100m (300ft) of other sediments that have to be removed before it can be mined.

### POLLUTION

The principal concern regarding the burning of fossil fuels is its effect on the climate. All fossil fuels release carbon dioxide into the atmosphere when they are burned. Carbon dioxide makes up only about 0.03 per cent of the air, but by absorbing infrared radiation, it has a huge effect on the heat balance of the atmosphere. With increased carbon dioxide in the atmosphere from burning fossil fuels, some of the Earth's heat, which normally escapes back into space, is trapped, creating the possibility of potentially deadly climate change.

**OIL FIELD**
*The search for oil relies heavily upon stratigraphy determined by the fossil record. Microfossils in drill cores and cuttings aid the search.*

**COAL SEAMS**
*Coal is the fossilized remains of plants accumulated over millions of years. Here coal seams form in Portuguese cliffs.*

**FOSSILIZED CORAL**
*Corals are relatively delicate life forms. Because of their sensitivity, they act as barometers of changes to climates and marine habitats.*

**BANDED JURASSIC LIMESTONE**
*This coastal cliff in Glamorgan, Wales, UK, consists of exposed beds of Jurassic limestone; the stratification is useful for age-dating.*

# PLANTS

THE FIRST FOSSIL EVIDENCE of land plants dates from the Ordovician Period (490 to 443 million years ago). It is believed that land plants evolved from marine algae that moved onto the land. The first plant fossils of a visible size were the psilophytes, which date from the Silurian Period (443 to 418 million years ago). Consisting of slender, forking tubes only a few centimetres long, they were little more than creeping root systems with no leaves.

The key evolutionary step during the Silurian was the development of vascular plants – those with an inner structure permitting the flow of water and nutrients through the entire plant. During the Early Devonian Period (418 to 354 million years ago), plants developed small clusters of light–capturing cells on their surfaces that eventually evolved into leaves. By the end of the Devonian, tree-sized plants called lycopods had evolved.

**MAGNOLIA**
*Fossilized magnolia leaves 20 million years old have been found in Idaho, USA.*

## ALGAE

ALGAE WERE some of the earliest organisms to appear on Earth. Their fossil remains date back to the Precambrian, although the fossil record is far from complete. Those that fossilized had structures impregnated with silica, calcite, or aragonite, or had thick-walled cysts (body cavities). Fossils range from more or less complete individuals to carbonized impressions. The two genera shown here belong to different phyla in the kingdom Protista.

*Bythotrephis gracilis*

*Impression*

**BYTHOTREPHIS SPECIMEN**

**BRANCHING**
*Bythotrephis had a simple branching structure and little or no calcification. This brown alga is mainly found as impressions. It can be mistaken for worm burrows.*

*Mastopora favus*

*Honeycomb pattern*

**GLOBULAR**
*Mastopora was a green alga with a globular structure covered by a calcified coating. A reef dweller, it lived from the Ordovician to the Silurian.*

**MASTOPORA SPECIMEN**

## HORSETAILS

THERE ARE LIVING SPECIES of horsetails, more properly referred to as sphenopsids. Their stems are conspicuously jointed, and the leaves form whorls about the stem. They grew in moist, rich soils; their tubers are found fossilized in fossil soils. Fossil sphenopsids are found as far back as the Late Devonian.

*Jointed stem*

*Asterophyllites equisetiformis*

**GIANT HORSETAIL**
Asterophyllites *was a giant horsetail, reaching 10m (33ft) in height. It lived from the Late Carboniferous to the Early Permian, in wetlands and swamps. Other horsetail species were closer to the size of the modern equivalents, reaching 50cm (20in) in height.*

**ASTEROPHYLLITES SPECIMEN**

**MODERN SPHENOPSID**

## FERNS

FERNS, OR PETROPHYTA, enter the fossil record in the Middle Devonian. They vary in size from tiny, gauzy plants to large tree ferns several metres tall. Although their fossils can be confused with those of sphenopsids, fossil fern leaves are flat and arranged on fronds.

**MODERN FERN**

**TREE FERN**
Pecopteris *was a large tree fern that lived from the Early Carboniferous to the Early Permian. It reached 3m (13ft) in height. It usually broke up before fossilization.*

**PECOPTERIS SPECIMEN**

*Impression*

*Pecopteris unita*

**MAZON CREEK**
*This site in Illinois, USA, is a rich area for fossilized ferns.*

# LYCOPODS

LYCOPODS ARE called club-mosses today. They first appeared in the Late Silurian, and were most abundant during the Late Carboniferous. Modern lycopods are small, herbaceous plants, but during the Carboniferous they were the size of large trees, with dense foliage and their spores contained in cones. Fossils of lycopod logs that measure up to 1m (3ft) in diameter have been found.

**MODERN CLUB-MOSS**

*Bark*

*Scale-like surface*

**LEPIDODENDRON SPECIMEN**

### CLUB-MOSS
Lepidodendron *is the most prolific fossilized club-moss, being found worldwide. It thrived during the Carboniferous, growing in hot, humid swampland. It reached 30m (100ft) in height.*

*Lepidodendron aculeatum*

# SEED FERNS

THESE PLANTS ORIGINATED in the late Devonian and were widespread towards the end of the Paleozoic Era. Known more properly as pteridosperms, they persisted into the Mesozoic, when they became extinct. They were small trees with fern-like leaves, some of which reached heights of up to 8m (26ft). Their foliage is not easily distinguished from that of other ferns of the period, but evidence of seeds puts the pteridosperms in their own group.

*Paripteris gigantea*

**PARIPTERIS SPECIMEN**

### POLLEN ORGAN
*Preserved in this fossil is the seed-like, pollen-bearing organ, or potoniea, of the Carboniferous* Paripteris; *it formed at the base of each leaf-frond. The fern grew to a height of 5m (16ft) and lived in higher altitude regions of hot swampland. Its fossils are found worldwide.*

### FORKED LEAVES
*This shrub-like fern reached a height of 4m (13ft) and is noted for its unusual "Y"-forked leaf. It is a Triassic fossil commonly found in the southern hemisphere, where it inhabited tropical forests of tree ferns.*

**DICROIDIUM SPECIMEN**

*Dicroidium species*

*Alethopteris serlii*

**ALETHOPTERIS SPECIMEN**

### FERN FROND
*This fossilized frond segment is from* Alethopteris, *which grew to 5m (16ft). A Lower Carboniferous to Early Permian seed fern, it has thick, strong-veined leaflets. It flourished in hot swamps worldwide.*

## WEGENER'S THEORY OF CONTINENTAL DRIFT

Born in Berlin in 1880, Alfred Wegener taught meteorology at Hamburg and geophysics at the University of Graz. In about 1910, noting the apparent close fit between the coastlines of Africa and South America, he developed the idea that in the Late Paleozoic Era (about 250 million years ago) all the present-day continents had formed a single large mass that had subsequently broken apart. He discovered that many closely related fossil organisms occurred in similar rock strata on widely separated continents, particularly those found in both the Americas and in Africa. The moving apart of the continents he termed Continental Drift. However, he was unable to explain satisfactorily the driving forces behind the continents' movement, and by 1930 his theory had been rejected by most geologists and had fallen into obscurity. It was not until the 1960s that sea-floor spreading was discovered, and his theory was resurrected as an important part of the now well-accepted theory of plate tectonics.

**WEGENER**
*Fifty years passed before evidence appeared to support Wegener's theories. He died in 1930.*

*Glossopteris species*

### KEY TO THEORY
*Glossopteris fossils found on different continents played an important part in Wegener's concept of Continental Drift.*

**GLOSSOPTERIS SPECIMEN**

### A SINGLE LANDMASS
*Alfred Wegener theorized that all of the continents were once connected in a single landmass that he named Pangea, and which subsequently broke apart.*

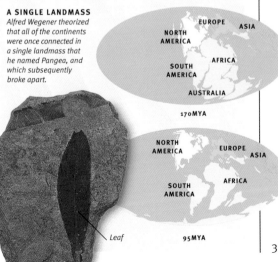

*Leaf*

EUROPE   ASIA
NORTH AMERICA
AFRICA
SOUTH AMERICA
AUSTRALIA

170MYA

NORTH AMERICA   EUROPE   ASIA
AFRICA
SOUTH AMERICA

95MYA

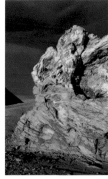

# THE ARIZONA PETRIFIED FOREST

COVERING A VAST AREA OF ARIZONA, THE FOSSILIZED WOOD OF THE PETRIFIED FOREST PRESERVES AN ENTIRE TRIASSIC ECOSYSTEM.

**ARGENTINIAN FOREST**
*There are a number of petrified forests worldwide. The Petrified Forest National Park (above) is in Santa Cruz province, southern Argentina.*

The Arizona Petrified Forest National Park occupies an area of 378 sq km (146 sq miles) approximately 30km (20 miles) east of Holbrook, Arizona. The area is high desert, and extends into the Painted Desert (see p.66). Erosion has exposed extensive areas of petrified, or fossil, wood, including whole logs and entire trees.

Formed by the invasion of minerals into cavities between and within the cells of natural wood, this fossil wood is preserved from the Triassic Period (200–250 million years ago), when the Petrified Forest area of Arizona was a large basin on the edge of the huge landmass known as "Pangea" (see p.333). Lush forests covered the landscape, with towering coniferous trees up to 3m (9ft) in diameter and almost 60m (200ft) tall. Ferns, cycads, and giant horsetails lined the waterways.

### THE PROCESS OF PETRIFICATION

As trees died or fell, some were deposited directly in river beds or on floodplains, while others were washed into the current area of the forest by flood water. Most of the trees decomposed, but others were buried in the accumulating sediments, which included a considerable amount of volcanic ash. Silica from the ash was dissolved by ground water, and as it percolated downwards, it filled or replaced cell walls, crystallizing as chalcedony – cryptocrystalline quartz. The ground water was also iron-rich, which combined with quartz during petrification, creating the rainbow of colours seen in the fossil wood of Arizona today. Eventually the forest area was buried by further accumulation of sediments. Many millions of years later, the whole of north-central Arizona, including the solidified sediments containing the fossilized logs, was uplifted, creating the Colorado Plateau. Erosion removed the overlying rock layers, and exposed the petrified logs.

**CROSS SECTION**
*The cell-by-cell replacement of original plant tissue by silica was so precise that in many instances the internal structure – sometimes even the cell structure – was faithfully reproduced, as well as the log's external shape.*

Preserved growth rings

Original cell structure replaced by chalcedony

**PETRIFIED FOREST LANDSCAPE**
*Brilliant reds and yellows predominate in the Forest today. Most of the fossilized logs are from three extinct trees: Araucarioxylon arizonicum, Woodworthia, and Schilderia.*

# CONIFERS

PLANTS THAT REPRODUCE by means of an exposed seed, including all conifers, are known as gymnosperms. Conifers are usually shrubs or trees that produce woody cones of various shapes and sizes. The leaves are often needle-shaped but can be flat and broad. Their woody trunks are often rich in resin. They made up a large part of Jurassic forests, although they appeared earlier, in the Late Carboniferous. Coniferous fossils can be needles or leaves, cones, or the wood itself. Most fossil wood, commonly called petrified wood, is from conifers. Large deposits of fossil conifer wood are referred to as petrified forests (see The Arizona Petrified Forest, pp.334–35).

**IDENTIFYING SIGNS**
*Fossil coniferous wood often shows immaculately preserved cell structure and growth rings. It is possible in some instances to determine individual species, but to do so usually requires close microscopic examination and detailed knowledge.*

Conifer species

Shale

Branch imprint

Wide growth ring

**CONIFEROUS WOOD SPECIMEN**

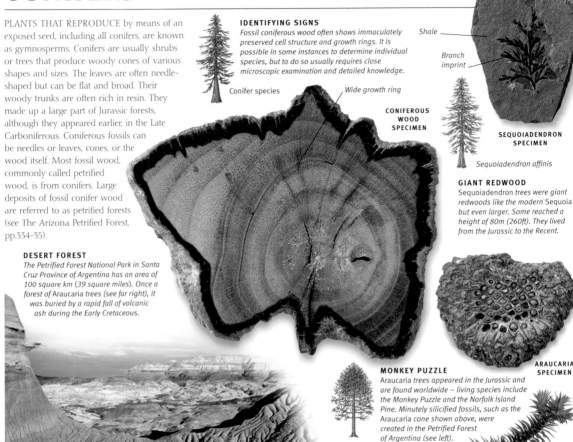

**SEQUOIADENDRON SPECIMEN**

*Sequoiadendron affinis*

**GIANT REDWOOD**
*Sequoiadendron trees were giant redwoods like the modern Sequoia but even larger. Some reached a height of 80m (260ft). They lived from the Jurassic to the Recent.*

**DESERT FOREST**
*The Petrified Forest National Park in Santa Cruz Province of Argentina has an area of 100 square km (39 square miles). Once a forest of Araucaria trees (see far right), it was buried by a rapid fall of volcanic ash during the Early Cretaceous.*

**ARAUCARIA SPECIMEN**

**MONKEY PUZZLE**
Araucaria *trees appeared in the Jurassic and are found worldwide – living species include the Monkey Puzzle and the Norfolk Island Pine. Minutely silicified fossils, such as the* Araucaria *cone shown above, were created in the Petrified Forest of Argentina (see left).*

*Araucaria miribilis*

**MODERN ARAUCARIA**

# ANGIOSPERMS

FLOWERS CHARACTERIZE angiosperm plants, of which there are over 250,000 living species. They first appeared in the Early Cretaceous, although the oldest fossil remains from that time are limited to pollen grains. The earliest angiosperm plant fossils are most similar to small bushes or small herbaceous plants. At the beginning of the Late Cretaceous, woody angiosperms began to evolve, including several modern groups, such as the magnolia, laurel, sycamore, and rose families. In general the woody angiosperms have broad leaves that are often fossilized, but fossil angiosperm wood is difficult to distinguish from fossil conifer wood. Fossil palm wood is the one exception (see *Palmoxylon*, right).

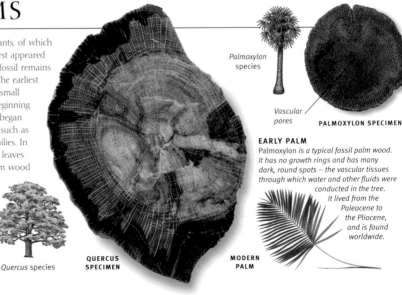

*Palmoxylon* species

Vascular pores

**PALMOXYLON SPECIMEN**

**EARLY PALM**
*Palmoxylon is a typical fossil palm wood. It has no growth rings and has many dark, round spots – the vascular tissues through which water and other fluids were conducted in the tree. It lived from the Paleocene to the Pliocene, and is found worldwide.*

**ANCESTRAL OAK**
*Quercus, the genus name for oak trees, appeared during the Eocene. Found worldwide, its fossilized wood exhibits the same characteristics as living oak.*

*Quercus species*

**QUERCUS SPECIMEN**

**MODERN PALM**

# INVERTEBRATES

EARLY FOSSIL EVIDENCE of invertebrates, the first multicellular animals, comes from the Pound Quartzite of the Ediacara Hills, north of Adelaide, South Australia. These fossils date from about 700 to 670 million years ago, during the Precambrian Period (which began about 3,800 million years ago). The fossils include soft corals, sponges, jellyfish, various sorts of worms, and yet unexplained forms. It was not until around the beginning of the Cambrian Period, 543 million years ago, that animals first began to develop hard shells. Because these are preserved more readily as fossils, we have a much clearer picture of the evolution of invertebrate life beyond that time. The rapid diversification of life during the Cambrian Period is referred to as the Cambrian Explosion, although some paleontologists think it may just represent an explosion of easily fossilized life. Evidence of land invertebrates – arthropod tracks – appears in the Ordovician.

**EDIACARAN FOSSIL**
*This Late Precambrian specimen from Australia is one of the earliest fossils of a complex animal life form.*

# FORAMINIFERA

FORAMINIFERA ARE MINUTE, sometimes microscopic, single-celled organisms, some of which secrete a shell made of calcite. They belong to the kingdom Protista. They have lived since the Cambrian in a wide range of ocean environments. There are many species of them, and they have existed in such large numbers that their shells can constitute an entire rock unit. Limestone principally made of these shells is called foraminiferal limestone, and may appear to be made from grains of rice.

**SHELLS OF THE PYRAMIDS**
*Nummulites is a relatively large foraminiferan with a shell made up of small internal chambers, rather like those of a nautilus. The pyramids of Egypt were built from Nummulitic limestone, which is found in Europe, Asia, and the Middle East.*

Shell

**ALVEOLINA SPECIMEN**

*Nummulites ghinensis*

**NUMMULITES SPECIMEN**

**GRAINS OF RICE**
Alveolina has the grain-of-rice shape typical of a number of foraminiferan species. It is a common species in foraminiferal limestones. The identification of individual foraminiferan species is difficult for the non-professional.

*Alveolina elliptica*

**ALVEOLINA ELLIPTICA**

# WORMS

MANY SOFT-BODIED invertebrate groups are loosely called worms, but fossils of worms rarely occur. What are preserved are their burrows and the tubes of calcite, secreted by some of them as dwellings, which in some areas were numerous enough to be reef-forming. Body details of some worms are preserved in the Middle Cambrian Burgess Shale of Canada, and the even earlier Ediacaran rocks of Australia.

*Rotularia bognoriensis*

*Glomerula plexus*

Coiled tube

**SEA SPIRAL**
*Rotularia was a ridged living tube coiled into a flat spiral. This worm was free-living on the sandy floors of shallow oceans. It lived during the Eocene, and is found mainly in Europe.*

**MODERN SEA WORM**

**ROTULARIA SPECIMEN**

**GLOMERULA SPECIMEN**

**TANGLED TUBES**
*The living tubes of Glomerula grew irregularly and twisted, with the tubes of a number of individuals intertwining. It was probably a filter-feeder, and the convoluted masses of worms lived unattached on the sea bed. Glomerula is found worldwide, and survived from the Early Jurassic to the Paleocene.*

# SPONGES

FOSSIL SPONGES APPEAR from the Cambrian onwards, and reached their peak during the Cretaceous. Sponges are animals that feed by filtering nutrients from water. Several classes of sponges have become extinct, but living sponges have evolved little from their earliest ancestors. As fossils, they can be mistaken for corals, but the surfaces of sponges only show the coarsest ornamentation, and their walls tend to be thicker than those of corals.

**ACTINOSTROMA SPECIMEN**

*Irregular shape*

*Concentric growth layers*

### GLASS SPONGE
Hydnoceras *represents the class of sponges sometimes called glass sponges. Their skeletons are composed of tiny, needle-like splinters of silica in a mesh-like network. Living examples are confined to deep water, but they were once found at all depths.*

**MODERN SPONGE**

*Main opening*

*Hydnoceras tuberosum*

**HYDNOCERAS SPECIMEN**

### REEF-BUILDER
Actinostroma *is an example of a stromatoporoid sponge. The skeleton was composed of calcareous splinters. They lived in large colonies and were major reef-builders. In the Silurian and Devonian they were concentrated in such numbers that their skeletons became important rock-formers.*

*Actinostroma clathratum*

# BRYOZOANS

BRYOZOANS FIRST APPEARED in the Ordovician, and more than 4,000 species are living today. These creatures appeared in the Ordovician and resemble miniature corals, and are easily confused with them. Like corals, they are colony animals but are, in fact, more closely related to brachiopods. They secrete calcareous skeletons, and each individual forms connections with its neighbours. They form free-standing colonies, or sheet-like encrustations on stones or other shells.

*Schizoretepora notopachys*

### COMPLEX FRONDS
*This* Schizoretepora *bryozoan is a colony of complexly folded fronds. It is one of several similar-appearing species, and is found worldwide.*

**SCHIZORETEPORA SPECIMEN**

### SEA MAT
Constelleria, *a sea mat, lived from the Ordovician to the Silurian, and is typical of a free-standing bryozoan, living on the sea bed. It is found in North America, Europe, and Asia.*

*Fenestella plebeia*

**FENESTELLA SPECIMEN**

### LACE CORAL
Fenestella, *a lace coral, is found both as a free-standing bryozoan and as a flat encrustation. It lived from the Silurian to the Permian and is found worldwide.*

*Constellaria antheloidea*

**CONSTELLARIA SPECIMEN**

# GRAPTOLITES

A GROUP OF SMALL, floating, aquatic, colonial animals that first appeared during the Cambrian Period, graptolites became extinct during the Carboniferous Period. They had a fingernail-like outer covering and no mineralized hard parts. They are most frequently preserved as carbonaceous impressions on black shales, although they are sometimes found in limestones. They had one or more branches, and are easily mistaken for fossil plants rather than animals. They underwent distinct evolutionary changes over time, making them very useful for age-dating rocks.

*Orthograptus intermedius*

### STRAIGHT BRANCH
Orthograptus *is a typical branch-like graptolite, exhibiting a single branch that is actually two branches back to back. It is found worldwide in rocks of the Lower to Middle Ordovician.*

**ORTHOGRAPTUS SPECIMEN**

*Triangular section*

*Coiled form*

### COILED BRANCH
Monograptus *is one of a group of single-branch graptolites with distinctive theca (the tubular structure that housed each individual). They evolved rapidly, were distributed worldwide, and are useful for dating and correlating Silurian rocks.*

**MONOGRAPTUS SPECIMEN**

*Monograptus convolutus*

# CORALS

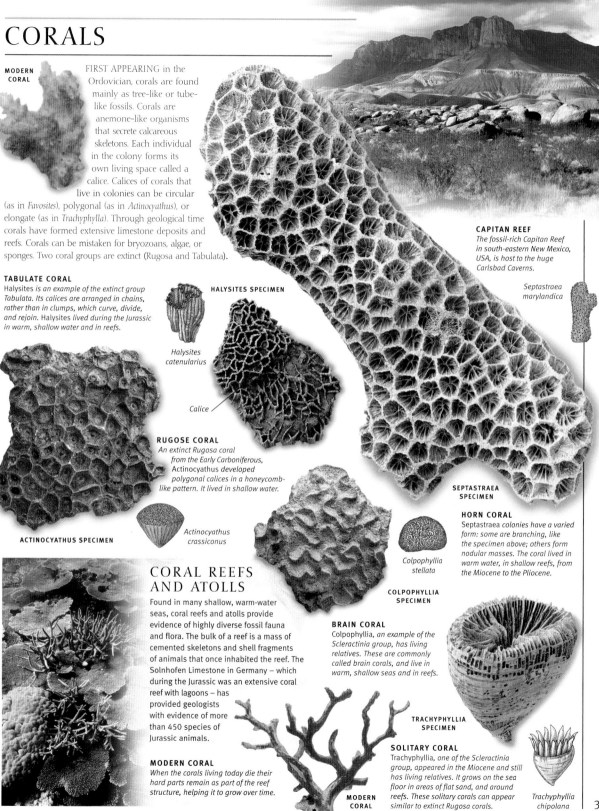

**MODERN CORAL**

FIRST APPEARING in the Ordovician, corals are found mainly as tree-like or tube-like fossils. Corals are anemone-like organisms that secrete calcareous skeletons. Each individual in the colony forms its own living space called a calice. Calices of corals that live in colonies can be circular (as in *Favosites*), polygonal (as in *Actinocyathus*), or elongate (as in *Trachyphylla*). Through geological time corals have formed extensive limestone deposits and reefs. Corals can be mistaken for bryozoans, algae, or sponges. Two coral groups are extinct (Rugosa and Tabulata).

**CAPITAN REEF**
*The fossil-rich Capitan Reef in south-eastern New Mexico, USA, is host to the huge Carlsbad Caverns.*

*Septastraea marylandica*

**TABULATE CORAL**
*Halysites is an example of the extinct group Tabulata. Its calices are arranged in chains, rather than in clumps, which curve, divide, and rejoin. Halysites lived during the Jurassic in warm, shallow water and in reefs.*

**HALYSITES SPECIMEN**

*Halysites catenularius*

Calice

**RUGOSE CORAL**
*An extinct Rugosa coral from the Early Carboniferous, Actinocyathus developed polygonal calices in a honeycomb-like pattern. It lived in shallow water.*

**ACTINOCYATHUS SPECIMEN**

*Actinocyathus crassiconus*

**SEPTASTRAEA SPECIMEN**

**HORN CORAL**
*Septastraea colonies have a varied form: some are branching, like the specimen above; others form nodular masses. The coral lived in warm water, in shallow reefs, from the Miocene to the Pliocene.*

## CORAL REEFS AND ATOLLS

Found in many shallow, warm-water seas, coral reefs and atolls provide evidence of highly diverse fossil fauna and flora. The bulk of a reef is a mass of cemented skeletons and shell fragments of animals that once inhabited the reef. The Solnhofen Limestone in Germany – which during the Jurassic was an extensive coral reef with lagoons – has provided geologists with evidence of more than 450 species of Jurassic animals.

*Colpophyllia stellata*

**COLPOPHYLLIA SPECIMEN**

**BRAIN CORAL**
*Colpophyllia, an example of the Scleractinia group, has living relatives. These are commonly called brain corals, and live in warm, shallow seas and in reefs.*

**TRACHYPHYLLIA SPECIMEN**

**MODERN CORAL**
*When the corals living today die their hard parts remain as part of the reef structure, helping it to grow over time.*

**MODERN CORAL**

**SOLITARY CORAL**
*Trachyphyllia, one of the Scleractinia group, appeared in the Miocene and still has living relatives. It grows on the sea floor in areas of flat sand, and around reefs. These solitary corals can appear similar to extinct Rugosa corals.*

*Trachyphyllia chipolana*

339

# CRUSTACEANS

CRUSTACEANS ARE carnivorous creatures belonging to the phylum Arthropoda (see panel, right). Crabs, lobsters, shrimps, and woodlice are modern crustaceans and are mainly aquatic, like their fossil ancestors. They differ from trilobites and other arthropods in that they have two pairs of appendages in front of the mouth, and three pairs of limbs near the mouth that function as jaws. The last abdominal appendage may be flattened to form a tail fan.

MODERN CRAB

*Large claw*

*Abdomen*

**PALEOCARPILIUS SPECIMEN**

### MUD CRAB
Paleocarpilius *is one of several species of crab that are found as fossils. It lived in Europe and Africa, but various other crab species are found across the world. Separated claws, legs, and shells are also common.*

*Paleocarpilius aquilinus*

### BARNACLE
*Appearing in the Eocene Period,* Balanus *still inhabits the world's oceans. Fossil barnacles are found worldwide and look much like their modern counterparts.*

*Calcareous plate*

*Balanus concavus*

**BALANUS SPECIMEN**

# TRILOBITES

ALTHOUGH NOW EXTINCT, trilobites were arthropods, a large and successful group of living organisms. Trilobites first appeared at the beginning of the Cambrian, and a few species persisted into the Permian. They had an external skeleton composed of material similar to that of the the shells of modern beetles. Most trilobites had a pair of compound eyes, although some of them were eyeless. Their sizes range from only a few millimetres to the huge *Paradoxides harlani*, which grew to more than 45cm (18in) in length. Trilobites are easily recognized by their distinctive three-segment form, consisting of a head, tail, and segmented thorax. Trilobites could roll themselves up for protection like the modern woodlouse (which is actually a crustacean), and *Phacops* is often found in a rolled-up position.

### DEFENCE POSITION
Phacops *had large eyes and lived in shallow, warm seas. This specimen was found in the defensive rolled-up position. This species is distributed worldwide.*

**MODERN WOODLOUSE**

*Phacops africanus*

*Rolled-up position*

**PHACOPS SPECIMEN**

*Thorax*

**ACADAGNOSTUS SPECIMEN**

*Acadagnostus exaratus*

### TINY TRILOBITE
*This trilobite was 8mm (⁵⁄₁₆in) long, blind, with two thoracic segments. It was found in America, Europe, and Australia, and lived in deep water.*

### STRAWBERRY HEAD
*The head shield of this trilobite earned it the name of the "strawberry-headed" trilobite. It lived worldwide in shallow seas during the Lower Ordovician to Silurian.*

**TRILOBITE BROOCH**
*This superbly fossilized trilobite has been mounted in a gold brooch.*

## ARTHROPODS

The bodies of all arthropods are covered by a shell, and it is this suit of armour that has allowed the group to become so successful. In addition, the shell fossilizes easily, providing a good fossil record. There are four subphyla of arthropods: *Trilobita*, *Uniramia* (including the class *Insecta*), *Crustacea*, and *Chelicerata*.

### BURGESS SHALE
*This shale in the Canadian Rockies is a rich source of trilobites and many other kinds of fauna.*

*Curved shell*

*Head tubercules*

*Encrinurus variolaris*

**ENCRINURUS SPECIMEN**

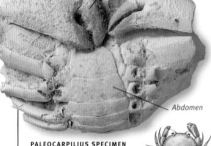

# CHELICERATES

LIKE OTHER ARTHROPODS (see panel, opposite), chelicerates have an external skeleton made of a material composed of chitin (a complex sugar) and protein. The body is segmented into a head and thorax shield, and an abdomen. The segments bear six pairs of jointed appendages – the first pair, the chelicerae, are claws for catching food; the second pair has various functions; and the remaining pairs are dedicated to movement. They lack the antennae of other arthropods. Chelicerates include eurypterids, horseshoe crabs, scorpions, and spiders. The latter two groups may have evolved from their sea-dwelling ancestors. All chelicerates are relatively uncommon in fossil form.

### HORSESHOE CRAB
Mesolimulus *lived from the Jurassic to the Cretaceous, and is an ancestor of modern horseshoe crabs. It was tolerant of changes in salinity, and laid its eggs in shallow mud-flats. This particular species is found in Europe and the Middle East.*

MESOLIMULUS SPECIMEN

*Mesolimulus walchii*

### SEA SCORPION
Paracarcinosoma *is a eurypterid, a scorpion-like sea creature from the Silurian, and a contemporary of trilobites. It has been suggested that the decline in trilobite species was due to the eurypterids. They are found in Europe, North America, and Asia, and lived in the sea, and in fresh to brackish water.*

*Paired appendages*

*Abdomen*

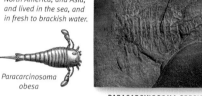

*Paracarcinosoma obesa*

PARACARCINOSOMA SPECIMEN

*Graeophonus analicus*

### EARLY SPIDER
Graeophonus *is a close relative of modern spiders. Its eyes were raised from its head on distinct knobs. It is found in North America and Europe, where it flourished in shallow-water habitats such as swamps and lagoons.*

GRAEOPHONUS SPECIMEN

# INSECTS

INSECTS LIVE BOTH ON land and in fresh water, and are distinguished from other arthropods (see panel, opposite) by having three pairs of legs. The earliest insects appeared in the Devonian, and were followed in the Permian by giant insects such as the dragonflies of the order Protodonata, some of which had a wing-span of more than 50cm (20in). Damselflies and mayflies evolved around this time, and still exist today. The first insects were wingless, but by the Upper Carboniferous wings had evolved, providing the first powered flight. Most insect fossils are identifiable if well preserved.

*Wing imprint*

### DRAGONFLY
Petalura *is a typical dragonfly, found in Europe and Australasia. It is a carnivorous insect – nymphs and adults feed on small aquatic and flying animals. Petalura is found in rocks from the Jurassic to the Recent.*

*Petalura species*

PETALURA SPECIMEN

*Abdomen*

MODERN DRAGONFLY

### WATER BEETLE
Hydrophilius *is a type of insect, commonly called a water beetle, that survives today. It evolved in the Pleistocene. The beetle feeds on plant and animal matter, and its larvae are carnivorous. This specimen is preserved in tar sand.*

### COCKROACH
Cockroaches have been around for a very long time. This specimen is from the Lower Carboniferous and was found in the UK.

*Embedded beetle*

*Tar sand*

*Hydrophilius species*

HYDROPHILIUS SPECIMEN

*Archimylacris eggintoni*

*Ironstone nodule*

ARCHIMYLACRIS SPECIMEN

341

# BRACHIOPODS

ABUNDANT AS FOSSILS, brachiopods appeared at the beginning of the Cambrian Period, and a few species are still living today. More than 35,000 species are known, with more being discovered each year. In some Paleozoic rocks they are the most abundant fauna. Brachiopods are bottom-feeding marine animals composed of two shells, or valves. Viewed from above, the shells are symmetrical. Most fossil brachiopods have a small hole at the central tip of the bottom valve. Here a fleshy stalk, called a pedicle, emerged, with which the brachiopod attached itself to the sea floor. There are two main groups within the phylum Brachiopoda: the Articulata, the shells of which are hinged, and the hingeless Inarticulata.

Convex shell

### SMOOTH SHELL
Pentamerus *inhabited shallow water, often in huge numbers. It is from the Silurian, and is found worldwide.*

Opening for pedicle

**PENTAMERUS SPECIMEN**

Pentamerus oblongus

### LAMP SHELL
Digonella *and brachiopods of similar appearance are sometimes referred to as lamp shells. Found worldwide, the animal is from the Middle Jurassic. It lived in shallow waters, where it attached itself to soft mud or fragments of other shells.*

**RIBBED SHELL**
Platystrophia *is an Ordovician brachiopod found worldwide. It lived on lime mud and sandy sea floors.*

Platystrophia biforata

**PLATYSTROPHIA SPECIMEN**

**DIGONELLA SPECIMEN**

Digonella digona

### LAMP SHELLS EMBEDDED IN ROCK
Digonella *and other lamp shells are often found fossilized in clusters, half-buried in fine-grained, muddy limestones.*

Sharp rib

### BUTTERFLY SHELL
*Perhaps the most recognizable of all brachiopods, the spirifers are sometimes called butterfly shells.* Mucrospirifer *is from the Devonian, lived on soft mud, and is found worldwide.*

Deep fold

Overlapping growth layers

**MUCROSPIRIFER SPECIMEN**

Mucrospirifer mucronata

Chonetes species

Short spines

### LITTLE RELATIVE
Chonetes *is a comparatively small brachiopod, with a thin shell. It lived from the Carboniferous to the Permian, and is distributed worldwide. It lived on soft lime mud.*

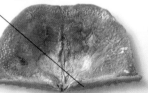

**CHONETES SPECIMEN**

**LEPTAENA SPECIMEN**

### WRINKLED SHELL
*An Ordovician brachiopod,* Leptaena *is usually found in fine-grained, limey shales. Its wrinkles helped to anchor it in sand. It is found worldwide.*

Leptaena species

**FARO BEACH, GOTLAND**
*Gotland, a Swedish island in the Baltic Sea, is a rich source of brachiopod fossils.*

# SCAPHOPODS AND CHITONS

*Helminthochiton turnacianus*

SCAPHOPODS AND CHITONS are two classes of mollusc (see panel, p.344). Scaphopods had tapering, tube-like shells, which were open at both ends. The head and foot of the animal occupied the larger end, which was buried in sea-floor sediment. Chitons had a shell of eight overlapping segments, similar to trilobites. Underneath they had a foot that exerted a powerful grip on hard surfaces.

**HELMINTHOCHITON SPECIMEN**

*Dentalium sexangulum*

**DENTALIUM SPECIMEN**

**TOOTH-LIKE SHELL**
*Dentalium is a typical scaphopod. It appeared in the Middle Triassic, but still has living relatives. It lived in a fairly wide range of ocean-floor environments.*

**COAT-OF-MAIL SHELL**
*Helminthochiton is an Ordovician to Carboniferous chiton that lived in warm seas on algae-covered debris. It is usually found as segments in Europe and North America.*

# BIVALVES

THESE MOLLUSCS (see panel, p.344) are composed of two valves connected by a hinge of organic material. The valves generally have interlocking teeth along the line of the hinge. They differ from brachiopods in that the valves are not symmetrical, and they lack a pedicle hole. There are many different kinds of bivalves, and they are classified by their hinges. Most bivalves were filter-feeders, and many burrowed into sediment, stone, or wood. Others attached themselves by means of their foot to other submerged objects.

*Pinna hartmannii*

**PEN SHELL**
*Also known as the pen shell or fan mussel, Pinna appeared in the Early Carboniferous and is still present. It lives in groups, with its pointed end buried in the sea-floor sediment.*

**PINNA SPECIMEN**

**FIMBRIA SPECIMEN**

*Net ornament*

**BASKET SHELL**
*Found worldwide, examples of Fimbria still live today. This bivalve appeared in the Middle Jurassic, living in warm, shallow seas. Its ornamentation gives its surface a net-like appearance.*

*Fimbria subpectunculus*

**PIERRE SHALE**
*The Cretaceous Pierre Shale of western USA contains countless numbers of bivalve fossils.*

**GRYPHEA SPECIMEN**

*Gryphea arcuata*

**DEVIL'S TOENAIL**
Gryphea *is popularly called the devil's toenail. It is a thick, curled shell, in which the smaller valve fits like a lid into the larger valve. It lived from the Lower Triassic to the Lower Jurassic, on muddy sea beds.*

**GLYCYMERIS SPECIMEN**

*Glycymeris brevirostris*

**BITTERSWEET CLAM**
Glycymeris *developed a thick shell, with weak growth ribs and ornamentation. The area of the hinge is grooved and triangular. It lived in shallow waters, burrowing in sand and mud.*

*Ligament pit*

**SCALLOP FOREBEAR**
*The strong radiating ribs and fan shape are the most noticeable features of Oxytoma. It lived from the Lower Triassic to the Lower Cretaceous, dwelled in a variety of marine environments, and is found worldwide.*

**OXYTOMA SPECIMEN**

*Oxytoma*

**PECTEN SPECIMEN**

*Pecten beudanti*

**SCALLOP**
Pecten *appeared in the Eocene, and today's scallop species are living relatives. It has a large, circular shell, with broad, radiating ribs. Its right valve is convex, while its left valve is concave. Scallops prefer moderately shallow waters, and a clean, sandy sea bed.*

343

# GASTROPODS

FIRST APPEARING IN the Cambrian Period, gastropods have developed into the most successful of all molluscs (see panel, below). There are over 60,000 living species, inhabiting marine, freshwater, and terrestrial environments. The living animal has a head with eyes and a mouth, and a large foot for crawling. All of these disappear during fossilization, so classification is based on shell characteristics. In general, gastropods have a single, spiral, asymmetric shell principally composed of aragonite, although some have calcite interlayering. A few species show no spiralling. Most gastropods are easily recognizable as such, although a few, such as *Platyceras*, can somewhat resemble certain bivalves.

**MODERN CONUS**

*Conus sauridens*

**CONUS SPECIMEN**

### CONICAL CARNIVORES
Cone shells, represented here by Conus, *first appeared in the Lower Cretaceous and are still living. They are carnivores, eating worms, other molluscs, and even fish. They live in a variety of ocean environments, and are found worldwide.*

*Platyceras haliotis*

### SEA SNAIL
Platyceras, *which lived worldwide from the Silurian to the Early Carboniferous, existed in close cooperation with crinoids, and has been found fossilized connected to them. Its shell is tightly wound for the first few coils, then rapidly spreads outwards.*

**PLATYCERAS SPECIMEN**

### TURRITELLA
*Commonly known as turritella, this gastropod first appeared in the Lower Eocene, and has modern relatives. A large deposit of silicified Archimediella occurs in Wyoming, USA, where it is called Turritella Agate.*

*Archimediella pontoni*

Square opening

**ARCHIMEDIELLA SPECIMEN**

**CALLISTOMA SPECIMEN**

**TOP SHELL**
*Cone-shaped Callistoma first appeared in the Early Cretaceous. It is informally called a top shell. It is extensively ornamented with granulated spiral cords. Callistoma remains an inhabitant of rocky shorelines worldwide.*

Granulated whorl

*Callistoma nodulosum*

Sharp-edged whorl

Flattened whorl

**EUOMPHALUS SPECIMEN**

*Euomphalus pentangulus*

### RIDGED DISC
Euomphalus *lived from the Silurian to the Middle Permian. It has a disc-shaped, coiled shell with a slightly raised spire. The whorls are stepped, creating a sharp ridge between them. This sea snail lived on marine vegetation, and is found worldwide.*

### ROUNDED SPIRE
Bourguetia *lived from the Middle Triassic to the Lower Jurassic. Although similar in appearance to land snails, it lived among coral reefs in shallow, warm seas. Fossils of this sea snail are found principally in Europe and New Zealand.*

*Bourguetia semanni*

Whorls regularly increasing in size.

**BOURGUETIA SPECIMEN**

Long spire

## MOLLUSCS

The phylum Mollusca comprises a large and varied group of animals that, for the most part, have lived in salt water. A few moved into fresh water, and an even smaller group, the snails, moved onto the land. Most molluscs have calcium carbonate shells. There are six main groups, or classes, of molluscs: monoplacophorans, chitons, scaphopods, bivalves, gastropods, and cephalopods. All mollusc groups appeared during the Cambrian,

apart from the scaphopods, which appeared during the Ordovician. By the end of the Cambrian, molluscs were to be found everywhere and in great variety. There are currently about 75,000 living species of molluscs, thriving in a great range of salt-water and freshwater habitats.

### MUSSELS
*For centuries people have collected mussels and consumed the soft-bodied animals inside them.*

# NAUTILOIDS AND AMMONOIDS

THESE TWO GROUPS OF CEPHALOPODS are among the most advanced molluscs (see panel, opposite). Ammonoids evolved from nautiloids, which appeared during the Ordovician. Only a single genus of nautiloid survives today, the pearly nautilus. Nautiloids and ammonoids are both characterized by a series of internal chambers that permitted the animal to control its own buoyancy. All the chambers were connected by a central stem called a siphuncle. Most were good swimmers, and were scavengers, carnivores, or both. In general, nautiloids can be distinguished from ammonoids by the pattern formed by their sutures, the joins between their chambers. In nautiloids, the line of the sutures is straight or gently curving. In ammonoids it is highly complex, often appearing fern-like. The ammonoids emerged in the Devonian, and became extinct at the end of the Cretaceous. They evolved rapidly, and are good fossils for dating Late Paleozoic and Mesozoic rocks.

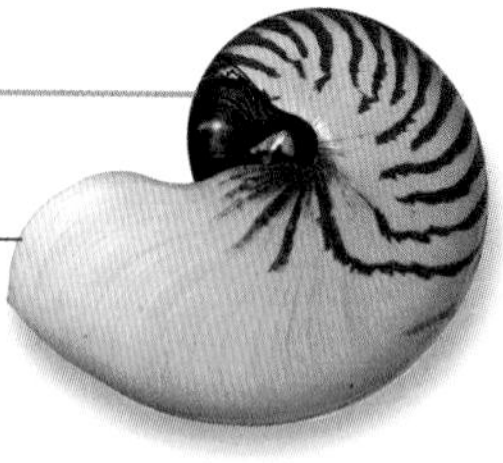

**CENOCERAS SPECIMEN**

*Cenoceras* species

**MODERN CHAMBERED NAUTILUS**

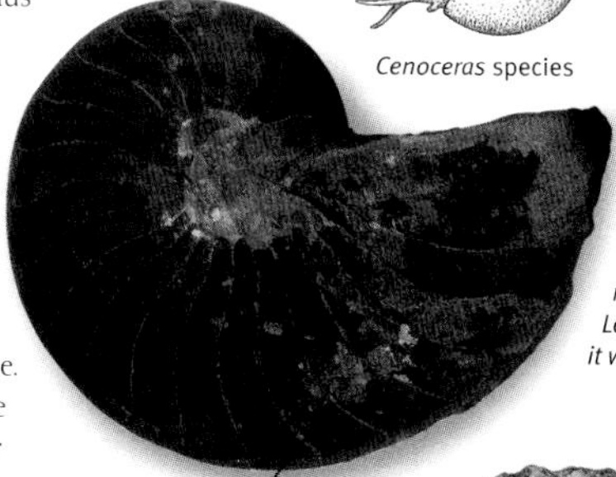

### NAUTILOID SIMPLICITY
Cenoceras *is a nautiloid, and exhibits the straight or gently curving lines of its sutures. A marine animal that lived from the Lower Triassic to the Middle Jurassic, it was a bottom-feeder.*

Rounded shell

*Mantelliceras* species

### PATTERN OF JOINS
*The ammonoid* Phylloceras *illustrates the elaborate suture patterns developed by the ammonoids. Its external ornamentation was simple, but the polished internal mould (right) shows the complex suturing of its chambers. It lived from the Early Jurassic to the Lower Cretaceous, and its streamlined shape suggests it was a good swimmer.*

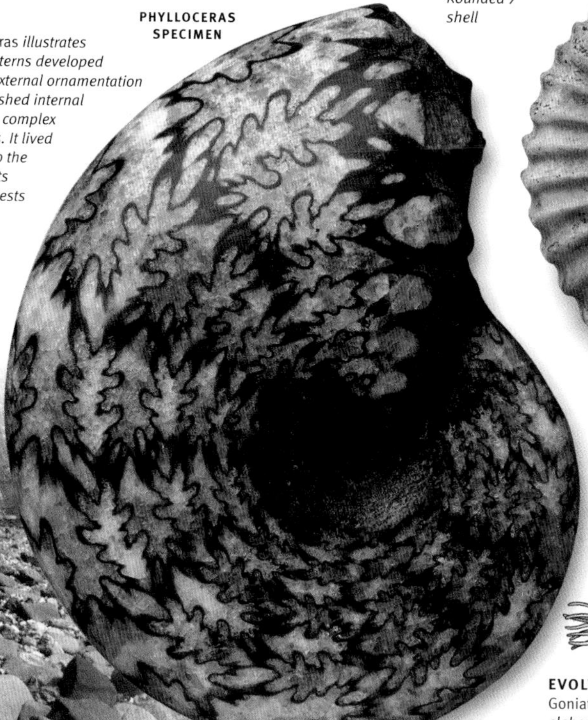

**PHYLLOCERAS SPECIMEN**

*Phylloceras* species

**MANTELLICERAS SPECIMEN**

### PROMINENT RIBS
Mantelliceras *shows the heavier ribbing developed by many ammonoids. A European species is shown above, but others of similar appearance, some with heavier and more elaborate ribbing, evolved in other parts of the world during the Cretaceous.*

*Goniatites crenistria*

### EVOLVED SUTURING
Goniatites *shows the increasing elaboration of suturing present in intermediates between nautiloids and ammonoids. It is distributed worldwide, and lived during the Early Carboniferous. It was a denizen of shallow, shelving seas and reefs.*

### POPULAR COLLECTION SITE
*The beaches at Lyme Regis, England, are littered with rocks bearing thousands of ammonite specimens.*

**GONIATITES SPECIMEN**

**BACCULITES SPECIMEN**

*Bacculites* species

### STRAIGHT AMMONOID
*In effect,* Bacculites *was an uncoiled ammonoid, since only the earliest portion of its shell was coiled. It occurred in huge numbers in some localities, almost to the exclusion of any other ammonoid species. Some individuals grew to more than 1m (3ft) in length. It lived during the Lower Cretaceous.*

345

# BELEMNOIDS AND SQUID

BELEMNOIDS AND SQUID are members of the cephalopod group of molluscs (see panel, p.344) that includes today's cuttlefish, octopus, and squid. Belemnoids are found particularly frequently as fossils because they had a hard internal shell made from calcite, known as the guard. While on rare occasions soft parts are found fossilized, the guard is usually all that remains. Belemnoids first appeared during the Early Carboniferous Period and became extinct during the Eocene.

*Pachyteuthis abbreviata*

U-shaped chamber

Guard

### PACHYTEUTHIS SPECIMENS

Pyrite replacement

### GUARDED PREDATOR
Pachyteuthis *is a typical belemnite, and its guard is found fossilized worldwide. It lived from the Middle to Late Jurassic, and was a scavenger or predator. The U-shaped hole at the bottom of the guard housed the chambered internal shell that helped to regulate buoyancy.*

### SLIM SCAVENGER
Belemnitella *is a Late Cretaceous belemnite that inhabited shallow waters. Its guard tapers gently, and has a pointed tip. It is often found grouped in large numbers.*

### MODERN SQUID
*A close living relative of the belemnites, the squid has an internal support, the pen, made of chitin, a horn-like substance.*

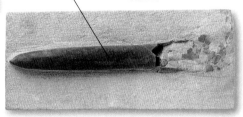

Impression of veins on guard surface

*Belemnitella mucronata*

**BELEMNITELLA SPECIMEN**

# ECHINOIDS

COMMONLY CALLED SEA URCHINS and sand dollars, echinoids are a group of echinoderms (see panel, right) with a rigid, globular skeleton made of thin calcite plates. Many possessed spines, also made of calcite, attached to the body by a ball-and-socket arrangement. The spines were used for defence and, in some cases, walking. Some echinoids were bottom-feeders, foraging on the sea bed; others tended to burrow into soft silt layers. Spines are often found fossilized separately.

*Temnocidaris sceptrifera*

Detached spine

### LAKE TEXOMA
*Many Lower Cretaceous echinoids are found around this lake on the border of Texas and Oklahoma, USA.*

Tubercle

## ECHINODERMS

Characterized by their hard, spiny covering or skin and their five-fold radial symmetry, the phylum Echinodermata appeared as fossils in the Lower Cambrian. They evolved rapidly into many species in eight different classes, including the echinoids, asteroids, crinoids, ophiuroids, blastoids, and cystoids. They are well represented in the fossil record, with approximately 13,000 fossil species so far described.

### MODERN ECHINOID
*Today's spiny sea urchin is a living relative of species now known only from fossils.*

### SPINY SEA URCHIN
Temnocidaris *is a Lower Cretaceous spiny echinoid found mainly in Europe. It has five radially arranged plates with large tubercles (the tubercles are the attachment points for the spines).*

### STARFISH
*Modern starfish species belong to one class of living echinoderms.*

### SEA URCHINS FOSSILIZED IN CHALK
*These superb examples of the Cretaceous sea urchin* Tylocidaris *retain many of their club-like spines.*

Attached spine

**TEMNOCIDARIS SPECIMEN**

**ENCOPE SPECIMEN**

*Encope micropora*

### SAND DOLLAR
Encope *appeared in the Miocene and is still a living species. Commonly called a sand dollar, it lives in tropical waters and shows the bilateral symmetry common in burrowing echinoids.*

**HEMIASTER SPECIMEN**

*Hemiaster batnensis*

### HEART URCHIN
Hemiaster *is representative of a group of echinoids commonly called heart urchins. It appeared in the Cretaceous, is found worldwide, and has living relatives. It buried itself in mud and lived by filter-feeding organic material from the mud.*

# ASTEROIDS

POPULARLY CALLED STARFISH, asteroids are a group of echinoderms (see panel, opposite), recorded as far back as the Ordivician. They usually have five arms, although a few species have more. The body has relatively little calcified material in its makeup, so entire examples preserved as fossils are very rarely found.

*Pentasteria cotteswoldiae*

**BURROWER**
*Pentasteria, an asteroid of the Jurassic to Middle Cretaceous, is found principally in Europe. It probably burrowed into shallow sand.*

**PENTASTERIA SPECIMEN**

# CRINOIDS

**SAGENOCRINITES SPECIMEN**

ALSO KNOWN AS SEA LILIES, crinoids are a group of echinoderms (see panel, opposite) that usually possess a cup-shaped body and five or more feathery arms. They are attached to the sea bed by a flexible stem made up of disc-like plates that are either circular or pentagonal in cross section. The cup is also made of a number of loosely connected plates. The cup and the stem tend to fall apart after death, so crinoid fossils are most often found as segments of stems and arms, and as disarticulated cup-plates. Crinoid skeletal components are made of calcite, and fossilize well. Complete crinoids are rare and beautiful fossils. Crinoids evolved rapidly into a large number of now-extinct species, and thus are good index fossils for dating rocks, particularly those of the Paleozoic Era.

**MODERN SEA LILY ON A SPONGE**

**SEA LILY**
*Sagenocrinites is found in Europe and North America. This Middle Silurian crinoid shows all the basic components of crinoid skeletons: the cup-shaped body, arms, and stem.*

*Sagenocrinites expansus*

# OPHIUROIDS

ALTHOUGH THEY ARE starfish-like in appearance, ophiuroids are more closely related to the echinoids. Commonly called brittle stars, their bodies are flat and disc-like, with five flexible arms different in structure to those of the asteroids. Living ophiuroids are found in huge masses on the sea floor, and fossil remains suggest that the same was true of earlier species. In areas with a strong current, they use some of their arms to link to each other, and other arms to capture food. Since their arms are long and fragile, it is rare to find complete fossilized specimens.

**BRITTLE STAR**
*Geocoma is a Jurassic ophiuroid found worldwide. Its habitat was lime mud. The arms are unusually long and covered by fine spines.*

*Geocoma carinata*

Long arms

Small central disc

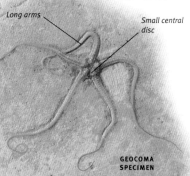

**GEOCOMA SPECIMEN**

# BLASTOIDS

AN EXTINCT CLASS of echinoderm (see panel, opposite), the blastoids existed from the Middle Ordovician to the Late Permian. They were sedentary animals that, rather like the crinoids, remained anchored to the sea floor by a stem consisting of a column of linked, circular plates. Blastoids had a highly regular structure; their body region consisted of three circles of five plates, made of calcium carbonate and shaped like a rosebud. Many blastoid species are useful as index fossils. While they can be mistaken for crinoids, blastoids lack attachments for arms, and their bodies are often found intact.

**SEDENTARY ANIMAL**
*Pentremites is one of the most common, widespread, and easily recognizable blastoids. It is abundant in Carboniferous marine sediments in the USA. It lived attached to hard layers on the sea-bed.*

Openings

**PENTREMITES SPECIMEN**

*Pentremites pyriformis*

Broad, biconical form

# CYSTOIDS

LIVING FROM the Ordovician to the Devonian, the cystoids are an extinct group of echinoderms (see panel, opposite). They bear a superficial resemblance to crinoids but lack the crinoids' true arms. They fed by filtering water through short limbs called brachioles. Anchorage to the sea floor was achieved by wrapping their short, tapering stems around any convenient fixed object. Useful index fossils for the Ordivician, cystoids are relatively uncommon fossils.

Body interior

*Pleurocystes rugeri*

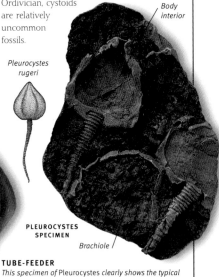

**PLEUROCYSTES SPECIMEN**

Brachiole

**TUBE-FEEDER**
*This specimen of Pleurocystes clearly shows the typical components of a cystoid: the short, tapered stem, the two brachioles (feeding tubes), and the large plates that make up the theca (body).*

# VERTEBRATES

**GEORGES CUVIER**
*This French naturalist (1769–1832) first identified vertebrates as a distinct group.*

THE EARLIEST FOSSILS of vertebrates (a sub–phylum of the phylum *Chordata*) are microscopic scale–bones belonging to the jawless fish (agnathans), which proliferated in the Silurian and Devonian. Bone, cartilage, and the enamel–like substance that forms teeth first occurred in fish. The modification of these substances in later fish contributed to the evolution of the first amphibians, able to live on dry land. Amphibians appeared in the Devonian, followed by the first full-time land dwellers, the reptiles,

early in the Carboniferous. In the Triassic, one major branch of reptiles gave rise to mammals, while another brought giant reptiles, flying and aquatic reptiles, lizards, and snakes. Mammals were therefore present concurrently with dinosaurs in the Cretaceous, but only became prominent after their demise at the end of that period. Fossils of land animals are less common than those of marine vertebrates because natural opportunities for fossilization, which depends on burial soon after death, occur much less often.

## FISH

THE FIRST SKELETAL REMAINS of fish occur in the Devonian. Fish fossils can range from entire skeletons to scale-bones or teeth. In some localities, such as the vast Eocene lakes of western North America, literally millions of fish were killed on different occasions as a result of volcanic ash falls. In some former lake-bottom sediments, a dozen or more complete sardine-sized fish skeletons can be recovered from a single square metre of shale. Fish teeth of various species are commonly found as fossils, such as the examples from *Ceratodus* illustrated below. As with other animal groups, the fossil record for fish is still incomplete.

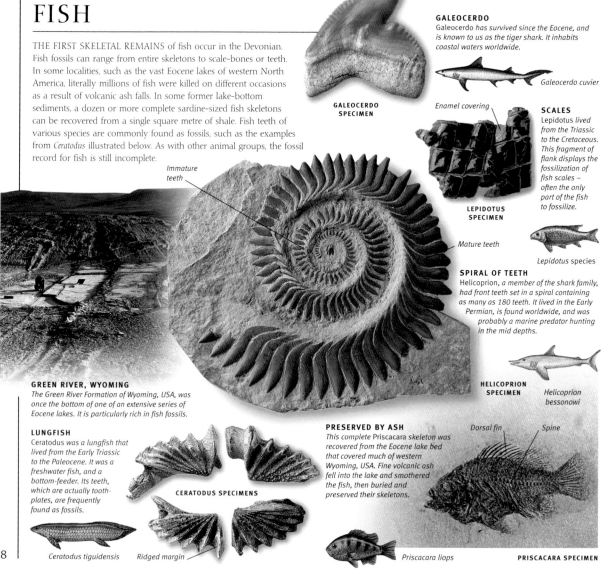

**GALEOCERDO SPECIMEN**

**GALEOCERDO**
*Galeocerdo has survived since the Eocene, and is known to us as the tiger shark. It inhabits coastal waters worldwide.*

*Galeocerdo cuvier*

*Enamel covering*

**SCALES**
*Lepidotus lived from the Triassic to the Cretaceous. This fragment of flank displays the fossilization of fish scales – often the only part of the fish to fossilize.*

**LEPIDOTUS SPECIMEN**

*Lepidotus species*

*Immature teeth*

*Mature teeth*

**SPIRAL OF TEETH**
Helicoprion, *a member of the shark family, had front teeth set in a spiral containing as many as 180 teeth. It lived in the Early Permian, is found worldwide, and was probably a marine predator hunting in the mid depths.*

**HELICOPRION SPECIMEN**

*Helicoprion bessonowi*

**GREEN RIVER, WYOMING**
*The Green River Formation of Wyoming, USA, was once the bottom of one of an extensive series of Eocene lakes. It is particularly rich in fish fossils.*

**LUNGFISH**
Ceratodus *was a lungfish that lived from the Early Triassic to the Paleocene. It was a freshwater fish, and a bottom-feeder. Its teeth, which are actually tooth-plates, are frequently found as fossils.*

**CERATODUS SPECIMENS**

*Ceratodus tiguidensis*

*Ridged margin*

**PRESERVED BY ASH**
*This complete* Priscacara *skeleton was recovered from the Eocene lake bed that covered much of western Wyoming, USA. Fine volcanic ash fell into the lake and smothered the fish, then buried and preserved their skeletons.*

*Dorsal fin*

*Spine*

*Priscacara liops*

**PRISCACARA SPECIMEN**

# AMPHIBIANS AND REPTILES

IN THE UPPER Devonian, amphibians became the first vertebrates to adapt for a living on the land. Resembling today's salamanders in appearance, these amphibians still relied on water for laying and hatching their eggs. The next evolutionary step occurred in the Carboniferous, with the emergence of reptiles. Having developed waterproof eggs, in addition to body scales that prevented dehydration, many reptiles were freed from the need to remain in a wet environment. Amphibian and reptile fossils are most likely to be fragmentary, consisting of solid bones or individual teeth, but in rare instances fairly complete skeletons or eggs are discovered.

### MODERN FROG
*Frogs first appeared in the Mesozoic; modern frogs were to evolve during the Eocene.*

### FROG FOREBEAR
*This Rana fossil is a rare, complete amphibian skeleton. Rana appeared in the Eocene, and is still living. It is found worldwide, except in Australasia.*

**RANA SPECIMEN**

*Rana pueyoi*

### TRICERATOPS SPECIMEN

### THREE HORNS
*One of the best known of all dinosaurs, the Lower Cretaceous, three-horned Triceratops was the most advanced of the horned dinosaurs.*

*Triceratops prorosus*

### TURTLESHELL
*Well-preserved fossil shells of turtles and tortoises are often found. The marine turtle Cimochelys lived during the Lower Cretaceous and fossils are found principally in Europe.*

*Cimochelys benstedi*

**CIMOCHELYS SPECIMEN**

# BIRDS

IT IS NOW GENERALLY agreed that birds evolved from one of the families of carnivorous dinosaurs, although exactly which is still unclear. The first identifiable bird remains are found in the Late Jurassic, although the main explosion in bird species did not occur until the Cretaceous, when many reptiles became extinct. Today there are more than 8,600 bird species. As bird bones are hollow, thin-walled, and fragile, fossils are relatively rare.

### DAWN BIRD
*Hesperornis, a flightless, diving, dawn bird, lived during the Late Cretaceous. It ate fish, and lived around warm seas.*

**HESPERORNIS SPECIMEN**

*Hesperornis regalis*

Nostril

### DODO
*Raphus, commonly called the dodo, became extinct in the 17th century.*

**RAPHUS SPECIMEN**

*Raphus cucullatus*

# MAMMALS

THE FIRST MAMMALS appeared during the Cretaceous Period, having evolved from therapsid reptiles. The explosion in mammalian species did not take place until after the extinction of the dinosaurs and other reptiles at the end of the Cretaceous. Fossilized mammal bones are hollow and generally thick-walled. Mammal teeth are multi-faceted, enamel-covered, and have bony roots. Many fossil mammals can be identified from teeth alone.

*Ursus speleus*

**URSUS SPECIMEN**

### BEAR BONES
*The name Ursus refers to a group that includes cave, brown, and grizzly bears. It evolved during the Pliocene and is still around today. Caves in Europe, in particular, are very rich in fossilized cave bear remains.*

*Homo habilis*

**HOMO HABILIS SPECIMEN**

### EARLY MAN
*Homo habilis represents a middle stage in the evolution of Homo sapiens, the modern human. Homo habilis appeared during the Lower Pliocene and disappeared during the Lower Pleistocene. Remains of this hominid are found in Africa.*

### OLDUVAI GORGE, TANZANIA
*Paleontologists have found the fossilized remains of more than 50 hominids in rocks of the Olduvai Gorge in northern Tanzania.*

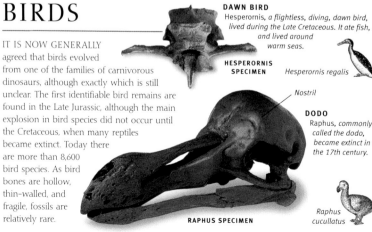

# THE DINOSAUR QUARRY

FIRST DISCOVERED IN 1909 AND DECLARED A NATIONAL MONUMENT IN 1915, THE DINOSAUR QUARRY IN THE WESTERN USA IS ONE OF THE RICHEST SOURCES OF DINOSAUR BONES IN THE WORLD, CONTAINING 1,600 BONES FROM 11 DIFFERENT DINOSAUR SPECIES.

**AERIAL VIEW OF THE QUARRY**
*The Quarry is located in the high desert country straddling the north-east border of the American state of Utah and the north-west border of Colorado. Part of the Morrison Formation, it was once a river channel in which the bones of dead dinosaurs were buried.*

On 17 August 1909, Earl Douglass, a paleontologist working for the Carnegie Museum of Pittsburgh, Pennsylvania, was walking along the small drainage channels of the Uinta Basin, searching for dinosaur skeletons. His search was rewarded. On that day he found eight still-articulated tail vertebrae of a herbivorous dinosaur called Apatosaurus, weathering from the rock. The excavation that followed revealed one of the most complete Apatosaurus skeletons ever discovered. Further excavation in the same area over the next 75 years uncovered some 1,600 bones from 11 different dinosaur species. This rock outcrop later became known as the Quarry, and it became the centrepiece of the Dinosaur National Monument.

The Quarry is an uptilted slab of sandstone belonging to the Morrison Formation. Dinosaur remains have been found in this formation in a number of localities in the American West. The rocks of the Morrison Formation were laid down as sediments about 150 million years ago, in the middle of the Jurassic Period, also known as the Age of Dinosaurs. The sediment in the Quarry area was sand and gravel from a fast-moving stream that flowed through the area. Dead dinosaurs were carried downstream and deposited in the river channel, where they were buried by rapidly accumulating sediments. Some of the dinosaur bodies were still intact, others were little more than scattered bones.

In the 150 million years that followed, the river disappeared, and the whole of what is now northern Colorado and Utah, including the dinosaur bone-bed, was buried under deep accumulations of sediment. The sand of the river channel solidified into rock, and the original bone material was replaced by mineral matter. Then, the sedimentary rock containing the fossil bones was tilted upwards by huge earth movements across the whole region. Eventually, weathering wore down and exposed the bone-bed, revealing the vertebrae at the surface for Douglass to discover.

**UNCOVERING DINOSAUR BONES IN THE ROCK WALL**
*The rock layer containing the dinosaur bones first discovered by Earl Douglass today forms one wall of the Quarry Visitor Centre. Paleontologists continue to chip away at the rock to uncover the bones.*

**DINOSAUR NATIONAL MONUMENT**
*Around three-quarters of all the dinosaur bones found at the Quarry belong to the long-necked, long-tailed herbivores, known as the Sauropods. The Apatosaurus belongs to this family. It was 21–23m (70–75ft) in length and weighed about 34 tonnes.*

# GLOSSARY

## A

**ACCESSORY MINERALS** The mineral constituents of a rock that occur in such small amounts that they are disregarded in its definition.

**ACCRETION** The drawing together through gravitational attraction of particles of various sizes to form planets.

**ACICULAR** Needle-like, referring to the needle-shaped habit of some minerals.

**ACID ROCK** See FELSIC ROCK.

**ADAMANTINE LUSTRE** A type of bright mineral lustre, similar to that of diamond.

**AMPHIBOLE GROUP** A group of common rock-forming minerals, often with complex composition but mostly ferro-magnesian silicates.

**AMYGDALE** A secondary in-filling of a void in an igneous rock. Minerals that occur as amygdales include quartz, calcite, and the zeolite group.

**ASSOCIATED MINERALS** Minerals found growing together, but not necessarily intergrown.

**AUREOLE** The area surrounding an igneous intrusion where contact metamorphism has occurred.

## B

**BASAL CLEAVAGE** Cleavage that occurs parallel to the basal crystal plane of a mineral.

**BASIC ROCK** See MAFIC ROCK.

**BATHOLITH** A huge, irregularly shaped mass of igneous rock formed from the intrusion of magma at depth.

**BEDDING** Layering of sedimentary rocks. Beds or strata are divided by bedding planes.

**BLADED** Wide and flat, describing the habit of some minerals.

**BRECCIA** A sedimentary rock made up of angular fragments.

**BRILLIANT CUT** A gemstone cut that is round with triangular facets top and bottom.

## C

**CABOCHON** A gemstone cut with a domed upper surface and a flat or domed under surface; stones cut in this way are said to be cut en cabochon.

**CAST** A fossil formed by the infilling of a mould.

**CHATOYANCY** The cat's-eye effect shown by some stones cut en cabochon.

**CHITIN** The horny substance that forms the shell of arthropods.

**CLAST** A fragment of rock, especially when incorporated into a sedimentary rock.

**CLAY** Mineral particles smaller than about 0.002mm (0.00008in).

**CLEAVAGE** The way certain minerals break along planes dictated by their atomic structure.

**COLUMNAR** Like columns, as in rocks formed by the growth of crystals in elongated prisms.

**CONCHOIDAL FRACTURE** A curved or shell-like fracture in many minerals and some rocks.

**CONCRETIONS** Commonly discrete, rounded, nodular rock masses formed in beds of shale or clay.

**CONVECTION** The movement and circulation of fluids (such as molten rock) in response to temperature differences.

**CRYPTOCRYSTALLINE** Crystalline, but very fine-grained. Individual components may be seen only under a microscope.

**CUSHION CUT** A gemstone cut that is roughly square, but with rounded sides and corners.

**CUT** The process of shaping a gemstone by grinding and polishing; the shape of the final gem, as, for example, in EMERALD CUT.

## D

**DENDRITIC** With a tree-like habit.

**DERMAL ARMOUR** The bony armour located in the skin of some vertebrates.

**DETRITAL** Referring to sedimentary rocks: formed essentially of fragments and grains derived from existing rocks.

**DIFFRACTION** The splitting of light into its component colours.

**DISCOVERY / TYPE LOCALITY** The site where a mineral was first recognized as a new mineral.

**DOUBLE REFRACTION** The splitting of light into two separate rays as it enters a stone.

**DULL LUSTRE** A shine that has minimal reflectiveness.

**DYKE** or **DIKE** A sheet-shaped igneous intrusion that cuts across existing rock structures.

## E

**EARTHY LUSTRE** A non-reflective, mineral lustre.

**EMERALD CUT** A gemstone cut that is an elongated octagon, with trapezoidal facets top and bottom.

**EROSION** The wearing away of the material on the surface of the Earth..

**ERUPTION** A discharge of lava, pyroclastic material, or gas from a volcanic cone or vent.

**EVAPORITE** A mineral or rock formed by the evaporation of saline water.

**EXTRUSIVE ROCK** A rock formed from lava that flowed onto the Earth's surface, or was ejected as pyroclastic material.

## F

**FACES** The external flat surfaces that make up a crystal's shape.

**FACET / FACETING** The cutting and polishing of multiple flat surfaces (facets) on a gemstone.

**FELSIC ROCK** An igneous rock comprising more than 65 per cent total silica and more than 20 per cent quartz. Also known as acidic rock.

**FOLIATION** The laminated, parallel orientation or segregation of different minerals.

**FOSSIL** Any record of past life preserved in rocks, including bones and shells, footprints, excrement, and borings.

**FOSSIL FUEL** A fuel such as coal or oil derived from once-living organisms buried underground.

## G

**FUMAROLE** In volcanic regions, an opening in the ground through which hot gases are emitted.

**GEODE** A hollow cavity within rock that is lined with crystals.

**GLASSY TEXTURE** The smooth consistency of an igneous rock in which glass formed due to rapid solidification.

**GRANULAR** Having grains, or being in the form of grains.

**GRAPHIC TEXTURE** The surface appearance of some igneous rocks in which quartz and feldspar have intergrown to produce an effect resembling written script.

**GROUNDMASS** A fine-grained rock in which larger crystals are set or upon which they rest. Also called MATRIX.

## H

**HABIT** The mode of growth and appearance of a crystal, a result of its molecular structure.

**HACKLY FRACTURE** A mineral fracture that has a rough surface with small protuberances, as on a piece of broken cast iron.

**HEMIMORPHIC** Referring to a crystal: having a different facial development at each end.

**HINGE** The linear area along which the two valves of a mollusc shell are joined.

**HYDROTHERMAL VEIN** A rock fracture in which minerals have been deposited by fluids from deep within the Earth's crust.

## I

**IGNEOUS ROCK** A rock that is formed through the solidification of molten rock.

**INCLUSION** A crystal or fragment of another substance enclosed in a crystal or rock.

**INTERMEDIATE ROCK** An igneous rock with a total silica content of 55–65 per cent, intermediate in composition between felsic and mafic rocks.

**INTRUSION** A body of igneous rock that invades older rock.

**IRIDESCENCE** The reflection of light from internal elements of a stone, yielding a rainbow-like play of colours.

# L

**LACCOLITH** A mass of intrusive igneous rock with a dome-shaped top and generally flat base.

**LAPIDARY** A craftsperson who cuts and polishes gemstones.

**LAMELLAR** In thin layers or scales, composed of plates or flakes.

**LAVA** Molten rock extruded onto the Earth's surface.

**LUSTRE** The shine of a mineral, caused by reflected light.

# M

**MAFIC ROCK** An igneous rock with a silica content of 45–55 per cent total silica. Such rocks have less than 10 per cent quartz and are rich in iron-magnesium minerals. Mafic rocks are also known as basic rocks.

**MAGMA** Molten rock that may crystallize beneath the Earth's surface or be erupted as lava.

**MASSIVE** Referring to a mineral: having no definite shape.

**MATRIX** See GROUNDMASS.

**METALLIC LUSTRE** A shine like that of polished metal.

**METAMORPHIC ROCK** A rock that has been transformed by heat or pressure (or both) into another rock.

**METASOMATISM** A process that changes the composition of a rock or mineral by the addition or replacement of chemicals.

**METEOR** A rock from space that completely vaporizes while passing through the atmosphere.

**METEORITE** A rock from space that reaches the Earth's surface.

**MICROCRYSTALLINE** Having crystals that are so miniscule that they are detectable only with the aid of a microscope.

**MINERALOGIST** A scientist who primarily studies minerals.

**MIXED CUT** A gemstone cut consisting of a mixture of trapezoidal and triangular facets.

**MOULD** The fossilized imprint of the body of an organism.

# N O

**NATIVE ELEMENT** A chemical element that is found in nature uncombined with other elements.

**OOLITHS** Individual, spherical sedimentary grains from which oolitic rocks are formed. Most ooliths are made of concentric layers of calcite.

**ORE** A rock or mineral from which a metal can be profitably extracted.

# P

**PALEOECOLOGY** The study of the ecology of fossil animals and plants.

**PETROLOGIST** A geologist who specializes in the study of rocks.

**PHENOCRYST** A relatively large crystal set into the groundmass of an igneous rock to give a porphyritic texture.

**PHYLUM** A diverse group of organisms with a fundamental characteristic in common.

**PLACER DEPOSIT** A deposit of minerals derived by weathering, and concentrated in streams or beaches because of their high specific gravity.

**PLANETESIMAL** Large rocky body, roughly the size of asteroids, that was a precursor to a planet.

**PLASTIC** Referring to a rock: susceptible to being easily folded when subjected to high temperature and pressure.

**PLATE BOUNDARY** The point at which two or more tectonic plates meet.

**PLATY HABIT** The growth habit shown by flat, thin crystals.

**PLAYA** A desert basin that is intermittently filled with a lake.

**PLEOCHROIC** Referring to a mineral or gem: presenting different colours to the eye when viewed from different directions.

**PLUTON** A mass of igneous (plutonic) rock that has formed beneath the surface of the Earth by solidification of magma.

**POLYMORPH** A substance that can exist in two or more crystalline forms; one crystalline form of such a substance.

**POLYP** An individual member of a coral colony.

**PORPHYRITIC TEXTURE** An igneous rock texture in which large crystals are set in a finer matrix.

**PORPHYROBLASTIC TEXTURE** A metamorphic rock texture characterized by relatively large crystals in a finer-grained matrix.

**PRECIPITATION** The condensation of a solid from a liquid or gas.

**PRISMATIC HABIT** A mineral habit in which parallel rectangular faces form prisms.

**PROTOLITH** The rock that existed prior to undergoing metamorphic transformation into a different rock type.

**PSEUDOMORPH** A crystal with the outward form of another species of mineral.

**PYROCLASTIC** Consisting of material ejected from a volcanic vent.

# R

**RADIOMETRIC DATING** The determination of absolute ages of minerals and rocks by measuring certain radioactive and radiogenic atoms in the minerals.

**RECRYSTALLIZATION** The redistribution of components to form new minerals or mineral crystals, and thus in some cases a new rock. Recrystallization occurs during lithification and metamorphism.

**RECTANGULAR CUT** A gemstone cut into a rectangle.

**REFRACTIVE INDEX** A measure of the slowing down and bending of light as it enters a stone, used for identifying cut gemstones and some minerals.

**RESINOUS LUSTRE** A shine having the reflectivity of resin.

**RETICULATED** Having a network or a net-like mode of crystallization.

**RIBS** Raised ornamental bands in an invertebrate's body.

**ROUGH** Uncut gem crystal.

# ST

**SALT DOME** A large, intrusive mass of salt, sometimes with petroleum trapped beneath.

**SCHILLER EFFECT** The brilliant play of bright colours in a crystal, often due to minute, rod-like inclusions.

**SCHISTOSITY** A foliation that occurs in coarse-grained metamorphic rocks, generally the result of platy mineral grains.

**SECONDARY MINERAL** A mineral that replaces another mineral as the result of a weathering or alteration process.

**SEDIMENTARY ROCK** A rock that originates on Earth's surface as an accumulation of sediments or precipitates from water.

**SEMIMETAL** A metal that is not malleable. Arsenic and bismuth are examples of semimetals.

**SILL** A sheet-shaped igneous intrusion that follows the bedding of existing rocks.

**SLATY CLEAVAGE** The tendency of a rock, such as slate, to break along very flat planes into thin flat sheets.

**SPECIFIC GRAVITY** The ratio of the mass of the mineral to the mass of an equal volume of water. Specific gravity is numerically equivalent to density (mass divided by volume) in grams per cubic centimetre.

**STRIATION** A parallel groove or line appearing on a crystal.

**SUTURE** The line where the whorls join on gastropod shells.

**TWINNED CRYSTALS** Crystals that are grown together, for example as mirror images or at 90 degrees to each other.

# UV

**ULTRAMAFIC ROCK** An igneous rock with a silica content of less than 45 per cent. Also known as an ultrabasic rock.

**VALVE** One of two parts of the shell of an animal, such as a mollusc or brachiopod, which are sometimes joined by a hinge.

**VEIN** A thin, sheet-shaped mass of rock that fills fractures in other rock.

**VESICLE** A small spherical or oval cavity produced by a bubble of gas or vapour in lava, left after the lava solidifies.

**VITREOUS LUSTRE** A shine resembling that of glass.

**VOLCANIC PIPE** A fissure through which lava flows.

# W

**WELL SORTED** Referring to a sediment or sedimentary rock: having grains or clasts that are all roughly the same size.

**WHORL** One complete turn in the structure of an animal's shell.

353

# ACKNOWLEDGMENTS

## Author's Acknowledgments

Many people have worked very hard to make this book possible. Jane Laing has put in endless hours over many months planning, editing, and coordinating all aspects of the book. Margaret Carruthers has done a meticulous and much appreciated job of oversight on behalf of the Smithsonian, and Liz Wheeler of Dorling Kindersley has brought this all about. Christine Lacey, Miranda Harvey and Alison Gardner have done a superb job on design, and Jane Simmonds on editing. Gary Ombler and Linda Burgess have provided excellent new photography. Assistant Curator Monica Price and the Oxford University Natural History Museum have provided superb cooperation and access to their mineral collection, as has the Gemmological Society of Great Britain. The Diamond Trading Company have been very helpful with diamond pictures. Gaea Crystals of Wendover, Buckinghamshire and Mineral Imports of Teddington, Middlesex, have been generous in allowing the use of their commercial stocks for photography.

I would also like to thank Lilian Verner-Bonds for keeping me afloat, and Guy and Meriel Ballard for their patience throughout the 18 months it took to create the book. Finally, I would like to acknowledge Professors Malcolm McCallum, Joseph Weitz, Ernest Wolff, David Harris, and John Campbell, of the (then) Colorado State University Department of Geology whose teaching is the ultimate foundation of this book.

## Dorling Kindersley and Grant Laing Partnership

would like to thank the following people for their help in the preparation of this book: Monica Price and the Oxford University Museum of Natural History for providing access to over 250 mineral specimens photographed at the museum, and advice with aspects of the text; Doug Garrod and Lorne Stather and the Gemmological Society of Great Britain for providing access to, and advice concerning, more than 50 gemstones photographed at their premises; Mineral Imports, Teddington, for the loan of over 40 mineral specimens photographed for the book; Gaea Crystals, Wendover, for allowing photography at their premises of five mineral specimens featured in the book; R. Holt and Co. Ltd, London, for supplying gemstones photographed for pp.222–23 and 314–15; Greg Ohanian and Wilde One's, London, for providing access to Native American jewellery and some mineral specimens photographed at the premises; Ronald Bonewitz for the loan of mineral specimens and gemstones photographed for the book; Anne-Marie Bulat, Jasmine Burgess, Linda Burgess, Linda Dare, Nicola Dawtry, Miranda Harvey, Christine Lacey, Betty Laing, Jane Laing, Frances Pearl, Monica Price, Liz Statham, Jean Taylor, Ulrica Terje, Lilian Vernor-Bonds, and Liz Wheeler for the loan of jewellery and artefacts photographed for the book; Chip Clark from the American National Museum of Natural History, Smithsonian Institution, for his photographs from the National Gem Collection that appear in the book; Margaret Curruthers and Richard Efthim for their advice on the content of the text; Katie Mann and Ellen Nanney at the Smithsonian Institution for their liaison work; Chris Alderman at The Diamond Trading Company for assistance in locating photographs of diamonds and diamond mining; David Waters at the Oxford University Museum of Natural History for providing 34 photographs of thin sections for use in the book, together with identification advice; Neil Fletcher for supplying 40 location photographs; Helen Stallion and Mariana Sonnenberg for additional picture research; Miezan van Zyl for editorial assistance; and Anne Thompson and Marilou Prokopiou for design assistance.

# Picture Credits

ACKNOWLEDGMENTS

Smithsonian Institution: cr. 137 Corbis: © Eric and David Hosking: tl; Geoff Dann © Dorling Kindersley, Courtesy of the Wallace Collection, London: ca. 138 Alamy Images: Pavel Filatov: cl; Corbis: tr, © Ludovic Maisant: tl; DK Images: © Judith Miller/Colin Baddiel: cr; Geophotos: crb. 139 Alamy Images: Leslie Garland/LGPL: tl. 140 Corbis: © Nick Rains; Cordaiy Photo Library Ltd: tl, © Brownie Harris: ca; Science Photo Library: br, Chris Knapton: crb. 141 Alamy Images: Natalie Tepper/Arcaid: clb, Andre Jenny: tl. 141 Corbis: © Lowell Georgia: clb. 142 Alamy Images: Leslie Garland/LGPL: cl; Corbis: © Charles O'Rear: cr. 143 akg-images: cb, bl; Alamy Images: Ethel Davies/ImageState: tr, Mediacolors : tl; Geophotos: clb. 144 Alamy Images: WilmarPhotography.com: br; Geophotos: cl. 145 Alamy Images: Bowmann/F1 online: tl, Van Hilversum: tc, J Schwanke: cb; Corbis: © Gerald French: clb, © Hulton Deutsch Collection: bca. 146 Jens C Andersen, Camborne School of Mines, School of Geography, Archaeology and Earth Resources, University of Exeter: cr; Corbis: cl. 147 Corbis: © Archivo Iconografico, S.A: tr, © Robert Holmes: tl. 148 Ron Bonewitz: tl; DK Images: Colin Keates, Courtesy of the Natural History Museum, London: bl; © 2004 Smithsonian Institution: cla. 149 Corbis: © Nathan Benn: br. 150 Ron Bonewitz: tl; Corbis: © Sheldan Collins: cra, © John Garrett: tr, © Araldo de Luca: clb; Michael Freeman: cl. 151 DK Images: Colin Keates © Natural History Museum, London, UK: cl; © 2004 Smithsonian Institution: cr, t; Harold & Erica Van Pelt: The American Museum of Natural History, gift of Mrs. George Bowen DeLong: br. 152 www.bridgeman.co.uk: Christie's Images, London, UK: br. 153 Courtesy John Harris, Gemlab: tc; © 2004 Smithsonian Institution: tr, cab, car, cbl, cbr; Harold & Erica Van Pelt: American Museum of Natural History: tl. 154 Ron Bonewitz: tl; Corbis: bl, © Frank Lane Picture Agency: clb; Science Photo Library: Mehau Kulyk: tr, David Nunuk: br. 155 Alamy Images: Bryan and Cherry Alexander Photography: br; Corbis: © Michael Freeman: crb, © Richard Hamilton Smith: tl, © Roger Ressmeyer: tr; Science Photo Library: British Antarctic Survey: cra. 156 Corbis: © Richard Klune: bla; Reg Grant: tl; Sterling Hill Mining Museum: Maureen Verbeek: tr. 157 Ron Bonewitz: tl; Crown © The Royal Collection © 2002 Her Majesty Queen Elizabeth II: c, tr. 158 Geophotos: cl; Science Photo Library: Photo Reseachers: tr; Sterling Hill Mining Museum: car, Maureen Verbeek: tl; Dave Waters: cb. 159 Corbis: © Hulton Deutsch Collection: cra; DK Images: © Judith Miller/H Bonnar: tr; Geophotos: tl; © 2004 Smithsonian Institution: cal, car; Courtesy of the Trustees of the V&A: cla. 162 Alamy Images: Walter Bibikow/Jon Arnold Images: clb; Corbis: © Bettmann: br, © Jason Burke; Eye Ubiquitous: car; DK Images: Colin Keates, Courtesy of the Natural History Museum, London; Geophotos: tl, crb; Science Photo Library: Arnold Fisher: cr. 163 Corbis: © George Hall: crb, © David Muench: clb; Geophotos: tl, tr. 164 Alamy Images: Pavel Filatov: cb, Christa Knijff: tl; Corbis: clb, © Bettmann: cr, © Roger Ressmeyer: tr. 165 Corbis: © Gallo Images: clb. 165 Geophotos: tl, tc, cra; Ron Bonewitz: cb. 166 Alamy Images: Pavel Filatov: cla; Sterling Hill Mining Museum: Maureen Verbeek: clb. 167 Corbis: © James L. Amos: bl, tr, © Bill Ross: br, © Paul A Souders: ccr, © Vince Streano: tl. 168 Alamy Images: Pavel Filatov: tl; Corbis: © Michael S. Yamashita: tr; Geophotos: clb. 169 Corbis: © Gianni Dagli Orti: c, © David Muench: tl. 170 Corbis: © Chris Lisle: crb, © Alinari Archives: bl; Geophotos: cl. 171 Alamy Images: Kevin Taylor: bl; Corbis: © Nick Rains; Cordaiy Photo Library: cl, © Fremont Stevens/Corbis Sygma: tc; Geophotos: tl. 172 © Christie's Images Ltd: clb, bc; Corbis: © Sandro Vannini: tl. 173 © The British Museum: cra; Geophotos: cr. 174 Corbis: © James L. Amos: br, © Bettmann: cr, cb; Geophotos: tl. 175 Corbis: © Dave G. Houser: clb, © Dewitt Jones: cr, © Alan Schein Photography: crb; Geophotos: tl. 176 Geophotos: cl, cr. 177 Alamy/Pictor International/ImageState: tl; Corbis: tr; Kevin Downey Photography, MA, USA: br. 178 Alamy Images: Tom Till: tl; Corbis: © Bettmann: cra; DK Images: Dave King, Courtesy of the University Museum of Archaeology and Anthropology, Cambridge; © 2004 Smithsonian Institution: crb, bl. 179 Alamy Images: Pictor International/ImageState: cl; © Christie's Images Ltd: c; Corbis: © Kevin Schafer: tl, © Michael S. Yamashita: cr; © 2004 Smithsonian Institution: bl. 180 Alamy Images: Markus Bassler/Bildarchiv Monheim GmbH: tl; DK Images: © Judith Miller/Barie Macey: tr, © The Wallace Collection, London: bl; © Smithsonian Instiution: bcl. 181 Alamy Images: Mediacolors: clb, Noel Yates: tcl; Corbis: © Alinari Archives: cb, © Sally A. Morgan/Ecoscene: tl; Steve Sleight: crb. 182 Ron Bonewitz: clb; www.bridgeman.co.uk: Bibliotheque Nationale, Paris, France: cr; Geophotos: tl; DK Images: Colin Keates © Natural History Museum, London, UK: ccr. 183 Corbis: © O. Alamany & E. Vicens: tl, © Nick Rains; Cordaiy Photo Library: crb, © Sally A. Morgan/Ecoscene: tr, © Macduff Everton: clb. 184 akg-images: Erich Lessing: bl; DK Images: © Judith Miller/Freeman's: tr. 185 www.bridgeman.co.uk: Private Collection: br; © The British Museum: cl, c; © 2004 Smithsonian Institution: tr. 186 akg-images: cl; Corbis: © Antoine Gyori: tl. 187 Corbis: © Steve Raymer. 188 Alamy Images: Witry Pascal: cl. 189 Alamy Images: Jacques Jangoux: tl; Corbis: © David Muench: cl; © Alan Towse: cra; Science Photo Library: Hank Morgan/University of Massachusetts, AT, Amherst: tr, US Navy: cr. 190 akg-images: Erich Lessing: bl; Geophotos: tl; Science Photo Library: cra. 191 Corbis: © Paul A Souders: tr; Geophotos: tl, clb; Courtesy Joe Marty. Photograph: Henry Barwood: bl; Jan McClellan www.designjewel.com: crb; © 2004 Smithsonian Institution: crb. 192 Alamy Images: Pavel Filatov: clb, crb; Corbis: © Danny Lehman: tl; Geophotos: tr. 193 Geophotos: tl, clb; © 2004 Smithsonian Institution: cr. 194 Corbis: © Nick Rains; Cordaiy Photo Library: tr; Reg Grant: tl; Geophotos: clb. 195 Art Directors/TRIP: N Rudakov: tl; The Natural History Museum, London: cla. 196 Corbis: © John Van Hasselt/Corbis Sygma: tcr, © Danny Lehman: tl; Heritage Image Partnership: The British Museum: tr; The Picture Desk: Art Archive: Dagli Orti: bl. 197 Corbis: © Dave G. Houser: cr, © Joseph Sohm/ChromoSohm Inc: br. 198 Corbis: © Randy Faris. 199 Corbis: © Randy Faris: tr; Ancient Art and Architecture: cr. 200 Corbis: © Nick Rains; Cordaiy Photo Library: tl, © Hulton Deutsch Collection: cl; Geophotos: clb; © 2004 Smithsonian Institution: cla. 201 Corbis: © Danny Lehman: tl; Geophotos: clb; © 2004 Smithsonian Institution: cb, crb. 202 akg-images: Erich Lessing: br; Geophotos: tl. 203 Alamy Images: Leslie Garland Picture Library: crb; www.bridgeman.co.uk: Iraq Museum, Baghdad, Iraq: car, Musee du Louvre, Paris, France: cra; © The British Museum: tr; Geophotos: tl, clb. 204 Geophotos: cl. 205 Corbis: © Danny Lehman: cl, © David Turnley: br; Geophotos: tl. 206 Corbis: © Tom Bean: clb, © Viviane Moos: crb; Geophotos: tl. 207 Alamy Images: Christa Knijff: tl; Corbis: © Peter Russell; The Military Picture Library: bl; Geophotos: cl.

208 Alamy Images: Terry Whitakker/Travel Ink: cl; Science Photo Library: John Howard: bc, bra. 209 Alamy Images: Tom Till: cl; Corbis: © Massimo Listri: tr; Geophotos: tl, crb. 210 Geophotos: tl; Science Photo Library: BSIP Dr Pichard T.: cr; © 2004 Smithsonian Institution: b. 211 Corbis: © Gallo Images: tl, © Ron Watts: tr. 212 Courtesy of Merlin L. Cohen: br, cr; Corbis: Digital Image © 1996 CORBIS; Original image courtesy of NASA: clb, © Joseph Sohm/ChromoSohm Inc.: bl; Science Photo Library: NASA: tl. 213 Corbis: tc; Kevin Downey Photography, MA, USA: tr, cr. 214 Alamy Images: Stephen Saks Photography: tl; Corbis: © Kevin R. Morris: clb, © Charles O'Rear: br; Science Photo Library: Jerry Mason: tr; Maximilian Stock Ltd: bc. 215 Geophotos: tl, clb, cb; The Wellcome Institute Library, London: tr. 216 Alamy Images: Iain Farley: cb; Geophotos: clb; © 2004 Smithsonian Institution: ca; Dave Waters: tc, tl. 217 Alamy Images: Leslie Garland Picture Library: tl; Geophotos: cl; Science Photo Library: NASA: cr, Peter Ryan: tr; © 2004 Smithsonian Institution: crb, bl. 218 Corbis: © David Muench: tr. 219 akg-images: Erich Lessing: cla; Ancient Art & Architecture Collection: tr; DK Images: Colin Keates © Natural History Museum, London, UK: br; Geophotos: tl; Science Photo Library: Sheila Terry: ccr; © 2004 Smithsonian Institution: c, cr. 220 Ron Bonewitz: tl; Corbis: © Charles O'Rear: crb; Geophotos: clb. 221 Corbis: © Dave G. Houser: clb, © Ludovic Maisant: tl; DK Images: Judith Miller/Dorling Kindersley/Joseph H Bonnar: bla; © 2004 Smithsonian Institution: bl, cr. 222 Ron Bonewitz: tl; © 2004 Smithsonian Institution: clb. 223 www.bridgeman.co.uk: Bonhams, London, UK: tl; © Christie's Images Ltd: crb; Corbis: © Elio Ciol: cr. 224 Alamy Images: Robert Harding Picture Library: clb; Corbis: © Sheldan Collins: bl; Geophotos: tl. 225 Alamy: Gondwana Photo Art: tl, Jacques Jangoux: clb. 226 Ron Bonewitz: cl; Corbis: © Francis G. Mayer: br; Geophotos: tl; © 2004 Smithsonian Institution: c, cr. 227 Alamy Images: Mr Yogesh More/Ephotocorp: cl; www.bridgeman.co.uk: Heini Schneebeli: c; DK Images: Dave King: tl © Judith Miller/Dreweatt Neate: crb; The Natural History Museum, London: tr. 228 Alamy Images: Tibor Bognar: clb, Mr Yogesh More/Ephotocorp: tl; www.bridgeman.co.uk: Musee du Louvre, Paris, France/Giraudon: cb; Museo Acheologico Nazionale, Naples, Italy: cal. 229 Alamy Images: Frantisek Staud: tl; www.bridgeman.co.uk: The Barber Institution of Fine Arts, University of Birmingham: tc, Bonhams, London, UK: tr; Hermitage, St Petersburg, Russia: cla; Museum of Antiquities, Newcastle upon Tyne: bc; Ron Bonewitz: clb. 230 Ron Bonewitz: tl; © 2004 Smithsonian Institution: r. 231 akg-images: br; Tourist-Information Idar-Oberstein: bl. 232 Corbis: © Dave G. Houser: tl, © Paul A Souders: tr; © 2004 Smithsonian Institution: cra, cr. 233 © 2004 Smithsonian Institution: cr, br, bcl. 234 Alamy Images: Paul Shawcross: cr. 235 Ron Bonewitz: tl, © 2004 Smithsonian Institution: tr, b. 236 Alamy Images: Witry Pascal: cl; Corbis: tl. 237 akg-images: cl; Ron Bonewitz: tl; Corbis: © Blaine Harrington III: clb. 238 Alamy Images: Pavel Filatov: tl; © Christie's Images Ltd: cr; Corbis: © The Cover Story: crb; Ron Bonewitz: clb; Courtesy of the Trustees of the V&A: Designer R Pearson: br. 239 Corbis: tl, © Ralph White: clb; DK Images: The British Museum: cra. Colin Keates, Courtesy of the Natural History Museum, London: br. 240 Alamy Images: Robert Harding Picture Library: bl, Tom Till: tr; Corbis: © Gavriel Jecan: tl. 241 Corbis: © The Cover Story: cra, © Wolfgang Kaehler: tl, © Hubert Stadler: cr; © 2004 Smithsonian Institution: tr. 242 Alamy Images: Christa Knijff: cl; Dave Waters: crb. 243 Alamy Images: Ronald Naar/ImageShop/Zefa: tl; Art Directors/TRIP: cl; © Christie's Images Ltd: c; Corbis: © Charles O'Rear: tr. 244 www.bridgeman.co.uk: Egyptian National Museum, Cairo, Egypt/Giraudon: cl; Corbis: © Reza; Webistan: tl; DK Images: © Judith Miller/Sloan's: br. 245 akg-images: Erich Lessing: tl; www.bridgeman.co.uk: Bargello, Florence, Italy: br, Private Collection: tr. 246 © The British Museum: tl, cl, b. 247 DK Images: The British Museum. 248 Alamy Images: Christa Knijff: cl, tl; DK Images: Alan Keohane: cb; Science Photo Library: tr. 249 Alamy Images: Ian Wallace: cl; Corbis: © Bettmann: br. 250 Alamy Images: Mr Yogesh More/Ephotocorp: clb, tl, WilmarPhotography.com: tc; Derwent Water Systems, Matlock, UK: bl; DK Images: Alan Williams, Courtesy of the National Trust: tl. 251 Alamy Images: Martin J Dignard: clb; Geophotos: tl, cb; Science Photo Library: Jeremy Walker: cr. 252 Corbis: © Jacques Langevin/CORBIS SYGMA: tr, © Tim Thompson: clb; Science Photo Library: car, Jon Wilson: crb; © 2004 Smithsonian Institution: cl. 253 Alamy Images: Bowmann/F1 online: tl, Andre Jenny: clb; Corbis: © George D Lepp: cla; Science Photo Library: Astrid & Hanns-Frieder Michler: br. 254 DK Images: Alan Keohane: tl. 255 www.bridgeman.co.uk: Chapel of the Planets, Tempio Malatestiano, Rimini, Italy: tc; Corbis: © Ric Ergenbright: tl; Science Photo Library: John Beatty: crb, Alexander Tsiaras: cl. 256 © The British Museum: cr; Reg Grant: cl. 257 Corbis: © Raymond Gehman: cb; www.gemstonesgifts.com: cr. 258 Alamy Images: Pavel Filatov: tl; Corbis: © Roger Garwood & Trish Ainslie: br, © Michael S. Yamashita: bl; Science Photo Library: Dr Jeremy Burgess: cl. 259 akg-images: CDA/Guillemot: c; www.bridgeman.co.uk: © Museum of Fine Arts, Houston, Texas, USA, Museum purchase with funds provided by the Museum Collectors: tr; Corbis: © Robert Estall: cl, © H. David Seawell: cb; Frank de Wit: tl. 260 Alamy Images: Pavel Filatov: tl, Jack Sullivan: crb; Corbis: © Danny Lehman: clb; DK Images: The British Museum: ca, © Judith Miller/Cooper Owen: cl. 261 Corbis: © Richard A. Cooke: cl, © Lowell Georgia: cla; DK Images: Alan Williams: clb; Ron Bonewitz: tl; Sterling Hill Mining Museum: Maureen Verbeek: cb. 262 Corbis: © Stefano Bianchetti: cla, © Bob Krist: crb, © David Muench: tl; Ron Bonewitz: clb. 263 Alamy Images: Leslie Garland Picture Library: tl, Pavel Filatov: clb; Nigel Lacey: c; Courtesy www.ohiotoxicmold.com: cr. 264 Corbis: © Bowers Museum of Cultural Art, photo Don Wiechec: tc, © Pat O'Hara: tl, © Enzo and Paolo Ragazzini: tr; The Natural History Museum, London: cb. 265 Corbis: © Gianni Dagli Orti: tl, © Wolfgang Kaehler: cl, © Sandro Vannini: br, © David H. Wells: cr; Geophotos: tr. 266 Alamy Images: Mr Yogesh More/Ephotocorp: tl, Andy Christodolo/Cephas Picture Library: tl; DK Images: Kim Sayer: clb. 267 Alamy Images: Jacques Jangoux: tl; Corbis: © Sheldan Collins: br, © Jon Sparks: cl. 268 Alamy Images: Leslie Garland Picture Library: crb, The PhotoLibrary Wales: clb; Geophotos: tl. 269 Alamy Images: J Schwanke: tl; Corbis: © Paul A Souders: clb. 270 Corbis: © Roger Ressmeyer: bl; 270 Jenny Reakes: cl; Dave Waters: c. 271 Alamy Images: Bowmann/F1 online: tl; Art Directors/TRIP: cl. 272 Corbis: © Richard Klune: crb; Geophotos: tl, clb; Ron Bonewitz: tl; Science Photo Library: Michael Dunning: tc, Detlev van Ravenswaay: tr. 273 Science Photo Library: Alexis Rosenfeld: bcl; © 2004 Smithsonian Institution: cla, cl. 274 Alamy Images: Robert Harding Picture Library: tl; www.bridgeman.co.uk: The British Museum, London, UK: bcl, Peabody Museum, Harvard University, Cambridge, MA, USA: bl; DK Images: Harry Taylor: bcl; Werner Forman Archive: Dallas Museum of Fine Art: c. 275 www.bridgeman.co.uk: Museo de America, Madrid, Spain: tr; DK Images: © Judith Miller/Sloan's: tl; © 2004 Smithsonian Institution: cr, tcl. 276 Science Photo Library: Chris Knapton: cr, Sheila Terry: bl. 277 Alamy Images: Pavel Filatov: tl; Ron Bonewitz: clb; © 2004 Smithsonian Institution: tr. 278 Corbis: © Jan Butchofsky-Houser: cl; Dave Waters: crb. 279 Corbis: © Nathan Benn: tl; Dave Waters: tc, crb. 280 Corbis: © Asian Art & Archaeology, Inc.: bl, © Royal Ontario Museum: clb, bc, © Tim Thompson: tl. 281 Kremlin Museums, Moscow, Russia/www.bridgeman.co.uk: tl; Corbis/ © Lowell Georgia: tr. 282 The Picture Desk: Art Archive/Genius of China Exhibition: tl, b. 283 National Geographic Image Collection: The Xuzhou Museum, China/O. Louis Mazzatenta: tc. 284 Alamy Images: Martin J Dignard: tl; Corbis: © Richard Cummins: crb, © Blaine Harrington III: cl. 285 Corbis: © Hulton Deutsch Collection: tr, © Roger Garwood & Trish Ainslie: tl, © David Muench: clb, © Hubert Stadler: br. 286 Alamy Images: P. S. Lahiri/Indiapicture: cr. 286 Geophotos: cl. 287 Alamy Images: Thaler Shmuel/Index Stock: cl; Geophotos: tl, cl; Division of Mines and Geology, California: cr. 288 www.bridgeman.co.uk: Ashmolean Museum, University of Oxford, UK: cl; Corbis: © Tom Bean: tl; Mount Mica Rarities: bl. 289 DK Images: cra, br, bc; © 2004 Smithsonian Institution: r. 290 Alamy Images: Pavel Filatov: tl; Courtesy Steven Hoffmeyer, Hoffmeyer.com Photography: cr; Science Photo Library: US Dept of Energy: cl; © 2004 Smithsonian Institution: tr. 291 Corbis: © William A. Bake: c, © Bettmann: tr; Geophotos: tl, cl. 292 Alamy Images: Pavel Filatov: tl; Corbis: © Gianni Dagli Orti: tr, © Ann Johansson: cra; © 2004 Smithsonian Institution: bl. 293 www.bridgeman.co.uk: Bonhams, London, UK: crb, Courtesy of the Warden and Scholars of New College, Oxford, UK: c; DK Images: © Judith Miller/N. Bloom & Son Ltd: b; Royal Geographical Society, London: tr; © 2004 Smithsonian Institution: tl. 294 Corbis: © Bob Krist: clb; Geophotos: tl. 295 Corbis: © Paul Almasy: clb, © Mimmo Jodice: cb. 296 © Christie's Images Ltd: cr; Geophotos: clb; Ron Bonewitz: tl. 297 Alamy Images: Pavel Filatov: clb, Helder Silva: crb; Corbis: © Paul Almasy: tr, © Charles Krebs: tl. 298 Courtesy of Dr Peter Bancroft: cl. 299 Ancient Art & Architecture Collection: cr; www.bridgeman.co.uk: Bonhams, London, UK: cl; DK Images: © Judith Miller/HY Duke and Son: cl. 300 Alamy Images: Leslie Garland Picture Library: cl; www.bridgeman.co.uk: The British Museum, London, UK: bl, © Christie's Images Ltd: cla; © 2004 Smithsonian Institution: br. 301 © Christie's Images Ltd: tr, cra, cbr; Corbis: © Nick Rains; Cordaiy Photo Library: clb; DK Images: © Judith Miller/Joseph H Bonnar: cb; Ron Bonewitz: tl; © 2004 Smithsonian Institution: cl. 302 Alamy Images: Robert Harding Picture Library: clb, Witry Pascal: tl; © Christie's Images Ltd: tr, cb, cbl; © 2004 Smithsonian Institution: b. 303 Alamy Images: Robert Harding Picture Library: clb, Witry Pascal: tl; © Christie's Images Ltd: cla, cl. 304 Alamy Images: Malie Rich-Griffith: tl; Corbis: © Jim Sugar: bl. 305 Alamy Images: Dan White: br. 306 Alamy Images: Pavel Filatov: tl; © 2004 Smithsonian Institution: cr, crb; Courtesy of Dr Peter Bancroft: cb. 307 Corbis: © Bettmann: tr; DK Images: © Judith Miller /Goodwins Antiques Ltd: cr. 308 © Christie's Images Ltd: tr; Corbis: © O. Alamany & E. Vicens: tl; DK Images: Harry Taylor: tcr. 309 Corbis: © Bettmann: crb, © John and Lisa Merrill: tl; Jenny Reakes: clb. 310 Alamy Images: Jacques Jangoux: tl; Corbis: © Bettmann: bc; Sterling Hill Mining Museum: Maureen Verbeek: clb. 311 Alamy Images: Pavel Filatov: cb; Ron Bonewitz: tl. 312 Geophotos: clb, cb; Sterling Hill Mining Museum: Maureen Verbeek: tl. 313 Alamy Images: Jacques Jangoux: tcl, Profimedia CZ s.r.o.: tl, Jon Sparks: clb; Novosti (London): cb. 314 Alamy Images: Mediacolors: tc; Science Photo Library: David Nunuk: bl. 315 www.bridgeman.co.uk: Palazzo Vecchio (Palazzo della Signoria) Florence, Italy: br; © Judith Miller/Dorling Kindersley/Woolley and Wallis: cr. 316 akg-images: cl; Corbis: © Antoine Gyori: bl. 317 akg-images. 318 Ron Bonewitz: tl; Mary Evans Picture Library: br; Getty Images: bla. 319 Ron Bonewitz: tl; Science Photo Library: Sinclair Stammers: clb. 320 Science Photo Library: Andrew J Martinez: crb. 321 Ron Bonewitz: tl; www.bridgeman.co.uk: Galleria degli Uffizi, Florence, Italy, Giraudon: br, Heini Schneebeli: bl; DK Images: tr, cl, cla, cra, c, bra © Judith Miller / Dorling Kindersley / Lyon and Turnbull: bc. 322 Corbis: © Anthony Redpath: tl; Colin Keates © Dorling Kindersley, Courtesy of the Natural History Museum, London: tr; Courtesy of the Victoria and Albert Museum, London: tc; Courtesy of the IDNR: br. 323 DK Images: Michel Zabe © CONACULTA-INAH-MEX. Authorized reproduction by the Instituto Nacional de Antropologia e Historia: br; SPL: Alexis Rosenfeld: tr. 324 Getty Images. 325 David Doubilet: br; The Picture Desk: Art Archive/The Bodleian Library Oxford: tr. 326 Alamy Images: Frank Chmura: t. 328 National Geographic Image Collection: Michael Melford: tl; Science Photo Library: WG: cb. 329 Corbis: © Tom Bean: cl, © Layne Kennedy: bl, © Reuters: br; Science Photo Library: Sinclair Stammers: ca. 330 Corbis: © Derek Hall; Frank Lane Picture Agency: tl; Science Photo Library: James King-Holmes: br. 331 Corbis: © Chinch Gryniewicz; Ecoscene: b; Science Photo Library: tr, Martin Bond: cr, Astrid and Hanns-Frieder Michler: tl, Sinclair Stammers: cb. 332 DK Images: The Natural History Museum: cl, cr, crb, bl. 333 DK Images: The Natural History Museum ca, cl, c, tr, bc; Science Photo Library: bl. 334 Corbis: © Hubert Stadle:r tl; DK Images: Photo Alan Keohane; Rainbow Forest Museum, Arizona: bl. 335 Corbis: © George H H Huey. 336 Corbis: © Hubert Stadler: cl; DK Images: The Natural History Museum: tc, crb. 337 Courtesy of Dr Peter Crimes: tr. 339 Corbis: © William A Bake: tr; Science Photo Library: Geogrette Douwma: bl. 340 Corbis: © Jonathan Blair: bc; Colin Keates © Dorling Kindersley, Courtesy of the Natural History Museum, London tr. 342 Alamy Images: Frank Chmura: b. 343 Alamy Images: Robert Harding Picture Library: cl. 344 DK Images: The Natural History Museum: cl. 345 Corbis: © Derek Hall; Frank Lane Picture Agency: bl; DK Images: The Natural History Museum: tc. 346 Corbis: bl. 347 Science Photo Library: Matthew Oldfield, Scubazoo: tc. 348 Corbis: © Bettmann: tl, © James L. Amos: cl; DK Images: The Natural History Museum: ca, cra, c, bl, br. 349 Corbis: © Anthony Bannister; Gallo Images: br; DK Images: The Natural History Museum: tr, c, bl, bl.

All other images © Dorling Kindersley.

**For further information see: www.dkimages.com**